TRIGONOMETRY IDENTITIES

1. $\sin A = \dfrac{1}{\csc A}$

2. $\cos A = \dfrac{1}{\sec A}$

3. $\tan A = \dfrac{1}{\cot A}$

4. $\tan A = \dfrac{\sin A}{\cos A}$

5. $\cot A = \dfrac{\cos A}{\sin A}$

6. $\sin^2 A + \cos^2 A = 1$

7. $1 + \tan^2 A = \sec^2 A$

8. $1 + \cot^2 A = \csc^2 A$

9. $\sin(A + B) = \sin A \cos B + \cos A \sin B$

10. $\sin(A - B) = \sin A \cos B - \cos A \sin B$

11. $\cos(A + B) = \cos A \cos B - \sin A \sin B$

12. $\cos(A - B) = \cos A \cos B + \sin A \sin B$

13. $\tan(A + B) = \dfrac{\tan A + \tan B}{1 - \tan A \tan B}$

14. $\tan(A - B) = \dfrac{\tan A - \tan B}{1 + \tan A \tan B}$

15. $\sin 2A = 2 \sin A \cos A$

16. $\cos 2A = \cos^2 A - \sin^2 A = 2 \cos^2 A - 1 = 1 - 2 \sin^2 A$

17. $\tan 2A = \dfrac{2 \tan A}{1 - \tan^2 A}$

18. $\sin \tfrac{1}{2} A = \pm \sqrt{(1 - \cos x)/2}$

19. $\cos \tfrac{1}{2} A = \pm \sqrt{(1 + \cos A)/2}$

20. $\tan \tfrac{1}{2} A = \dfrac{\sin A}{1 + \cos A}$

21. $\sin A + \sin B = 2 \sin \tfrac{1}{2}(A + B)\cos \tfrac{1}{2}(A - B)$

22. $\sin A - \sin B = 2 \cos \tfrac{1}{2}(A + B)\sin \tfrac{1}{2}(A - B)$

23. $\cos A + \cos B = 2 \cos \tfrac{1}{2}(A + B)\cos \tfrac{1}{2}(A - B)$

24. $\cos A - \cos B = -2 \sin \tfrac{1}{2}(A + B)\sin \tfrac{1}{2}(A - B)$

25. $\sin A \cos B = \tfrac{1}{2}\{\sin(A + B) + \sin(A - B)\}$

26. $\cos A \sin B = \tfrac{1}{2}\{\sin(A + B) - \sin(A - B)\}$

27. $\cos A \cos B = \tfrac{1}{2}\{\cos(A + B) + \cos(A - B)\}$

28. $\sin A \sin B = \tfrac{1}{2}\{\cos(A - B) - \cos(A + B)\}$

CALCULUS AND ANALYTIC GEOMETRY
FOR ENGINEERING TECHNOLOGY

BERNARD J. RICE
University of Dayton

JERRY D. STRANGE
University of Dayton

Delmar Publishers Inc.®

For information, address Delmar Publishers Inc.
3 Columbia Circle Drive, PO Box 15-015
Albany, NY 12212-5015

10 9 8 7 6 5 4 3 2

Printed in the United States of America
Published simultaneously in Canada
by Nelson Canada
A Division of The Thomson Corporation

ISBN 0-8273-3868-6

■ PREFACE

This book is an outgrowth of our *Technical Mathematics with Calculus*. It is intended for students enrolled in an engineering technology curriculum. The material is presented under the assumption that the student has a working knowledge of college-level algebra and trigonometry.

The primary objective of a technical calculus book is to relate mathematical concepts to practical engineering technology problems. We have attempted to attain this goal by using physical problems that are well within the scope of the beginning student as a springboard to mathematical concepts. Our approach is exemplified in Chapters 2 and 6 in connection with the concepts of the derivative and the definite integral, respectively. Both chapters are designed to lead the student to the desired mathematical concept by identifying the common thread in several apparently unrelated physical problems. For example, in Chapter 2, the similarity in the solutions to the slope of a tangent line, the velocity of a particle, and the current in a circuit is used to motivate the definition of the derivative.

The scope of topics covered in the text is standard for technical calculus books. Algebraic functions are discussed in the preliminary chapters and are followed by the transcendental functions. Selected topics from differential equations and infinite series are presented in the concluding chapters.

Analytic geometry is presented as it is needed to facilitate an understanding of calculus applications. Except in Chapters 1 and 11, the two subjects are interwoven. The conic sections of Chapter 11 can be taught at any time after Chapter 1 is covered, but they are delayed in the arrangement of topics to allow earlier coverage of the transcendental functions. The book is designed to satisfy the needs of a two-semester course; however, those desiring a one-semester coverage of calculus will find the first seven chapters ideal for this purpose.

The presentation features a variety of learning aids. Numerous worked examples are included in each section, many of which have step-by-step comments to lead students through the solution. Because repetition is an important part of the learning process, we have included an abundant supply of graded problems in the end-of-section exercise sets. In addition, each chapter concludes with a set of review exercises. Exercises with application to a specific technology are "called-out" with a distinctive logo. (A key to the application symbols follows the Table of Contents.) Answers to the odd-numbered exercises are provided at the end of the book.

Comment and *Warning* statements are included throughout the book to alert the student to important ideas and processes and to warn of common pitfalls. Also, many of the mathematical procedures are presented in an easy-to-follow, step-by-step format. These procedural steps are boxed—as are key formulas, equations, and definitions—for emphasis and for easy reference. An additional learning aid is the Glossary of Important Terms at the back of the book.

Several important concepts in the book are reinforced through the use of simple computer programs. These interactive programs are written in BASIC language for the Apple IIe computer. They occur at the ends of selected sections and are highlighted with the application symbol shown at the left. The programs will help the student understand the process or concept being presented in the section. We are grateful to Professor J. W. Friel for permitting us to use these programs. Anyone interested in further information on these programs may contact Professor Friel at the University of Dayton.

Finally, we wish to acknowledge the contributions of the following individuals who reviewed the manuscript for this text: Henry D. Davison, St. Petersburg Junior College; David Sherren, Fairmont State College; Donald W. Sibrel, Nashville State Technical Institute; Lawrence A. Trivieri, Mohawk Valley Community College; Roman Voronka, New Jersey Institute of Technology. Thanks also to our editor at Breton Publishers, George J. Horesta, and to Sylvia Dovner and her staff at Technical Texts, Inc., for their efforts on behalf of this project.

Bernard J. Rice
Jerry D. Strange

■ CONTENTS

CHAPTER 12
Calculus of Functions of Two Variables   397

CHAPTER 13
Differential Equations   433

■ APPLICATION SYMBOLS

 MECHANICAL ENGINEERING

 ELECTRICAL ENGINEERING

 CIVIL ENGINEERING

 CHEMICAL ENGINEERING

 INDUSTRIAL ENGINEERING

 ARCHITECTURE

 AEROSPACE

 ECOLOGY/BIOLOGY

 PHYSICS

 BUSINESS/ECONOMICS

 NAVIGATION

 MINING/GEOLOGY

CHAPTER 1

Prerequisites for Calculus

Many problems that arise in science and technology can be solved by the methods of algebra and trigonometry. However, there are many other applied problems that can only be adequately understood with the concepts of calculus, the topic of this book. Some of the ideas from algebra, trigonometry, and analytic geometry that are used frequently in the study of calculus are reviewed in this chapter.

1.1 THE REAL NUMBER LINE

The discussion in this book is limited to real numbers. We presume familiarity with the fundamental operations of addition, subtraction, multiplication, division, and exponentiation.

The magnitude of a real number a, independent of its sign, is called the **absolute value** of the number and is denoted by $|a|$. We define the absolute value of a real number a by

$$|a| = \begin{cases} a, & \text{if } a \geq 0 \\ -a, & \text{if } a < 0 \end{cases}$$

Thus, $|a|$ is positive for every nonzero number, because if $a > 0$, $|a| = a$, which is positive; and if $a < 0$, $|a| = -a$, which is also positive. For example, $|3| = 3$ and $|-3| = -(-3) = 3$.

The real numbers can be represented geometrically by associating each number with a point on a line. The line is called the **real number line**, or simply the **number line**. We begin by choosing any point on the line. This point is called the **origin**. With the origin, we associate the real number 0. Then, we select any other point to the right of 0 and with it associate the real number 1. The segment determined by the origin and the point corresponding to 1 is called a **unit distance**. (Note that this unit distance can be any distance that we choose.)

Proceeding from 1, we mark 2 units, 3 units, 4 units, and so on. With these points, we associate the numbers 2, 3, 4, ... Thus, the entire set of natural numbers is associated with equally spaced points on the line. Similarly, unit distances to the left of 0 associate points with negative integers. See Figure 1.1.

■ Figure 1.1

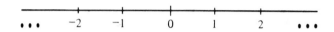

To locate rational numbers on the number line, divide the unit distance into an appropriate number of equal parts. For example, if the unit distance is divided into thirds, each of the points is labeled as shown in Figure 1.2.

■ Figure 1.2

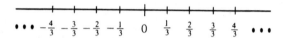

When a point on the number line is associated with a real number, the number is called the **coordinate** of the point. The point is called the **graph** of the number. In other words, the point represents the real number. The two concepts are often used interchangeably. That is, we say "the point 2" instead of "the point whose coordinate is 2."

The distance between any two points on the number line is given by the absolute value of their difference.

DEFINITION

Distance between Two Points   The distance between two points a and b, denoted by $d(a, b)$, is

$$d(a, b) = |a - b|$$

From our understanding of the absolute value of a number, we see that

$$d(a, b) = d(b, a)$$

EXAMPLE 1   ■ **Problem**

Find the distance between the indicated points in Figure 1.3.

■ **Solution**

We have

$$d(-3, 2) = |-3 - 2| = |-5| = 5$$

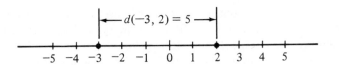

■ **Figure 1.3**

The distance from the origin of the number line to the point a is $d(a, 0) = |a - 0| = |a|$. That is, $|a|$ is the distance from the origin to point a.

An **interval** is the set of all real numbers between two given real numbers. Special notation is used to indicate intervals. The set of real numbers greater than a and less than b, called an **open interval**, is denoted by (a, b) and defined by

$$(a, b) = \{x \mid a < x < b\}$$

The notation $\{x \mid a < x < b\}$ is read "the set of all x such that x is greater than a and less than b." The notation (a, b) presumes that $a < b$. The set of real numbers greater than or equal to a and less than or equal to b, called a **closed interval**, is denoted by $[a, b]$ and defined by

$$[a, b] = \{x \mid a \le x \le b\}$$

The points corresponding to a and b are called **endpoints** of the interval. Thus, the distinction between an open and a closed interval is the exclusion or inclusion of the endpoints. Graphically, an endpoint to be included in the interval is represented by a closed dot. Otherwise, an open dot is used. See Figure 1.4.

■ **Figure 1.4**   Open interval (a, b) Closed interval $[a, b]$

If only one of the endpoints is included in an interval, the interval is said to be **half-open**. The half-open intervals $(a, b]$ and $[a, b)$ are depicted in Figure 1.5.

■ **Figure 1.5** $(a, b]$ $[a, b)$

The set of real numbers is **unbounded** in the sense that there is no largest real number. We use the symbol ∞ (**infinity**) to convey the idea of unboundedness. Thus, the set of all real numbers greater than or equal to a is represented in interval notation by $[a, \infty)$. Notice that a right parenthesis is used to indicate that the interval is open on the right. Similarly, $(-\infty, a]$ denotes the set of all real numbers less than or equal to a. These intervals are shown in Figure 1.6.

■ **Figure 1.6** $(-\infty, a]$ $[a, \infty)$

Comment: The concept of infinity is useful in many situations where large numbers are involved. However, in using the symbol ∞, you must realize that it is not a number but a symbol representing the idea of unboundedness. Because ∞ is not a number in the ordinary sense of the word, it cannot be combined arithmetically with real numbers.

The absolute inequalities $|x| < b$ and $|x| > b$ both describe sets of real numbers that can be represented by intervals on the real line. From the definition of absolute value, $|x| < b$ and $|x| > b$ have the following interpretations:

$$|x| < b \quad \text{means that} \quad -b < x < b$$
and
$$|x| > b \quad \text{means that} \quad x < -b \quad \text{or} \quad x > b$$

Figure 1.7 pictures these two intervals.

■ **Figure 1.7**

$$|x| < b \qquad\qquad |x| > b$$

1.2 THE CARTESIAN COORDINATE SYSTEM

Just as we form a correspondence between real numbers and points on a line, we associate ordered pairs of numbers with points in a plane. The basic tool used to accomplish this association is called the **rectangular coordinate system** or the **Cartesian* coordinate system**. To construct such a system, draw a pair of mutually perpendicular number lines to intersect at the zero point of each line, as shown in Figure 1.8.

Normally, the horizontal line is called the ***x*-axis**, the vertical line the ***y*-axis**, and their intersection the **origin**. Considered together, the two axes are called the coordinate axes. As you can see, the coordinate axes divide the plane into four zones, or **quadrants**. The upper right quadrant is called the first quadrant, and the others are numbered consecutively counterclockwise as in Figure 1.8. The coordinate axes are not considered to be in any quadrant.

Graphing

To locate points in the plane, use the origin as a reference point. The displacement of a point in the plane to the right or left of the *y*-axis is called the ***x*-coordinate**, or **abscissa**, of the point and is denoted by *x*. Values of *x* measured to the right of the *y*-axis are considered *positive* and to the left, *negative*. The displacement of a point

---

*Named in honor of René Descartes (1596–1650), the first person to systematically use ordered pairs of numbers.

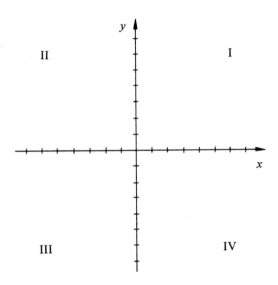

■ **Figure 1.8**

in the plane above or below the *x*-axis is called the **y-coordinate**, or **ordinate**, of the point and is denoted by *y*. Values of *y* above the *x*-axis are considered *positive* and below the *x*-axis, *negative*. Together, the abscissa and the ordinate of a point are called the **coordinates** of the point, conventionally written in parentheses, with the abscissa written first and separated from the ordinate by a comma, that is, (*x*, *y*).

We observe the following:

- A point (*x*, *y*) lies in quadrant I if both coordinates are positive.
- A point (*x*, *y*) lies in quadrant II if the *x*-coordinate is negative and the *y*-coordinate is positive.
- A point (*x*, *y*) lies in quadrant III if both coordinates are negative.
- A point (*x*, *y*) lies in quadrant IV if the *x*-coordinate is positive and the *y*-coordinate is negative.

Since the first number represents the horizontal displacement and the second the vertical displacement, we see the significance of order. For example, the ordered pair (3, 5) represents a point displaced 3 units to the right of the origin and 5 units above it, while (5, 3) represents a point that is 5 units to the right and 3 units up. See Figure 1.8.

To be precise, we should distinguish between the point and the ordered pair. However, the distinction is usually blurred, and we say "the point (*x*, *y*)" instead of "the point whose coordinates are (*x*, *y*)."

Use of the rectangular coordinate system establishes a one-to-one correspondence between the points in a plane and all possible ordered pairs of real numbers (*x*, *y*). That is, each point in the plane can be described by a unique ordered pair of numbers (x, y), and each ordered pair of numbers (x, y) can be represented by a unique point in the plane called the **graph** of the ordered pair.

EXAMPLE 1   ■ **Problem**

Locate the points $P(-1, 2)$, $Q(2, 3)$, $R(-3, -4)$, $S(3, -5)$, and $T(\pi, 0)$ in the plane.

■ **Solution**

a.   Point $P(-1, 2)$ is in quadrant II because the x-coordinate is negative and the y-coordinate is positive.

b.   Point $Q(2, 3)$ is in quadrant I because both coordinates are positive.

c.   Point $R(-3, -4)$ is in quadrant III because both coordinates are negative.

d.   Point $S(3, -5)$ is in quadrant IV because the x-coordinate is positive and the y-coordinate is negative.

e.   Point $T(\pi, 0)$ is not in any quadrant but lies on the positive x-axis.

The points are plotted in Figure 1.9.

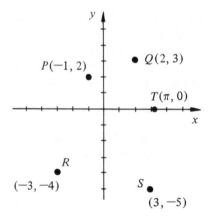

■ **Figure 1.9**

In the plot of an entire *set* of ordered pairs, the corresponding set of points in the plane is called the **graph of the set**. Sometimes, the points in the graph form a recognizable pattern, such as a straight line or a circle.

EXAMPLE 2   ■ **Problem**

Graph the set of points in the plane whose ordinate y is 3.

■ **Solution**

This set is defined by the equation $y = 3$. The graph is shown in Figure 1.10.

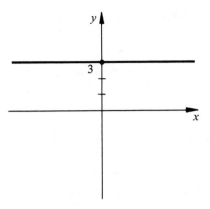

■ Figure 1.10   ■

The Distance between Two Points

Consider two points $P_1(x_1, y_1)$ and $P_2(x_2, y_2)$ that determine a slant line, as shown in Figure 1.11. Draw a line through P_1 parallel to the x-axis and a line through P_2 parallel to the y-axis. These two lines intersect at the point $M(x_2, y_1)$. Hence, by the Pythagorean theorem,

$$[d(P_1, P_2)]^2 = [d(P_1, M)]^2 + [d(M, P_2)]^2 \qquad (1.1)$$

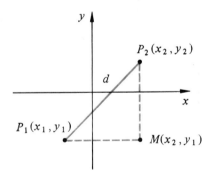

■ Figure 1.11

We see from Figure 1.11 that $d(P_1, M)$ is the horizontal distance between P_1 and P_2. Therefore, the distance $d(P_1, M)$ is given by

$$d(P_1, M) = |x_2 - x_1|$$

Likewise, the vertical distance $d(M, P_2)$ is given by

$$d(M, P_2) = |y_2 - y_1|$$

Making these substitutions into Equation (1.1) and denoting $d(P_1, P_2)$ by d gives

$$d^2 = (x_2 - x_1)^2 + (y_2 - y_1)^2$$

Taking the square root yields the distance formula.

FORMULA

> **The Distance Formula**   The length of the line segment from $P_1(x_1, y_1)$ to $P_2(x_2, y_2)$ is
>
> $$d = \sqrt{(x_2 - x_1)^2 + (y_2 - y_1)^2} \qquad\qquad (1.2)$$

Equation (1.2) is used to find the distance between two points in the plane directly from their coordinates. The order in which the two points are labeled is immaterial, since

$$(x_2 - x_1)^2 = (x_1 - x_2)^2 \qquad \text{and} \qquad (y_2 - y_1)^2 = (y_1 - y_2)^2$$

EXAMPLE 3   ■ **Problem**

Find the distance between $(-3, -6)$ and $(5, -2)$. (See Figure 1.12.)

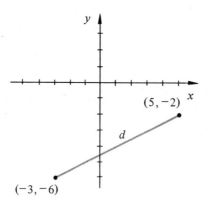

■ **Figure 1.12**

■ **Solution**

Let $(x_1, y_1) = (-3, -6)$ and $(x_2, y_2) = (5, -2)$. Substituting these values into the distance formula, we get

$$d = \sqrt{(x_2 - x_1)^2 + (y_2 - y_1)^2} = \sqrt{[5 - (-3)]^2 + [-2 - (-6)]^2}$$
$$= \sqrt{64 + 16} = \sqrt{80} = 4\sqrt{5}$$

When making the substitutions, do not forget to include the signs of the co-ordinates.   ■

The midpoint of the line segment connecting $P_1(x_1, y_1)$ and $P_2(x_2, y_2)$ is indicated by M in Figure 1.13. To obtain the coordinates of M, draw lines through P_1, M, and P_2 parallel to the y-axis so that they intersect the x-axis at Q_1, N, and Q_2. We know from geometry that if M is a midpoint of the line segment from P_1 to P_2, then N must be the midpoint of the line segment from Q_1 to Q_2. The midpoint of the line segment from Q_1 to Q_2, which corresponds to the abscissa of M, is given in Section 1.1 to be $(x_1 + x_2)/2$. A similar analysis shows that the ordinate of M is given by $(y_1 + y_2)/2$.

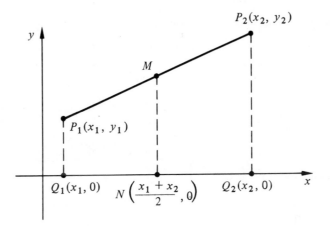

■ Figure 1.13

Midpoint Formula The midpoint of the line segment from $P_1(x_1, y_1)$ and $P_2(x_2, y_2)$ is located at

$$\left(\frac{x_1 + x_2}{2}, \frac{y_1 + y_2}{2}\right) \tag{1.3}$$

EXAMPLE 4 ■ **Problem**

Find the midpoint of the line segment from $(2, -3)$ to $(-1, -5)$.

■ **Solution**

By Expression (1.3), the coordinates of the midpoint are

$$\left(\frac{2 + (-1)}{2}, \frac{-3 + (-5)}{2}\right)$$

This pair simplifies to $(\frac{1}{2}, -4)$. Check this result by drawing the indicated line segment and locating the midpoint. ■

EXERCISES FOR SECTIONS 1.1–1.2

In Exercises 1–8, find the distance between the indicated points on the number line.

1. $-3, 1$ 2. $5, 9$

3. $-5, -1$ 4. $-2, 7$

5. $-0.1, 7$ 6. $-0.5, -0.3$

7. $-6, 0$ 8. $-8, -1$

Sketch the graph of each interval in Exercises 9–20.

9. $(-2, 3)$ 10. $[-2, 3]$

11. $(-\infty, 0]$ 12. $[1, \infty)$

13. $(-\infty, \infty)$ 14. $(-3, 3]$

15. $[2, 3]$ 16. $[\frac{1}{4}, \frac{1}{2})$

17. $|x| < 2$ 18. $|x| < 4$

19. $|x| \geq 1$ 20. $|x| \geq 2$

Plot the ordered pairs in Exercises 21–26.

21. $(3, 2)$ 22. $(4, 6)$

23. $(-2, \frac{1}{2})$ 24. $(-6, -5)$

25. $(\frac{1}{4}, \frac{1}{2})$ 26. $(-2.5, 1.7)$

27. In which two quadrants do the points have positive abscissas?

28. In which two quadrants do the points have negative ordinates?

29. In which quadrant are both the abscissa and the ordinate negative?

30. In which quadrants is the ratio y/x negative?

31. What is the ordinate of a point on the x-axis?

In Exercises 32–36, graph the given set.

32. $\{(0, 0), (1, 1), (2, 2), (3,3)\}$

33. $\{(0, 5), (1, 4), (2, 3), (3, 2), (4, 1), (5, 0)\}$

34. $\{(-2, 7), (0, 7), (3, 7), (7, 7)\}$

35. $\{(\sqrt{2}, 5), (\sqrt{2}, 2), (\sqrt{2}, 0), (\sqrt{2}, -\sqrt{3})\}$

36. $\{(-2, -1), (-1, -1), (0, -1), (1, \frac{1}{2}),$
     $(2, 1), (3, \frac{3}{2})\}$

Graph the set of points defined by the given equations or inequalities in Exercises 37–41.

37. $y \geq 0$ 38. $x = 0$

39. $x = -2$ 40. $x \geq 0$

41. $y = 2$

Plot the pairs of points in Exercises 42–49, and find the distance between the points. Compute the coordinates of the midpoint of the line segment connecting the points.

42. $(1, 2), (5, 4)$

43. $(0, 4), (-1, 3)$

44. $(-1, 5), (-1, -6)$

45. $(\frac{1}{2}, \frac{1}{2}), (\frac{1}{2}, -\frac{3}{4})$

46. $(-5, 3), (2, -1)$

47. $(0.5, 1.6), (6.2, 7.5)$

48. $(-3, 4), (0, 4)$

49. $(2, -6), (-\sqrt{3}, -3)$

50. The point $(x, 3)$ is 4 units from $(5, 1)$. Find x.

51. Find the distance between the points $(\sqrt{x}, \sqrt{y})$ and $(-\sqrt{x}, -\sqrt{y})$.

52. Find the distance between the points (x, y) and $(-x, y)$.

53. Find the distance between the points (x, y) and $(x, -y)$.

54. Find the point on the x-axis that is equidistant from $(0, -1)$ and $(3, 2)$.

55. Find the point on the y-axis that is equidistant from $(-4, -1)$ and $(-1, 2)$.

 The following BASIC program for computing the distance between two points in the plane is written for the APPLE IIe microcomputer. If you have another type of microcomputer, you may have to modify certain statements. Use this program to find the distance between the points listed in Exercises 42–49.

```
10   REM   DISTANCE FORMULA
20   HOME
30   PRINT "THIS PROGRAM WILL FIND THE DISTANCE"
40   PRINT "BETWEEN TWO POINTS SUPPLIED BY THE USER."
50   PRINT
60   PRINT "ENTER X,Y FOR THE FIRST POINT"
70   INPUT X1,Y1
80   PRINT "ENTER X,Y FOR THE SECOND POINT"
90   INPUT X2,Y2
100  D =   SQR ((X1 - X2) ^ 2 + (Y1 - Y2) ^ 2)
105  D =   INT (1000 * D) / 1000
110  PRINT "THE DISTANCE BETWEEN (";X1;",";Y1;") AND"
120  PRINT "(";X2;",";Y2;") IS ";D;" UNITS"
130  PRINT
140  PRINT "DO YOU WISH TO FIND ANOTHER DISTANCE? Y/N"
150  INPUT A$
160  HOME
170  IF A$ = "Y" THEN 50
180  END
```

1.3 FUNCTIONS

An important idea in mathematics is that of pairing numbers by some specified rule called a **rule of correspondence.** We use this idea when we pair a number y with each number x by the rule of correspondence $y = 3x$. The pairings in the following table were generated by using this rule:

x	-2	-1	0	1	2	3
y	-6	-3	0	3	6	9

Rules of correspondence that assign a unique value of y to each value of x are of particular interest to us, and we make the following definition.

DEFINITION

> **Function**   If the rule of correspondence between two variables x and y is such that there is exactly one value of y for each value x, then we say that y is a *function* of x.

Functions are frequently defined by some formula or expression involving x and y. The following remarks should clarify the important points of this definition.

- The expression $y = 3x^2$ defines a function since each value of x determines only one value of y. For instance, if $x = -2$, then y is 12.
- The expression $y = 8$ defines a function since y has the value 8 for any value of x that we choose. The definition does not require that y have a different value for each x. It only requires that the value of y be unique.
- The expression $y^2 = x$ does not define a function since there are two values of y for each positive value of x. For instance, if $x = 9$, then y can be either 3 or -3.
- The expression $y > x$ does not define a function since there are many values of y for each value of x. For instance, if $x = 2$, then y can be any number greater than 2.

Expressions such as $y^2 = x$ and $y > x$ that do not satisfy the condition in the preceding definition are called **relations**.

A function consists of three parts: a **rule of correspondence** that relates the dependent variable to the independent variable, the values that can be taken on by the independent variable, called the **domain** of the function, and the resulting values taken on by the dependent variable, called the **range**. If the rule of correspondence is given by an equation and the domain is not specified, we assume that the domain contains all real numbers for which the equation makes sense (as a real number).

EXAMPLE 1   The equation $y = x^2 + 5$ defines a function, since each value of x determines only one value of y. The domain consists of all real numbers, and the range consists only of those real numbers greater than or equal to 5 (since the smallest possible value of x^2 is 0).   ■

EXAMPLE 2   ■ **Problem**

Find the domain and range of $y = \sqrt{x}$.

■ **Solution**

If we substitute a negative real number for x in $y = \sqrt{x}$, we do not get a real number for y. However, each nonnegative real number substituted for x yields a nonnegative real number for y. Therefore, both the domain and the range of this function consist of all nonnegative real numbers.   ■

EXAMPLE 3   Consider the rule of correspondence $y = 4/(x - 3)$. Here, any value except 3 can be substituted for x (remember that division by 0 is not allowed). Therefore, the domain consists of all real numbers other than 3. The range consists of all real numbers other than 0, since $4/(x - 3)$ can be any value other than 0. We verify this result by solving for x in terms of y and noting any restrictions on y. Thus,

$$y = \frac{4}{x - 3}$$

$$xy - 3y = 4 \qquad \text{multiplying both sides by } x - 3$$

$$xy = 3y + 4 \qquad \text{adding } 3y \text{ to both sides}$$

$$x = \frac{3y + 4}{y} \qquad \text{dividing both sides by } y$$

The only limitation on y is that it cannot equal zero. Therefore, the range consists of all real numbers except 0.   ■

Functional rules are usually given by a formula, but they may be given in other ways such as diagrams, tables, and graphs. The next two examples show rules given by a table and a diagram.

EXAMPLE 4   In Figure 1.14, we have a rule of correspondence given by a diagram. The arrows point from values of the independent variable to values of the dependent variable. Since each value of the independent variable determines just one value of the dependent variable, we have a function. The domain consists of 1, 2, 3, and 4. The range consists of $\sqrt{6}$, 4, and 5. Note that neither 2 nor 12 belongs to the range. Also, note that both 1 and 4 are associated with the same value in the range. Nothing in the definition of a function excludes that possibility.

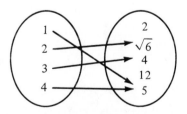

■ **Figure 1.14**   A function   ■

EXAMPLE 5   The following table of values defines a function:

x	-3	-2	-1	0	1	2	3
y	4	0	1	2	0	5	-4

In this case, the domain consists of the numbers -3, -2, -1, 0, 1, 2, 3. The range consists of the numbers -4, 0, 1, 2, 4, 5. The table is the rule of correspondence for this function.   ■

EXAMPLE 6   Although a functional relation can be given in a variety of forms, it must never assign more than one range element to the same domain element. Figure 1.15 does not describe a function, since 3 is assigned to the two different elements 5 and 6.

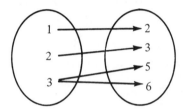

■ **Figure 1.15**   ■

Functions are usually denoted by a single letter, such as f, F, g, G, h, or H. If x represents an element of the domain of a function f, then the corresponding value in the range is denoted by $f(x)$, which is read, "f of x." Note that the symbol $f(x)$ does *not* indicate the product of f and x. The symbol $f(x)$ represents the value in the range of f that corresponds to the value of x in the domain of f. For example, if we let f represent the function indicated in Figure 1.14, we have

$$f(1) = 5 \qquad f(2) = \sqrt{6} \qquad f(3) = 4 \qquad f(4) = 5$$

In general, if y is a function of x, we write

$$y = f(x)$$

EXAMPLE 7   ■ **Problem**

Evaluate $f(x) = x^2 + 3x$ at $x = 2$ and at $x = a - 4$.

■ **Solution**

Substitute 2 for x, so that

$$f(2) = 2^2 + 3 \times 2 = 10$$

Now substitute $a - 4$ for x, so that

$$f(a - 4) = (a - 4)^2 + 3(a - 4) = a^2 - 8a + 16 + 3a - 12$$
$$= a^2 - 5a + 4$$

■

The Graph of a Function

By the **graph of a function** f, we mean the set of points in the plane whose coordinates satisfy the equation $y = f(x)$. Domain values are plotted along the horizontal axis, and range values are plotted along the vertical axis. Most functions have infinitely many ordered pairs. Since we cannot plot infinitely many ordered pairs, we plot some selected ordered pairs and then draw a smooth curve through the points.

EXAMPLE 8 ■ **Problem**

Graph the function $y = 3x + 4$. (The domain consists of all real numbers.)

■ **Solution**

Choose some convenient values for x and compute the corresponding values of y as in the following table:

x	-3	-2	-1	0	1	2	3
y	-5	-2	1	4	7	10	13

Then, plot the ordered pairs, and connect the points with a smooth curve. The graph turns out to be a straight line, as shown in Figure 1.16.

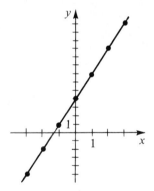

■ **Figure 1.16**

Comment: Functions of the form $y = ax + b$, where a and b are constants, are called **linear functions**. The name comes from the fact that functions of this form have straight-line graphs.

EXAMPLE 9 ■ **Problem**

Graph the function $y = x^2$. (The domain consists of all real numbers.)

■ **Solution**

Choose some convenient values of x and compute the corresponding values of y, as shown in the following table:

x	-3	-2	-1	0	1	2	3
y	9	4	1	0	1	4	9

Plot the ordered pairs. Then, observing the trend indicated by the points, connect them with a smooth curve. See the graph in Figure 1.17.

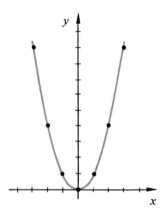

■ Figure 1.17

Comment: Functions of the form $y = ax^2 + bx + c$, where a, b, and c are constants and $a \neq 0$, are called **quadratic functions** because of the "squared" term. Quadratic functions have a U-shaped graph called a **parabola**. The graph of $y = x^2$ shown in Figure 1.17 is typical of the parabolic shape.

EXAMPLE 10 ■ **Problem**

Graph the function $y = \sqrt{4 - x}$.

■ **Solution**

The domain consists of all real numbers x for which $4 - x \geq 0$. (Otherwise, the range values would not be real numbers.) Hence, the domain is the set of all real numbers x such that $x \leq 4$. So, the graph must be on or to the left of the vertical line $x = 4$. See Figure 1.18.

x	y
4	0
2	$\sqrt{2}$
0	2
-2	$\sqrt{6}$

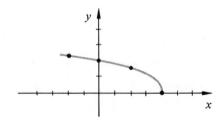

■ Figure 1.18

Increasing and Decreasing Functions

A function f whose graph rises from left to right is said to be an **increasing function**. See Figure 1.19(a). Thus, as x increases, $f(x)$ also increases. A function f whose graph falls from left to right is said to be a **decreasing function**. See Figure 1.19(b). In this case, $f(x)$ decreases as x increases. In general, we speak of a function increasing or decreasing on some interval.

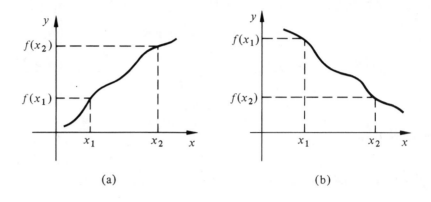

■ **Figure 1.19**   (a)   (b)

EXAMPLE 11 Increasing and decreasing functions are illustrated in Figure 1.20.

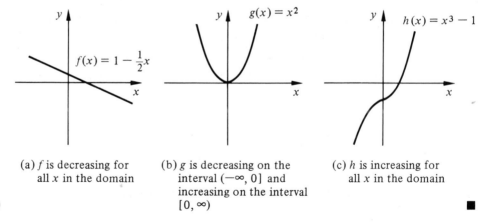

(a) f is decreasing for all x in the domain

(b) g is decreasing on the interval $(-\infty, 0]$ and increasing on the interval $[0, \infty)$

(c) h is increasing for all x in the domain

■ **Figure 1.20**   ■

Defining a Function by a Graph

Every graph consists of ordered pairs of real numbers. If no two points of the graph are on the same vertical line, then the set of points is the graph of a function. The sets of points in Figures 1.21(a) and 1.21(b) are graphs of functions. The sets of points in Figures 1.21(c) and 1.21(d) are graphs of relations that are *not* functions.

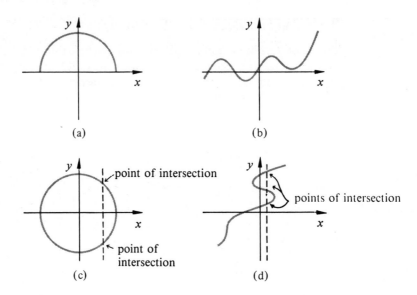

■ **Figure 1.21**

EXERCISES FOR SECTION 1.3

In Exercises 1–9, tell whether the given equation defines a function. (Assume that y is the dependent variable.)

1. $y = 2x + 5$ 2. $y = x^2$

3. $y = 10$ 4. $y < 3x$

5. $y = \sqrt[3]{x}$ 6. $y^2 = x^3$

7. $y = 5x + 1$ 8. $y = \dfrac{1}{x}$

9. $y = \pm\sqrt{x}$

Tell whether the table defines a function in Exercises 10–13.

10.

x	1	1	2	3
y	2	3	4	7

11.

x	2	7	8
y	3	9	3

12.

x	1	3	7	12
y	2	5	1	-2

13.

x	-1	1	1
y	6	2	4

Tell whether the diagram defines a function in Exercises 14–17.

14.

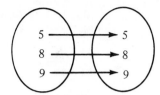

15.

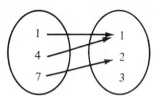

16.

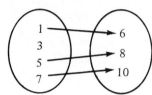

17.

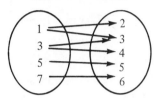

In Exercises 18–24, find the domain and range of the function.

18. $f(x) = 2x$

19. $g(x) = 4 - 3x$

20. $G(x) = \dfrac{1}{x^2}$

21. $y = \dfrac{4}{x + 3}$

22. $y = \dfrac{5}{x - 2}$

23. $f(x) = \sqrt{-x}$

24. $f(x) = \sqrt{1 - x}$

25. For the function $f(x) = 3x + 1$, find the following:

   (a) $f(3)$

   (b) $f(\pi)$

   (c) $f(z)$

   (d) $f(x - h)$

   (e) The element in the domain for which $f(x) = 10$.

   (f) The range of the function.

   (g) $f(f(3))$

26. For the function $G(t) = t^2 - 2t + 1$, find the following:

   (a) $G(2)$

   (b) $G(-1)$

   (c) $G(x^2)$

   (d) $G(x - h)$

   (e) $G(\sqrt{t})$

   (f) $G(G(0))$

Graph the functions in Exercises 27–42.

27. $\{(-2, 3), (0, 4), (2, 5)\}$ 28. $\{(-1, 0), (0, 0), (1, 2), (3, -4)\}$

29. $f(x) = 3x + 5$ 30. $y = 2 - \frac{1}{2}x$

31. $z = t^2 + 4$ 32. $f(x) = -x^2$

33. $y(x) = \sqrt{x}$ 34. $\phi = w^2/2$

35. $p = z^2 - z - 6$ 36. $v = 10 + 2t$

37. $y = \sqrt{16 - 4x^2}$ 38. $y = \sqrt{25 - x^2}$

39. $y = \sqrt{x - 2}$ 40. $y = \sqrt[3]{2 - x}$

41. $y = -x^3$ 42. $\beta = -\sqrt{\alpha}$

43. The work w done in moving an object varies with the distance s according to $w = \sqrt[3]{2s}$. Show this relationship graphically.

44. The path of a certain projectile is described by the function $h = 100x - 2x^2$, where h is the vertical height (in feet) and x is the horizontal displacement (in feet). Draw the path of the projectile.

45. An office machine is supposed to be serviced once a month. If it is not serviced, the cost of repairs is $20 plus five times the square of the number of months the machine goes unserviced. Express the cost of repairs as a function of the number of months the machine goes unserviced, and draw the graph.

46. An archway is to be built in the form of a semiellipse. Draw the shape of the archway if it is described by the function $y = \sqrt{64 - 4x^2}$, where y is the vertical height of the arch and x is the horizontal distance from the centerline of the arch. Both x and y are measured in feet.

In Exercises 47–52, indicate if the given graph determines a function.

47.
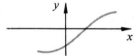

48.
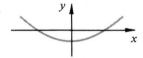

49.

50.
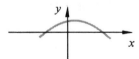

51.
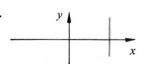

52.

Use the following BASIC program to compute values of the functions given in Exercises 29–42.

```
10   REM   EVALUATION OF A FUNCTION
20   HOME
30   PRINT "THIS PROGRAM WILL EVALUATE ANY"
40   PRINT "CONTINUOUS FUNCTION ON ANY"
50   PRINT "INTERVAL CHOSEN BY THE USER"
60   PRINT
70   PRINT "THE PROGRAM CONTAINS THE FUNCTION"
80   PRINT "     F(X) = 25 - X^2"
90   PRINT "DO YOU WISH TO CHANGE THIS FUNCTION? Y/N"
100  INPUT A$
110  IF A$ = "Y" THEN 310
120  PRINT "ENTER THE ENDPOINTS OF THE INTERVAL YOU WISH TO USE"
130  INPUT A,B
140  PRINT "AT HOW MANY POINTS DO YOU WISH EVALUATION?"
150  INPUT N
160 N = N - 1
170 H = (B - A) / N
180  DEF  FN F(X) = 25 - X ^ 2
190  PRINT "X                F(X)"
200  FOR I = 0 TO N
210 X = A + I * H
220 Y =  FN F(X)
224 X =  INT (1000 * X) / 1000
226 Y =  INT (1000 * Y) / 1000
230  PRINT X,Y
240  NEXT I
250  PRINT "DO YOU WISH TO RUN THIS PROGRAM AGAIN? Y/N"
260  INPUT B$
270  HOME
280  IF B$ = "Y" THEN 90
290  GOTO 340
310  PRINT "TYPE"
320  INVERSE : PRINT "180 DEF FNF(X) = (YOUR FUNCTION) <RET>"
330  PRINT "RUN 120 <RET>": NORMAL
340  END
```

▬▬▬ 1.4 LINEAR FUNCTIONS: STRAIGHT LINES

Functions of the form

$$y = ax + b, \qquad a \neq 0$$

where *a* and *b* are constants, are called **linear functions.** Expressions of this type are also called **linear equations** in two variables. We will use the term *linear function* if we are stressing the functional concept of pairing *x* and *y* and *linear equation* otherwise. As noted in the previous section, the graph of a linear function is a straight line. In this section, we look at some of the important properties of straight lines.

Slope of a Straight Line

An important property for describing the orientation of a straight line in the plane is its inclination with respect to the x-axis. The inclination of a line is described by giving either an angle of inclination or a slope.

The **angle of inclination** of a straight line is defined as the smallest positive angle between the line and the positive x-axis. Hence, the angle of inclination, denoted by α in Figure 1.22 is always less than 180°. If a line is parallel to the x-axis, its angle of inclination is defined to be zero.

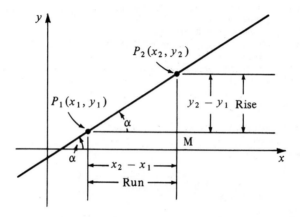

■ **Figure 1.22**

The inclination of a straight line is also described by giving its slope. We define the **slope** of a line as the ratio of the vertical rise of the line to the corresponding horizontal run. That is,

$$\text{slope} = \frac{\text{vertical rise}}{\text{horizontal run}}$$

To find the slope of a straight line in the coordinate plane, we use the coordinates of any two points on the line. If the coordinates of two points on a line are known, the slope of the line is defined as the difference in the ordinates of the two points divided by the difference in the corresponding abscissas. When we apply this definition to the line segment $P_1 P_2$ in Figure 1.22, the slope m of the line is expressed by the following formula:

FORMULA

> **Slope Formula**   The slope of a line through (x_1, y_1) and (x_2, y_2) is
>
> $$m = \frac{y_2 - y_1}{x_2 - x_1}$$
>
> (1.4)

EXAMPLE 1   ■ **Problem**

Find the slope of the straight line passing through the points $(-5, 1)$ and $(2, -3)$. See Figure 1.23.

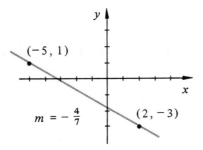

■ **Figure 1.23**

■ **Solution**

Letting $(x_1, y_1) = (-5, 1)$, and $(x_2, y_2) = (2, -3)$, and using Equation (1.4), we find that the slope is

$$m = \frac{y_2 - y_1}{x_2 - x_1} = \frac{-3 - 1}{2 - (-5)} = \frac{-4}{7} = -\frac{4}{7}$$

If we interchange the labels of the given points and let $(x_1, y_1) = (2, -3)$ and $(x_2, y_2) = (-5, 1)$, the result is

$$m = \frac{y_2 - y_1}{x_2 - x_1} = \frac{1 - (-3)}{-5 - 2} = \frac{4}{-7} = -\frac{4}{7}$$

Thus, the order in labeling the given points does not matter. We interpret a slope of $-\frac{4}{7}$ to mean that for every seven units moved to the right in the direction of the x-axis, the straight line moves down four units.   ■

An interesting and important relationship exists between the angle of inclination of a line and its slope. Referring to Figure 1.22, we see that α is also an angle of the right triangle P_1MP_2. The side opposite angle α is $y_2 - y_1$, and the side adjacent to angle α is $x_2 - x_1$. From trigonometry, the ratio of the side opposite an angle to the side adjacent to the angle is the tangent of the angle. Thus,

$$\tan \alpha = \frac{y_2 - y_1}{x_2 - x_1} \tag{1.5}$$

But the ratio on the right side of Equation (1.5) is also the slope of the line. Therefore, the slope of a line is equal to the tangent of the angle of inclination, and we write

$$m = \tan \alpha \tag{1.6}$$

From Equation (1.6), we see that lines that have angles of inclination greater than 90° and less than 180° have negative slopes, because the tangent of an obtuse angle is negative. Also, the slope of a line parallel to the y-axis is undefined because tan 90° is undefined.

EXAMPLE 2   ▪ **Problem**

Find the angle of inclination of the straight line shown in Figure 1.23.

▪ **Solution**

The slope of the line is $m = -\frac{4}{7}$. Therefore, by Equation (1.6), we have

$$\tan \alpha = m = -\tfrac{4}{7} = -0.5714$$

The angle α is an obtuse angle since $\tan \alpha$ is negative. Using a calculator, we find

$$\text{Arctan}(0.5714) = 29.7°$$

Hence, the angle of inclination is

$$\alpha = 180° - 29.7° = 150.3°$$ ▪

In working with straight lines, the following facts about slopes are useful:

- The slope of a straight line is positive if, as you follow the curve from left to right, you move up on the coordinate axes. See Figure 1.24(a). Thus, the ratio of rise to run is always positive when the straight line is in this position.
- The slope of a straight line parallel to the x-axis is zero, because the rise is zero for any run. See Figure 1.24(b).
- The slope of a straight line is negative if, as you follow the curve from left to right, you move down on the coordinate axes. From Figure 1.24(c), we see that when the run is positive, the rise is negative, and when the run is negative, the rise is positive. The ratio of rise to run (that is, the slope) in this case must be negative.

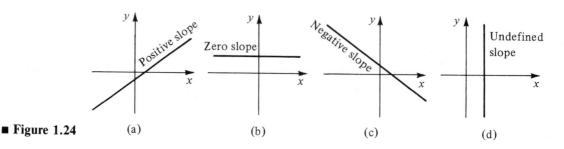

▪ **Figure 1.24**   (a)   (b)   (c)   (d)

- The slope of a straight line parallel to the y-axis is undefined, because the run is zero for any rise. Therefore, to apply Formula (1.4), we would have to divide by zero. (Remember, division by zero is an undefined operation.) See Figure 1.24(d).
- Parallel straight lines have equal slopes.
- Perpendicular straight lines have slopes that are negative reciprocals. That is, $m_1 = -1/m_2$ ($m_2 \neq 0$ and neither line is vertical).

The last statement may not be completely obvious. In Figure 1.25, we show two perpendicular lines with slope $m_1 = \tan \alpha_1$ and $m_2 = \tan \alpha_2$. Then, since α_1 and α_2 differ by 90°, we have $\tan \alpha_2 = \tan(\alpha_1 + 90°) = -\cot \alpha_1 = -1/\tan \alpha_1$, or

$$m_2 = -\frac{1}{m_1}$$

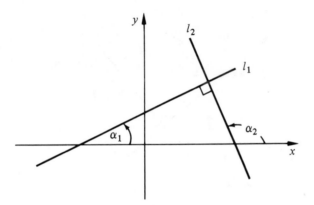

■ **Figure 1.25**

EXAMPLE 3   ■ **Problem**

Draw the graph of the equation $y = 3$. What is its slope?

■ **Solution**

The graph of $y = 3$ is a straight line passing through the point $(0, 3)$ parallel to the x-axis; (see Figure 1.26). The slope of this line is 0 since it is parallel to the x-axis.

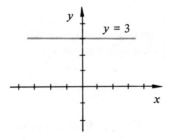

■ **Figure 1.26**

EXAMPLE 4   ▪ **Problem**

Find the slope of a line passing through $(2, 1)$ and $(-4, 6)$. Then, find the slope of a line drawn perpendicular to the given line at $(2, 1)$. See Figure 1.27.

▪ **Solution**

The slope m of the given line is

$$m = \frac{y_2 - y_1}{x_2 - x_1} = \frac{6 - 1}{-4 - 2} = -\frac{5}{6}$$

The slope m of a line perpendicular to the given line is

$$m = -\frac{1}{m} = -\frac{1}{-5/6} = \frac{6}{5}$$

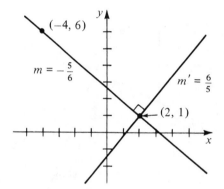

▪ **Figure 1.27**

■

Since the slope is the ratio of a change in y to the corresponding change in x, it is also referred to as the rate of change of y with respect to x. Rates of change that occur in applied mathematics have interpretations other than the slope of a line. The following example illustrates one such interpretation.

EXAMPLE 5   ▪ **Problem**

The average velocity of an object is defined as the distance traveled divided by the corresponding elapsed time. If the object has traveled s_1 units at time t_1, and s_2 units at time t_2, the average velocity of the object is given by

$$V_{avg} = \frac{s_2 - s_1}{t_2 - t_1}$$

A car travels a distance of 100 mi in 2.5 hr. What is the average velocity of the car?

■ **Solution**

In this problem the distance traveled is

$$s_2 - s_1 = 100$$

The corresponding time is

$$t_2 - t_1 = 2.5$$

Therefore, the average velocity of the car is

$$V_{avg} = \frac{100}{2.5} = 40 \text{ mi/hr}$$ ■

Methods of Describing a Line

Whereas every straight line can be represented by a linear equation, sometimes certain *forms* of representation are important. Two important forms of straight lines are the point-slope form and the slope-intercept form. Each form shows two properties of a line that can be determined by inspection.

   The **point-slope form** of a straight line is used when we know the slope and one point on the line. See Figure 1.28. To obtain the point-slope form of a straight line, we choose an arbitrary point $P(x, y)$ different from $P_1(x_1, y_1)$ on the straight line. Then, by the definition of the slope of a straight line, we have

$$\frac{y - y_1}{x - x_1} = m$$

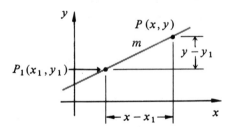

■ **Figure 1.28**

Rearranging the terms yields the point-slope formula.

FORMULA

The Point-Slope Form of a Straight Line   The equation of a straight line passing through (x_1, y_1) with slope m is

$$y - y_1 = m(x - x_1) \qquad\qquad (1.7)$$

Note that since *any* fixed point (x_1, y_1) may be used in (1.7), the equation will *appear* different for various choices of the point. But the resulting equations are equivalent.

EXAMPLE 6   The point-slope form of the equation representing the line that passes through (2, 1) with slope $\frac{1}{3}$ is

$$y - 1 = \tfrac{1}{3}(x - 2)$$

However, the two points $(0, \frac{1}{3})$ and $(-1, 0)$ are also on the line. So two other point-slope forms of the same line are

$$y - \tfrac{1}{3} = \tfrac{1}{3}(x - 0) \qquad \text{and} \qquad y = \tfrac{1}{3}(x + 1)$$

Infinitely many other point-slope forms of this equation are possible.   ■

EXAMPLE 7   ■ **Problem**

The amount that a spring stretches is directly proportional to the applied force. See Figure 1.29. This relation is known as Hooke's law. Find the equation relating the length d of a spring to applied force f, if the length of the spring is 5 in. when a 6-lb weight is applied and 7 in. when a 14-lb weight is applied. Assume $f \geq 0$.

■ **Solution**

Since spring length is directly proportional to applied force, the desired equation is linear. Figure 1.30 shows this relation graphically. The slope of the straight line between these two points is then given by

$$m = \frac{d_2 - d_1}{f_2 - f_1} = \frac{7 - 5}{14 - 6} = \frac{2}{8} = \frac{1}{4}$$

■ **Figure 1.29**

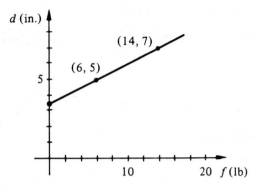

The point-slope form of this straight line can be written in the form

$$d - d_1 = m(f - f_1)$$

Now letting $(f_1, d_1) = (6, 5)$ and $m = \frac{1}{4}$, we have

$$d - 5 = \tfrac{1}{4}(f - 6)$$

or

$$4d - f = 14$$

Since $d = \frac{7}{2}$ when $f = 0$, the free length of the spring is 3.5 in.   ■

If the point used in the point-slope form is $(0, b)$ we have the special case of the **slope-intercept form.** Substituting $x_1 = 0$ and $y_1 = b$ into the point-slope form yields

$$\frac{y - b}{x} = m$$

Solving for y we have the slope-intercept formula.

FORMULA

> **The Slope-Intercept Form of a Straight Line**   The equation of a straight line passing through $(0, b)$ with slope m is
>
> $$y = mx + b \qquad\qquad (1.8)$$

To obtain the slope-intercept form of a linear equation, solve for the variable y. Then, the constant on the right-hand side is the y-intercept, and the coefficient of x is the slope.

EXAMPLE 8   ■ **Problem**

Rearrange the linear equation $3x + 2y = -5$ into slope-intercept form and draw its graph.

■ **Solution** .

To write in slope-intercept form, solve for y:

$$2y = -3x - 5$$

$$y = \underbrace{-\frac{3}{2}x}_{m} + \underbrace{\left(-\frac{5}{2}\right)}_{b}$$

The slope is $-\frac{3}{2}$ and the y-intercept is $-\frac{5}{2}$. See Figure 1.31.

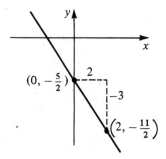

■ **Figure 1.31**   ■

EXAMPLE 9   ■ **Problem**

Suppose that the cost of insurance, maintenance, and parking for your car is $1100 per year and the cost for gas and oil is $0.07 per mile. Then, the total annual cost C of operating your car is given by the linear function

$$C = 1100 + 0.07m$$

where m is the miles driven during the year. Show this function graphically.

■ **Solution**

We consider C as the dependent variable. Hence, the graph of the function intersects the C axis at $(0, 1100)$. The line has slope $m = 0.07$, as shown in Figure 1.32. The graph is not extended to the left of the C-axis because a negative amount of mileage is meaningless. In this case, the slope of the line indicates the actual cost per mile over and above a certain fixed cost of $1100 that you spend to maintain and operate your car.

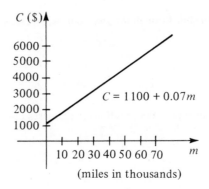

■ Figure 1.32

$C = 1100 + 0.07m$

(miles in thousands)

EXERCISES FOR SECTION 1.4

Sketch the straight line determined by the pairs of points given in Exercises 1–4, and compute the slope of the line.

1. $(1, 2), (5, 4)$

2. $(-1, -1), (3, -6)$

3. $(-5, 2), (3, -7)$

4. $(7, 3), (0, 5)$

5. What is the slope of a 25-ft ladder that is leaning against a building if the foot of the ladder is 12 ft from the building?

6. What is the slope of a 120-ft support cable for a broadcasting antenna if the cable is anchored 75 ft from the base of the antenna?

7. The current (in amperes, A) in a resistor (units in ohms, Ω) is defined as the charge transferred (in coulombs, C) divided by the corresponding time (in seconds). What is the current in a 5-Ω resistor if the charge varies 15 C in 3 sec?

8. The acceleration of a particle is defined as the change in velocity divided by the corresponding change in time. If the initial velocity is 25 ft/sec and the velocity is 40 ft/sec at the end of 10 sec, what is the acceleration? What units does acceleration have?

9. The velocity of a car is 22.5 ft/sec when $t = 0.5$ sec, and 31.4 ft/sec when $t = 0.8$ sec. What is the acceleration of the car? See Exercise 8 for the definition of acceleration.

10. A charged particle is 3.2 cm from a reference point when $t = 2$ sec and 10.7 cm from the point when $t = 5$ sec. What is the average velocity of the particle?

In Exercises 11–14, draw the line passing through the given point with the given slope, and then determine its equation.

11. $(2, 5), m = \frac{1}{2}$

12. $(-1, -3), m = 3$

13. $(5, -2), m = -7$

14. $(3, 4), m = -\frac{2}{5}$

In Exercises 15–20, draw the line through the given points, and then find its equation.

15. $(1, 3), (6, 2)$

16. $(2, 5), (-3, -7)$

17. $(-1, -1), (1, 2)$

18. $(0, 0), (3, -2)$

19. $(0, 2), (-5, 0)$

20. $(\frac{1}{2}, \frac{1}{3}), (-\frac{1}{2}, \frac{1}{3})$

Find the slope and the y-intercept of the equations in Exercises 21–32, and then draw the line.

21. $2x - 3y = 5$

22. $3x + 4y = 0$

23. $x + y = 2$

24. $5x + 2y = -3$

25. $4y - 2x + 8 = 0$

26. $-5x - y - 2 = 0$

27. $2y = x + 5$

28. $3x + 6 = 2y$

29. $y = 5$

30. $x + y = 1$

31. $x - 5y + 7 = 0$

32. $y = 1 - x$

In Exercises 33–42, write the equation of the line perpendicular to the given line at the indicated point.

33. $3x + 2y = 7$ at $(1, 2)$

34. $x + 3y = 11$ at $(2, 3)$

35. $x - y = 2$ at $(5, 3)$

36. $2x - 5y = 2$ at $(-4, -2)$

37. $2y - 3x = 1$ at $(1, 2)$

38. $y = x$ at $(0, 0)$

39. $5x - 7y = 0$ at $(0, 0)$

40. $5x - 7y = 3$ at $(2, 1)$

41. $-x + 3y = 5$ at $(-2, 1)$

42. $-2y - x = 3$ at $(-1, -1)$

43. Show that the equation of the line perpendicular to $Ax + By = C$ at (x_1, y_1) is $Bx - Ay = Bx_1 - Ay_1$. *Hint:* Use the fact that the product of the slopes equals -1.

44. Express degrees Celsius as a linear function of degrees Fahrenheit if $0°C$ corresponds to $32°F$ and $100°C$ corresponds to $212°F$, as indicated in the following figure.

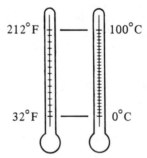

45. The current I in a resistor is a linear function of the applied voltage V (in volts, V). Find the equation relating the current to the voltage if the current is $\frac{1}{2}A$ when the voltage is 6 V and $\frac{2}{3}A$ when the voltage is 8 V.

46. In an experiment to determine the coefficient of friction, it is found that a 10-lb block has a frictional force of 3 lb and a 25-lb block has a frictional force of 7.5 lb. What is the equation relating frictional force to weight if the frictional force is a linear function of weight?

47. *Linear depreciation* is one of several methods approved by the Internal Revenue Service for depreciating business property. If the original cost of the property is C dollars and if it is depreciated linearly over N years, its value V remaining at the end of n years is given by

$$V = C - \frac{n}{N}C$$

Find the value after 5 years of a typewriter whose initial cost of $300 is to be depreciated over 20 years.

48. If you borrow P dollars at the simple interest rate i, the annual interest is $P \cdot i$. Hence, the amount A owed at the end of n years is given by

$$A = P(1 + i \cdot n)$$

Find the amount you owe after 5 years if you borrow $2000 at 8% simple interest.

49. Mr. Smith wants to borrow $4000 to buy a new car. He wishes to pay off the loan with monthly payments stretching over 3 years. If he is charged 12% simple interest and if he computes the monthly payments by dividing the total amount due in 3 years by 36, how much will Mr. Smith have to pay each month?

50. A manufacturer of fountain pens can expect to sell 21,000 felt-tip pens if she charges $0.80 per pen but only 10,000 if she raises the price to $1.00. Assuming that the relationship is linear, find the equation of the line relating the number of pens to their price. How many felt-tip pens can she sell if she charges $0.90?

The following BASIC program finds the equation of the straight line passing through two given points. Use this program for Exercises 15–20.

```
10   REM  EQUATION OF A LINE
20   HOME
30   PRINT "THIS PROGRAM WILL FIND THE EQUATION OF"
40   PRINT "A LINE AFTER THE USER INPUTS TWO POINTS.
50   PRINT
60   PRINT "ENTER X,Y FOR THE FIRST POINT"
70   INPUT X1,Y1
80   PRINT "ENTER X,Y FOR THE SECOND POINT"
90   INPUT X2,Y2
100  D = X2 - X1
110   IF D = 0 THEN 170
120  M = (Y2 - Y1) / D
130  B = Y2 - M * X2
140   PRINT "THE EQUATION OF THE LINE IS"
144  M =  INT (1000 * M) / 1000
146  B =   INT (1000 * B) / 1000
150   PRINT "Y = ";M;"*X + ";B
160   GOTO 190
170   PRINT "THE EQUATION OF THE LINE IS"
180   PRINT "X = ";X1
190   PRINT
200   PRINT "DO YOU WISH TO FIND ANOTHER LINE? Y/N"
```

```
210   INPUT A$
220   HOME
230   IF A$ = "Y" THEN 60
240   END
```

1.5 QUADRATIC FUNCTIONS: PARABOLIC GRAPHS

Functions that can be put into the form

$$f(x) = ax^2 + bx + c$$

where a, b, and c are constants and $a \neq 0$, are called **quadratic functions**. Functions of this type arise naturally in describing physical quantities such as the path of a projectile, the shape of a radar antenna, and the area of a circle.

EXAMPLE 1   **a.** $f(x) = x^2 - 2x + 1$ is a quadratic function with $a = 1$, $b = -2$, and $c = 1$.
b. $g(x) = 9 - x^2$ is a quadratic function with $a = -1$, $b = 0$, and $c = 9$.
c. $h(x) = x^2 + 3x$ is a quadratic function with $a = 1$, $b = 3$, and $c = 0$.   ■

The graph of a quadratic function is a U-shaped curve called a **parabola**. The parabola is the graph of the quadratic function, just as the straight line is the graph of the linear function. The graphs of $y = x^2$ and $y = -x^2$, which are shown in Figure 1.33, typify the parabolic shape. The parabola in Figure 1.33(a) opens upward, and the one in 1.33(b) opens downward. The low (or high) point of the parabola is called the **vertex**. The parabola is **symmetric** about a vertical line through its vertex.

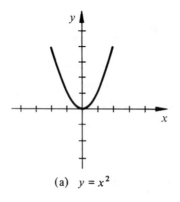

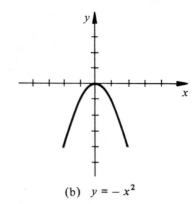

■ **Figure 1.33**   (a) $y = x^2$ (b) $y = -x^2$

The constants a, b, and c in the quadratic function $ax^2 + bx + c$ determine the location and the shape of the parabola. As a matter of fact, the vertex of the parabola is at the origin only if b and c are zero. If b or c is a nonzero coefficient, we can show that the coordinates of the vertex of the parabola are as follows:

FORMULA

> **The Coordinates of the Vertex of a Parabola**   The vertex of the parabola for $y = ax^2 + bx + c$ is located at
>
> $$x = -\frac{b}{2a}, \quad y = f\left(-\frac{b}{2a}\right)$$

For example, the vertex of the graph of $y = 2x^2 - 6x$ is located at

$$x = \frac{-(-6)}{2(2)} = \frac{3}{2}, \quad y = 2\left(\frac{3}{2}\right)^2 - 6\left(\frac{3}{2}\right) = -\frac{9}{2}$$

See Figure 1.34.

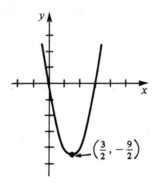

$$\left(\frac{3}{2}, -\frac{9}{2}\right)$$

■ **Figure 1.34**

To sketch the graph of a quadratic function, use the following procedure:

PROCEDURE

> **Sketching the Graph of $f(x) = ax^2 + bx + c$**
>
> 1. Determine the sign of the x^2 coefficient. If $a > 0$, the parabola opens upward; if $a < 0$, it opens downward.
> 2. Determine the y-intercept by letting $x = 0$. That is, $f(0) = c$ is the y-intercept.
> 3. Determine the x-intercepts by letting $f(x) = 0$. The x-intercepts are given by
>
> $$x = \frac{-b \pm \sqrt{b^2 - 4ac}}{2a}$$

If $b^2 - 4ac > 0$, the graph has two x-intercepts. If $b^2 - 4ac = 0$, the graph has one x-intercept. If $b^2 - 4ac < 0$, the graph has no x-intercepts and, therefore, does not cross the x-axis. If $f(x)$ is factorable, finding the x-intercepts from the factors is sometimes easier.

4. Determine the location of the vertex by
$$x = -b/2a \quad \text{and} \quad y = f(-b/2a).$$

EXAMPLE 2   ▪ **Problem**

Sketch the graph of $f(x) = x^2 - x - 6$.

▪ **Solution**

Note that $a = 1, b = -1$, and $c = -6$.

1. The parabola opens upward since $a > 0$.
2. The y-intercept is $f(0) = -6$.
3. The x-intercepts are found by letting $f(x) = 0$. Thus,
$$x^2 - x - 6 = 0$$
    Observing that the quadratic is factorable, we have
$$(x - 3)(x + 2) = 0$$
    Therefore, the x-intercepts are $x = 3$ and $x = -2$.
4. To locate the vertex, note that
$$x = -\frac{b}{2a} = -\frac{-1}{2(1)} = \frac{1}{2}$$
and
$$y = f\left(\frac{-b}{2a}\right) = \left(\frac{1}{2}\right)^2 - \frac{1}{2} - 6 = -\frac{25}{4}$$

Hence, the vertex is located at $(\frac{1}{2}, -\frac{25}{4})$. The graph of $f(x) = x^2 - x - 6$ is shown in Figure 1.35.

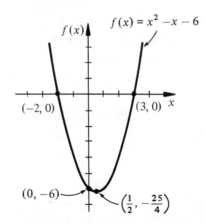

▪ **Figure 1.35**

EXAMPLE 3   ■ **Problem**

Sketch the graph of $y = x^2 + 3x$.

■ **Solution**

Here $a = 1, b = 3$, and $c = 0$.

1.   The parabola opens upward since $a > 0$.
2.   The y-intercept is $y(0) = 0$.
3.   The x-intercepts are obtained from $x^2 + 3x = 0$. Since $x^2 + 3x$ factors into $x(x + 3)$, we conclude that the x-intercepts are $x = 0$ and $x = -3$.
4.   The vertex is at

$$x = -\frac{3}{2(1)} = -\frac{3}{2}$$

and

$$y\left(-\frac{3}{2}\right) = \left(-\frac{3}{2}\right)^2 + 3\left(-\frac{3}{2}\right) = \frac{9}{4} - \frac{9}{2} = -\frac{9}{4}$$

These points and our knowledge of the parabolic shape yield the graph in Figure 1.36.

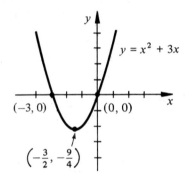

y = x² + 3x

(−3, 0)   (0, 0)

$\left(-\frac{3}{2}, -\frac{9}{4}\right)$

■ **Figure 1.36**   ■

Sometimes, we want to find a quadratic function whose graph has x-intercepts at $x = a$ and $x = b$. Then, the function has linear factors $(x - a)$ and $(x - b)$. Hence, the quadratic function has the form $f(x) = k(x - a)(x - b)$, where k is a constant. The value of k can be determined if an additional point on the parabola is known.

EXAMPLE 4   ■ **Problem**

Determine the form of the quadratic function whose graph is shown in Figure 1.37.

■ **Solution**

1.   The x-intercepts are 5 and -3. Hence, the form of the function is

$$y = k(x - 5)(x + 3)$$

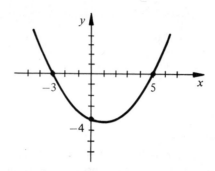

■ **Figure 1.37**

2. Since the y-intercept is -4, substitute $x = 0$ and $y = -4$ into the functional expression to get

$$-4 = k(-5)(3)$$

or

$$k = \frac{4}{15}$$

The desired function is then

$$y = \frac{4}{15}(x - 5)(x + 3) = \frac{4}{15}(x^2 - 2x - 15)$$

$$= \frac{4}{15}x^2 - \frac{8}{15}x - 4$$

■

EXERCISES FOR SECTION 1.5

In Exercises 1–20 sketch the graph of the given quadratic function. Give the domain, range, and x-intercepts of each function. Also, indicate the y-intercept and the vertex of each.

1. $f(x) = x^2 - 4$
2. $y = 2x^2 - 6$
3. $y = x^2 + 1$
4. $g(t) = t^2 + 5$
5. $f(x) = 9 - x^2$
6. $y = x^2 + 2x$
7. $x(t) = t^2 - 3t$
8. $s = 3t - 16t^2$
9. $y = x^2 - 5x - 6$
10. $y = x^2 + 7x + 12$
11. $y = 3x^2 + 2x + 1$
12. $y = x^2 + 7x + 13$
13. $y = 2 - x^2 + x$
14. $y = x - x^2 - 1$
15. $f(x) = 2x^2 - 3x - 5$
16. $f(t) = -2t^2 - 4t + 5$
17. $y = 3 + x - x^2$
18. $g(x) = 3 + x + x^2$
19. $G(x) = 1 + x + x^2$
20. $f(x) = \frac{1}{2}x^2 - x - 2$

In Exercises 21–30, determine the quadratic function whose graph is given.

21.

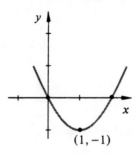

$(1, -1)$

22.

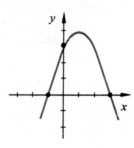

23.

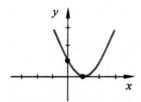

24.

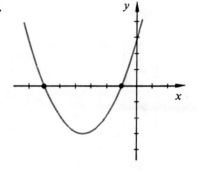

25.

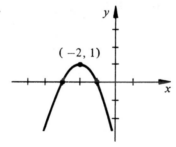

$(-2, 1)$

26.

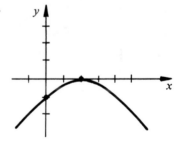

27.

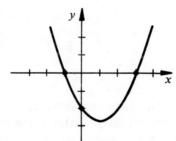

28.

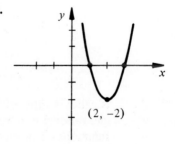

$(2, -2)$

29.   **30.**

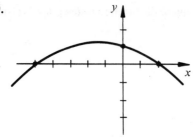   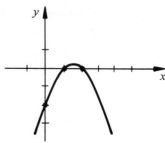

31. If $f(x) = x^2$, how are the graphs of $f(x)$ and $-f(x)$ related?

32. If $f(x) = x^2$, how are the graphs of $f(x)$ and $f(-x)$ related?

 33. A radar antenna is to be built with a cross section that is described by the function $y = 0.05x^2$, where x and y are in feet. Draw the cross section of the antenna if it has a diameter of 15 ft.

 34. When an object is thrown into the air, its height above the ground is a quadratic function of time. Sketch the graph of height versus time from $t = 0$ to $t = 3$ if the functional relationship is $h = 48t - 16t^2$. What is the object's maximum height?

 35. Sketch the graph of electric current versus time, given that current varies according to $i = t^2 - 5t + 6$ (in amperes). When is the current zero? Assume the domain is all $t \geq 0$.

 36. The load being applied to a cantilever beam varies with the distance from the fixed end according to $F = 32 - 2x^2$ (in pounds). Sketch the graph of this quadratic function from $x = 0$ to $x = 4$. Where does the maximum load occur?

 37. The velocity of a certain rocket varies with time according to $v = 100 + 200t - 25t^2$, where v is in feet per second when t is in seconds. Sketch the graph of the velocity function for $t \geq 0$. What is the maximum velocity of the rocket? At what time is the velocity equal to zero?

 38. Using the graph from Exercise 37, estimate the time it takes for the velocity of the rocket to equal its initial velocity.

▬▬▬ 1.6 RATIONAL FUNCTIONS

Functions that are the ratio of two polynomials are called **rational functions.** Specifically, if $N(x)$ and $D(x)$ are polynomials without common factors, then

$$f(x) = \frac{N(x)}{D(x)}$$

is called a rational function of x. Notice the similarity between rational functions and rational numbers. Rational numbers are quotients of integers, whereas rational functions are ratios of polynomials. Some examples of rational functions are

$$\frac{3x^5 + 5}{20x^3 + x - 16}, \quad \frac{4}{x^2 - 4}, \quad \frac{3x^2 + 5x - 6}{(x - 5)(x + 3)}$$

We will use

$$f(x) = \frac{1}{x}$$

to illustrate the typical characteristics of graphs of rational functions. The function $f(x) = 1/x$ is not defined at $x = 0$; however, it is defined for all other values of x. To determine the behavior of the graph, we note that as x approaches zero from the right, $f(x)$ is positive and increases indefinitely. We say that $f(x)$ **increases without bound.** Similarly, as x approaches zero from the left, $f(x)$ is negative and decreases indefinitely. Using this knowledge and computing some points on either side of $x = 0$ yields the graph shown in Figure 1.38. The vertical line at $x = 0$ is called a **vertical asymptote** of the graph.

x	y
-3	$-\frac{1}{3}$
-2	$-\frac{1}{2}$
-1	-1
$-\frac{1}{2}$	-2
$-\frac{1}{4}$	-4
$\frac{1}{4}$	4
$\frac{1}{2}$	2
1	1
2	$\frac{1}{2}$
3	$\frac{1}{3}$

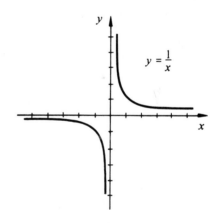

■ **Figure 1.38**

Figure 1.38 also shows that the graph comes closer to the x-axis as x increases. Similarly, as x decreases, the graph comes closer to the x-axis, although negative. In fact, as x increases, the value of $1/x$ approaches zero. (But no matter how large x becomes, $1/x$ will not equal zero.) The same observation holds as x decreases. The line $y = 0$ is called a **horizontal asymptote** of the graph.

From the foregoing discussion, you can see that the asymptotes of a rational function are very helpful in sketching its graph. Use the following definition to determine the vertical asymptotes of a rational function in which $N(x)$ and $D(x)$ have no common factors.

RULE

> **Finding the Vertical Asymptote**   If $D(x)$ represents the denominator of a rational function and if $D(k) = 0$, then $x = k$ is a vertical asymptote.

A rational function may have more than one vertical asymptote, determined by the values of x for which the denominator is zero. Thus, the rational function $(2x^2 + 5)/(x^2 - 9)$ has two vertical asymptotes, the lines $x = 3$ and $x = -3$.

The graph of a rational function may have at most one horizontal asymptote. To find the horizontal asymptote, or to determine that one does not exist, use the following rule.

RULE

> **Finding the Horizontal Asymptote**   Divide the numerator and the denominator of the function by the highest power of x occurring in either. If the resulting expression approaches a constant c as $|x|$ becomes large, then $y = c$ is a horizontal asymptote. If the expression becomes large as $|x|$ becomes large, there is no horizontal asymptote.

To illustrate the technique, we consider $y = (2x^2 + 5)/(x^2 - 9)$. Dividing the numerator and the denominator of this function by x^2 yields

$$y = \frac{2 + 5/x^2}{1 - 9/x^2}$$

Now as $|x|$ becomes large, both $5/x^2$ and $9/x^2$ approach zero. Hence, the value of the function approaches 2 as $|x|$ becomes large, and we conclude that $y = 2$ is a horizontal asymptote.

As an example of a rational function that does not have a horizontal asymptote, consider

$$y = \frac{x^3 - 6}{5x + 1}$$

Dividing the numerator and the denominator by x^3 yields

$$y = \frac{1 - 6/x^3}{5/x^2 + 1/x^3}$$

As $|x|$ becomes large, both terms in the denominator approach zero and the numerator gets close to one. Therefore $|y|$ becomes large.

PROCEDURE	**Sketching Graphs of Rational Functions**
	1. Locate any vertical and horizontal asymptotes by the methods just described. Draw the asymptotes as dashed lines on the coordinate axes.
	2. Plot any x-intercepts of the graph. The x-intercepts are found by equating the numerator to zero and solving for x.
	3. Plot any y-intercept of the graph. The y-intercept is found by equating x to zero.
	4. Plot a point or two for values of x near any vertical asymptotes of the graph.
	5. Draw the graph through the selected points so that it approaches the asymptotes identified in step 1.

EXAMPLE 1   ■ **Problem**

Sketch the graph of $y = \dfrac{3x}{x - 2}$.

■ **Solution**

Intercepts: $x = 0, y = 0$.

Vertical asymptote: The denominator of the function is zero for $x = 2$. Since the numerator and denominator have no common factors, the line $x = 2$ is a vertical asymptote.

Horizontal asymptote: To find the horizontal asymptote, divide the numerator and the denominator by x to get

$$y = \frac{3}{1 - 2/x}$$

As x becomes large, the fraction $2/x$ approaches zero, and therefore, the value of y approaches 3. The line $y = 3$ is the horizontal asymptote.

Finally, determine some additional points in the vicinity of the vertical asymptote, as shown in the table in Figure 1.39. The graph is now readily determined. See Figure 1.39.

x	y
-2	$\frac{3}{2}$
-1	1
0	0
1	-3
2	undef.
3	9
4	6

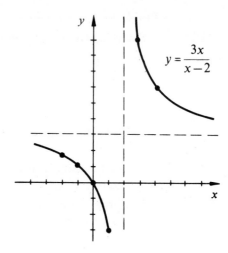

$y = \dfrac{3x}{x-2}$

■ Figure 1.39

EXAMPLE 2   ■ **Problem**

Sketch the graph of $y = \dfrac{2x}{x^2 - 4}$.

■ **Solution**

Vertical asymptotes: The graph of this function has vertical asymptotes at $x = 2$ and $x = -2$, since these are the zeros of the demoninator.

Horizontal asymptotes: Divide the numerator and denominator by x^2 to get

$$y = \frac{2/x}{1 - 4/x^2}$$

As x gets very large, the numerator approaches zero and the denominator approaches one. Therefore, $y = 0$ is the horizontal asymptote. The graph is then determined by plotting a few additional points, as shown in Figure 1.40.

x	y
-4	$-\frac{2}{3}$
-3	$-\frac{6}{5}$
-1	$\frac{2}{3}$
0	0
1	$-\frac{2}{3}$
3	$\frac{6}{5}$
4	$\frac{2}{3}$

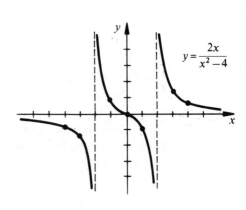

$y = \dfrac{2x}{x^2 - 4}$

■ Figure 1.40

EXAMPLE 3   ▪ **Problem**

The heat generated by a certain chemical process varies with time according to

$$H = \frac{t}{t^2 + 4}$$

where H is the heat (in calories) and t is the time (in seconds). Draw the graph of this function, and find the maximum heat output.

▪ **Solution**

There are no vertical asymptotes since there is no real value of t for which $t^2 + 4 = 0$. However, the t-axis is a horizontal asymptote. Since a negative time has no physical significance, we consider only positive values of t. Plotting a few additional points and drawing the curve so that it approaches zero as t increases, we obtain the graph in Figure 1.41. The graph illustrates that the heat output increases rapidly at first, reaching a peak of 0.25 cal at about 2.0 sec. From this point, the heat output decreases gradually toward zero.

$$H = \frac{t}{t^2 + 4}$$

t	H
0	0
2	0.25
4	0.20
6	0.15
8	0.12
10	0.09

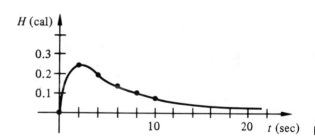

▪ **Figure 1.41**

EXERCISES FOR SECTION 1.6

Compute intercepts, and locate asymptotes (if any). Then, graph the rational functions in Exercises 1–18.

1. $y = -\dfrac{1}{x}$

2. $y = \dfrac{5x + 2}{2 - x}$

3. $y = \dfrac{3x - 2}{x - 4}$

4. $z = \dfrac{2}{t - 1}$

5. $p = \dfrac{1}{w + 1}$

6. $v = \dfrac{t^2 - 1}{t^2 + 1}$

7. $s = \dfrac{5w^2}{4 - w^2}$

8. $m = \dfrac{u + 1}{u^2 - 1}$

9. $y = \dfrac{x^2}{x^2 - 9}$

10. $y = \dfrac{-2x}{x^2 - 4}$

11. $s = \dfrac{(t + 2)^2}{t^2 + 2t}$

12. $w = \dfrac{3x^2}{x^2 - 3x}$

13. $y = \dfrac{3}{w^2 - 9w}$

14. $r = \dfrac{5z + 10}{3z - z^2}$

15. $y = \dfrac{1}{p^2 - 3p - 4}$

16. $n = \dfrac{1 + s^2}{s^2 - s - 2}$

17. $y = \dfrac{5x + 3}{x^2 + x}$

18. $y = \dfrac{x^2 + 3x + 2}{x^2 + x}$

19. The force between two unit masses varies with the distance between them according to the formula $F = 1/d^2$. The force is in dynes when the distance is measured in centimeters. Show this relationship graphically.

20. Owing to leakage, the pressure in a hydraulic system varies with time according to $P = 10/(t^2 + 1)$, in pounds per square inch (psi) when t is the time in seconds. Show this pressure variation graphically.

21. The coefficient of friction μ of a plastic block sliding on an aluminum table varies according to $\mu = (v^2 + 3)/(4v^2 + 5)$, where v is the velocity of the block (in centimeters per second). Draw the graph of this function.

■ 1.7 MULTIRULE FUNCTIONS

Functions that require more than one rule to completely describe their behavior are called **multirule functions**. To write a multirule function, we express each rule and indicate the interval for which it applies.

EXAMPLE 1   ■ **Problem**

Draw the graph of f if the function is given by

$$f(x) = \begin{cases} 4, & \text{if } x < 0 \\ -2x + 4, & \text{if } 0 \le x \le 3 \\ -2, & \text{if } x > 3 \end{cases}$$

■ **Solution**

Note that the analytic description given here describes *one* function, even though the rule of correspondence is given in *three* parts. From Section 1.4 we know that the individual rules each describe straight lines; however, these lines must be restricted to their intervals of definition. Figure 1.42 shows the graph of the function. The graphs of $y = 4$, $y = -2x + 4$, and $y = -2$ are first made as dashed lines, and then the portions relevant to this function are made solid, as shown in Figures 1.42(a) and 1.42(b).

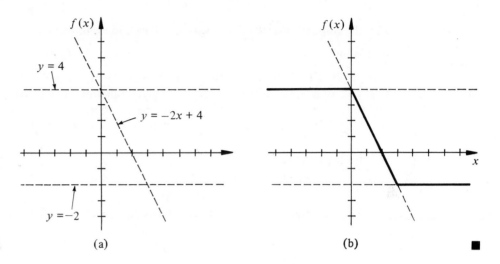

■ **Figure 1.42**   (a)   (b)   ■

EXAMPLE 2   ■ **Problem**

Sketch the graph of the function whose definition is

$$f(x) = \begin{cases} 1, & \text{if } x < 0 \\ x^2 + 1, & \text{if } 0 \le x < 2 \\ -x + 7, & \text{if } 2 \ge x \end{cases}$$

■ **Solution**

The graphs of $y = 1$, $y = x^2 + 1$, and $y = -x + 7$ are drawn as dashed lines in Figure 1.43. The portions of these curves that comprise the given function are shown as a solid line.

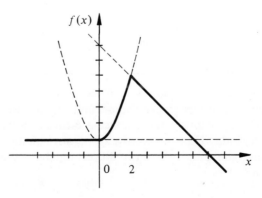

■ **Figure 1.43**   ■

EXAMPLE 3 ■ **Problem**

The yearly maintenance cost for a water purification system is expected to vary with time according to

$$C(t) = \begin{cases} 20t, & \text{if } 0 \le t < 10 \\ 30t - 100, & \text{if } 10 \le t \le 25 \end{cases}$$

Plot the graph of this function if C is in dollars and t is in years.

■ **Solution**

The graph of $C = 20t$ is a straight line through the origin, but only the portion for $0 \le t < 10$ is a part of the graph of the given function. The equation $C = 30t - 100$ is another straight line and represents the function for $10 \le t \le 25$. The function is undefined for $t < 0$ and for $t > 25$. Its graph is shown in Figure 1.44.

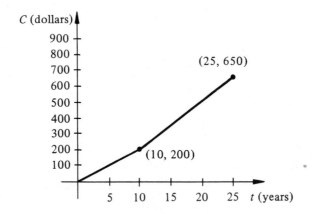

■ **Figure 1.44**

One of the most useful multirule functions is the **absolute value function** $f(x) = |x|$, where $|x|$ is called the absolute value of x and is defined as follows:

DEFINITION

> **Absolute Value Function**   The *absolute value* of x, denoted by $|x|$, is given by
>
> $$|x| = \begin{cases} x, & \text{if } x \ge 0 \\ -x, & \text{if } x < 0 \end{cases}$$

When we consider $|x|$, we sometimes say that we are "taking the absolute value of x." Quite obviously, this statement means that we consider the numerical value of x independently of the positive or negative quality. Thus, since the domain of the absolute value of x is the set of all real numbers, the range is the set of non-negative real numbers.

The graph of $f(x) = |x|$ consists of a combination of the graphs of $y = -x$ (for $x < 0$) and $y = x$ (for $x \geq 0$). These two graphs are shown in Figure 1.45 with the portion relevant to the given function drawn as solid lines.

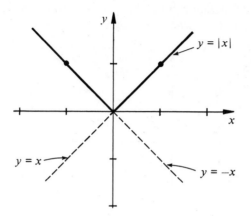

■ **Figure 1.45**

Slight variations of the absolute value function are also important. Examples 4 and 5 graphically show two variations.

EXAMPLE 4   ■ **Problem**

Draw the graph of $y = |x - 2|$.

■ **Solution**

This function is given by the two-part rule

$$y = \begin{cases} x - 2, & \text{if } x \geq 2 \\ -(x - 2), & \text{if } x < 2 \end{cases}$$

See Figure 1.46. This graph consists of a part of $y = x - 2$ for $x \geq 2$ and a part of $y = -x + 2$ for $x < 2$.

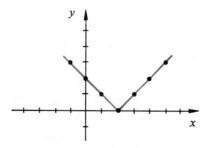

■ **Figure 1.46**   ■

EXAMPLE 5 ▪ **Problem**

Graph the function $y = |x| - 2$.

▪ **Solution**

The graph of this function can be obtained by moving the graph of $y = |x|$ down two units. See Figure 1.47, where the graph of $y = |x|$ is shown as a dashed curve.

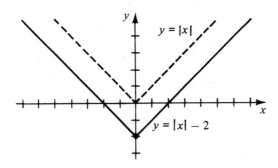

▪ **Figure 1.47**

Comment: If the equation had been $y = |x| + 2$, the graph would have been obtained by *raising* the graph of $y = |x|$ two units. ▪

EXERCISES FOR SECTION 1.7

Graph each function in Exercises 1–20.

1. $f(x) = \begin{cases} 2, & \text{if } x < 0 \\ 3, & \text{if } x \geq 0 \end{cases}$

2. $g(x) = \begin{cases} -1, & \text{if } x < 0 \\ 1, & \text{if } x \geq 0 \end{cases}$

3. $F(t) = \begin{cases} 0, & \text{if } t \leq 0 \\ t, & \text{if } t > 0 \end{cases}$

4. $f(x) = \begin{cases} x, & \text{if } x < 0 \\ -2, & \text{if } x \geq 0 \end{cases}$

5. $h(x) = \begin{cases} 0, & \text{if } x < 2 \\ 2x - 4, & \text{if } x \geq 2 \end{cases}$

6. $H(t) = \begin{cases} -2, & \text{if } t < 0 \\ t - 2, & \text{if } t \geq 0 \end{cases}$

7. $G(x) = \begin{cases} 0, & \text{if } x < 0 \\ 1, & \text{if } 0 < x < 2 \\ 0, & \text{if } x \geq 2 \end{cases}$

8. $f(t) = \begin{cases} 0, & \text{if } t < 0 \\ 2t, & \text{if } 0 \leq t < 3 \\ 6, & \text{if } t \geq 3 \end{cases}$

9. $y(x) = \begin{cases} 0, & \text{if } x < -2 \\ x^2 + 2, & \text{if } -2 < x < 0 \\ 0, & \text{if } x > 0 \end{cases}$

10. $z(x) = \begin{cases} x^2, & \text{if } x < 0 \\ -x^2, & \text{if } x \geq 0 \end{cases}$

11. $f(x) = \begin{cases} x, & \text{if } x < 1 \\ -x, & \text{if } x \geq 1 \end{cases}$

12. $f(x) = \begin{cases} 0, & x \text{ an integer} \\ 1, & x \text{ not an integer} \end{cases}$

13. $f(x) = |x + 1|$

14. $y = |x - 4|$

15. $g(t) = |t - 3|$

16. $F(x) = |x + \frac{1}{2}|$

17. $H(x) = |2x|$

18. $Z = |0.25t|$

19. $y = |x| + 3$

20. $y = |x| - 5$

21. The load (in kilograms) applied to a tensile specimen varied with time according to

$$F(t) = \begin{cases} 1000t, & \text{if } 0 \le t \le 5 \text{ min} \\ 5000, & \text{if } 5 < t \le 60 \text{ min} \end{cases}$$

Sketch the load-time curve.

22. The voltage applied to a given circuit varied with time according to

$$v(t) = \begin{cases} 0 \text{ V}, & \text{if } t < 2 \\ 6 \text{ V}, & \text{if } t \ge 2 \end{cases}$$

Sketch the voltage-time curve.

23. The cost of concrete varied with the number of yards delivered according to

$$c(n) = \begin{cases} 60 \text{ dollars}, & \text{if } 0 \le n \le 3 \\ 20n \text{ dollars}, & \text{if } 3 < n \le 10 \end{cases}$$

Sketch the graph.

24. A chemical supply house charges $1.50/gal for muriatic acid on orders up to 10 gal. For orders from 10 to 20 gal, the price is $1.25/gal; and from 20 to 30 gal, the price is $1.10/gal. Orders in excess of 30 gal cost $1.00/gal. Write the rule for the function that relates the cost per gallon to the size of the order. Sketch the graph.

25. Let $f(x) = |x|$. Find two real numbers a and b for which $f(a + b) \ne f(a) + f(b)$.

26. Compare $f(x) = |x|$ and $g(x) = |x - 2|$ by sketching their graphs in the same coordinate plane. How do the two graphs differ?

27. Compare $f(x) = |x|$ with $g(x) = |x| + 2$ by sketching their graphs in the same coordinate plane. How do the two graphs differ?

28. Compare $f(x) = |x|$, $g(x) = |2x|$, and $h(x) = 2|x|$ by sketching their graphs in the same coordinate plane. How do the graphs differ?

REVIEW EXERCISES FOR CHAPTER 1

In Exercises 1–6, find the distance between the given points on the number line.

1. 14, 5

2. $-4, 6$

3. $-2, 0$

4. $-5, -3$

5. $-7, -10$

6. 0.9, 0.2

In Exercises 7–14, sketch the graph of each interval.

7. $(3, 7]$

8. $[-2, \infty)$

9. $[-5, -1]$

10. $(9, 12)$

11. $[2, \infty)$ **12.** $(-\infty, -1]$

13. $(-\infty, -4)$ **14.** $[0, \infty)$

In Exercises 15–20, sketch the interval (or intervals) represented by the given absolute value.

15. $|x| < 4$ **16.** $|x| < \frac{1}{2}$

17. $|x| \leq 2$ **18.** $|x| \leq 3$

19. $|x| > 10$ **20.** $|x| > 7$

21. Find the distance between the points $(3, 7)$ and $(-2, -1)$. Find the midpoint of the line segment.

22. Find the distance between the points $(2, -2)$ and $(7, -5)$. Find the midpoint of the line segment.

In Exercises 23–32, tell whether the given equation, table, or diagram defines a function.

23. $y = x^4$ **24.** $y = x \pm 5$

25. $y < 2$ **26.** $y = 1/x^2$

27. $y = x - x^{-1}$

28.

x	2	-1	4	7
y	3	0	7	-3

29.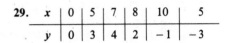

x	0	5	7	8	10	5
y	0	3	4	2	-1	-3

30.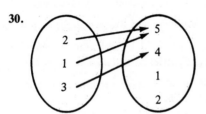

31.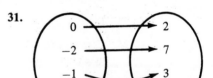

32.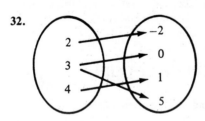

33. Find the range and domain of $f(x) = 2x^2$.

34. Find the range and domain of $y = 7 + 2x$.

35. Find the range and domain of $g(x) = \sqrt{x - 3}$.

36. Find the range and domain of $h(x) = 3/(x + 5)$.

In Exercises 37–40, draw the graph of the function.

37. $F(x) = -\sqrt{x} + 2$ **38.** $s = t^2 - 3t$

39. $y = \sqrt{-x}$ **40.** $M = \sqrt{2(t - 3)}$

Sketch the straight line determined by the pairs of points in Exercises 41–44, and compute the slope of the line.

41. $(-1, 3), (0, -2)$ **42.** $(2, -7), (-1, -4)$

43. $(3, 2), (1, -5)$ **44.** $(8, 0), (0, -3)$

45. Draw the straight line passing through $(-1, 2)$ with a slope of $\frac{2}{5}$.

46. Draw the straight line passing through $(0, -3)$ with a slope of 4.

47. Find the slope of a line drawn perpendicular to the line through $(5, 4)$ and $(-1, -1)$.

48. Find the slope of a line drawn perpendicular to the line through $(-2, 0)$ and $(-2, 5)$.

49. Determine the equation of the line passing through $(3, -1)$ with a slope of -2.

50. Determine the equation of the line passing through the point $(-5, 2)$ with a slope of $\frac{1}{3}$.

51. Determine the equation of the line passing through $(7, 2)$ and $(-1, 4)$.

52. Determine the equation of the line passing through $(1, 3)$ and $(-1, -2)$.

In Exercises 53–56, find the equation of the line perpendicular to the given line at the indicated point.

53. $x - 2y = -5$ at $(0, \frac{5}{2})$ **54.** $-4y - 3 = 3x$ at $(1, -\frac{3}{2})$

55. $x - 3y = 7$ at $(1, -2)$ **56.** $5x - 2y = 4$ at $(2, 3)$

In Exercises 57–58, sketch the graph of the given quadratic function. Give the domain and the range. Indicate the intercepts and the vertex of each.

57. $y = 2x^2 - 3x - 9$ **58.** $y = x^2 - 5x + 4$

Sketch the graph of each function in Exercises 59–68.

59. $y = \dfrac{x}{x^2 - 4}$ **60.** $y = \dfrac{5}{x(x + 3)}$

61. $y = \begin{cases} 1, & \text{if } x < 0 \\ -1, & \text{if } x \geq 0 \end{cases}$ **62.** $f(x) = \begin{cases} -1, & \text{if } x < 0 \\ x - 1, & \text{if } x \geq 0 \end{cases}$

63. $g(x) = \begin{cases} x, & \text{if } x < 0 \\ x^2, & \text{if } x \geq 0 \end{cases}$ **64.** $y = \begin{cases} 2 - x, & \text{if } x < 0 \\ 2, & \text{if } x \geq 0 \end{cases}$

65. $F(x) = |x - \frac{1}{2}|$ **66.** $y = |2 - x|$

67. $y = |t| + 2$ **68.** $m(t) = |t| - 0.5$

69. The cross section of a parabolic radar antenna is described by the equation $y = 0.3x^2$, where x and y are measured in feet. Sketch the cross section of an antenna having a diameter of 10 ft.

 70. Sketch the path of a rocket if the height of the rocket is related to its horizontal displacement by $h = 72x - 3x^2$, where h and x are measured in miles. Estimate the maximum height obtained by the rocket.

 71. The power (in watts) dissipated by a certain resistive circuit is given by

$$P = \begin{cases} 0, & \text{if } t \leq 2 \text{ sec} \\ 10(t-2), & \text{if } t > 2 \text{ sec} \end{cases}$$

Sketch the graph of power versus time.

 72. A unit step function defined by

$$u(t) = \begin{cases} 0, & \text{if } 0 \leq t < 1 \text{ millisecond (msec)} \\ 1, & \text{if } 1 \leq t \leq 3 \text{ msec} \\ 0, & \text{if } t > 3 \text{ msec} \end{cases}$$

is used to check the performance of a linear-feedback circuit in an automated welder. Sketch the graph of this function.

CHAPTER 2

The Rate of Change of a Function

In this chapter, we begin our discussion of a branch of mathematics called *calculus*.* Although some of the notions of calculus were known and used as early as 250 B.C., the subject was not clearly understood until the latter part of the seventeenth century when Sir Isaac Newton and Gottfried von Leibniz independently and almost simultaneously developed it. We begin our study of calculus by introducing the concept of a *limit*, an idea that is the foundation of calculus and sets it off from other branches of elementary mathematics.

▬ 2.1 LIMITS AND CONTINUITY

Limits

In analytic geometry, we draw the graph of a function by connecting a few selected points with a smooth curve. No attempt is made to examine the behavior of the function near the plotted points; we just assume that they can be connected by a smooth curve. The analysis of the behavior of a function near a point is fundamental to calculus and important to our understanding of the real world. The following example give some idea of how to examine carefully the behavior of a function in the neighborhood of a given point.

EXAMPLE 1 ■ **Problem**

Determine the behavior of $f(x) = x^2$ when x is close to 2. Do not evaluate $f(2)$.

■ **Solution**

Notice first that we are not to determine the value of x^2 when $x = 2$; the problem is to describe the behavior of x^2 when x is close to 2. We proceed numerically by evalu-

*The word *calculus* means "a method of calculation."

ating x^2 for values of x that are successively closer and closer to 2. The following table shows the result of some typical computations.

	$x \rightarrow$			2.0			$\leftarrow x$
x	1.9	1.99	1.999		2.001	2.01	2.1
x^2	3.61	3.960	3.9960		4.0040	4.040	4.41

Inspection of the values in this table suggests that as x gets closer and closer to 2, x^2 gets closer and closer to 4. Therefore, we state that when x is near 2, the value of x^2 is near 4. Figure 2.1 is a picture of the situation.

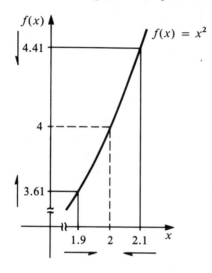

■ Figure 2.1

You may wonder why we expended so much effort to show that $f(x) = x^2$ is near 4 when x is near 2 since it is obvious that $f(2) = 4$. The reason, as the next example shows, is that the same technique can be used to study the behavior of a function near a point for which the function itself is undefined.

EXAMPLE 2   **■ Problem**

Let $g(x) = \dfrac{x^2 - 9}{x - 3}$, $x \neq 3$. Examine the behavior of the function for x near 3.

■ Solution

Although the given function is undefined at $x = 3$, we can investigate its behavior for other values of x in the neighborhood of 3 by proceeding as in the previous example. Evaluating $g(x) = (x^2 - 9)/(x - 3)$ for some selected values of x close to 3, we get the following table:

	$x \rightarrow$			3.0			$\leftarrow x$
x	3.10	3.01	3.001		2.999	2.990	2.90
$g(x)$	6.10	6.01	6.001		5.999	5.990	5.90

The table suggests that $g(x)$ is close to 6 when x is close to 3.

Another approach to this problem involves the algebraic simplification of $g(x)$ as follows:

$$g(x) = \frac{x^2 - 9}{x - 3} = \frac{(x + 3)(x - 3)}{x - 3} = x + 3 \qquad \text{(provided } x \neq 3)$$

With the function expressed in this form, we see that $g(x)$ is close to 6 when x is close to 3. Keep in mind that $g(3)$ is undefined, and therefore, *the function is never actually equal to 6.* Using the fact that $g(x) = x + 3$ provided $x \neq 3$, we conclude that the graph of the given function is a straight line with a "hole" at the point (3, 6). See Figure 2.2.

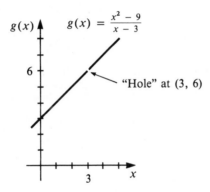

■ **Figure 2.2**

In Example 2, an algebraic procedure facilitated our understanding of the behavior of $(x^2 - 9)/(x - 3)$ near $x = 3$. Such simplifications are not always possible, as the next example shows.

EXAMPLE 3 ■ **Problem**

The function $(\sin x)/x$ is undefined at $x = 0$. Use a calculator to determine the behavior of this function for x close to zero.

■ **Solution**

A calculator with a $\sin x$ button is necessary for this problem. Furthermore, to get the numbers in the following table, the calculator must be in the radian mode. We observe that $(\sin x)/x$ appears to be close to 1 when x is close to 0.

	$x \to 0$			
x(radians)	0.5	0.1	0.05	0.01
$(\sin x)/x$	0.95885	0.99833	0.99958	0.99998

The numerical procedures used in the preceding examples are tedious, but the analysis is typical of what we must do to determine the behavior of a function near a given point. These examples are intended to motivate the following definition.

DEFINITION

Limit of a Function   We say that L is the *limit of a function* $f(x)$ as x approaches a if whenever x is close to a, then $f(x)$ is close to a fixed number L, $x \neq a$. We write

$$\lim_{x \to a} f(x) = L$$

If such a number L does not exist, we say that the limit does not exist.

Using the notation introduced in the definition of a limit to express the results of Examples 1 through 3, we have the following expressions:

Example 1: $\lim_{x \to 2} x^2 = 4$

Example 2: $\lim_{x \to 3} \dfrac{x^2 - 9}{x - 3} = 6$

Example 3: $\lim_{x \to 0} \dfrac{\sin x}{x} = 1$

EXAMPLE 4   ■ **Problem**

Show that $\lim_{x \to 0} f(x)$ does not exist for

$$f(x) = \begin{cases} x + 2, & \text{if } x \geq 0 \\ x - 2, & \text{if } x < 0 \end{cases}$$

■ **Solution**

For this function, the limit of $f(x)$ depends on *how* x approaches zero. Notice that when x approaches zero from the right (denoted by $x \to 0^+$), we get $\lim_{x \to 0^+} (x + 2)$ = 2. However, when x approaches zero from the left (denoted by $x \to 0^-$), we get $\lim_{x \to 0^-} (x - 2) = -2$. We conclude that $\lim_{x \to 0} f(x)$ does not exist since $f(x)$ does not get close to a single number when x is close to zero. The graph of the function is shown in Figure 2.3.

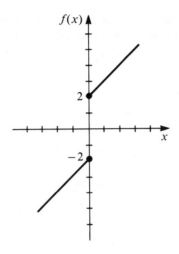

■ **Figure 2.3**

Before presenting any more examples, we state some theorems that make the computation of limits easier.

THEOREM

Limit Calculations   If $\lim\limits_{x \to a} f(x)$ and $\lim\limits_{x \to a} g(x)$ exist and c is a constant, then the following statements are true:

1. $\lim\limits_{x \to a} c = c$

Expressed in words, the limit of a constant is that constant.

2. $\lim\limits_{x \to a} cf(x) = c \lim\limits_{x \to a} f(x)$

Expressed in words, the limit of a constant times a function is equal to the constant times the limit of the function.

3. $\lim\limits_{x \to a} [f(x) + g(x)] = \lim\limits_{x \to a} f(x) + \lim\limits_{x \to a} g(x)$

Expressed in words, the limit of a sum is the sum of limits. This result may be extended to any finite sum.

4. $\lim\limits_{x \to a} [f(x)g(x)] = \lim\limits_{x \to a} f(x) \lim\limits_{x \to a} g(x)$

Expressed in words, the limit of a product is the product of the limits.

5. $\lim\limits_{x \to a} \dfrac{f(x)}{g(x)} = \dfrac{\lim\limits_{x \to a} f(x)}{\lim\limits_{x \to a} g(x)}$ *provided* $\lim\limits_{x \to a} g(x) \ne 0$

Expressed in words, the limit of a quotient is the quotient of the limits.

Warning: Beginning students often tend to misuse this theorem because they forget that **none of the statements are valid unless both limits exist.**

EXAMPLE 5   ■ **Problem**

Find $\lim\limits_{x \to 0} (x + 3)(x - 2)$.

■ **Solution**

Applying statement 4 of the theorem, we have

$$\lim_{x \to 0} (x + 3)(x - 2) = \lim_{x \to 0} (x + 3) \lim_{x \to 0} (x - 2)$$

Since $x + 3$ is near 3 when x is near 0, we have

$$\lim_{x \to 0} (x + 3) = 3$$

Similarly,

$$\lim_{x \to 0} (x - 2) = -2$$

Hence,

$$\lim_{x \to 0} (x + 3)(x - 2) = (3)(-2) = -6 \qquad\blacksquare$$

EXAMPLE 6   ■ **Problem**

Find $\lim\limits_{x \to 4} \dfrac{3x^2 - x}{\sqrt{x}}$.

■ **Solution**

Applying statement 5 of the theorem, we have

$$\lim_{x \to 4} \frac{3x^2 - x}{\sqrt{x}} = \frac{\lim\limits_{x \to 4} (3x^2 - x)}{\lim\limits_{x \to 4} \sqrt{x}} = \frac{\lim\limits_{x \to 4} 3x^2 - \lim\limits_{x \to 4} x}{\lim\limits_{x \to 4} \sqrt{x}}$$

$$= \frac{3 \lim\limits_{x \to 4} x^2 - \lim\limits_{x \to 4} x}{\lim\limits_{x \to 4} \sqrt{x}} = \frac{3(16) - 4}{2} = 22 \qquad\blacksquare$$

EXAMPLE 7   ■ **Problem**

Show that $\lim\limits_{x \to 2} \dfrac{x^2 + x - 6}{x - 2} = 5$.

■ **Solution**

Note that the limit of the denominator as x approaches 2 is equal to 0, and hence, statement 5 of the theorem may *not* be applied. However, if $x \neq 2$, we may perform the following algebraic simplification:

$$\frac{x^2 + x - 6}{x - 2} = \frac{(x + 3)(x - 2)}{x - 2} = x + 3$$

The desired limit is then more obviously

$$\lim_{x \to 2} (x + 3) = 5$$

Keep in mind that in finding such a limit we are *not* finding the value of $(x^2 + x - 6)/(x - 2)$ when $x = 2$ (that value does not exist) but the value that it approaches as x approaches arbitrarily close to 2. ■

Comment: The technique of the previous example carries with it the general hint that you are usually better off to perform algebraic simplifications *before* computing a limit.

Continuity

When you sketch the graph of the function, you usually plot several points and then connect those points with a smooth, connected curve. In drawing such a curve, you are assuming that the function can be represented by a continuous set of points. Intuitively, we say that a function is **continuous** if we can sketch its graph without removing our pencil from the paper as we draw it. In Figure 2.4, the graphs in parts (a) and (b) are continuous, while those in parts (c) and (d) are not. Points where a function is not continuous are called **discontinuities.** The discontinuity in Figure 2.4(c) is a *finite* discontinuity, and that in Figure 2.4(d) is an *infinite* discontinuity.

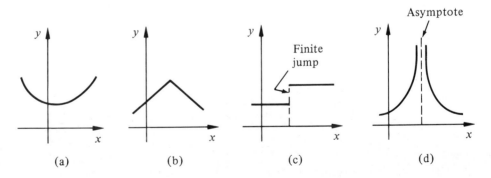

■ **Figure 2.4**   (a)   (b)   (c)   (d)

The continuity of a function at a point is established by applying the following definition.

DEFINITION

Continuity at a Point   A function f is *continuous* at $x = a$ if

$$\lim_{x \to a} f(x) = f(a)$$

EXAMPLE 8   The function $F(x) = x^2$ is continuous at $x = 3$ since $F(3) = 9$ and

$$\lim_{x \to 3} x^2 = 9$$

■

EXAMPLE 9   The function

$$f(x) = \begin{cases} x + 2, & \text{if } x \geq 0 \\ x - 2, & \text{if } x < 0 \end{cases}$$

has a finite discontinuity at $x = 0$ since $f(0) = 2$. But as we approach zero from the left, we get

$$\lim_{x \to 0^-} f(x) = \lim_{x \to 0^-} (x - 2) = -2$$

The graph of this function is shown in Figure 2.3.

■

EXERCISES FOR SECTION 2.1

In Exercises 1–20, find the indicated limit.

1. $\displaystyle \lim_{x \to 3} \frac{3x}{2x + 1}$

2. $\displaystyle \lim_{t \to 0} \frac{t - 3}{t^2 + t + 4}$

3. $\displaystyle \lim_{t \to 0} \frac{t^3 - 3t^2 + 4t}{t^2 + 2t}$

4. $\displaystyle \lim_{x \to 5} (-5)$

5. $\displaystyle \lim_{t \to 0} \frac{3t^2 + t}{t}$

6. $\displaystyle \lim_{x \to 0} \frac{x^3 + 2x}{x^4 + 5x^2 + 3}$

7. $\displaystyle \lim_{x \to 0} \frac{x^2 + x + 3}{x - 1}$

8. $\displaystyle \lim_{x \to 2} \frac{7 - x^2}{3 + x - 3x^2}$

9. $\displaystyle \lim_{i \to 0} \frac{4i + 5i^3}{3i + 4i^2}$

10. $\displaystyle \lim_{x \to 1} \frac{x^2 - 1}{x - 1}$

11. $\displaystyle \lim_{x \to 3} \frac{x^2 - 2x - 3}{x - 3}$

12. $\displaystyle \lim_{x \to 5} \frac{2x^2 - 7x - 15}{x - 5}$

13. $\displaystyle \lim_{x \to 3} \frac{3x^2 - 2x - 21}{x - 3}$

14. $\displaystyle \lim_{x \to 4} \frac{\sqrt{x} - 2}{x - 4}$

15. $\displaystyle \lim_{x \to 1} \frac{\sqrt{x} - 1}{x - 1}$

16. $\displaystyle \lim_{x \to 1} \frac{x^3 - 1}{x - 1}$

17. $\lim\limits_{x\to 1} \dfrac{(1/x) - 1}{x - 1}$

18. $\lim\limits_{x\to 1/2} \dfrac{(1/x) - 2}{2x - 1}$

19. $\lim\limits_{x\to 1/3} \dfrac{(1/x) - 3}{x - (1/3)}$

20. $\lim\limits_{x\to 1/3} \dfrac{6x^2 + 13x - 5}{3x - 1}$

21. Let

$$f(x) = \begin{cases} x^2 - 1, & \text{if } x \le 0 \\ x, & \text{if } x > 0 \end{cases}$$

Show that $f(x)$ is discontinuous at $x = 0$.

22. Let

$$f(x) = \begin{cases} 3x, & \text{if } x < -1 \\ 2, & \text{if } -1 \le x \le 1 \\ 2x^2, & \text{if } 1 < x \end{cases}$$

Show that $f(x)$ is discontinuous at $x = -1$ and continuous at $x = 1$.

23. Let

$$f(x) = \begin{cases} x^2 + c, & \text{if } x \le 1 \\ 3x + 5, & \text{if } x > 1 \end{cases}$$

Assign a value to c so that $f(x)$ is continuous at $x = 1$.

24. Let

$$f(x) = \begin{cases} 2x + c, & \text{if } x < 2 \\ k, & \text{if } x = 2 \\ x^2, & \text{if } x > 2 \end{cases}$$

Assign values to c and k so that $f(x)$ is continuous at $x = 2$.

In Exercises 25–30, use a calculator to help you evaluate the limits. (*Note: x* is a real number, so be sure your calculator is in the radian mode for sin *x* and cos *x*.

25. $\lim\limits_{x\to 0} \dfrac{\sin^2 x}{x}$

26. $\lim\limits_{x\to 0} \dfrac{\cos x - 1}{x}$

27. $\lim\limits_{x\to 0} \dfrac{e^x - 1}{x}$

28. $\lim\limits_{x\to 0} \dfrac{\sin 2x}{2x}$

29. $\lim\limits_{x\to 0} (1 + x)^{1/x}$

30. $\lim\limits_{x\to 0} (1 + x)^{2/x}$

31. The current in a circuit is given by

$$i = \dfrac{t^3 - 8}{t - 2}$$

Find $\lim\limits_{t\to 2} i$.

32. In the study of radioactivity, the following expression occurs:

$$\frac{1 - 2e^{-t} + e^{-2t}}{t}$$

Find the limiting value of this expression as t approaches zero.

The following BASIC program computes the limit of a function $f(x)$ as x becomes close to a number a. Use this program to compute the limit of the functions in this exercise set.

```
10   REM  EVALUATION OF A LIMIT
20   HOME
30   PRINT "THIS PROGRAM WILL FIND THE LIMIT"
40   PRINT "OF A FUNCTION WHEN
50   PRINT "THE USER SUPPLIES THE X-VALUE"
60   PRINT : PRINT "THE FUNCTION IN THIS PROGRAM"
70   PRINT "IS    (X^2-4)/(X-2)"
80   PRINT "IF THE USER CHANGES THE FUNCTION"
90   PRINT "THE NEW FUNCTION WILL REMAIN UNTIL"
100   PRINT "THE USER CHANGES IT AGAIN, OR"
110   PRINT "RELOADS THE PROGRAM."
120   PRINT : PRINT "DO YOU WISH TO CHANGE THIS FUNCTION? Y/N"
130   INPUT A$
140   IF A$ = "Y" THEN 510
150 X1 = 2
160   PRINT : PRINT "THE X VALUE IS   ";X1
170   PRINT "DO YOU WISH TO CHANGE THIS VALUE? Y/N"
180   INPUT B$
190   IF B$ = "N" THEN 230
200   IF B$ < > "Y" THEN 170
210   PRINT "WHAT X-VALUE DO YOU WISH TO USE?"
220   INPUT X1
230   PRINT "DO YOU WISH TO APPROACH FROM"
240   PRINT "FROM THE LEFT OR THE RIGHT? L/R"
250   INPUT R$
260 K = 1
270   IF R$ = "L" THEN K =  - 1
280   DEF  FN F(X) = (X ^ 2 - 4) / (X - 2)
290 XN = X1 + K
300 Y1 =  FN F(XN)
310   PRINT "X              F(X)"
320   PRINT XN,Y1
330 Y0 = Y1
340 ERR = 0.0005
350   FOR N = 1 TO 15
360 XN = X1 + K / 2 ^ N
370 Y1 =  FN F(XN)
380   PRINT XN,Y1
390   IF  ABS (Y0 - Y1) < ERR THEN 440
400 Y0 = Y1
410   NEXT N
420   PRINT "NO CONVERGENCE AFTER 15 ITERATIONS"
430   GOTO 460
440   PRINT : PRINT "AS X APPROACHES ";X1
445 Y1 =  INT (1000 * Y1 + .1) / 1000
450   PRINT : PRINT "THE APPARENT LIMIT IS ";Y1
460   PRINT : PRINT "DO YOU WISH TO RUN THIS PROGRAM AGAIN? Y/N"
470   INPUT C$
```

```
480  HOME
490  IF C$ = "Y" THEN 60
500  GOTO 540
510  PRINT "TYPE"
520  INVERSE : PRINT "280 DEF FN F(X) = (YOUR FUNCTION) <RET>"
530  PRINT "RUN 150": NORMAL
540  END
```

■■■■■ 2.2 MORE ON LIMITS

What happens to the value of

$$f(x) = \frac{6x + 5}{3x + 1}$$

as x becomes very large? To discover the answer to this question, we can make a table showing the values of $f(x)$ for increasingly larger values of x:

$$x \rightarrow$$

x	0	1	10	100	1000
$f(x)$	5	2.8	2.1	2.01	2.001

From the table, we see that the values of $f(x)$ are decreasing and seem to be getting closer and closer to 2 as the values of x increase. We use the idea of a limit to describe this behavior.

DEFINITION

> **Limit of a Function for Large x**   We say that L is the *limit of a function* $f(x)$ as x increases if $f(x)$ is near a fixed number L when x is very large. We denote this result by
>
> $$\lim_{x \to \infty} f(x) = L$$
>
> which is read, "the limit of $f(x)$ as x increases without bound is L."

The infinity symbol is used to characterize limits in which either x or $f(x)$ becomes unbounded. For instance, $\lim_{x \to a} f(x) = -\infty$ means that $f(x)$ becomes negatively unbounded as x approaches a. Technically, a function like this one does not have a limit, but we use the ∞ symbol to indicate that the values of the function become unbounded near a.

Figure 2.5 shows the graphical representation of $\lim_{x \to \infty} f(x) = L$ and $\lim_{x \to a} g(x) = \infty$. Notice that $\lim_{x \to \infty} f(x) = L$ corresponds to a *horizontal asymptote* of $f(x)$, and $\lim_{x \to a} g(x) = \infty$ corresponds to a *vertical asymptote* of $g(x)$. (See Section 1.6.)

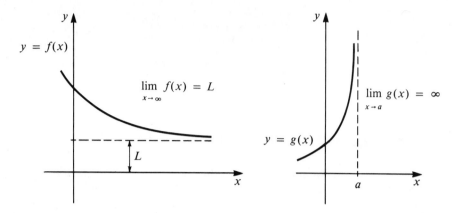

■ Figure 2.5

EXAMPLE 1   **■ Problem**

Evaluate $\lim\limits_{x \to \infty} \dfrac{1}{x}$.

■ Solution

The following table shows that $1/x$ tends toward zero as x increases. We conclude that $\lim\limits_{x \to \infty} 1/x = 0$ since we can make the value of $1/x$ as close to zero as we please by choosing the value of x to be sufficiently large. Thus, to make $1/x < 0.00001$, choose $x > 100{,}000$. Figure 2.6 shows this limit as a horizontal asymptote at $y = 0$.

x	1	10	100	1000
$1/x$	1	0.1	0.01	0.001

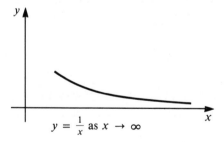

■ Figure 2.6

$$y = \frac{1}{x} \text{ as } x \to \infty$$

■

EXAMPLE 2   **■ Problem**

Evaluate $\lim\limits_{x \to \infty} \dfrac{x^2 + 5x + 3}{4x - 3x^2}$.

■ Solution

Neither the limit of the numerator nor the limit of the denominator exists. (Both increase indefinitely as $x \to \infty$). Therefore, we change algebraically the form of the given expression. The trick is to divide numerator and denominator by x^2, to obtain

$$\lim_{x \to \infty} \frac{x^2 + 5x + 3}{4x - 3x^2} \lim_{x \to \infty} = \frac{1 + (5/x) + (3/x^2)}{(4/x) - 3}$$

$$= \frac{1 + 0 + 0}{0 - 3} = -\frac{1}{3}$$

∎

Comment: The technique used in Example 2 to identify the limit of the given function can be used anytime the given function is a rational function. In general, **before taking the limit of a rational function as $x \to \infty$, divide numerator and denominator by the highest power of x occurring in either.**

EXAMPLE 3

■ **Problem**

Evaluate $\lim_{x \to \infty} \dfrac{5x^3 - 4}{x^2 + 2x}$.

■ **Solution**

Dividing numerator and denominator by x^3, we write

$$\lim_{x \to \infty} \frac{5x^3 - 4}{x^2 + 2x} = \lim_{x \to \infty} \frac{5 - (4/x^3)}{(1/x) + (2/x^2)}$$

As $x \to \infty$, the numerator approaches 5 and the denominator approaches 0. Consequently, the ratio increases without bound as $x \to \infty$, so the limit does not exist. ∎

EXAMPLE 4

■ **Problem**

Evaluate $\lim_{x \to 2} \dfrac{x^2 + 9}{x - 2}$.

■ **Solution**

When x is close to 2, the numerator is close to 13. The denominator is close to 0 when x is close to 2. The closer the denominator is to 0, the larger the absolute value of the quotient becomes. As x approaches 2 from the right, we have

$$\lim_{x \to 2^+} \frac{x^2 + 9}{x - 2} = \infty$$

Likewise, as x approaches 2 from the left, we have

$$\lim_{x \to 2^-} \frac{x^2 + 9}{x - 2} = -\infty$$

The unbounded nature of the function near $x = 2$ is represented by a vertical asymptote in Figure 2.7.

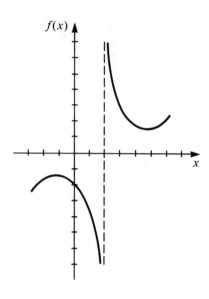

■ **Figure 2.7**   ■

EXAMPLE 5   ■ **Problem**

The electric current in a circuit varies with time according to

$$i = \frac{3t^3 - t^2 + 5t}{t^3 + 10}$$

The current is in amperes, and the time is in seconds. What happens to the current as $t \to \infty$? (This limiting value is called the *steady-state current*.)

■ **Solution**

Here we note that

$$i = \frac{3t^3 - t^2 + 5t}{t^3 + 10} = \frac{3 - (1/t) + (5/t^2)}{1 + (10/t^3)}$$

From this expression, we have

$$\lim_{t \to \infty} i = 3\,\text{A}$$

See Figure 2.8.

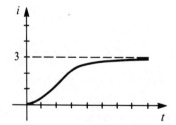

■ **Figure 2.8**   ■

Sometimes, a limit does not exist because the function fails to approach any fixed number. As the next example shows, sin x is a function of this type.

EXAMPLE 6   ■ **Problem**

Evaluate $\lim\limits_{x \to \infty}$ sin x (x a real number).

■ **Solution**

The sine function is periodic with a period of 2π. Therefore, sin x takes on all values between 1 and -1 as $x \to \infty$. Since sin x does not get close to a specific number as $x \to \infty$, we conclude that the limit does not exist.   ■

EXERCISES FOR SECTION 2.2

In Exercises 1–22, evaluate the indicated limit.

1. $\lim\limits_{x \to \infty} \dfrac{1}{x + 2}$

2. $\lim\limits_{t \to \infty} \dfrac{2}{1 - t}$

3. $\lim\limits_{x \to \infty} \dfrac{3x}{2x + 1}$

4. $\lim\limits_{x \to \infty} \dfrac{2x^3 + 7x + 2}{x^3 + 5}$

5. $\lim\limits_{x \to \infty} \dfrac{x^2 + 4}{3x^2 + 7}$

6. $\lim\limits_{x \to \infty} \dfrac{7 - x^2}{3 + x - 3x^2}$

7. $\lim\limits_{x \to \infty} \dfrac{10 + 3x - x^3}{2x^2 - 5}$

8. $\lim\limits_{x \to \infty} \dfrac{x^4 - 9}{x + 3}$

9. $\lim\limits_{x \to -\infty} \dfrac{x^2 + 3x - 1}{x(x - 1)}$

10. $\lim\limits_{x \to -\infty} \dfrac{3x^5 - 5x^2}{x^5 + 4x^4 - 3}$

11. $\lim\limits_{x \to \infty} \log x$

12. $\lim\limits_{x \to -\infty} \cos x$

13. $\lim\limits_{x \to \infty} \dfrac{x^2}{x^3}$

14. $\lim\limits_{x \to -\infty} e^x$

15. $\lim\limits_{x \to \infty} e^{-x}$

16. $\lim\limits_{x \to 1} \dfrac{x + 1}{(x - 1)^2}$

17. $\lim\limits_{x \to 4} \dfrac{x - 3}{x^2 - 7x + 12}$

18. $\lim\limits_{x \to 0} \log x$

19. $\lim\limits_{x \to 2} \dfrac{x^2 + x - 6}{x^2 - 4}$

20. $\lim\limits_{x \to 0} \dfrac{\cos x}{x}$

21. $\lim\limits_{x \to \pi/2} \dfrac{\sin x}{x - (\pi/2)}$

22. $\lim\limits_{x \to 3} \dfrac{x^2 + 8}{(x - 3)^2}$

 23. The mass of a particle m is known to change with its velocity v according to the formula

$$m = \dfrac{m_0}{[1 - (v^2/c^2)]^{1/2}}$$

where m_0 is the rest mass and c is the velocity of light. Find the limit of m as the velocity approaches c.

24. The value of the current in a circuit is given at any time t by the expression

$$i = e^{-t} + \frac{3t - 1}{t}$$

Find the steady-state current as t increases.

25. The current in a particular RL circuit is given by

$$i = \frac{V}{R}(1 - e^{-Rt/L})$$

where V is the applied voltage, R is the resistance, L is the inductance, and t is time. Find the steady-state current as t increases, if V, R, and L are constant.

26. The terminal velocity of a particle is defined as the limiting velocity as t becomes large. Find the terminal velocity of a particle whose velocity varies as

$$v = \frac{\sin t}{t} + \frac{3t - 1}{t + 1} + e^{-t}$$

2.3 APPLICATIONS OF THE LIMIT IDEA

The limit idea is the root of calculus. In this section, we show how this idea is used to solve two important problems, the tangent problem and the rate problem.

The Tangent Problem

The graph of a function $y = g(x)$ is shown in Figure 2.9. The straight line that touches the graph at point P is called a **tangent line** of the graph. The problem is to find the slope of a tangent line to a curve at a given point. While this problem may seem of minor importance, many of the concepts of science and technology are related to the tangent problem.

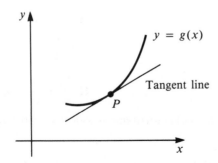

■ **Figure 2.9**

As a preliminary step to solving for the slope of a tangent line, we define a secant line. A straight line that passes through two distinct points on a curve is called a *secant line* to the curve. A secant line through the points P and Q is shown in Figure 2.10(a). From Chapter 1, we know that the slope of the secant line is

$$m_{\text{sec}} = \frac{g(x) - g(x_0)}{x - x_0}$$

If we move point Q along the curve toward point P [see Figure 2.10(b)], the secant line rotates about P, and its slope gets closer to the slope of the tangent line. Since the slope of the secant line approaches the slope of the tangent line as $Q \rightarrow P$, we write

$$m_{\text{tan}} = \lim_{Q \rightarrow P} m_{\text{sec}} = \lim_{Q \rightarrow P} \frac{g(x) - g(x_0)}{x - x_0}$$

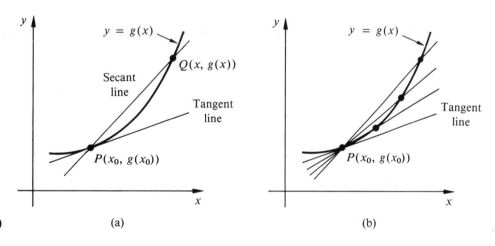

■ **Figure 2.10**   (a)   (b)

Since $Q \rightarrow P$ implies that $x \rightarrow x_0$, we define the slope of a tangent line to $y = g(x)$ at $P(x_0, y_0)$ as follows:

DEFINITION

> **The Slope of a Tangent Line**   The *slope $m(x_0)$ of the tangent line* to the graph of $y = g(x)$ at the point $P(x_0, y_0)$ is
>
> $$m(x_0) = \lim_{x \rightarrow x_0} \frac{g(x) - g(x_0)}{x - x_0}$$

EXAMPLE 1   ▪ **Problem**

Use the definition to find the slope of the tangent line to $y = x^2 - 3x$ at $x_0 = 2$.

▪ **Solution**

Here, $g(x) = x^2 - 3x$ and $g(x_0) = g(2) = 2^2 - 3(2) = -2$. Therefore,

$$m(2) = \lim_{x \to 2} \frac{(x^2 - 3x) - (-2)}{x - 2} \qquad \text{applying definition of slope of tangent}$$

$$= \lim_{x \to 2} \frac{x^2 - 3x + 2}{x - 2} \qquad \text{limit has form } \frac{0}{0}$$

$$= \lim_{x \to 2} \frac{(x - 1)(x - 2)}{x - 2} \qquad \text{factoring numerator}$$

$$= \lim_{x \to 2} (x - 1) = 1 \qquad \text{canceling and taking limit}$$

The tangent line is shown in Figure 2.11.

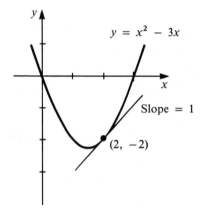

▪ **Figure 2.11**   ▪

EXAMPLE 2   ▪ **Problem**

Use the definition to find the slope of the tangent line to $y = x^3$ at $x_0 = -1$.

▪ **Solution**

Since $g(x) = x^3$ and $g(-1) = -1$, we have

$$m(-1) = \lim_{x \to -1} \frac{x^3 - (-1)}{x - (-1)} \qquad \text{applying definition}$$

$$= \lim_{x \to -1} \frac{x^3 + 1}{x + 1} \qquad \text{limit has form } \frac{0}{0}$$

$$= \lim_{x \to -1} \frac{(x + 1)(x^2 - x + 1)}{x + 1} \qquad \text{factoring numerator}$$

$$= \lim_{x \to -1} (x^2 - x + 1) = 3 \qquad \text{canceling and taking limit}$$

This result is shown graphically in Figure 2.12.

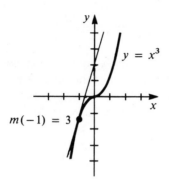

■ **Figure 2.12**

The Rate Problem

The next problem is concerned with the rate of change of a variable. Some common examples of rates of change are velocity, acceleration, electric current, and power. We approach the rate problem in the context of finding the velocity of a particle moving in a straight line.

The ratio of the distance traveled by a particle to the elapsed time is called the **average velocity** of the particle. The average velocity is denoted and defined by

$$v_{\text{avg}} = \frac{s_2 - s_1}{t_2 - t_1} \qquad \text{(2.1)}$$

To illustrate the use of this formula, we consider a particle whose displacement from a fixed point varies with time according to $s = 3t^2$ (in feet). That is, after 1 sec, the particle has moved 3 ft; after 2 sec, it has moved 12 ft; and after 3 sec, it has moved 27 ft. See Figure 2.13. The average velocity over any two consecutive times can be calculated by Formula (2.1). For instance, the average velocity from $t = 1$ to $t = 2$ is

$$v_{\text{avg}} = \frac{12 - 3}{2 - 1} = 9 \text{ ft/sec}$$

■ **Figure 2.13**

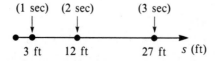

Now, suppose that instead of finding the average velocity from $t = 1$ to $t = 2$ sec, we wish to find the velocity at the instant $t_0 = 2$ sec. The formula for v_{avg} fails here because there is no elapsed time, making the denominator zero. However, we can use the average velocity to estimate the velocity at $t_0 = 2$ sec by choosing a second point close to 2. For instance, if we let $t = 2.01$ sec, we have

$$v_{avg} = \frac{3(2.01)^2 - 3(2)^2}{2.01 - 2} = \frac{0.1203}{0.01} = 12.03 \text{ ft/sec}$$

If this value isn't close enough, we choose another time closer to $t_0 = 2$ sec until the average velocity is an acceptable estimate of the velocity at $t_0 = 2$ sec.

We can make the average velocity as close as we please to the velocity at t_0 by choosing t sufficiently close to t_0. Therefore, we conclude that the velocity at t_0 is the limit of the average velocity as $t \to t_0$. Using $v(t_0)$ to denote the velocity at t_0, we make the following definition:

DEFINITION

Instantaneous Velocity   If the displacement of a particle varies with time according to $s = s(t)$, then the *velocity* of the particle at t_0 is

$$v(t_0) = \lim_{t \to t_0} \frac{s(t) - s(t_0)}{t - t_0}$$

EXAMPLE 3   ▪ **Problem**

An object is thrown vertically upward with a velocity of 64 ft/sec. The height of the object at any time is known from physics to be given by the formula $s = 64t - 16t^2$ (in feet). What is the velocity of the object 1 sec after it is thrown upward?

▪ **Solution**

Since $s(t) = 64t - 16t^2$, then $s(1) = 64(1) - 16(1)^2 = 48$. Therefore, the velocity of the object at $t_0 = 1$ is

$$v(1) = \lim_{t \to 1} \frac{(64t - 16t^2) - 48}{t - 1} \qquad \text{applying definition of velocity at } t_0 = 1$$

$$= \lim_{t \to 1} \frac{-16(t^2 - 4t + 3)}{t - 1} \qquad \text{limit has form } \frac{0}{0}$$

$$= \lim_{t \to 1} \frac{-16(t - 3)(t - 1)}{t - 1} \qquad \text{factoring}$$

$$= \lim_{t \to 1} -16(t - 3) = 32 \text{ ft/sec} \qquad \text{canceling and taking limit} \qquad ▪$$

EXAMPLE 4   ▪ **Problem**

Find the velocity of an object at $t_0 = 2$ sec given that the displacement of the object varies with time according to $s = 1/(t + 3)$ (in meters).

▪ **Solution**

Since $s(t) = 1/(t + 3)$, then $s(2) = \frac{1}{5}$. So the velocity at $t_0 = 2$ sec is

$$v(2) = \lim_{t \to 2} \frac{\dfrac{1}{t + 3} - \dfrac{1}{5}}{t - 2} \qquad \text{limit has form } \frac{0}{0}$$

$$= \lim_{t \to 2} \frac{(2 - t)/5(t + 3)}{t - 2} \qquad \text{adding fractions in numerator}$$

$$= \lim_{t \to 2} \frac{-1}{5(t + 3)} = \frac{-1}{25} \text{ m/sec} \qquad \text{canceling and taking limit}$$

The minus sign indicates that the object is moving to the left.   ▪

The approach used to define instantaneous velocity of a particle is typical of rate problems in general. That is, other instantaneous rates may be defined in the same way. For example, the electric current in a circuit is the time rate of transferring electric charge past a point in a circuit. If $q(t)$ denotes the charge, then the average current i_{avg} in the circuit over some time interval $t - t_0$ is

$$i_{avg} = \frac{q(t) - q(t_0)}{t - t_0}$$

The unit of current is the ampere when the charge is measured in coulombs and the time in seconds. The value of an instantaneous current is approximated at t_0 by computing the average current over very small time intervals about t_0. Making these time intervals smaller and smaller gives the following definition of current at some time t_0:

DEFINITION

> **Instantaneous Current**   If the electric charge transferred through a circuit varies with time according to the formula $q = q(t)$, then the *current* in the circuit at time t_0 is given by
>
> $$i(t_0) = \lim_{t \to t_0} \frac{q(t) - q(t_0)}{t - t_0}$$

EXAMPLE 5   ▪ **Problem**

The electric charge in a circuit varies with time according to the formula $q(t) = 4t^2$. What is the current in the circuit 2 sec after the switch is closed?

■ **Solution**

Using the definition of current with $q(t) = 4t^2$ and $t_0 = 2$, we have

$$i(2) = \lim_{t \to 2} \frac{q(t) - q(2)}{t - 2} = \lim_{t \to 2} \frac{4t^2 - 16}{t - 2} = \lim_{t \to 2} \frac{4(t - 2)(t + 2)}{t - 2}$$

$$= \lim_{t \to 2} 4(t + 2) = 16 \text{ A}$$

■

EXERCISES FOR SECTION 2.3

In Exercises 1■14, use the definition of the slope of a tangent line to compute the slope at the indicated point.

1. $y = 3x^2$ at $x_0 = -1$

2. $y = 3x^2$ at $x_0 = 2$

3. $y = x^2 + 5$ at $x_0 = 1$

4. $y = 9 - x^2$ at $x_0 = -4$

5. $y = x^2 - 2x$ at $x_0 = 0$

6. $g(x) = x - 4x^2$ at $x_0 = \frac{1}{4}$

7. $y = x^2 - 6x + 5$ at $x_0 = 3$

8. $y = x^2 + 2x + 1$ at $x_0 = 0$

9. $y = x^3$ at $x_0 = 2$

10. $y = \frac{1}{2}x^3$ at $x_0 = 1$

11. $y = \frac{1}{x}$ at $x_0 = 2$

12. $y = \frac{1}{x^2}$ at $x_0 = 1$

13. $y = \sqrt{x}$ at $x_0 = 1$

14. $y = \sqrt{x}$ at $x_0 = 4$

15. Write the equation of the tangent line to the curve $y = x^2$ at $x_0 = 1$.

16. Write the equation of the tangent line to the curve $y = x^2$ at $x_0 = \frac{1}{2}$.

In Exercises 17–22, find the velocity $v(t_0)$ at the given instant for the given distance-time equation.

17. $s = 6 + 9t$; $t_0 = 2$ sec

18. $s = t^2 + 2t$; $t_0 = 1$

19. $s = t^2 + 2t$; $t_0 = 2$

20. $s = \frac{3t^2}{2}$; $t_0 = 3$

21. $s = \frac{1}{t^2}$; $t_0 = 2$

22. $s = \frac{4}{t + 1}$; $t_0 = 5$

In Exercises 23–28, find the current $i(t_0)$ given the charge-time equation.

23. $q = 3t^2 - t$; $t_0 = 2$ sec

24. $q = 3t^2 - t + 1$; $t_0 = 1$

25. $q = t^2$; $t_0 = 0.1$

26. $q = 2t^3$; $t_0 = 1$

27. $q = \frac{5}{t}$; $t_0 = 0.5$

28. $q = \frac{1}{2t}$; $t_0 = 0.2$

29. Acceleration is defined as the time rate of change of velocity. Write the expression for the instantaneous acceleration of a particle at $t = t_0$, given that its velocity varies with time according to $v = V(t)$.

30. Using the result from Exercise 29, find the acceleration of an object at $t = 5$ sec, given that its velocity equation is $v = 3t^2 + 2$ (in feet per second).

31. Using the result from Exercise 29, find the acceleration of an object at $t = 0.8$ sec, given that its velocity is expressed by $v = t^2 - 3t$ (in centimeters per second).

2.4 THE DERIVATIVE

We have defined the concepts of slope of a tangent line, instantaneous velocity, and electric current by taking the limit of average quantities over small intervals. We write these three averages here in order to point out their similarities.

Slope of a secant line: $m_{\text{sec}} = \dfrac{g(x) - g(x_0)}{x - x_0}$

Average velocity of a particle: $v_{\text{avg}} = \dfrac{s(t) - s(t_0)}{t - t_0}$

Average current in a circuit: $i_{\text{avg}} = \dfrac{q(t) - q(t_0)}{t - t_0}$

Notice that these three quantities all have the same mathematical form; the only difference is one of physical interpretation. By recognizing this similarity, we can study all three problems at once. That is, the slope problem, the velocity problem, and the electric current problem are mathematically the same. To avoid reference to specific applications, we replace $g(x)$, $s(t)$, and $q(t)$ by the function $f(x)$ and study the quotient

$$\frac{f(x) - f(x_0)}{x - x_0}$$

This quotient is called the **difference quotient** of $f(x)$ at x_0.

The limit (if it exists) of the difference quotient as $x \to x_0$ is called the **derivative** of $f(x)$ at x_0. We denote the derivative of $f(x)$ at x_0 by $f'(x_0)$ and define it by

$$f'(x_0) = \lim_{x \to x_0} \frac{f(x) - f(x_0)}{x - x_0}$$

In this form, you can see that the derivative can be interpreted as the slope of a tangent line, the velocity of a particle, or the electric current in a circuit. The context of the problem tells you how to interpret the derivative.

While the derivative of $f(x)$ at x_0 may be defined as above, it is more commonly defined with $x - x_0 = \Delta x$. The notation Δx is read as "delta x" and is intended to suggest a small change in the variable x. Observing that if $x - x_0 = \Delta x$, then $x = x_0 + \Delta x$ and $\Delta x \to 0$ as $x \to x_0$, we make the following definition:

DEFINITION

> **Derivative of a Function at a Specific Point** The *derivative* of $f(x)$ at x_0 is given by
>
> $$f'(x_0) = \lim_{\Delta x \to 0} \frac{f(x_0 + \Delta x) - f(x_0)}{\Delta x}$$

The derivative of a function $f(x)$ is defined at all points for which the limit of the difference quotient exists. Consequently, a new function is derived from $f(x)$ by evaluating the limit of the difference quotient at each point. This derived function is denoted by $f'(x)$ and is called the **derivative function** of $f(x)$, or more commonly the derivative of $f(x)$. Practically, the formula for $f'(x)$ is the formula for $f'(x_0)$ with x_0 replaced by x.

DEFINITION

> **Derivative of a Function at Any Point** The *derivative* of $f(x)$ at any x is given by
>
> $$f'(x) = \lim_{\Delta x \to 0} \frac{f(x + \Delta x) - f(x)}{\Delta x}$$

A function is said to be **differentiable** at those points where the derivative exists. The process of finding the derivative of a function is called **differentiation.**

EXAMPLE 1   ■ **Problem**

Find the derivative of $y = x^2$.

■ **Solution**

Here, $f(x) = x^2$; therefore, $f(x + \Delta x) = (x + \Delta x)^2 = x^2 + 2x(\Delta x) + (\Delta x)^2$. Using the definition of a derivative, we find

$$f'(x) = \lim_{\Delta x \to 0} \frac{x^2 + 2x(\Delta x) + (\Delta x)^2 - x^2}{\Delta x} = \lim_{\Delta x \to 0} \frac{2x(\Delta x) + (\Delta x)^2}{\Delta x}$$

$$= \lim_{\Delta x \to 0} (2x + \Delta x) = 2x \qquad ■$$

The Delta Process

The process of finding the derivative of $f(x)$ is sometimes thought of as a four-step procedure called the **delta process.** In this approach, the limit of the difference quotient is determined in the following four steps.

PROCEDURE	**The Delta Process**
	1. Evaluate $f(x)$ at $x + \Delta x$.
	2. Subtract the value of $f(x)$ from $f(x + \Delta x)$ to obtain Δy, the increment of change in the value of the function.
	3. Divide the Δy obtained in step 2 by Δx, to obtain $\Delta y / \Delta x$. Simplify this expression by canceling any common factors from numerator and denominator.
	4. Take the limit of $\Delta y / \Delta x$ as Δx approaches zero. This step gives the derivative; that is, $$f'(x) = \lim_{\Delta x \to 0} \frac{\Delta y}{\Delta x}$$

To illustrate the use of the delta process to find the derivative of a function, we choose the function $f(x) = x^2$, which is the same function used in Example 1. In this way, you will be able to see the similarities in the two methods. In fact, the delta process is simply the process of applying the definition of the derivative in four separate steps.

EXAMPLE 2   ■ **Problem**

Use the delta process to find the derivative of $y = x^2$.

■ **Solution**

The four steps in the delta process are as follows:

1. Evaluate $f(x)$ at $x + \Delta x$. Here, $f(x) = x^2$, so

$$f(x + \Delta x) = (x + \Delta x)^2 = x^2 + 2x(\Delta x) + (\Delta x)^2$$

2. Compute Δy.

$$\Delta y = f(x + \Delta x) - f(x) = [x^2 + 2x(\Delta x) + (\Delta x)^2] - x^2$$
$$= 2x(\Delta x) + (\Delta x)^2$$

3. Divide Δy by Δx. Using Δy from step 2, we get

$$\frac{\Delta y}{\Delta x} = \frac{2x(\Delta x) + (\Delta x)^2}{\Delta x} = 2x + \Delta x$$

4. The derivative is the limit of $\Delta y/\Delta x$ as Δx approaches zero. Thus,

$$f'(x) = \lim_{\Delta x \to 0} \frac{\Delta y}{\Delta x} = \lim_{\Delta x \to 0} (2x + \Delta x) = 2x$$

This answer agrees with the result obtained in Example 1.   ■

EXAMPLE 3   ■ **Problem**

Find the equation of the line tangent to the curve $y = x^2$ at $x = -\frac{2}{3}$. See Figure 2.14.

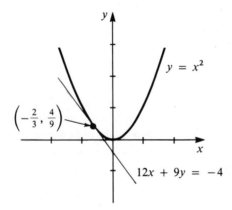

■ **Figure 2.14**

■ **Solution**

From Example 1, we know that the derivative is $f'(x) = 2x$. Hence, the slope of the tangent line when $x = -\frac{2}{3}$ is

$$f'(-\tfrac{2}{3}) = 2(-\tfrac{2}{3}) = -\tfrac{4}{3}$$

The equation of the line is found by using the point-slope form of a straight line:

$$y - \tfrac{4}{9} = -\tfrac{4}{3}(x + \tfrac{2}{3})$$
$$12x + 9y = -4$$

    ■

Physical Interpretations of the Derivative

As noted earlier, the derivative is the generalization of the physical concept of rate of change, and consequently, it can represent any number of physical quantities. For instance, the derivative of the displacement function of an object is an expression for the velocity of the object. The derivative of the function describing the electric charge flowing past a point is an expression for the electric current. There are many other applications of the derivative, and we will discuss many of them in this book. The following table lists just a few of these applications.

If the function represents	The derivative represents
A graph	The slope of a tangent line
Displacement	Velocity
Electric charge	Electric current

EXAMPLE 4 ■ **Problem**

The displacement (in centimeters) of a charged particle varies with time (in seconds) according to the formula $s = t^2 + 3t$. Find a formula for the instantaneous velocity of the particle.

■ **Solution**

Velocity is the rate of change of displacement with respect to time. Thus, the velocity of the particle is given by the derivative of its displacement function, that is, $v = s'(t)$. So,

$$v = \lim_{\Delta t \to 0} \frac{s(t + \Delta t) - s(t)}{\Delta t} = \lim_{\Delta t \to 0} \frac{(t + \Delta t)^2 + 3(t + \Delta t) - (t^2 + 3t)}{\Delta t}$$

$$= \lim_{\Delta t \to 0} \frac{t^2 + 2t(\Delta t) + (\Delta t)^2 + 3t + 3(\Delta t) - t^2 - 3t}{\Delta t}$$

$$= \lim_{\Delta t \to 0} \frac{2t(\Delta t) + (\Delta t)^2 + 3(\Delta t)}{\Delta t} = \lim_{\Delta t \to 0} (2t + \Delta t + 3) = 2t + 3$$

Therefore, the velocity (in centimeters per second) of the particle at any time t is given by

$$v = 2t + 3$$ ■

EXAMPLE 5 ■ **Problem**

Find a formula for the current (in amperes) in a circuit if the charge transferred (in coulombs) varies with time according to $q = 1/(t + 1)$.

■ **Solution**

Electric current is the rate of change of charge with respect to time. Thus, $i = q'(t)$, and we write

$$i = \lim_{\Delta t \to 0} \frac{q(t + \Delta t) - q(t)}{\Delta t}$$

Here, $q(t) = 1/(t + 1)$, and $q(t + \Delta t) = 1/(t + \Delta t + 1)$. The formula for the current, then, is given by

$$i = \lim_{\Delta t \to 0} \frac{\dfrac{1}{t + \Delta t + 1} - \dfrac{1}{t + 1}}{\Delta t} = \lim_{\Delta t \to 0} \frac{\dfrac{(t + 1) - (t + \Delta t + 1)}{(t + 1)(t + \Delta t + 1)}}{\Delta t}$$

$$= \lim_{\Delta t \to 0} \frac{-1}{(t + 1)(t + \Delta t + 1)}$$

$$= \frac{-1}{(t + 1)(t + 1)} = \frac{-1}{(t + 1)^2}$$ ■

EXAMPLE 6   ■ **Problem**

What is the current in the circuit of Example 5 when $t = 2$ sec?

■ **Solution**

The current at any time t is given by $i = -1/(t + 1)^2$. Letting $t = 2$ in this formula, we obtain

$$i(2) = \frac{-1}{(2 + 1)^2} = -\frac{1}{9} = -0.111 \text{ A}$$

The negative sign indicates that the flow of electric charge is decreasing.   ■

Comment: In addition to the $(')$ notation for the derivative of a function, another commonly used symbol for the derivative of $y = f(x)$ is

$$\frac{dy}{dx}$$

This notation comes from the use of

$$\frac{\Delta y}{\Delta x}$$

for the slope of a secant line of $y = f(x)$. See Figure 2.15. Then, the slope of a tangent line is denoted as dy/dx and is defined as

$$\frac{dy}{dx} = \lim_{\Delta x \to 0} \frac{\Delta y}{\Delta x}$$

The derivative of $y = f(x)$ may thus be represented by y', $f'(x)$, or dy/dx.

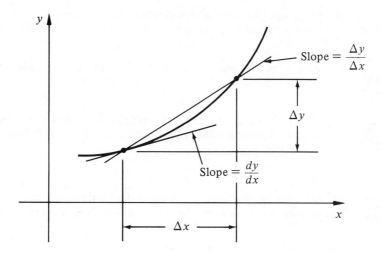

$$\text{Slope} = \frac{\Delta y}{\Delta x}$$

$$\Delta y$$

$$\text{Slope} = \frac{dy}{dx}$$

$$\Delta x$$

■ **Figure 2.15**

EXERCISES FOR SECTION 2.4

Find the derivative of the function in Exercises 1–16.

1. $y = \frac{1}{3}x^2$

2. $f(x) = 5x^2$

3. $y = x^2 + 3x$

4. $y = 2x - x^2$

5. $f(x) = \frac{1}{2}x^2 + \frac{1}{3}x$

6. $f(x) = x^2 - 2x + 5$

7. $y = 7 + 5x + 3x^2$

8. $y(x) = -\frac{1}{x}$

9. $y = \frac{5}{x}$

10. $y = \frac{3}{x}$

11. $f(x) = \frac{2}{x^2}$

12. $f(x) = x^3$

13. $y = x^3 - 2x + 3$

14. $y = x^3 - x^2$

15. $y = \frac{1}{x + 5}$

16. $y = \frac{2}{x - 2}$

Use the results obtained in Exercises 1–16 to help you solve the remaining problems.

17. Find the slope of the curve $y = \frac{1}{3}x^2$ at (a) $x = -1$, (b) $x = 0$, and (c) $x = 1$.

18. Find the slope of the curve $y = 7 + 5x + 3x^2$ at (a) $x = \frac{1}{2}$, (b) $x = 1$, and (c) $x = 2$.

19. The equation of motion of a particle along a straight line varies with time (in seconds) in accordance with $s = t^2 + 3t$ (in feet). What is the velocity of the particle when (a) $t = 0.1$ and (b) $t = 0.5$ sec?

20. The charge (in coulombs) transferred in a circuit varies with time (in seconds) as $q = 3/t$. What current (in amperes) flows in the circuit when (a) $t = 0.5$ and (b) $t = 1.5$ sec?

21. What is the equation for the velocity of a particle if its displacement (in feet) varies with time (in seconds) according to $s = 2/(t - 2)$?

22. What is the equation for the acceleration of an object if its velocity (in feet per second) varies with time (in seconds) according to $v = t^2 - 2t + 3$? See Exercise 29 of the previous exerise set.

The following BASIC program calculates the derivative of $f(x)$ at a specified value of x. Use this program to find the derivative of the following functions at the indicated points.

23. Given $f(x) = \frac{1}{2}x^{1/3}$, find $f'(-2), f'(\frac{1}{2}), f'(2)$.

24. Given $f(x) = 5x + x^2$, find $f'(-2), f'(0), f'(1.5)$.

25. Given $f(x) = 3x^{-2}$, find $f'(-3), f'(0.2), f'(2)$.

```
10   REM  PROGRAM TO FIND THE DERIVATIVE AT A POINT
20   HOME
30   PRINT "THIS PROGRAM WILL FIND THE DERIVATIVE"
40   PRINT "AT A POINT SUPPLIED BY THE USER."
50   PRINT : PRINT "THE FUNCTION IN THIS PROGRAM"
60   PRINT "IS (X - 10)/(X +10)
70   PRINT "IF THE USER CHANGES THE FUNCTION,"
80   PRINT "THE NEW FUNCTION WILL REMAIN UNTIL"
90   PRINT "THE USER CHANGES IT AGAIN, OR"
100  PRINT "RELOADS THE PROGRAM."
110  PRINT : PRINT "DO YOU WISH TO CHANGE THE FUNCTION? Y/N"
120  INPUT A$
130  IF A$ = "Y" THEN 420
140  PRINT "ENTER THE X-VALUE WHERE YOU WOULD LIKE"
150  PRINT "TO FIND THE DERIVATIVE."
160  INPUT X1
170 EPS = 0.00005
180  HTAB 12: INVERSE : PRINT "PATIENCE PLEASE"
190  HTAB 12: PRINT "CALCULATING!!!": NORMAL
200  DEF  FN F(X) = (X - 10) / (X + 10)
210  PRINT "X            F(X)          F(X)          X"
     BL$ = "     ": REM  FIVE BLANKS
230  FOR N = 1 TO 15
240 H = 1 / 2 ^ N
250 XL = X1 - H
260 XR = X1 + H
270 YL = ( FN F(X1) -  FN F(XL)) / H
280 YR = ( FN F(XR) -  FN F(X1)) / H
282 XL =   INT (10000 * XL) / 10000
284 YL =   INT (10000 * YL) / 10000
286 YR =   INT (10000 * YR) / 10000
288 XR =   INT (10000 * XR) / 10000
290  PRINT XL;BL$;YL;BL$;YR;BL$;XR
300  IF  ABS (YR - YL) < EPS THEN 350
```

```
310   NEXT N
320   PRINT "AFTER 15 ITERATIONS THERE IS NO"
330   PRINT "APPARENT CONVERGENCE!"
340   GOTO 380
350 YA = (YL + YR) / 2
355   PRINT
360   PRINT "AT X = ";X1;" THE APPARENT VALUE OF"
370   PRINT "THE DERIVATIVE IS ";YA
380   PRINT : PRINT "DO YOU WANT TO RUN THIS PROGRAM AGAIN? Y/N"
390   INPUT B$
395   HOME
400   IF B$ = "Y" THEN 50
410   GOTO 450
420   PRINT "TYPE"
430   INVERSE : PRINT "200 DEF FN F(X) = (YOUR FUNCTION) <RET>"
440   PRINT "RUN 140": NORMAL
450   END
```

REVIEW EXERCISES FOR CHAPTER 2

In Exercises 1–22, find the indicated limit.

1. $\lim\limits_{x \to 3} x^3$

2. $\lim\limits_{x \to 2} \sqrt{2x}$

3. $\lim\limits_{k \to 1} (5k + 1)$

4. $\lim\limits_{t \to 0} (3 - 2\sqrt{t})$

5. $\lim\limits_{x \to 0} (3x^3 - 5x)$

6. $\lim\limits_{x \to 9} \sqrt{x - 9}$

7. $\lim\limits_{x \to 4} \sqrt{4 - x}$

8. $\lim\limits_{x \to 1} (3 - 5x)$

9. $\lim\limits_{t \to 0} \dfrac{t + 4t^2}{t}$

10. $\lim\limits_{t \to 0} \dfrac{2t + 5t^2}{t}$

11. $\lim\limits_{x \to 3} \dfrac{x^2 + x - 12}{x - 3}$

12. $\lim\limits_{x \to 2} \dfrac{2x^2 - x - 6}{x - 2}$

13. $\lim\limits_{y \to 9} \dfrac{\sqrt{y} - 3}{y - 9}$

14. $\lim\limits_{x \to 1} \dfrac{3x^2 - 5x + 2}{x - 1}$

15. $\lim\limits_{x \to 2} \dfrac{x + 3}{x^2 - x - 2}$

16. $\lim\limits_{x \to 1} \dfrac{x^2 + 1}{x^2 - 1}$

17. $\lim \dfrac{x + 4}{3x - 1}$

18. $\lim\limits_{x \to \infty} \dfrac{3x^2 + x}{2x - x^2}$

19. $\lim\limits_{x \to \infty} \sin x$

20. $\lim\limits_{x \to \infty} \dfrac{5 + 7x}{4x^3 - 2x + 1}$

21. $\lim\limits_{x \to 0} \dfrac{\cos x}{x}$

22. $\lim\limits_{n \to 2} \dfrac{n}{n - 2}$

23. Use the definition of $m(x_0)$ to find the slope of the tangent line to $y = x^2 + x$ at $x_0 = 2$.

24. Use the definition of $m(x_0)$ to find the slope of the tangent line to $y = 3x - x^2$ at $x_0 = 2$.

25. The displacement (in feet) of a car varies with time (in seconds) according to $s = 3t^2 + t$. Use the definition of $v(t_0)$ to find the velocity of the car at $t_0 = 5$ sec.

26. The displacement (in meters) of an object varies with time (in seconds) according to $s = \sqrt{t}$. Use the definition to find the velocity of the object at $t_0 = 4$ sec.

In Exercise 27–34, use the definition to find the derivative function of the given function.

27. $y = \frac{1}{2}x^2$

28. $f(x) = 2 - 3x^2$

29. $f(x) = x^2 + 5x - 1$

30. $y = 2x^2 - 2x + 3$

31. $g(t) = t^3 + t$

32. $s(t) = \dfrac{5}{t^2}$

33. $y = \dfrac{1}{x + 1}$

34. $y = 1 + \dfrac{1}{x}$

CHAPTER 3

Formulas for Finding the Derivative

If the four-step process were always used to compute derivatives, the variety of applications would be limited since the process is unwieldy and time-consuming. Furthermore, a certain amount of mathematical sophistication is required to evaluate the limit even if the functions are relatively simple. In this chapter, we will develop techniques that make the derivative easy to use. In fact, these formulas will permit us to largely ignore the definition of the derivative in terms of limits and to have an algorithmic approach to calculus. These easy-to-remember and easy-to-use formulas have made the calculus available to everyone, not just the individuals skilled in higher mathematics. We begin by limiting the discussion to the simple but important case of finding derivatives of polynomial functions.

3.1 DERIVATIVES OF POLYNOMIALS

In this section we introduce some new notation for the derivative and develop formulas that simplify the process of finding the derivative of a function. We note that the derivative of a function $f(x)$ is frequently written as

$$f'(x) = \frac{d}{dx}[f(x)]$$

where the symbol d/dx is called the **derivative operator** and indicates that $f(x)$ is to be differentiated with respect to the variable x. Thus, the derivative of the function $f(x) = x^3 + 4$ may be denoted by

$$f'(x) = \frac{d}{dx}(x^3 + 4)$$

Two useful properties of the derivative operator d/dx follow. The letters u and v represent any differentiable functions of the same variable, and c represents any constant. Note also that

$$\frac{d}{dx}[u(x)] \quad \text{and} \quad \frac{du}{dx}$$

both mean the derivative of the function u with respect to x.

PROPERTY 1

Derivative of a Constant Times a Function

$$\frac{d}{dx}[cu(x)] = c\frac{d}{dx}[u(x)] = c\frac{du}{dx}$$

PROPERTY 2

Derivative of a Sum of Functions

$$\frac{d}{dx}[u(x) + v(x)] = \frac{d}{dx}[u(x)] + \frac{d}{dx}[v(x)] = \frac{du}{dx} + \frac{dv}{dx}$$

Expressed in words, Property 1 states that *the derivative of a constant times a function is equal to the constant times the derivative of the function.* Property 2 states that *the derivative of a sum of two or more functions is equal to the sum of the derivatives of the individual functions.* The use of these properties decreases the number of specific differentiation formulas that must be developed.

EXAMPLE 1 ▪ **Problem**

Apply Properties 1 and 2 to $\dfrac{d}{dx}(3x^5 + 7x^2 + 2)$.

▪ **Solution**

We have

$$\frac{d}{dx}(3x^5 + 7x^2 + 2) = \frac{d}{dx}(3x^5) + \frac{d}{dx}(7x^2) + \frac{d}{dx}(2)$$

$$= 3\frac{d}{dx}(x^5) + 7\frac{d}{dx}(x^2) + \frac{d}{dx}(2)$$

All that is required to complete this problem is a knowledge of the differentiation formulas for $(d/dx)(x^n)$ and $(d/dx)(c)$. ■

EXAMPLE 2 ▪ **Problem**

Write an expression for $f'(x)$ if $f(x) = 2\sin x - \log 3x$.

■ **Solution**

Applying Properties 1 and 2, we have

$$f'(x) = 2 \frac{d}{dx} (\sin x) - \frac{d}{dx} (\log 3x)$$

Here, we need the specific formulas for $(d/dx)(\sin x)$ and $(d/dx)(\log 3x)$ to complete the problem. These formulas are introduced in later chapters. ■

The preceding examples demonstrate the need for *specific* differentiation formulas to complement Properties 1 and 2. In this section, we will derive formulas for $(d/dx)(x^n)$ and $(d/dx)(c)$.

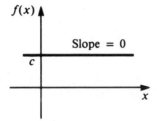

■ **Figure 3.1**

We know that the graph of $f(x) = c$ is a straight line parallel to the x-axis. Since the slope of a line parallel to the x-axis is zero, it follows from the geometric interpretation of the derivative as the slope of a curve at a point that the derivative must be zero for all points on the line. See Figure 3.1. We state this conclusion in the following formula.

FORMULA

The Derivative of a Constant

$$\frac{d}{dx} (c) = 0 \tag{3.1}$$

where c is any constant.

EXAMPLE 3 The derivatives of 200, $-\sqrt{5}$, and 10^7 are as follows:

$$\frac{d}{dx} (200) = 0$$

$$\frac{d}{dx} (-\sqrt{5}) = 0$$

$$\frac{d}{dx} (10^7) = 0$$

■

The formula for differentiating polynomial functions of the form $f(x) = x^n$ is given next.

FORMULA

The Derivative of x^n

$$\frac{d}{dx}(x^n) = nx^{n-1}$$

(3.2)

where n is a real number.

For $n = 1$, Formula (3.2) gives $(d/dx)(x) = x^0 = 1$. To demonstrate the validity of this result, we use the definition of the derivative to get

$$\frac{d}{dx}(x) = \lim_{\Delta x \to 0} \frac{x + \Delta x - x}{\Delta x}$$

$$= \lim_{\Delta x \to 0} 1 = 1$$

which agrees with Formula (3.2). The graph of $f(x) = x$ is shown in Figure 3.2. The graph is a straight line with a slope of one.

■ **Figure 3.2**

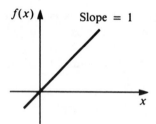

Next, consider $f(x) = x^2$. Then, by Formula (3.2), we have

$$\frac{d}{dx}(x^2) = 2x^{2-1} = 2x$$

Also, from the definition of the derivative, we have

$$\frac{d}{dx}(x^2) = \lim_{\Delta x \to 0} \frac{x^2 + 2x(\Delta x) + (\Delta x)^2 - x^2}{\Delta x} = \lim_{\Delta x \to 0} (2x + \Delta x) = 2x$$

This result agrees with the result obtained by using Formula (3.2).

Finally, if $f(x) = x^3$, then $(d/dx)(x^3) = 3x^2$. This result is verified by

$$\frac{d}{dx}(x^3) = \lim_{\Delta x \to 0} \frac{x^3 + 3x^2(\Delta x) + 3x(\Delta x)^2 + (\Delta x)^3 - x^3}{\Delta x}$$

$$= \lim_{\Delta x \to 0} [3x^2 + 3x(\Delta x) + (\Delta x)^2] = 3x^2$$

By similar methods, we could show that $(d/dx)(x^4) = 4x^3$, and so on. In fact, the formula $(d/dx)(x^n) = nx^{n-1}$ is valid for any real number n. For instance, $(d/dx)(x^\pi) = \pi x^{\pi-1}$. We will use Formula (3.2) in this expanded interpretation throughout the book.

EXAMPLE 4 The derivatives of x^7, $x^{1/3}$, $t^{2.3}$, and m^{-2} are as follows:

$$\frac{d}{dx}(x^7) = 7x^6$$

$$\frac{d}{dx}(x^{1/3}) = \frac{1}{3}x^{-2/3}$$

$$\frac{d}{dt}(t^{2.3}) = 2.3t^{1.3}$$

$$\frac{d}{dm}(m^{-2}) = -2m^{-3}$$
∎

EXAMPLE 5 ∎ **Problem**

Evaluate $\dfrac{d}{dx}(3x^5 + x^{1/4} - 15)$.

∎ **Solution**

We have

$$\frac{d}{dx}(3x^5 + x^{1/4} - 15)$$

$$= 3\frac{d}{dx}(x^5) + \frac{d}{dx}(x^{1/4}) - \frac{d}{dx}(15) \qquad \text{Properties 1 and 2}$$

$$= 3(5x^4) + \frac{1}{4}x^{-3/4} - 0 \qquad \text{Formulas (3.1) and (3.2)}$$

$$= 15x^4 + \frac{1}{4}x^{-3/4}$$
∎

Frequently, we must manipulate an expression before we can apply the appropriate differentiation formula. The following examples show this situation.

EXAMPLE 6 ∎ **Problem**

Find $f'(x)$ if $f(x) = \dfrac{21}{x^3}$.

■ **Solution**

Before we can use Formula (3.2), we write the function in the form $f(x) = 21x^{-3}$. Once we have the function in this form, we can differentiate it to get

$$f'(x) = 21 \frac{d}{dx}(x^{-3}) = 21(-3x^{-4}) = -63x^{-4}$$

■

EXAMPLE 7 ■ **Problem**

Differentiate $p = \sqrt[3]{5t^7}$ with respect to t.

■ **Solution**

Writing the given expression as $p = \sqrt[3]{5}\,t^{7/3}$, we have

$$\frac{dp}{dt} = \sqrt[3]{5}\frac{d}{dt}(t^{7/3}) = \frac{7\sqrt[3]{5}}{3}t^{4/3}$$

■

EXAMPLE 8 ■ **Problem**

Find $g'(x)$ if $g(x) = (2x^2 + 5)(x - 7)$.

■ **Solution**

To evaluate this derivative, we first expand the indicated product so that we can use Properties 1 and 2. Thus,

$$g(x) = (2x^2 + 5)(x - 7) = 2x^3 - 14x^2 + 5x - 35$$

The desired derivative is then

$$g'(x) = \frac{d}{dx}(2x^3 - 14x^2 + 5x - 35) = 6x^2 - 28x + 5$$

■

EXAMPLE 9 ■ **Problem**

Find the equation of the straight line drawn tangent to the graph of $y = x^3$ at the point $(2, 8)$. See Figure 3.3.

■ **Solution**

Recall that we can write the equation of a line if we know the slope of the line and any point through which it passes. Here, we know the line must pass through $(2, 8)$, and we can find the slope by evaluating the derivative of $y = x^3$ at $x = 2$. Since the derivative of $y = x^3$ is $y' = 3x^2$, the slope of the tangent line at $x = 2$ is 12. Now, using the point-slope form of a straight line, we have

$$y - 8 = 12(x - 2)$$

which, when simplified, becomes

$$y = 12x - 16$$

■

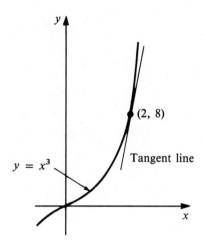

■ **Figure 3.3**

Differentiate the functions in Exercises 1–39.

1. $y = x^9$

2. $y = x^5$

3. $f(x) = 3x^{12}$

4. $g(x) = 6x^8 - 17$

5. $p = t^{-2} + 1$

6. $m = 5 + x^{-10}$

7. $y = 4x^{1/2}$

8. $y = 3x^{2/5}$

9. $h(t) = 2^7$

10. $y = \sqrt{13}$

11. $y = 3x + 5$

12. $y = ax^2 + bx^3$

13. $\theta = 3\phi + 4\phi^2$

14. $g(s) = \sqrt{s} - \sqrt[3]{s}$

15. $y = \dfrac{1}{x^3}$

16. $y = \left(\dfrac{2}{x}\right)^4 + \left(\dfrac{2}{x}\right)^2$

17. $z = \sqrt{t^3} - 5$

18. $f(x) = \dfrac{3}{4x}$

19. $p = \sqrt{3t}$

20. $h = \sqrt[3]{4k^2}$

21. $v = at^{5/2} - bt^{3/5}$

22. $y = \dfrac{1}{\sqrt{x^7}}$

23. $r = \dfrac{4}{\sqrt[3]{s}} + \dfrac{5}{\sqrt{s}}$

24. $s = \dfrac{10}{t}$

25. $y = x^4 - 2x^3 + 5x^2 - 7x - 1$

26. $y = \sqrt[5]{w} - 6\sqrt[5]{w^2}$

27. $p = \dfrac{1}{\sqrt{2t^3}}$

28. $m = \sqrt{4k^6}$

29. $y = \sqrt{x^2 + 2x + 1}$

30. $v = \sqrt[3]{\sqrt{x}}$

31. $y = (x - 2)^2$

32. $r = (s^{1/3} + 2)^2$

33. $w = 3t^{1/2} - 4t^{1/3} + 2$

34. $f(x) = \dfrac{2x^3 - 3x^2 + 5x}{x^3}$

35. $i = \dfrac{3 + t^2}{\sqrt{t}}$

36. $y = 15x^{1.39}$

37. $v = 0.9t^{2.6}$

38. $g(t) = 125t^{-1.42}$

39. $p = (12 + w)w^{-0.2}$

Find the equation of the line drawn tangent to the graph of the function at the indicated point in Exercises 40–43.

40. $y = x^3 - 3x^2 - 2x$, at $x = 3$

41. $y = x^2 + \dfrac{1}{x^2}$, at $x = \dfrac{1}{2}$

42. $y = x^{3/2} - x^{4/3}$, at $x = 1$

43. $y = x + 3x^{-1}$, at $x = -1$

In Exercises 44–47, find the values of x for which the graph has a slope of zero.

44. $y = x^2 - 6x + 13$

45. $y = x^3 + 9x^2 + 7$

46. $y = x^3 + 4x^2 + 4x$

47. $y = x - \dfrac{8}{x}$

48. To find the voltage across the terminals of an amplifier output, an engineer must differentiate the expression $2t^{3/2} + 3t^{5/2}$. Find the value of the derivative of this expression when $t = 4$.

49. To maximize cost, an economist must differentiate the expression $C(t) = 3t^2 + 2t^\pi$. Find $C'(3)$.

50. Show that

$$\frac{d}{dx}(x^2 + 3x)(2x - 3) \neq \frac{d}{dx}(x^2 + 3x) \cdot \frac{d}{dx}(2x - 3)$$

thus establishing that the derivative of a product is not the product of the derivatives.

■■■■ 3.2 COMPOSITE FUNCTIONS

Properties 1 and 2 of Section 3.1 allow us to compute the derivative of functions of the type $af(x) + bg(x)$, where a and b are constants, if the specific derivatives of f and g are known. In the preceding section, we limited our discussion to linear combinations of functions of the form x^n. Now, we introduce a fundamental rule of differentiation that shows how to differentiate functions that are composed of other more elementary functions. This rule is variously called the composite function rule or, more popularly, the **chain rule.** Rather than go through the formality of a definition of composition, we will give several examples to show the general idea.

EXAMPLE 1   (a) The function $y = (x^5 + x^4 + 1)^{10}$ may be thought of as the *composition* of the function $y = u^{10}$ with the function $u = x^5 + x^4 + 1$.
(b) The function $y = \sin x^2$ may be thought of as the composition of the function $y = \sin u$ with the function $u = x^2$.
(c) The function $y = \log e^{3x}$ may be thought of as the composition of the *three* functions $y = \log u$, $u = e^v$, and $v = 3x$.   ∎

EXAMPLE 2   Suppose the velocity of a missile varies with displacement s according to $v = s^{1/2}$, and the displacement, in turn, is given as a function of elapsed time to be $s = t^3 - 4$. Then, the velocity can be written as

$$v = (t^3 - 4)^{1/2}$$

We say that v is a composite function of t through the displacement function s. Alternatively, we can say that v as a function of time can be decomposed into functions of s and t.   ∎

The Chain Rule

In general, let $y = f(u)$ and $u = g(x)$. Then, we may regard y as a composite function of the variable x. The following succession of steps will at least motivate the formula for the chain rule. We begin by writing

$$\frac{\Delta y}{\Delta x} = \frac{\Delta y}{\Delta x} \cdot \frac{\Delta u}{\Delta u} \quad \text{if} \quad \Delta u \ne 0$$

which may be rewritten as

$$\frac{\Delta y}{\Delta x} = \frac{\Delta y}{\Delta u} \cdot \frac{\Delta u}{\Delta x}$$

We now apply the limit process to this equation. Figure 3.4 gives a typical relationship between y and x through u. As you can see, $\Delta u \to 0$ as $\Delta x \to 0$. Assuming both dy/du and du/dx exist, we have the limit

$$\lim_{\Delta x \to 0} \frac{\Delta y}{\Delta x} = \lim_{\Delta x \to 0} \frac{\Delta y}{\Delta u} \cdot \lim_{\Delta x \to 0} \frac{\Delta u}{\Delta x}$$

and therefore,

$$\frac{dy}{dx} = \frac{dy}{du} \cdot \frac{du}{dx}$$

which is the formula ordinarily called the chain rule.

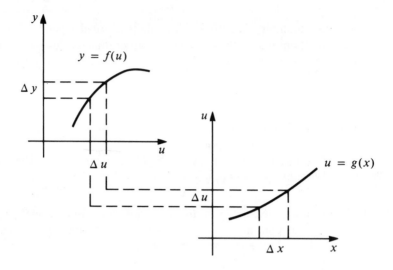

■ **Figure 3.4**

RULE

The Chain Rule Let $y = f(u)$ and $u = g(x)$ be differentiable functions. Then,

$$\frac{d}{dx}[f(u)] = \frac{d}{du}[f(u)] \cdot \frac{d}{dx}[g(x)] \tag{3.3}$$

or more briefly,

$$\frac{dy}{dx} = \frac{dy}{du} \cdot \frac{du}{dx}$$

Expressed in words, the chain rule says that *the derivative of a function formed by composition is the **product** of the derivatives of the composing functions.* When applying the chain rule, you must decompose the given function into two or more elementary functions. Most of the time, this decomposition can be done mentally, but it does take practice.

EXAMPLE 3   ■ **Problem**

Show how the chain rule is used to differentiate the function

$$y = (x^5 + x^4 + 1)^{10}$$

■ **Solution**

To differentiate $y = (x^5 + x^4 + 1)^{10}$, we let $u = x^5 + x^4 + 1$. Then, by (3.3)

$$\frac{d}{dx}(u^{10}) = 10u^9 \frac{du}{dx}$$

and

$$\frac{du}{dx} = 5x^4 + 4x^3$$

Therefore,

$$\frac{d}{dx}(x^5 + x^4 + 1)^{10} = 10(x^5 + x^4 + 1)^9(5x^4 + 4x^3)$$ ∎

The Derivative of u^n

Now, we will derive a formula for the derivative of composite functions of the form $y = u^n$, where u is any differentiable function of x. By the chain rule, we can write

$$\frac{dy}{dx} = \frac{dy}{du} \cdot \frac{du}{dx}$$

Replacing y with u^n, we get

$$\frac{d}{dx}(u^n) = \frac{d}{du}(u^n)\frac{du}{dx}$$

Finally, since $(d/du)(u^n) = nu^{n-1}$, we have the following rule.

RULE

> **The Power Rule**   Let u represent a differentiable function of x. Then
>
> $$\frac{d}{dx}(u^n) = nu^{n-1}\frac{du}{dx}$$ (3.4)

The mistake most beginners make in applying this formula is to forget to identify the independent variable. For example, if y is the independent variable of the function y^4, the derivative is given by

$$\frac{d}{dy}(y^4) = 4y^3$$

However, if y is a composite function of the variable x, by Formula (3.4), we get

$$\frac{d}{dx}(y^4) = 4y^3 \frac{dy}{dx}$$

EXAMPLE 4   ▪ **Problem**

Let $y = (3x + 5)^2$. Find $\dfrac{dy}{dx}$ in two different ways.

▪ **Solution**

First, apply Formula (3.4) with $u = 3x + 5$ and $n = 2$, to obtain

$$\frac{dy}{dx} = 2(3x + 5)^1 \cdot \frac{d}{dx}(3x + 5) = 2(3x + 5)(3) = 6(3x + 5)$$

Second, instead of using the chain rule, we could first expand the given binomial and then differentiate term by term. Thus,

$$y = (3x + 5)^2 = 9x^2 + 30x + 25$$

and

$$\frac{dy}{dx} = 18x + 30 = 6(3x + 5)$$

This result agrees with the result obtained by using Formula (3.4).   ▪

EXAMPLE 5   ▪ **Problem**

Find the derivative of $p = \sqrt[5]{4 - t}$ with respect to t.

▪ **Solution**

If we write the fifth root as a fractional exponent, the given equation has the form

$$p = (4 - t)^{1/5}$$

Using Formula (3.4) with $u = 4 - t$ and $n = \frac{1}{5}$, we have

$$\frac{dp}{dt} = \frac{1}{5}(4 - t)^{-4/5} \cdot \frac{d}{dt}(4 - t) = \frac{1}{5}(4 - t)^{-4/5}(-1)$$

$$= -\frac{1}{5}(4 - t)^{-4/5}$$ ▪

EXAMPLE 6   ▪ **Problem**

Differentiate $y = \dfrac{1}{(x^2 + x)^4}$ with respect to x.

▪ **Solution**

This expression may be written in the form u^n by use of a negative exponent. Thus,

$$y = \frac{1}{(x^2 + x)^4} = (x^2 + x)^{-4}$$

Formula (3.2) may now be used if we let $u = x^2 + x$ and $n = -4$. Hence,

$$\frac{dy}{dx} = -4(x^2 + x)^{-5} \cdot \frac{d}{dx}(x^2 + x) = -4(x^2 + x)^{-5}(2x + 1)$$

$$= -\frac{4(2x + 1)}{(x^2 + x)^5}$$

    ■

EXAMPLE 7   ■ **Problem**

Find the expression for the slope of the graph of $y^3 = x^2 + 4$.

■ **Solution**

The given equation may be rearranged in the form $y = (x^2 + 4)^{1/3}$. We then use Formula (3.4) to find the derivative of the given function. (Remember that the derivative of a function is the expression for the slope of the graph of the function.) Letting $u = x^2 + 4$ and $n = \frac{1}{3}$, we have

$$\frac{dy}{dx} = \frac{1}{3}(x^2 + 4)^{-2/3} \cdot \frac{d}{dx}(x^2 + 4) = \frac{1}{3}(x^2 + 4)^{-2/3}(2x)$$

$$= \frac{2}{3}x(x^2 + 4)^{-2/3}$$

    ■

EXAMPLE 8   ■ **Problem**

The light intensity I from a carbon filament is found to vary with the applied voltage v according to $I = \sqrt{\rho R v}$, where ρ is a constant and R is the resistance of the filament. Find the rate of change of light intensity with respect to voltage for a given filament.

■ **Solution**

The resistance R is considered to be fixed since we are concerned with a particular filament. The given equation can be written as

$$I = (\rho R v)^{1/2}$$

To find dI/dv, we let $u = \rho R v$ and $n = \frac{1}{2}$. Hence,

$$\frac{dI}{dv} = \frac{1}{2}(\rho R v)^{-1/2} \cdot \frac{d}{dv}(\rho R v) = \frac{1}{2}(\rho R v)^{-1/2}(\rho R) = \frac{\rho R}{2\sqrt{\rho R v}} = \frac{1}{2}\sqrt{\frac{\rho R}{v}}$$

    ■

EXERCISES FOR SECTION 3.2

Find the derivative of the function in Exercises 1–22 with respect to the independent variable by applying Formula (3.4).

1. $y = (4 - 3x)^5$ **2.** $y = (2x - 3)^3$

3. $s = (t^2 + 4)^4$ **4.** $p = (w^3 - 1)^9$

5. $y = (5x + 1)^{1/5}$

6. $z = (s^2 + 4)^{3/2}$

7. $w = \sqrt{4t^2 - 3t}$

8. $q = 2\sqrt[3]{1 - 3t}$

9. $i = (3t + 1)^{-2}$

10. $y = (x^2 + 3x + 4)^{-1}$

11. $y = \dfrac{2}{(2 + 5x)^3}$

12. $s = \dfrac{1}{(t^3 - t^2)}$

13. $L = \dfrac{3}{\sqrt{x^2 + 1}}$

14. $y = \dfrac{1}{\sqrt[5]{(1 - x)^3}}$

15. $s = \left(2 - \dfrac{1}{x}\right)^3$

16. $m = (2 + 3\sqrt{s})^4$

17. $r = \sqrt{1 + \pi\sqrt{y}}$

18. $z = [t - (1/t^2)]^3$

19. $y = \dfrac{1}{\sqrt[3]{4 + 3x - 2x^2}}$

20. $y = \sqrt[3]{\sqrt{3x - 2}}$

21. $y = \sqrt[3]{2 - 3x} + \dfrac{5}{\sqrt[3]{(2 - 3x)^2}}$

22. $s = (2t + 1)^3 + \dfrac{2}{(2t + 1)^3}$

23. The distance traveled (in centimeters) by a charged particle in a linear accelerator is given by $s = (3t + 4)^{3/2}$, where t is time (in seconds). Find the expression for the velocity of the particle as a function of time.

24. The velocity (in feet per second) of discharge of a horizontal jet of water issuing from an orifice that is h feet below the surface of the water is $v = \sqrt{2gh}$, where $g = 32$ ft/sec^2. This expression is *Torricelli's theorem.* How fast does v change with respect to h?

25. The fundamental frequency f of a vibrating string is $f = (1/2L)\sqrt{T/u}$, where L is length, u is the mass per unit length, and T is the tension in the string. Find the expression for df/dT. Assume u and L are constants.

26. The impedance Z of a series RL circuit is given by $Z = \sqrt{R^2 + X_L^2}$, where R is the resistance and X_L is the inductive reactance. What is dZ/dX_L if R is constant?

27. The voltage across the plates of a capacitor varies with time according to $v = (3t + 2)^{-1/3}$. Find the expression for the rate of change of voltage with respect to time.

28. The stagnation pressure (in pounds per square inch) at the nose of an object moving with a velocity V through a resisting medium is given by $P = \frac{1}{2}\rho V^2$, where ρ is the density of the medium. Determine the stagnation pressure, as a function of time, of an experimental rocket given that the displacement (in feet) of the rocket is known to vary with time according to $s = (2t + 4)^{3/2}$.

Exercises 29–33

29. The modulus of rigidity G of a circular shaft is related to the modulus of elasticity E and Poisson's ratio μ by the equation $G = (E/2)(1 + \mu)^{-1}$. Assuming E to be constant, determine the expression for $dG/d\mu$.

30. In Exercise 29, let $E = 30 \times 10^6$ and $\mu = 0.250$. Find the value of $dG/d\mu$.

31. The bending moment M of a simply supported beam is a function of the distance x (see the accompanying figure) measured from the left end of the beam. The change in bending moment M with respect to the distance x is called *shear* and is denoted by $s = dM/dx$. Determine the equation for the shear if the bending moment of a beam is given by $M = (2 - 3x)^2$. When the unit of bending moment is foot-pounds (ft-lb), the unit of shear is pounds.

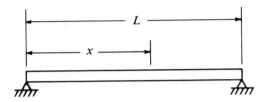

32. Evaluate both the bending moment and the shear force 2 ft from the left end of the beam in Exercise 31.

33. Find the shear equation for a cantilever beam if its bending moment is given by $M = (10 - x^2)^3$.

34. In the study of heat transfer, the rate of heat conduction q through a material is given by

$$q = -kA \frac{dT}{dx}$$

where k is a constant called the thermal conductivity, A is the area normal to the direction of flow, and dT/dx is the rate of change of temperature with respect to the thickness of the material. If the temperature of a concrete wall varies with thickness according to $T = (3x + 2)^{-2}$, find the expression for the rate of heat conduction.

3.3 DERIVATIVES OF PRODUCTS AND QUOTIENTS

The Derivative of a Product

Consider the problem of finding the derivative of

$$y = (3x^2 + 1)(2x + 3)$$

The derivative y may be found by first expanding this product and then differentiating each term of the resulting expression. Thus,

$$y = (3x^2 + 1)(2x + 3) = 6x^3 + 9x^2 + 2x + 3$$

and

$$y' = 18x^2 + 18x + 2$$

Expanding a given product is usually either unwieldy, as in the product $(x^2 + 1)^{20}(x - 1)^5$, or impossible, as in the product $(x^2 + 3)^{1/2}(x + 1)^{2/3}$. So, we look for a formula for the derivative of the product of two functions. Our initial inclination might be to assume that the derivative of a product is equal to the product of the derivatives of the individual factors. If we apply this "rule" to the product

$$y = (3x^2 + 1)(2x + 3)$$

we obtain the value

$$y' = \frac{d}{dx}(3x^2 + 1) \cdot \frac{d}{dx}(2x + 3) = (6x)(2) = 12x$$

which is clearly not in agreement with the correct expression for y' as derived above. Hence, we conclude that the derivative of a product is *not* the product of the derivatives.

If h, f, and g are functions such that $h(x) = f(x) \cdot g(x)$, then the correct formula for finding the derivative of a product of two elementary functions is found by applying the definition of the derivative. Thus,

$$h'(x) = \lim_{\Delta x \to 0} \frac{h(x + \Delta x) - h(x)}{\Delta x}$$

$$= \lim_{\Delta x \to 0} \frac{f(x + \Delta x)g(x + \Delta x) - f(x)g(x)}{\Delta x}$$

$$= \lim_{\Delta x \to 0} \frac{f(x + \Delta x)g(x + \Delta x) - f(x + \Delta x)g(x) + f(x + \Delta x)g(x) - f(x)g(x)}{\Delta x}$$

$$= \lim_{\Delta x \to 0} \frac{f(x + \Delta x)[g(x + \Delta x) - g(x)] + g(x)[f(x + \Delta x) - f(x)]}{\Delta x}$$

$$= \lim_{\Delta x \to 0} f(x + \Delta x) \lim_{\Delta x \to 0} \frac{g(x + \Delta x) - g(x)}{\Delta x} + \lim_{\Delta x \to 0} g(x) \lim_{\Delta x \to 0} \frac{f(x + \Delta x) - f(x)}{\Delta x}$$

$$= f(x)g'(x) + g(x)f'(x)$$

We see that *the derivative of a product of two functions is equal to the first function times the derivative of the second plus the second function times the derivative of the first*. This result is formalized in the following product rule.

RULE

The Product Rule   Let u and v represent differentiable functions of x. Then,

$$\frac{d}{dx}(u \cdot v) = u\frac{dv}{dx} + v\frac{du}{dx} \qquad\qquad (3.5)$$

EXAMPLE 1   ■ **Problem**

Find the derivative of $y = (3x^2 + 1)(2x + 3)$ by using Formula (3.5).

■ **Solution**

Letting $u = 3x^2 + 1$ and $v = 2x + 3$, we have

$$\frac{dy}{dx} = (3x^2 + 1) \cdot \frac{d}{dx}(2x + 3) + (2x + 3) \cdot \frac{d}{dx}(3x^2 + 1)$$

$$= (3x^2 + 1)(2) + (2x + 3)(6x) = 18x^2 + 18x + 2$$

Notice that this result agrees with the result for y' that we derived at the beginning of this section.   ■

EXAMPLE 2   ■ **Problem**

Differentiate $y = 3x^4\sqrt{2x - 3}$.

■ **Solution**

Using the product formula with $u = 3x^4$ and $v = (2x - 3)^{1/2}$, we have

$$\frac{dy}{dx} = 3x^4 \cdot \frac{d}{dx}[(2x - 3)^{1/2}] + (2x - 3)^{1/2} \cdot \frac{d}{dx}(3x^4)$$

$$= 3x^4[\tfrac{1}{2}(2x - 3)^{1/2}(2)] + (2x - 3)^{-1/2}(12x^3)$$

This step completes the differentiation process. The result can be simplified considerably by factoring $3x^3(2x - 3)^{-1/2}$ from each term. Thus,

$$\frac{dy}{dx} = 3x^3(2x - 3)^{-1/2}[x + 4(2x - 3)] = \frac{3x^3(9x - 12)}{\sqrt{2x - 3}}$$ ■

EXAMPLE 3   ■ **Problem**

Differentiate $y = (5x + 6)^3(2x - 1)^4$.

■ **Solution**

Applying the product rule, we have

$$\frac{dy}{dx} = (5x + 6)^3 \cdot \frac{d}{dx}[(2x - 1)^4] + (2x - 1)^4 \cdot \frac{d}{dx}[(5x + 6)^3]$$

$$= (5x + 6)^3[4(2x - 1)^3(2)] + (2x - 1)^4[3(5x + 6)^2(5)]$$

Simplifying algebraically by factoring $(5x + 6)^2(2x - 1)^3$, we have

$$\frac{dy}{dx} = (5x + 6)^2(2x - 1)^3[8(5x + 6) + 15(2x - 1)]$$

$$= (5x + 6)^2(2x - 1)^3(70x + 33)$$ ∎

The Derivative of a Quotient

Let $h(x)$ be a quotient of two functions, specifically, $h(x) = f(x)/g(x)$. To derive a formula for the derivative of $h(x)$, we consider it as a product of $f(x)$ and $[g(x)]^{-1}$. Using the product rule gives

$$\frac{d}{dx} h(x) = \frac{d}{dx} [f(x)[g(x)]^{-1}] = f'(x)[g(x)]^{-1} + f(x) \frac{d}{dx} [g(x)]^{-1}$$

$$= \frac{f'(x)}{g(x)} - \frac{f(x)g'(x)}{[g(x)]^2} = \frac{g(x)f'(x) - f(x)g'(x)}{[g(x)]^2}$$

This result proves the following quotient rule.

RULE

The Quotient Rule   Let u and v represent differentiable functions of x. Then,

$$\frac{d}{dx}\left(\frac{u}{v}\right) = \frac{v \dfrac{du}{dx} - u \dfrac{dv}{dx}}{v^2}$$ (3.6)

for all x for which $v \neq 0$.

Expressed in words, *the derivative of a quotient is the denominator function times the derivative of the numerator minus the numerator function times the derivative of the denominator, all divided by the square of the denominator function.*

EXAMPLE 4   ∎ **Problem**

Differentiate $y = \dfrac{x^2}{(2 - x)^3}$.

∎ **Solution**

The application of Formula (3.6) to the given quotient yields

$$\frac{dy}{dx} = \frac{(2 - x)^3(2x) - x^2[3(2 - x)^2(-1)]}{[(2 - x)^3]^2}$$

$$= \frac{2x(2 - x)^3 + 3x^2(2 - x)^2}{(2 - x)^6} = \frac{x(2 - x)^2[2(2 - x) + 3x]}{(2 - x)^6}$$

$$= \frac{x(4 + x)}{(2 - x)^4}$$ ∎

EXAMPLE 5   ■ **Problem**

Differentiate $y = (x - 1)/x^2$. Then, find the value of x for which the slope of the tangent to the graph of this function is equal to zero.

■ **Solution**

Using the quotient rule, we have

$$y' = \frac{x^2(1) - (x - 1)(2x)}{(x^2)^2} = \frac{2 - x}{x^3}$$

The derivative is an expression for the slope of the tangent to the graph of a function. So, to find the value of x for which the slope of the tangent line is zero, we let $y' = 0$ and solve for x. Thus,

$$y' = \frac{2 - x}{x^3} = 0$$

from which we get $x = 2$. This result is shown graphically in Figure 3.5.

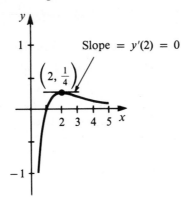

■ **Figure 3.5**

■

EXERCISES FOR SECTION 3.3

Find the derivative in Exercises 1–20.

1. $y = (2x + 3)(3x - 1)$

2. $s = (3t^2 + 2)(5t + 4)$

3. $y = \dfrac{3 + 2x}{3 - 2x}$

4. $p = \dfrac{w^2 - 1}{w^2 + 1}$

5. $q = t^3(t + 1)^2$

6. $f(x) = x^5(x^2 - 1)^3$

7. $m = \dfrac{3s^2}{s^2 - 4}$

8. $g(t) = t\sqrt{2 + 3t}$

9. $y = (3x + 2)^3(5x + 1)^4$

10. $z = (2 - x)^2(2 + x)^5$

11. $p(z) = \dfrac{(4z + 2)^3}{(3z + 1)^2}$

12. $s = \left(\dfrac{t^2 + 2}{t^2 - 2}\right)^3$

13. $y = \dfrac{(1 - x)^2}{4 - x^2}$

14. $B = (q^2 + q)^2 \sqrt{q^2 - q}$

15. $y = \sqrt[5]{\dfrac{m}{m - 1}}$

16. $f(t) = \sqrt{\dfrac{1 - 3t}{1 + 3t}}$

17. $y = \sqrt[3]{\dfrac{t^2 + 1}{t^2 - 1}}$

18. $E = \sqrt{\dfrac{\pi - v}{\pi + v}}$

19. $G = \dfrac{1}{(2x - 1)^3(3x + 5)^2}$

20. $h = \dfrac{1}{3m^4 \sqrt{m^2 + 1}}$

In Exercises 21–24, find the values of x for which the graph of the indicated function has a slope of zero.

21. $y = (2x - 3)^2(x + 4)$

22. $y = x\sqrt{x + 1}$

23. $y = \dfrac{x^2}{x - 1}$

24. $y = \dfrac{x}{x^2 + 1}$

25. Find the equation of the line drawn tangent to the graph of $y = x\sqrt[3]{x} + 3$ at $x = 5$.

26. Find the equation of the line drawn tangent to the graph of $y = (x + 3)/(x - 4)$ at $x = 2$.

27. The displacement (in centimeters) of an object varies with time according to $s = 3t(t^2 + 4)^{1/3}$, when t is in seconds. Find the velocity equation for this object. Calculate the displacement and velocity of the object when $t = 2$ sec.

28. The bending moment of a certain I beam is given by $M = 3x^2/4(x - 2)$. Find the equation for the shear. See Exercise 31, Section 3.2.

29. The voltage output (in volts) of a laboratory thermocouple varies with temperature according to

$$v = \dfrac{3T^2}{\sqrt{T^2 - T}}$$

Find the rate of change of voltage output with respect to temperature.

30. The charge transferred in an ionic solution varies with time as

$$q = \sqrt{\dfrac{2t - 3}{2t + 3}}$$

Find the expression for the electrical current in the solution. Recall that $i = dq/dt$.

■■■■■■ 3.4 HIGHER-ORDER DERIVATIVES

The derivative of $f(x) = x^4$ is $f'(x) = 4x^3$. We observe that since $f'(x)$ is a function of x, it may be differentiated with respect to x. Thus, we can write

$$\frac{d}{dx}[f'(x)] = \frac{d}{dx}[4x^3] = 12x^2$$

Derivatives of derivative functions are referred to as *higher derivatives* of the given function. Generally, the derivative of $f'(x)$ is called the **second derivative** of $f(x)$ and is denoted by

$$\frac{d}{dx}[f'(x)] = f''(x)$$

In terms of operator notation, the symbol d^2/dx^2 is used to indicate the second derivative of a function. Thus, the derivative of f' is also denoted by

$$\frac{d}{dx}[f'(x)] = \frac{d^2}{dx^2}[f(x)]$$

Notice that the 2 in d^2/dx^2 has nothing to do with "squaring."

Higher-order derivatives are defined in a manner analogous to that of the second derivative. The derivative of the second derivative of f is called the **third derivative** of f and is denoted by

$$\frac{d}{dx}[f''(x)] = f'''(x) = \frac{d^3}{dx^3}[f(x)]$$

Similarly, the derivative of the third derivative is called the **fourth derivative** of f and is denoted by

$$\frac{d}{dx}[f'''(x)] = f^{iv}(x) = \frac{d^4}{dx^4}[f(x)]$$

The scheme for additional derivatives is the same as for those indicated above.

EXAMPLE 1 ■ **Problem**

Find the first, second, and third derivatives for $g(x) = x^5$.

■ **Solution**

The derivatives are

$$g'(x) = 5x^4 \qquad g''(x) = 20x^3 \qquad g'''(x) = 60x^2$$

■

EXAMPLE 2   ■ **Problem**

Find the first, second, and third derivatives of the function defined by

$$y = 6x^2 + \sqrt[3]{x} + 9$$

■ **Solution**

We write the function as $y = 6x^2 + x^{1/3} + 9$. Then,

$$\frac{dy}{dx} = 12x + \frac{1}{3}x^{-2/3}$$

$$\frac{d^2y}{dx^2} = 12 - \frac{2}{9}x^{-5/3}$$

$$\frac{d^3y}{dx^3} = \frac{10}{27}x^{-8/3}$$

■

EXAMPLE 3   ■ **Problem**

Find the second derivative of $s = (t^2 + 5)^4$.

■ **Solution**

The first derivative of the given function is

$$\frac{ds}{dt} = 4(t^2 + 5)^3(2t) = 8t(t^2 + 5)^3$$

Since ds/dt is the product of $8t$ and $(t^2 + 5)^3$, we use the product rule to find d^2s/dt^2. Thus,

$$\frac{d^2s}{dt^2} = 8t[3(t^2 + 5)^2(2t)] + (t^2 + 5)^3(8) = 8(t^2 + 5)^2[6t^2 + (t^2 + 5)]$$

$$= 8(t^2 + 5)^2(7t^2 + 5)$$

■

Just as the first derivative gives an expression for the slope of the graph of a given function, the second derivative gives an expression for the slope of the graph of the first derivative, the third derivative gives an expression for the slope of the graph of the second derivative, and so on. The graphical relationship between $f(x)$, $f'(x)$, and $f''(x)$ is shown in Figure 3.6 for the function $f(x) = x^3 - 6x^2$. In general, the ordinate of the lower curve is numerically equal to the slope of the curve above it at the same point. Notice that the x-intercepts of the graph of $f'(x)$ correspond to the points on the graph of $f(x)$ that have horizontal slopes, and the x-intercept of the graph of $f''(x)$ corresponds to the horizontal slope of $f'(x)$.

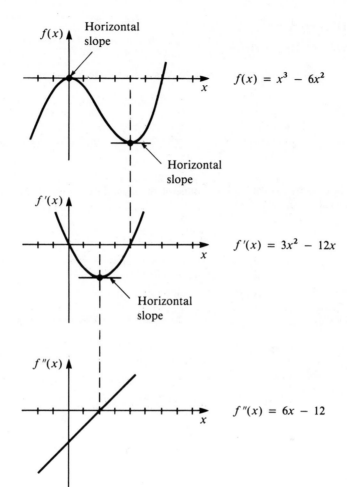

$$f(x) = x^3 - 6x^2$$

$$f'(x) = 3x^2 - 12x$$

$$f''(x) = 6x - 12$$

■ **Figure 3.6**   Note: Vertical axes are not to scale.

EXERCISES FOR SECTION 3.4

In Exercises 1–20, find the second derivative.

1. $y = 5x^3 - 3x^7$

2. $y = 7x^5 + 17$

3. $F(t) = t^{1/2} - 3t$

4. $s = \sqrt{t} + t^2$

5. $y = 3x^{2.7}$

6. $g(y) = 5y^{0.3}$

7. $r = \dfrac{6}{\sqrt[3]{w}} + \dfrac{9}{w^3}$

8. $y = x^2(x - 6x^3)$

9. $a = t^{1/2}(2t^{1/2} - 3t^{1/4})$

10. $v = (r^2 - 3r)^2$

11. $y = \sqrt{2x - 5}$

12. $y = (3 - 2x)^7$

13. $s = 5(3x + 7)^{-1}$

14. $r = 20\sqrt{1 - x}$

15. $y = (x^3 - 8)^5$

16. $y = \dfrac{1}{(x^2 + 4)^{1/3}}$

17. $v = \dfrac{2}{\sqrt{t^2 + 1}}$

18. $i = (3x^2 - 5x)^4$

19. $y = x(5x + 4)^3$

20. $y = x^2(3x + 2)^2$

In Exercises 21–24, find the value of the second derivative of the given function at the indicated point.

21. $y(x) = x^2 - 3x + 12$; find $y''(3)$.

22. $g(x) = x^3 - \frac{3}{2}x^2 - 6x + 7$; find $g''(0)$.

23. $h(z) = 6z^2 - 2z^3$; find $h''(1)$.

24. $s(t) = t^{-1}$; find $s''(2)$.

Sketch the graphs of $f(x)$, $f'(x)$, and $f''(x)$ for the functions in Exercises 25–28.

25. $F(x) = x^2$

26. $y = x^2 + 2$

27. $y = 4x - x^2$

28. $g(x) = x^2 + 4x$

29. For certain functions, the expression $t^2y'' + ty' - y$ is equal to zero. Show that it is equal to zero for $y = 3t + (2/t)$.

30. An electrical potential is given by $v(t) = 10/t$. Find the first and second derivatives of this potential.

3.5 IMPLICIT DIFFERENTIATION

Sometimes, an equation involves both x and y such that a given value of x determines none, one, or more real values of y. For example, the equation $x + y + 1 = 0$ determines a value of y for each value of x. The equation $x^2 + y^2 = 1$ determines *two* values of y for each x between -1 and 1 and *no* real values of y otherwise. The equation $x^2 + y^2 = -1$ determines no real value of y.

When an equation such as $x + y + 1 = 0$ or $x^2 + y^2 = 1$ is used to determine values of y for specific values of x, we say that y is defined **implicitly** as a function of x. When y is defined implicitly, we may be able to solve explicitly for y in terms of x. Thus, the equation $x + y + 1 = 0$ yields the explicit expression $y = -x - 1$. The equation $x^2 + y^2 = 1$ yields the two explicit functions $y = (1 - x^2)^{1/2}$ and $y = -(1 - x^2)^{1/2}$. We usually prefer to work with explicit functions, but often, because of the nature of the equation defining the implicit function, we cannot solve for y. In this case, we find the derivative directly from the implicit function. The next few examples illustrate the technique of **implicit differentiation.**

EXAMPLE 1   ■ **Problem**

Using implicit differentiation, find $\dfrac{dy}{dx}$ for $x^5 + y^4 = 25$.

■ **Solution**

In this case, y can be found explicitly, but we choose to differentiate it as an implicit function in order to show the technique. To find the derivative of y implicitly, we take the derivative of both sides of the given equation with respect to x. Thus,

$$\frac{d}{dx}(x^5 + y^4) = \frac{d}{dx}(25)$$

The derivative of a sum is the sum of the derivatives. So,

$$\frac{d}{dx}(x^5) + \frac{d}{dx}(y^4) = \frac{d}{dx}(25)$$

Starting with the first term on the left, we get

$$\frac{d}{dx}(x^5) = 5x^4$$

To differentiate the second term, we note that y is an implicit function of x. (This step is the key to implicit differentiation.) Thus,

$$\frac{d}{dx}(y^4) = 4y^3 \frac{dy}{dx}$$

Differentiating the constant term yields

$$\frac{d}{dx}(25) = 0$$

These steps combine to form the equation

$$5x^4 + 4y^3 \frac{dy}{dx} = 0$$

Solving for dy/dx, we have

$$\frac{dy}{dx} = -\frac{5x^4}{4y^3}$$

■

EXAMPLE 2   ■ **Problem**

Given $y^3 + x^2y^4 + x^3 = 1$, find $\dfrac{dy}{dx}$.

■ **Solution**

To find dy/dx, we differentiate each term with respect to x. Notice that the derivative of the middle term is

$$\frac{d}{dx}(x^2y^4) = x^2\frac{d}{dx}(y^4) + y^4\frac{d}{dx}(x^2) = x^2\left(4y^3\frac{dy}{dx}\right) + y^4(2x)$$

$$= 4x^2y^3\frac{dy}{dx} + 2xy^4$$

Completing the differentiation of the other terms with respect to x, we have

$$3y^2\frac{dy}{dx} + \left(4x^2y^3\frac{dy}{dx} + 2xy^4\right) + 3x^2 = 0$$

Solving for dy/dx gives

$$\frac{dy}{dx} = -\frac{3x^2 + 2xy^4}{3y^2 + 4x^2y^3}$$
   ■

EXAMPLE 3   ■ **Problem**

Find $\dfrac{d^2s}{dt^2}$ for $s^2 - 9t^2 = 16$.

■ **Solution**

Differentiating each term with respect to t, we have

$$\frac{d}{dt}(s^2) - \frac{d}{dt}(9t^2) = \frac{d}{dt}(16)$$

and

$$2s\frac{ds}{dt} - 18t = 0$$

Solving for ds/dt, we get

$$\frac{ds}{dt} = \frac{9t}{s}$$

Therefore, the second derivative is

$$\frac{d^2s}{dt^2} = \frac{d}{dt}\left(\frac{9t}{s}\right) = \frac{s(d/dt)(9t) - 9t(d/dt)(s)}{s^2} = \frac{9s - 9ts'}{s^2}$$

By substituting $s' = 9t/s$, we may write the result as

$$\frac{d^2s}{dt^2} = \frac{9s - 9t(9t/s)}{s^2} = \frac{9(s^2 - 9t^2)}{s^3} = \frac{9(16)}{s^3} = \frac{144}{s^3}$$
   ■

EXAMPLE 4   ■ **Problem**

Find the equation of the line normal (perpendicular) to $x^5 + 3y^4 = 3xy^2 + 1$ at the point $(1, 1)$.

■ **Solution**

The normal line is perpendicular to the tangent line at $(1, 1)$. To obtain the slope of the tangent line, we find y'. Differentiating the given function, we get

$$5x^4 + 12y^3y' = 3y^2 + 6xyy'$$

Substituting $x = 1$ and $y = 1$ into this relation, we get

$$5 + 12y' = 3 + 6y'$$

$$6y' = -2$$

$$y' = -\frac{1}{3}.$$

Using m_n to designate the slope of the normal line, we have

$$m_n = -\frac{1}{y'} = -\frac{1}{-\frac{1}{3}} = 3$$

Finally, in the point-slope form of a straight line, the equation of the normal line is

$$y - 1 = 3(x - 1)$$
$$y = 3x - 2$$

■

EXERCISES FOR SECTION 3.5

In Exercises 1–14, find dy/dx.

1. $x^2 + y^2 = 25$

2. $4x^2 - 9y^2 = 1$

3. $x^4 + y^3 = 15$

4. $\sqrt{x} + \sqrt{y} = 16$

5. $x + x^2y^2 + y = 1$

6. $x + \sqrt{xy} + y = 25$

7. $x^{2/3} + y^{2/3} = 3^{2/3}$

8. $2xy + y^2 = 2x + y$

9. $x^3 + xy + y^3 = 0$

10. $x = x^2y^2 + y$

11. $\frac{1}{x} + \frac{1}{y} = 1$

12. $\frac{1}{x^2} + \frac{1}{y^2} = 1$

13. $\frac{x}{y} - 2 = y$

14. $2x^{2/3} + y^{4/3} = 10$

Find d^2y/dx^2 **in Exercises 15–20.**

15. $x^2 + y^2 = 16$

16. $x^3 + y^3 = 4$

17. $3y^2 + xy = x$

18. $x^3 + 3xy = 9$

19. $6xy^2 + y^3 = 2$

20. $xy^2 + y = 3$

21. Find the equation of the straight line tangent to the graph of $x^2 + xy + y^2 = 19$ at the point (2, 3).

22. Find the equation of the line tangent to $x^{1/3} + 2y^{1/3} = 3$ at the point (1, 1).

23. Find the equation of the straight line normal to the graph of $\sqrt{x} + \sqrt{y} = 6$ at the point (4, 16).

24. The relationship between displacement (in centimeters) of a particle and time in seconds is $s^3 + t^2 = 31$. Find the velocity of the particle when $t = 2$ sec.

25. Find the current, when $t = 8$ sec, in a circuit whose charge-time relation is given by $q^2t + t^{2/3} = 12$. (Recall that $i = dq/dt$.)

REVIEW EXERCISES FOR CHAPTER 3

Find the derivative of the functions in Exercises 1–30.

1. $y = 2x^5 - 3x^3 + x$

2. $y = 2x^3 - x^2 + 4x$

3. $y = \sqrt{x} + 4\sqrt[3]{x}$

4. $s = 3x^{-2} + x^{-1/2}$

5. $i = \dfrac{15}{t^3} - \dfrac{8}{\sqrt[5]{t}}$

6. $m = \dfrac{11}{r} + \dfrac{2}{\sqrt[3]{r}}$

7. $z = w^2(\sqrt[3]{w} - 3w^5 + 7)$

8. $y = (x^3 + 2x)\sqrt{x}$

9. $y = \dfrac{x - 3}{x^2}$

10. $s = \dfrac{x^3 + 5x}{x^3}$

11. $F(x) = (2 - 3x)^7$

12. $s = (5 - t)^9$

13. $y = (x^2 + 2)^{-3}$

14. $y = 3(x^2 - 5)^{-1}$

15. $i = 2\sqrt[3]{t} - t^2$

16. $g(x) = 10\sqrt[4]{15x^3 - 7}$

17. $y = \dfrac{8}{\sqrt{3x + 7}}$

18. $y = \dfrac{1}{\sqrt{2t + 3}}$

19. $v = 3t(2t + 5)^3$

20. $m = r(r^2 - 4)^5$

21. $y = (x^2 + 1)^2(x^2 - 1)^2$

22. $f(t) = (3t + 5)^4(3t - 5)^7$

23. $y = (3x^3 + 1)^4(x^2 + 7x)^3$

24. $y = \sqrt{5x^2 - 7}\,\sqrt{5x^2 + 7}$

25. $w = \dfrac{s^5}{(s + 5)^3}$

26. $i = \dfrac{3t^2 - 5t}{t^3 - 1}$

27. $v = \dfrac{(5t - 3)^{1/5}}{(3t + 1)^{1/3}}$

28. $y = \dfrac{x}{\sqrt{2x - 1}}$

29. $p = (2t - 1)^2(5t + 4)^3(4t - 1)$

30. $q = (s - 3)(s^3 + 1)(2s - 1)^2$

31–40. Find the second and third derivatives of the functions given in Exercises 1–10.

41. Find y' given that $xy^2 - x^3 + 3y = 25$.

42. Find y' given that $y^2 + 4x^2y^2 - 5y = 0$.

43. Find the derivative function for $y = x - \frac{1}{2}x^2$. What is the slope of the tangent line to the graph of the given function at $x = -1$? At $x = 1$?

44. Find the derivative function for $y = 2x^3 - 9x^2 + 12x + 2$. Where is the slope of the tangent line equal to zero?

45. Find the equation of the line tangent to $(x^2 + y^2)^2 - 3y = 1$ at the point $(1, 1)$.

46. Find the equation of the line normal to $(2xy^2 - x)^2 + y = 2$ at the point $(1, 1)$.

 47. The displacement (in centimeters) of an object is given by $s = 5t^2 + 12t$ when time is in seconds. Use the derivative function to find the velocity of the object at $t = 3$ sec.

 48. The displacement (in feet) of an object is given by $s = 10 + 15t^3$. Use the derivative function to find the velocity of the object at $t = 0.5$ sec.

 49. The charge (in coulombs) transferred in a given electric circuit varies with time (in seconds) as $q = t\sqrt{t + 1}$. Find the expression for the current in the circuit.

 50. Find the current in the circuit in Exercise 49 at $t = 3$ sec.

CHAPTER 4

Applications of the Derivative

The importance of calculus in technology rests with its wide variety of uses in solving applied problems. In this chapter, we highlight applications of calculus in mechanics and electricity, as well as the more classic problem of locating maximum and minimum values of functions.

◼ 4.1 APPLICATIONS TO MECHANICS

In mechanics, we encounter problems in which the displacement of an object is given as a function of elapsed time. For example, the distance s (in feet) that an object falls from rest in t seconds is given by $s = 16t^2$. The derivative offers a convenient means of finding the velocity and acceleration of an object when its displacement-time function is known.

Recall from Chapter 2 that the **velocity** of an object is the derivative of the displacement-time function. Symbolically, we write the following formula.

FORMULA

> **Velocity Formula**
>
> $$v = \frac{ds}{dt} \tag{4.1}$$

EXAMPLE 1 ◼ **Problem**

Determine the displacement and the velocity of an object at $t = 3$ sec if its displacement (in feet) is given by $s = 16t^2$.

■ **Solution**

The displacement of the object at $t = 3$ sec is

$$s = 16(3)^2 = 144 \text{ ft}$$

To find the velocity-time function, we differentiate $s = 16t^2$ with respect to t. Thus,

$$v = \frac{d}{dt}(16t^2) = 32t$$

Substituting $t = 3$ into this expression, we get

$$v = 32(3) = 96 \text{ ft/sec} \qquad ■$$

The **acceleration** of an object is defined to be the time rate of change of velocity. Thus, acceleration is the derivative with respect to time of the velocity-time function, which is the same thing as the second derivative of the distance-time function. The acceleration is then represented symbolically as follows:

FORMULA

Acceleration Formula

$$a = \frac{dv}{dt} = \frac{d^2s}{dt^2} \qquad\qquad\qquad (4.2)$$

EXAMPLE 2   ■ **Problem**

Find the equations for the velocity and the acceleration of a rocket having an equation of motion described by $s = 3t^3 + 5t$. Find the velocity and the acceleration of the projectile after 1 sec. Assume s is measured in feet and time in seconds.

■ **Solution**

We find the equation for the velocity by differentiating the given equation with respect to time. Thus,

$$v = \frac{ds}{dt} = 9t^2 + 5$$

The expression for the acceleration is the derivative of the velocity function with respect to time. So,

$$a = \frac{dv}{dt} = 18t$$

We can now evaluate the velocity and the acceleration at $t = 1$ sec:

$$v = 9(1)^2 + 5 = 14 \text{ ft/sec} \qquad a = 18(1) = 18 \text{ ft/sec}^2 \qquad ■$$

EXAMPLE 3   ■ **Problem**

A projectile is fired vertically upward. The altitude (in feet) of the projectile at any time t (in seconds) is given by the function

$$h(t) = -16t^2 + 1280\,t$$

a. Find the velocity and acceleration for any time t.
b. Compute the velocity and acceleration at $t = 20$ and $t = 50$ sec.
c. Determine the maximum altitude obtained.
d. Sketch the graph of $h(t)$ versus t.

■ **Solution**

a. The velocity is the derivative of the altitude function. Thus,

$$v(t) = -32t + 1280$$

Also, the acceleration is the derivative of the velocity. Thus,

$$a(t) = -32$$

b. At $t = 20$ and $t = 50$ sec, we have

$$v(20) = 640 \text{ ft/sec}, \qquad v(50) = -320 \text{ ft/sec}$$
$$a(20) = a(50) = -32 \text{ ft/sec}^2$$

A positive velocity indicates that the projectile is rising; a negative velocity indicates it is falling. Thus, the projectile is rising at $t = 20$ sec and falling at $t = 50$ sec.

c. At the instant the projectile reaches its maximum altitude, $v(t) = 0$. To determine this instant of time, we set the expression for the velocity equal to zero and solve for t. Thus,

$$-32t + 1280 = 0$$

from which $t = 40$ sec. Hence, the maximum altitude reached is

$$h(40) = 25,600 \text{ ft}$$

d. A sketch of $h(t)$ versus t is shown in Figure 4.1. Note that when $t = 80$, the projectile is back at ground level.   ■

Comment: The distance-time graph given in Example 3 does not represent the path traveled by the object but, rather, gives the height above ground level at any time t. The actual path of the projectile is straight up and straight down.

In some practical situations, a graphical representation of the distance-time relationship is given instead of an equation. Since the first derivative is the slope of

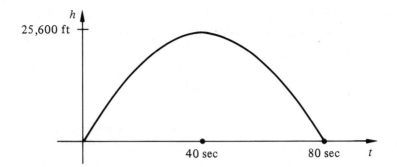

■ **Figure 4.1**

the graph at every point, the velocity of an object at time t is equal to the slope of the distance-time graph at time t. Similarly, the acceleration of an object at time t is numerically equal to the slope of the velocity-time graph at time t. Example 4 shows how the velocity can be obtained from an accurate graph of the displacement equation.

EXAMPLE 4 ■ **Problem**

What is the velocity of an object when $t = 1.0$ sec if the equation of motion is given by the following table?

Time (sec)	0	0.5	1.0	1.5	2.0	2.5	3.0	3.5	4.0	4.5
Displacement (ft)	0	2.0	3.3	3.9	3.8	2.3	1.5	1.9	1.3	1.0

■ **Solution**

We make a careful sketch of the distance-time relationship, as shown in Figure 4.2. Drawing a tangent line at (1.0, 3.3), we use its slope to estimate the velocity of the object at $t = 1.0$ sec. Our estimate is 1.2/0.5 = 2.4, and therefore, the velocity is 2.4 ft/sec.

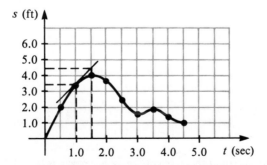

■ **Figure 4.2**

We remark again that the motion being considered here is *linear*; that is, the object is assumed to be traveling in a straight line. The distance-time graph does *not* represent the path traveled by the object; rather $s(t)$ gives the distance of the particle from some fixed reference point. ∎

Power p is defined as the time rate of change of doing work w. Symbolically, we write the following formula.

FORMULA

> **Power Formula**
>
> $$p = \frac{dw}{dt} \qquad\qquad (4.3)$$
>
> If work is given in foot-pounds (ft-lb) and time in seconds, the unit of power is foot-pounds per second (ft-lb/sec).

EXAMPLE 5   ∎ **Problem**

Find the expression for the power of a mechanical system if the system has the capacity to work (in foot-pounds) according to

$$w = 3t^2 + t^4$$

∎ **Solution**

From Formula (4.3), the power (in foot-pounds per second) is

$$p = \frac{dw}{dt} = \frac{d}{dt}(3t^2 + t^4) = 6t + 4t^3$$

∎

EXERCISES FOR SECTION 4.1

1. The displacement (in feet) of a projectile varies with time (in seconds) according to $s = 5 + 100t - 16t^2$.
   (a) Find the velocity-time function for the projectile.
   (b) What is the projectile's velocity after 1, 2, and 3 sec?

2. The displacement (in feet) of a car that starts from rest is given by $s = 6t^{3/2} + 10$, where t is time (in seconds). Determine the expression for the velocity of the car.

3. Find the velocity and the acceleration of a particle at $t = 8$ sec if its displacement (in centimeters) is given by $s = t^{4/3} + 5t$.

4. Determine the velocity of a missile 3 sec after launch if its displacement (in feet) is given by $s = 3t^3 - 4t$.

5. The plunger of a solenoid has a velocity $v = 3t^{1/2}$ (in centimeters per second). Find the acceleration of the plunger when $t = 0.01$ sec.

6. The distance-time function of a car is $s = \sqrt{2t^3 + t^2}$ (s is in feet and t is in seconds). What is the velocity of the car after 6 sec? Give your answer in miles per hour. (*Note:* 88 ft/sec = 60 mi/hr.)

7. An electric motor does work (in foot-pounds) according to $w = 3(2t + 5)^{-2}$. What is the expression for the power developed by the motor?

8. Find the power being developed by a motor after 1 sec, given that it does work (in foot-pounds) according to

$$w = \frac{1}{3 + 2t}$$

9. A cannon is fired vertically upward. The altitude (in feet) of the projectile varies with time from launch according to $h = 100 + 640t - 16t^2$. What is the maximum altitude obtained? (*Hint:* The velocity of the projectile is zero when it reaches its maximum altitude.)

10. Find the maximum height of a projectile whose height (in feet) above ground level is given by

$$h = 150t - 16t^2$$

11. In a testing of a new braking system for an airplane, the velocity (in feet per second) is found to be given by $v = \sqrt{50 - 0.4t^2}$. Find the deceleration function for this system.

12. The height of a projectile is given as a function of time to be

$$h = h_0 + v_0 t + \tfrac{1}{2}gt^2$$

where h_0, v_0, and g are constants. Find the equations for the velocity and the acceleration of the projectile.

13. The distance-time function for a charged particle in an electric field is

$$s = 0.56t^{1.3} + 2.61t$$

where s is in centimeters. What are the velocity and the acceleration of the particle when $t = 2.5$ sec? (You will need a calculator here.)

14. The work (in foot-pounds) being done on a system varies with time according to

$$w = 15.7 + 2.8t^2 - 1.8t^4$$

What is the power function?

15. During an aerobic exercise experiment, the work (in foot-pounds) being done by the subject varies with time according to

$$w = 1.5\sqrt{2t + 1} + 10$$

Find the power function for this subject.

16. The displacement (in feet) of a rocket sled varies with time (in seconds) according to $s = (t^{1/2} + 10)^3$. Determine the expressions for the velocity and the acceleration of the sled.

17. As a charged particle moves through an electric field, its distance (in centimeters) from a reference point increases with time (in seconds) according to $s = \sqrt{t^3 + 8}$.

(a) Determine expressions for the velocity and the acceleration of the particle.
(b) Determine the displacement, velocity, and acceleration of the particle at $t = 2$ sec.

18. In the metric system, the common unit of work is the joule (J). The unit of power is then joules per second (J/sec). A machine works according to $w = 3t^4 - 4t$. What power is being developed by the machine when $t = 1.1$ sec? (*Note:* 1 J/sec = 1 watt, W.)

19. A particle is observed in a laboratory experiment, and its displacement-time coordinates are recorded as follows:

t (sec)	0	1	2	3	4	5
s (cm)	0	0.1	0.3	0.6	1.0	1.5

Estimate the velocity of the particle at $t = 3$ sec from the graph obtained by connecting the points with a smooth curve.

20. The following data were obtained from the radar tracking of a weather balloon, where R is the slant range (in kilometers, km). At what rate is the balloon moving away from the radar when $t = 2$ sec?

t (sec)	0	1	2	3	4
R (km)	5.75	5.79	5.85	5.93	6.05

21. The following data were collected during an experiment in which the displacement of an object is measured as a function of time. Estimate the time at which the velocity of the object is zero.

t (sec)	0	2.0	4.0	6.0	8.0
h (ft)	5.0	11.4	13.1	13.5	12.0

22. The velocity-time relationship for an experimental rocket is given in the following table. Estimate the acceleration of the rocket at $t = 1.0$ and $t = 2.0$ sec.

t (sec)	0	0.5	1.0	1.5	2.0	2.5	3.0
v (ft/sec)	0	110	250	475	750	1100	1650

▌▌ 4.2 APPLICATIONS TO ELECTRICITY

In Chapter 2, we defined electric current as follows:

FORMULA

> **Current Formula**
>
> $$i = \frac{dq}{dt} \qquad\qquad\qquad (4.4)$$
>
> where q is the quantity of electric charge (in coulombs, C) and t is time (in seconds). The current is measured in amperes (A).

EXAMPLE 1 ▪ **Problem**

What is the current when $t = 0.1$ sec if the charge (in coulombs) varies according to $q = 2t - 3t^2$?

▪ **Solution**

The equation for the current (in amperes) is

$$i = \frac{dq}{dt} = \frac{d}{dt}(2t - 3t^2) = 2 - 6t$$

And therefore, the current when $t = 0.1$ sec is

$$i(0.1) = 2 - 6(0.1) = 2 - 0.6 = 1.4\,\text{A} \qquad\qquad ■$$

Voltage Induced in a Coil of Wire

When an electric current in a coil of wire changes, a voltage is induced in the coil that opposes the change in current. An inductance coil is represented schematically as shown in Figure 4.3. The formula for the induced voltage is as follows:

FORMULA

> **Induced Voltage Formula**
>
> $$v = L\frac{di}{dt} \qquad\qquad\qquad (4.5)$$
>
> where L is a constant of the coil of wire called the *inductance* of the coil. The unit of inductance is the henry (H).

■ Figure 4.3   Inductance coil

EXAMPLE 2   **■ Problem**

Find the equation for the voltage induced in a coil if $L = 10$ H and the current (in amperes) varies according to $i = 2t^3$.

■ Solution

The voltage (in volts) is

$$v = L\frac{di}{dt} = 10\frac{d}{dt}(2t^3) = 10(6t^2) = 60t^2$$

■

EXAMPLE 3   **■ Problem**

Express the induced voltage in a coil of wire in terms of the charge being transferred through the coil.

■ Solution

Assuming the charge q is a function of time, we have

$$i = \frac{dq}{dt}$$

Then, from Equation (4.5), we get

$$v = L\frac{di}{dt} = L\frac{d}{dt}\left(\frac{dq}{dt}\right) = L\frac{d^2q}{dt^2}$$

■

Current in a Capacitor

A rate of change is also used to describe the current in a capacitor. A schematic representation of a capacitor is shown in Figure 4.4.

■ Figure 4.4   Capacitor

If the capacitance (in farads, F) is denoted by C, then the current is given by the following formula.

FORMULA

> **Current in a Capacitor**
>
> $$i = C \frac{dv}{dt} \tag{4.6}$$

The current is in amperes if the voltage is measured in volts.

EXAMPLE 4 ■ **Problem**

Determine the equation for the current in a capacitor if $C = 10^{-6}$ F and the voltage (in volts) varies as $v = (t^2 + 5t)^{1/3}$.

■ **Solution**

Since $i = C(dv/dt)$, we have

$$i = 10^{-6} \frac{d}{dt} (t^2 + 5t)^{1/3} = 10^{-6} \left[\frac{1}{3} (t^2 + 5t)^{-2/3} (2t + 5) \right] \qquad ■$$

EXAMPLE 5 ■ **Problem**

The work done (in joules) by a certain electrical system varies with current according to

$$w = (i^2 + 2.5i)^{1/2}$$

The current (in amperes) is given by $i = t^2 - 7$. What is the power in the circuit when $t = 3$? Power is in watts.

■ **Solution**

Power is the derivative of the work-time equation. Since work is given as a function of current and current as a function of time, we can use the chain rule to find dw/dt, as follows:

$$p = \frac{dw}{dt} = \frac{dw}{di} \cdot \frac{di}{dt} = \frac{1}{2} (i^2 + 2.5i)^{-1/2} (2i + 2.5)(2t)$$

When $t = 3$, $i = 2$. So, the power at $t = 3$ is

$$p = \frac{1}{2}[(2)^2 + 2.5(2)]^{-1/2}[2(2) + 2.5][2(3)] = 6.5 \text{ W} \qquad ■$$

EXERCISES FOR SECTION 4.2

In Exercises 1–8, find the current-time equation from the given charge-time equation.

1. $q = 5.7t^3$

2. $q = 0.705t^2 - 0.023t$

3. $q = t^{0.3}(3.2 - 6.6t)$

4. $q = \dfrac{7}{1.4t + 2}$

5. $q = \dfrac{15}{(2t - 1)^{1.6}}$

6. $q = 0.4t^{-0.25}$

7. $q = \dfrac{5 - t^{0.2}}{t^{0.1}}$

8. $q = \dfrac{25t - 3}{t^{0.6}}$

Exercises 9–23

9. What is the current in a resistor when $t = 0.2$ sec if the charge (in coulombs) varies according to $q = 4t^{4/3}$.

10. The charge (in coulombs) is $q = 3t - 2t^2$. What charge has been transferred by the time the current becomes equal to zero?

11. A coil of copper wire has an inductance $L = 15$ H. What is the expression for the voltage induced in the coil if the current (in amperes) is $i = (4t^3 + 5)^2$?

12. Find the equation for the voltage induced in a coil if $L = 50$ H and the current (in amperes) varies according to $i = 0.01(t^2 + 2)^{1.3}$.

13. What is the induced voltage in the coil in Exercise 12 when $t = 2$ sec?

14. The electric power loss in a resistor of R ohms (Ω) when a current of i amperes flows in the resistor is given by $p = i^2R$. What is the power loss after 4 sec in a 1000-Ω resistor if the charge varies according to $q = t^{3/2}$? The unit of power is the watt when R is in ohms and i is in amperes.

15. What is the electric power loss in a 200-Ω resistor when $t = 8$ sec if the charge (in coulombs) transferred is $q = 3t^{4/3}$? See Exercise 14.

16. A capacitor of capacitance $C = 2$ F has a voltage (in volts) of $v = 25 + t^2$ applied to its terminals. What is the expression for the current in the capacitor?

17. In Exercise 16, what is the magnitude of the current when $t = 0.1$ sec?

18. A coil of wire has a current of $i = 4t^2 - 6t$ passing through it. The induced voltage in the coil is 0.4 V when $t = 2$ sec. What is the inductance of the coil?

19. The charge (in coulombs) transferred in an inductance coil is $q = 0.5t^3$. The inductance of the coil is 7 H. What is the expression for the induced voltage?

20. The charge (in coulombs) being transferred through a 1.5-H inductance coil is given by the equation $q = 0.35(3t + 1)^{1.2}$. What is the expression for the induced voltage?

21. The charge (in coulombs) transferred through a 1.2-H inductance coil is given by the equation $q = 0.82t^{0.3} + 0.25$. What is the induced voltage in the coil when $t = 0.02$ sec?

22. A certain electrical system expends energy (in joules) according to $w = (i^3 + 5)^{-1}$, where

i is the current (in amperes). What is the power in the circuit when $t = 0.5$ if the current varies with time according to $i = 2t + t^2$?

23. The energy output (in joules) of an electric circuit is given by $w = (i^2 + 6i)^{2/3}$, where i is the current (in amperes). What is the power in the circuit at $t = 3$ if the current varies with time as

$$i = \sqrt{t + 1}?$$

4.3 RELATED RATES

The use of the derivative to describe a rate of change was discussed previously for functions of a single variable. Often, the rate of change of a quantity depends on two or more quantities, each of which is a function of time. For instance, if a car is moving due north with a velocity v_N and another car is moving due east with a velocity v_E, the velocity at which the two cars are moving apart depends on both v_N and v_E. Problems in which a relationship exists between the time rates of change of several variables are called **related rate problems.** Usually, one of the rates can be determined in terms of the others.

EXAMPLE 1 ▪ **Problem**

A 20-ft ladder leans against a vertical wall. The top of the ladder slides down the wall at a rate of 4 ft/sec. How fast is the bottom of the ladder moving along the ground when it is 16 ft from the wall? See Figure 4.5.

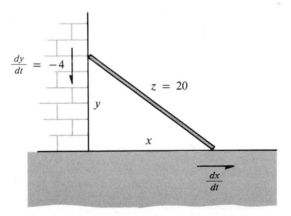

▪ **Figure 4.5**

▪ **Solution**

We recognize this problem as a rate problem since the question asked is, "How fast?" We can establish the relationship between the pertinent variables by referring to Figure 4.5. Let y be the height to the top of the ladder, and let x be the corresponding distance from the wall to the bottom of the ladder. From the Pythagorean theorem, we get

$$x^2 + y^2 = 400 \tag{4.7}$$

As you can see from Figure 4.5, $-dy/dt$ represents the rate at which the top of the ladder is moving downward, and dx/dt represents the rate at which the bottom of the ladder is moving away from the wall. Since x and y are both functions of the time t, they may be differentiated with respect to t. The relationship between dx/dt and dy/dt is then found by differentiating both sides of Equation (4.7) with respect to t.

$$\frac{d}{dt}(x^2) + \frac{d}{dt}(y^2) = \frac{d}{dt}(400)$$

or

$$2x\frac{dx}{dt} + 2y\frac{dy}{dt} = 0$$

Solving for dx/dt gives

$$\frac{dx}{dt} = -\frac{2y(dy/dt)}{2x} = -\frac{y}{x}\frac{dy}{dt} \tag{4.8}$$

To find the value of dx/dt at the specific instant in question, we note that the statement of the problem gives $x = 16$ and $dy/dt = -4$. The value $y = 12$ is found from Equation (4.7). Hence, from Equation (4.8) we have

$$\frac{dx}{dt} = -\left(\frac{12}{16}\right)(-4) = 3 \text{ ft/sec} \qquad\blacksquare$$

EXAMPLE 2   ■ **Problem**

When a stone is dropped into a pond, the ripples form concentric circles of increasing radius. What is the rate at which the area of one of these circles is increasing when the radius is 4 ft if the radius is increasing at a rate of 0.5 ft/sec?

■ **Solution**

We determine the equation that relates dA/dt and dr/dt. Note that area is related to radius by the formula

$$A = \pi r^2$$

Since A and r vary with time in this problem, we may differentiate both sides with respect to t to get

$$\frac{d}{dt}(A) = \frac{d}{dt}(\pi r^2)$$

or

$$\frac{dA}{dt} = 2\pi r \frac{dr}{dt}$$

Substituting $r = 4$ and $dr/dt = 0.5$, we get

$$\frac{dA}{dt} = 2\pi(4)(0.5) \approx 12.56 \text{ ft}^2/\text{sec}$$

■

EXAMPLE 3   ■ **Problem**

At 12:00 noon airplane A is 500 mi south of airplane B. Plane A is traveling west at 300 mi/hr, and B is traveling south at 400 mi/hr. What is the rate at which the distance between the two airplanes is changing at 1:00 P.M.?

■ **Solution**

Let x and y denote the miles covered by A and B, respectively, in t hours after noon. Let z be the distance between the two planes. See Figure 4.6.

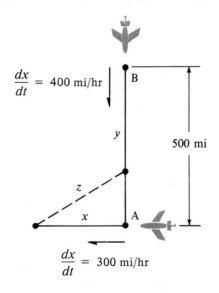

■ **Figure 4.6**

From the Pythagorean theorem,

$$z^2 = x^2 + (500 - y)^2$$

Differentiating both sides with respect to t gives

$$2z \frac{dz}{dt} = 2x \frac{dx}{dt} - 2(500 - y) \frac{dy}{dt}$$

or

$$z \frac{dz}{dt} = x \frac{dx}{dt} - (500 - y) \frac{dy}{dt}$$

By 1:00 P.M., A has gone 300 mi and B has gone 400 mi. Therefore, at that time,

$$z^2 = 300^2 + 100^2 = 100{,}000$$

Hence, $z = 316$ mi (approximately).

Using this value of z along with the other known values of x, y, dx/dt, and dy/dt, we have

$$316 \frac{dz}{dt} = 300(300) - (500 - 400)(400)$$

so that

$$\frac{dz}{dt} = \frac{90{,}000 - 40{,}000}{316} = \frac{50{,}000}{316} \approx 158 \text{ mi/hr} \qquad \blacksquare$$

The previous examples exhibit the following procedure for solving related rate problems.

PROCEDURE

> **Solving Related Rate Problems**
>
> 1. Carefully identify those quantities that remain constant and those that change with time.
> 2. Establish a *general* relationship between the variables and the constants.
> 3. Differentiate this implicit relationship with respect to time.
> 4. Evaluate the unknown rate at the specific time.

EXERCISES FOR SECTION 4.3

1. Two cars, A and B, leave a common point at 3 P.M. Car A travels east at 80 mi/hr, and car B moves north at 60 mi/hr. How fast are the two cars moving apart at 5 P.M.?

2. A man walks 200 ft north and then turns west. He walks 4 ft/sec. At what rate is the distance between the man and the starting point increasing 1 min after he turns west?

3. When a square plate is heated, the side length increases at a rate of 0.05 cm/min. How fast is the area increasing when the side is 10 cm long?

4. A spherical balloon is filled with gas at a rate of 5 ft^3/sec. Determine the rate at which the radius of the balloon is increasing when the radius is 2 ft.

5. The force of attraction between two unit charges, separated by a distance s, is $F = 1/s^2$ (in dynes, dyn). What is the rate at which the force is changing when $s = 2$ cm if the two charges are moving closer together at a rate of 3 cm/sec?

6. The power (in watts) in a resistor is given by $p = Ri^2$. Determine the relationship between dp/dt and di/dt. In a 50-Ω resistor, the current changes at a rate of 0.1 A/sec. At what rate is the power changing when the current is 2 A?

7. The stagnation pressure (in pounds per square inch) at the nose of a ballistic missile is given by $P = \frac{1}{2}\rho V^2$, where ρ is the density of the air and V is the velocity of the missile. At what rate is the stagnation pressure changing when the missile has a velocity of 800 ft/sec and an acceleration of 50 ft/sec^2? Assume $\rho = 0.015$.

8. In an experiment, the relationship between the tensile strength (in pounds per square inch) of a specimen and its temperature was found to be $s = 1000 - \sqrt{2T + 30}$. The temperature is increasing at a rate of 10°/min. At what rate is the tensile strength changing when $T = 35°$?

9. The electric resistance (in ohms) of a wire varies with temperature according to

$$R = 50 - 0.3T + 0.001T^2$$

What is the rate at which the resistance is changing when $T = 260°$ if the temperature is changing at a rate of 5°/min?

10. In AC circuit theory, the reactance X_L of an inductance coil is given by $X_L = 2\pi fL$, where f is the frequency of the alternating current and L is the inductance of the coil. What is the relationship between dX_L/dt and df/dt? At what rate is the reactance of a 10-H inductance coil changing if the frequency is changing at a rate of 0.5 hertz per second (Hz/sec)?

11. A meteorological balloon is released at a point 5000 ft downrange from a tracking radar. The balloon rises vertically at a rate of 20 ft/sec. How fast is the distance (slant range) between the balloon and the radar increasing 3 min after the balloon is released?

12. Water is flowing into an 8-ft-diameter conical tank at the rate of 20 ft^3/min. What is the rate of water level rise when the depth of the water is 4 ft if the tank is 12 ft deep?

13. If $y = (5x + x^3)^{1.3}$, and x is increasing at the rate of 0.5 unit per second, how fast is the slope of the graph changing when $x = 2$?

14. If $y = (2x + 3)^{2.7}$, and x is increasing at the rate of 0.3 unit per second, how fast is the slope of the graph changing when $x = 1.5$?

Exercises 15–18

15. The gas in a cylinder is being compressed by a piston so that the pressure is increasing at a rate of 3 psi/sec. At what rate is the volume of the gas decreasing when the pressure is 50 psi and the volume is 0.53 ft^3? Assume that $PV = C$ (*Boyle's law*), where C is a constant, holds for this situation.

16. A cylinder containing 0.8 ft³ of air at 25 psi is being compressed at a rate of 2.3 psi/sec. At what rate is the volume of gas decreasing if Boyle's law is valid?

17. The expansion of a gas under adiabatic conditions is given by $PV^{1.4} = C$, where P is pressure, V is volume, and C is a constant. What is the rate at which the pressure of a gas is increasing if $P = 15$ psi, $V = 5.6$ ft³, and the volume of the gas is decreasing at a rate of 0.08 ft³/sec?

18. A cylinder with a radius of 3.2 in. has one end sealed, with a freely moving piston at the other. What is the rate at which the piston is moving if the water is being pumped into the cylinder at 15 in.³/sec?

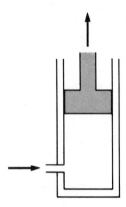

19. A point $P(x,y)$ moves on the graph of $x^2 + 3y^2 = 4$ such that $dx/dt = x + 1$. Find dy/dt at the point $(1, 1)$.

20. A metal rod has the shape of a right circular cylinder. As it is heated, its length increases by 30 microns/minute and its diameter by 20 microns/minute. At what rate is the volume changing when the rod has length 30 cm and diameter 5 cm?

21. Using a straw, a student drinks soda from a cup in the shape of an inverted cone that is 12 cm across the top and 12 cm deep. The soda is being consumed at a rate of 2 cm³/sec. How fast is the area of the surface of the soda decreasing when the depth of the soda left in the cup is 6 cm?

22. A 10-ft horizontal trough has a vertical cross section that is an isosceles right triangle. Water is poured into the trough at a rate of 8 ft³/min. How fast is the surface of the water rising when the trough is filled to a depth of 2 ft?

4.4 MAXIMUM AND MINIMUM VALUES OF A FUNCTION

Recall from Section 1.3 that a function $y = f(x)$ whose graph rises from left to right is said to be an **increasing function.** As shown in Figure 4.7(a), the tangent to the graph of an increasing function has a positive slope and, therefore, a positive derivative. A function whose graph falls from left to right is a **decreasing function.** A decreasing function has a negative derivative since the slope of a tangent line to its graph is negative. See Figure 4.7(b).

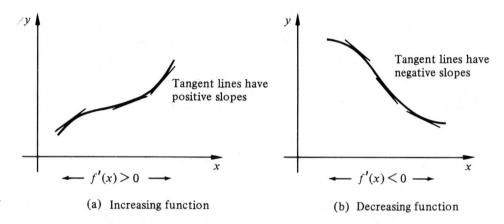

■ **Figure 4.7**   (a) Increasing function   (b) Decreasing function

EXAMPLE 1   ■ **Problem**

Use the first derivative to determine where $y = \frac{1}{3}x^3 - 2x^2 + 3x + 1$ is increasing and where it is decreasing.

■ **Solution**

The derivative of the given function is

$$y' = x^2 - 4x + 3 = (x - 1)(x - 3)$$

From this expression, we see that $y' = 0$ for $x = 1$ and $x = 3$. Therefore, we investigate the sign of y' on the intervals $x < 1$, $1 < x < 3$, and $x > 3$.

    If x is in the interval $x < 1$, then both factors of y' are negative; so their product is positive. If x is in the interval $1 < x < 3$, then the factor $x - 1$ is positive and the factor $x - 3$ is negative; so the product is negative. If x is in the interval $x > 3$, then both factors are positive, and so is their product. We summarize this discussion as follows:

$$x < 1 \qquad y' = (-)(-) > 0$$
$$1 < x < 3 \qquad y' = (+)(-) < 0$$
$$x > 3 \qquad y' = (+)(+) > 0$$

Therefore, the function is increasing on $x < 1$ and $x > 3$ and is decreasing on $1 < x < 3$. These relationships are shown in Figure 4.8.

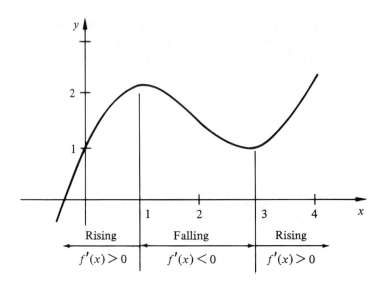

■ **Figure 4.8**

Figure 4.9 shows the graph of a typical function of x on which certain points on the interval [a, b] are indicated. The points labeled A, C, and E in the figure are called **relative maximum points** of $f(x)$ because each has a greater y value than any other point near it. Similarly, the points B, D, and F are called **relative minimum points** since the y value at each of these points is less than any nearby point. The points A and F are endpoints of the interval.

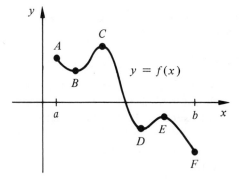

■ **Figure 4.9**

At point C, the functional value is clearly greater than that at A or E; therefore, this maximum is called an **absolute maximum point** on the interval [a, b]. For a similar reason, the point F is called the **absolute minimum** on the interval.

Referring to Figure 4.9, we note the following features:

1. The first derivative is negative between points A and B since the curve is falling on this interval.

2.   The first derivative is positive between B and C since the curve is rising on this interval.
3.   Since $f'(x) < 0$ between A and B and $f'(x) > 0$ between B and $C,$ the first derivative at point B is zero; that is, $f'(x_B) = 0.$
4.   For similar reasons, $f'(x) = 0$ at $C,$ $D,$ and $E.$

The implications of the preceding observations are generalized in the following theorem.

THEOREM 1

Relative Extremum Theorem   Let f be a differentiable function on the interval $[a, b]$. If a relative maximum (or minimum) value of the function occurs at a point X that is not an endpoint of the interval, then $f'(X)$ must equal zero.

Two important points should be made about this theorem.

■ The converse is *not* true; that is, $f'(X) = 0$ does not ensure a relative extreme at $x = X$. As an illustration, the function f defined by $f(x) = 1 - x^3$ has a zero derivative at $x = 0$, since $f'(x) = -3x^2$. However, referring to Figure 4.10, we see that the curve does not have a relative maximum or minimum at the point $(0, 1)$.

■ The theorem does not tell us how to locate extreme values that occur at endpoints of the interval or at points for which $f'(x)$ does not exist.

Collectively, the values of x for which $f'(x) = 0$, the values of x for which $f'(x)$ does not exist, and the endpoints are called **critical values.** Once the critical values have been identified, the following test may be used to determine the nature of the extrema at the points for which $f'(x) = 0$.

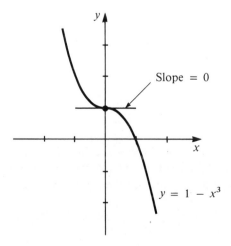

■ **Figure 4.10**

TEST

The First Derivative Test   If $f'(X) = 0$, then

1. $f(X)$ is a relative maximum if $f'(x)$ changes from $+$ to $-$ as x increases through X. See Figure 4.11.

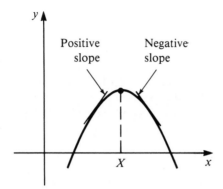

Positive slope   Negative slope

■ **Figure 4.11**

2. $f(X)$ is a relative minimum if $f'(x)$ changes from $-$ to $+$ as x increases through X. See Figure 4.12.

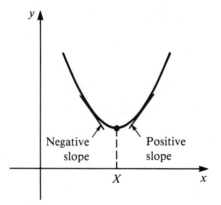

Negative slope   Positive slope

■ **Figure 4.12**

3. If $f'(x)$ does not change sign as x increases through X, then $f(X)$ is neither a maximum nor a minimum.

The change in sign of $f'(x)$ is determined by evaluating the derivative slightly to the left and then slightly to the right of the point. The procedure is demonstrated in the following examples.

EXAMPLE 2   ■ **Problem**

Use the first derivative test to show that $y = x^2 + 6x - 2$ has a relative minimum at $x = -3$.

■ **Solution**

The derivative of $y = x^2 + 6x - 2$ is $y' = 2x + 6 = 2(x + 3)$. We find the critical point by letting $y' = 0$. Thus,

$$2(x + 3) = 0$$
$$x = -3$$

To use the first derivative test, we first consider the sign of y' when $x < -3$. In this case, the factor $x + 3$ is negative. Next, we note that when $x > -3$, the factor $x + 3$ is positive. Since the derivative changes from negative to positive, we conclude that a relative minimum occurs at $x = -3$. The value of the minimum is $y = (-3)^2 + 6(-3) - 2 = -11$. See Figure 4.13.

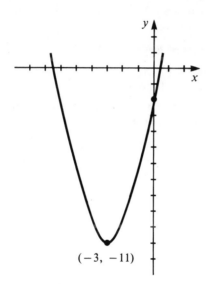

$(-3, -11)$

■ **Figure 4.13**

EXAMPLE 3   ■ **Problem**

Show that $y = x^3 - 3x^2 - 45x$ has a relative maximum at $x = -3$ and a relative minimum at $x = 5$.

■ **Solution**

The derivative of $y = x^3 - 3x^2 - 45x$ is

$$y' = 3x^2 - 6x - 45 = 3(x - 5)(x + 3)$$

Since the function is everywhere differentiable and there are no endpoint consider-ations, the only critical points are obtained by setting $y'(x) = 0$. Thus,

$$3(x - 5)(x + 3) = 0$$
$$x = -3, 5$$

Therefore, the critical values are -3 and $+5$.

To test $x = -3$, we substitute values into y' to the left of -3. In this case, the signs of the factors $(x - 5)$ and $(x + 3)$ are both negative, and hence, y' is positive to the left of -3. By substituting values of x to the right of -3, the factor $(x + 3)$ becomes positive, so y' is negative. Thus, $y(-3)$ is a relative maximum. You may easily check that $y(-3) = 81$. To test $x = 5$, we note that to the left of 5, the derivative is negative, and to the right, it is positive. Hence, $y(5) = -175$ is a rela-tive minimum.   ∎

PROCEDURE

Finding Absolute Extremes   To find the absolute extremes of a function on $[a, b]$, we proceed as follows:

1. Compute the values of $f(a)$ and $f(b)$.
2. Locate the values at which $f'(x) = 0$ or does not exist. Compute the corresponding values of the function.
3. The *absolute maximum* is the largest of the y values, and the *absolute minimum* is the smallest of the y values, as computed in steps 1 and 2.

EXAMPLE 4   ▪ **Problem**

Find the absolute maximum and minimum for

$$f(x) = x^3 - \frac{3}{2}x^2 - 6x + 2$$

on the interval $[0, 3]$.

▪ **Solution**

For this function,

$$f(0) = 2 \quad \text{and} \quad f(3) = (3)^3 - \tfrac{3}{2}(3)^2 - 6(3) + 2 = -\tfrac{5}{2}$$

are the functional values at the endpoints of the interval. The derivative of $f(x)$ is

$$f'(x) = 3x^2 - 3x - 6 = 3(x^2 - x - 2) = 3(x - 2)(x + 1)$$

From this expression, we see that the critical points are $x = -1$ and $x = 2$. Thus, since $x = 2$ is the only critical point on $[0, 3]$, we compute the value of $f(2)$ to be $(2)^3 - \tfrac{3}{2}(2)^2 - 6(2) + 2 = -8$. We make the following conclusions:

1. The absolute maximum is 2 and occurs when $x = 0$.

2. The absolute minimum is -8 and occurs when $x = 2$.

The graph of the function on [0, 3] is shown in Figure 4.14. ■

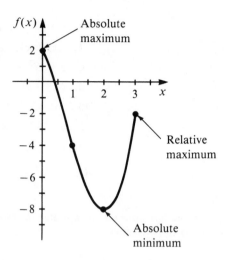

■ **Figure 4.14**

EXAMPLE 5 ■ **Problem**

Find the absolute maximum and minimum for $y = \sqrt{x}$ on [0, 4].

■ **Solution**

The values of y at the endpoints of the interval are $y = 0$ and $y = 2$. The derivative of the function is

$$y' = \frac{1}{2}x^{-1/2}$$

Equating y' to zero, we find that there is no solution to $\frac{1}{2}x^{-1/2} = 0$. Consequently, $y = \sqrt{x}$ has no relative maximum or relative minimum values within the interval [0, 4]. We conclude that $y = 0$ is the absolute minimum and $y = 2$ is the absolute maximum. See Figure 4.15.

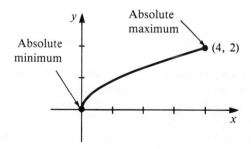

■ **Figure 4.15** ■

EXERCISES FOR SECTION 4.4

Use the first derivative test to locate any relative maximum and minimum points of the following functions. Find the numerical value of each maximum and minimum in Exercises 1–16.

1. $y = x^2 - 6x + 5$

2. $s = 3t^2 + 12t$

3. $i = t^3 - 3t^2 + 2$

4. $f(x) = x^4 - 8x^2$

5. $y = x^3 + 4$

6. $v = 6t^2 - t^3$

7. $p = x^3 - 3x$

8. $g(s) = s^4 + 10s^2$

9. $y = 3x^2 + 24x - 30$

10. $y = 4x - x^2$

11. $m = 3r^4 - 4r^3 + 1$

12. $s = 2t^3 - 18t^2 + 48t - 20$

13. $w = t^3 + 3t^2 + 3t + 5$

14. $f(x) = 2 + 5x - x^5$

15. $y = x + \dfrac{10}{x}$

16. $v = x^2 + \dfrac{16}{x}$

Find the absolute extremes for the functions on the indicated interval in Exercises 17–24. Use the first derivative test to identify relative extremes.

17. $y = 8x - x^2; 3 \le x \le 10$

18. $y = x^2 + x - 2; -3 \le x \le 1$

19. $y = x^3 + 3x^2; [-4, 0]$

20. $y = x^3 - 4x^2; [-2, 5]$

21. $y = \frac{1}{3}x^3 - \frac{3}{2}x^2 + 2x + 1; [0, 3]$

22. $y = x^3 - 6x^2 + 7; [0, 5]$

23. $y = x^{1/3}; 0 \le x \le 8$

24. $y = x^{4/3}; [-1, 1]$

25. A ball thrown vertically upward has an equation of motion given by $s(t) = 1000t - 32t^2$ (in feet), $t \ge 0$ sec. Find the maximum height that the ball rises above the ground.

26. The current (in amperes) in a circuit is given by $I(t) = \frac{1}{3}t^3 - \frac{7}{2}t^2 + 10t$, $t \ge 0$. Find the maximum value of the current for $0 \le t \le 5$.

27. Find the minimum power (in watts) developed by a motor from $t = 0$ to $t = 2$ if $p(t) = 3t^2 - 3t + 4$.

4.5 THE SECOND DERIVATIVE TEST

In the previous section, we explained how the first derivative of a function is used to identify any relative maximum or minimum of the function. In this section, we will discuss the relationship between the relative extremes of a function and its second derivative.

Consider the two cup-shaped graphs in Figure 4.16. The graph in Figure

4.16(a) opens upward and is said to be **concave up.** The graph in Figure 4.16(b) opens downward and is said to be **concave down.** We make the following observations about these two graphs.

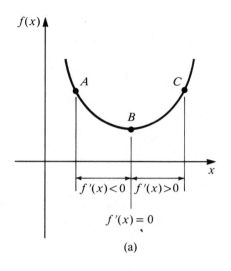

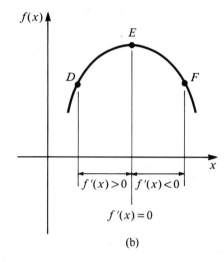

■ **Figure 4.16**   (a)   (b)

1. The slope of the graph in Figure 4.16(a) is negative to the left of *B* and positive to the right of *B.* Furthermore, the slope is *increasing* on the segment from *A* to *C* since it is negative at *A,* zero at *B,* and positive at *C.* If the first derivative is increasing, the second derivative is positive [$f''(x)$ is the rate of change of $f'(x)$]. We conclude that **a function whose graph is concave up has a positive second derivative.**

2. The concave-down graph in Figure 4.16(b) has a positive slope at *D,* a zero slope at *E,* and a negative slope at *F;* that is, the slope is *decreasing* on the segment from *D* to *F.* Therefore, we conclude that **a function whose graph is concave down has a negative second derivative.**

The relationship between concavity and the second derivative leads to another test for the relative extremes of a function. This test is based on the facts that a relative minimum is the low point of a concave-up graph and a relative maximum is the high point of a concave-down graph.

TEST

> **The Second Derivative Test**
>
> 1. If $f'(X) = 0$ and $f''(X) > 0$, then $f(X)$ is a relative minimum.
> 2. If $f'(X) = 0$ and $f''(X) < 0$, then $f(X)$ is a relative maximum.
> 3. The second derivative test fails if $f''(X) = 0$ or does not exist at X.

EXAMPLE 1   ▪ **Problem**

Use the second derivative test to show that $y = 7x - 2x^2$ has a relative maximum at $x = \frac{7}{4}$.

▪ **Solution**

To verify that $x = \frac{7}{4}$ is a critical value, we set the first derivative equal to zero and solve for x. Thus,

$$y' = 7 - 4x$$

Letting $y' = 0$, we get

$$7 - 4x = 0$$

$$x = \frac{7}{4}$$

The second derivative is $y'' = -4$. Since y'' is always negative, a relative maximum occurs at $x = \frac{7}{4}$. The value of the maximum is $y = 7(\frac{7}{4}) - 2(\frac{7}{4})^2 = \frac{49}{8}$. The graph is shown in Figure 4.17.

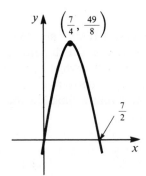

▪ **Figure 4.17**

EXAMPLE 2   ▪ **Problem**

Test $y(x) = x^2 + \dfrac{54}{x}$ for relative maximum and minimum values.

▪ **Solution**

From the given function, we obtain

$$y'(x) = \frac{dy}{dx} = 2x - 54x^{-2} = 2x - \frac{54}{x^2}$$

Setting $dy/dx = 0$, we obtain

$$2x - \frac{54}{x^2} = 0$$

Multiplying both sides by x^2 yields

$$2x^3 - 54 = 0$$
$$x^3 = 27$$
$$x = 3 \qquad (\textit{critical value})$$

The second derivative is

$$y''(x) = 2 + 108x^{-3} = 2 + \frac{108}{x^3}$$

Using the second derivative test at $x = 3$, we obtain

$$y''(3) = 2 + \frac{108}{3^3} = 6$$

Since $y''(3)$ is positive, $y(3) = 27$ is a relative minimum. See Figure 4.18.

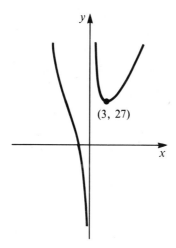

(3, 27)

■ **Figure 4.18**   ■

Many graphs contain both concave-up and concave-down segments. Points on the curve at which the sense of the concavity changes are called **points of inflection.** Since a change in concavity means that the sign of f'' changes, the value of f'' at a point of inflection is either zero or undefined.

Figure 4.19 shows the graph of a function $f(x)$ having four changes in the sense of concavity. At points A and C the second derivative is zero, while at B and D the second derivative is undefined. There is no point on the curve at D, so only points A, B, and C are points of inflection.

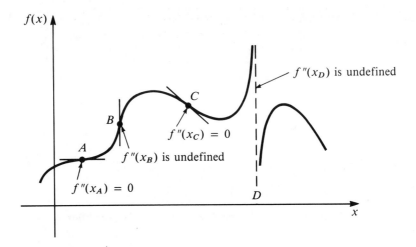

■ **Figure 4.19**

| DEFINITION | **Point of Inflection** A point $(X, f(X))$ is a *point of inflection* if $f''(x)$ changes sign as x passes through X. At a point of inflection, $f''(X)$ will be either zero or undefined. |

EXAMPLE 3 ■ **Problem**

Sketch the graph of $f(x) = x^3 + 3x^2 - 24x$.

■ **Solution**

The first and second derivatives of $f(x)$ are

$$f'(x) = 3x^2 + 6x - 24 = 3(x - 2)(x + 4)$$
$$f''(x) = 6x + 6 = 6(x + 1)$$

Note that $f'(x) = 0$ for $x = -4$ and 2. By the second derivative test, we note the following features:

1. Since $f''(-4) = -18$, a relative maximum occurs at $x = -4$.
2. Since $f''(2) = 18$, relative minimum occurs at $x = 2$.

Furthermore, a point of inflection occurs at $x = -1$, since $f''(-1) = 0$ and $f''(x)$ changes sign as x passes through -1.

Summarizing the results, we sketch the curve in Figure 4.20, using only these three points and our knowledge of concavity. The accompanying table summarizes the results.

x	$f(x)$	$f''(x)$	Remarks
-4	80	$-$	Maximum
-1	26	0	Point of inflection
2	-28	$+$	Minimum

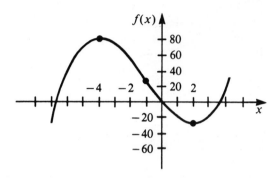

■ **Figure 4.20**   ■

EXAMPLE 4   ■ **Problem**

Locate any relative extremes and points of inflection for $y = 2 - x^{3/5}$.

■ **Solution**

Here, we note that $y' = -\frac{3}{5}x^{-2/5}$ is never equal to zero and is undefined when $x = 0$. Likewise, $y'' = \frac{6}{25}x^{-7/5}$ cannot equal zero and is undefined when $x = 0$. The sign of y' is negative for both $x < 0$ and $x > 0$. Hence, there are no relative extremes. The sign of y'' is negative for $x < 0$ (curve is concave down) and positive for $x > 0$ (curve is concave up). The graph in Figure 4.21 shows a vertical tangent at $x = 0$, along with the point of inflection at $(0, 2)$.

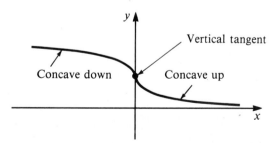

■ **Figure 4.21**   ■

EXAMPLE 5   A point of inflection on a distance-time graph corresponds to the time at which the velocity of the object is either a maximum or a minimum. A distance-time graph of a particle is shown in Figure 4.22. From this graph we can conclude that the particle has either a maximum or a minimum velocity at $t = 2$ sec. Notice that the point of inflection does not give the velocity; it only indicates where the maximum or minimum velocity will occur.

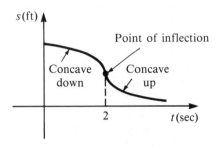

■ **Figure 4.22**

■

EXERCISES FOR SECTION 4.5

Use the second derivative test to find any relative extremes of the functions in Exercises 1–12.

1. $y = x^3$

2. $f(x) = 4 - 2x^3$

3. $f(x) = x^2 + 3x + 2$

4. $y = x^3 + 9x^2$

5. $y = x^{1/3}$

6. $f(x) = 6x - 3x^2$

7. $y = x^3 - 9x^2 + 24x - 10$

8. $y = x^4 - 4x^3$

9. $y = 4x + \dfrac{16}{x}$

10. $y = x - \dfrac{1}{x}$

11. $f(x) = x + \dfrac{1}{x^2}$

12. $f(x) = 2x^5 - 5x^2$

Locate any relative extremes and points of inflection on the graphs of the functions in Exercises 13–22. Use this information and your understanding of concavity to sketch the graph of the given function.

13. $f(x) = x^2 + x - 2$

14. $f(x) = -x^2 - 8x - 7$

15. $f(x) = x^3 + 3x^2$

16. $f(x) = x^3 - 4x^2$

17. $f(x) = \dfrac{x^3}{3} - \dfrac{3x^2}{2} + 2x + 1$

18. $f(x) = x^3 - 6x^2 + 7$

19. $f(x) = x^{1/3} + 2x$

20. $f(x) = x^{2/3} + 1$

21. $f(x) = x^{4/3}$

22. $f(x) = -x^{2/3}$

23. Determine the time ($t > 0$) when the velocity is an extreme given that the distance-time relation is expressed as $s(t) = t^5 - 3t^3 + t - 1$.

24. Let the work-time relation for a solenoid be given by $w(t) = -t^4 + t^3 + 5t + 5$. Find the time ($t > 0$) of maximum power transfer. (Recall that $p = dw/dt$.)

■■■■■ 4.6 CURVE SKETCHING

In this section, we briefly review and combine the procedures of curve sketching from analytic geometry and calculus. The following table summarizes the effect of the signs of f, f', and f'' on geometric properties of the curve $y = f(x)$.

Function	Positive	Zero	Negative
$f(x)$	Curve is above x-axis	x-intercept	Curve is below x-axis
$f'(x)$	Curve is rising	Horizontal tangent	Curve is falling
$f''(x)$	Curve is concave-up	Possible inflection point	Curve is concave-down

With this table in mind, the following steps will be a useful guide in making a rather complete graph of a function $y = f(x)$.

PROCEDURE

Curve Sketching   Use the following steps to sketch the graph of $y = f(x)$.

1. For the y-intercept: Let $x = 0$ and evaluate y.

2. For the x-intercepts: Let $y = 0$ and solve for x.

3. For relative extremes: Let $y' = 0$ and solve for x. Use the first or second derivative tests to check for relative maximum and minimum values.

4. For points of inflection: Let $y'' = 0$ and solve for x.

5. For vertical tangents or asymptotes: Note the values of x at which y, y', and y'' are undefined.

6. For trends for large x: Note what happens to y as x becomes large.

EXAMPLE 1   ■ **Problem**

Sketch the graph of $y = x^3 - 6x^2$.

■ **Solution**

We note that $y' = 3x^2 - 12x$ and $y'' = 6x - 12$.

1. The y-intercept is $y = (0)^3 - 6(0)^2 = 0$.

2. Letting $y = 0$, we get $x^3 - 6x^2 = 0$. The x-intercepts are $x = 0$ and 6.

3. $y' = 3x^2 - 12x = 0$ when $x = 0$ and 4. The second derivative test shows that a relative maximum of 0 occurs at $x = 0$, and a relative minimum of -32 occurs at $x = 4$.

4. $y'' = 6x - 12 = 0$ when $x = 2$. Since y'' changes sign as x passes through 2, then $(2, -16)$ is a point of inflection.

5. The function and its derivatives are defined everywhere. Hence, there are no vertical asymptotes or vertical tangents.

6. The absolute value of the functional values get very large as x moves away from the origin. The graph of this function is shown in Figure 4.23.

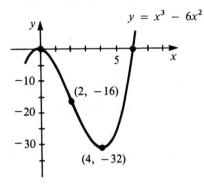

$y = x^3 - 6x^2$

$(2, -16)$

$(4, -32)$

■ **Figure 4.23**

■

EXAMPLE 2   ■ **Problem**

Sketch the graph of $y = 2 + x^{1/3}$.

■ **Solution**

Note that $y' = \frac{1}{3}x^{-2/3}$ and $y'' = -\frac{2}{9}x^{-5/3}$.

1. The y-intercept is $y = 2 + (0)^{1/3} = 2$.
2. Letting $y = 0$, we get $2 + x^{1/3} = 0$, which yields $x = -8$ as the x-intercept.
3. y' is never zero, so there are no relative extremes except possibly for y' undefined.
4. y'' is never zero, so there are no points of inflection without vertical tangents.
5. y' is undefined at $x = 0$ and positive for all other values of x. Note y'' is also undefined at $x = 0$ and changes from $+$ (concave-up) to $-$ (concave-down) as x passes through 0. This result, in conjunction with the fact that y is defined at $x = 0$, leads us to conclude that there is a vertical tangent at $(0, 2)$. The graph is concave up to the left of zero, and concave down to the right of zero, which means $(0, 2)$ is a point of inflection.
6. $y = 2 + x^{1/3}$ behaves approximately as $x^{1/3}$ when x is large. The graph is shown in Figure 4.24.

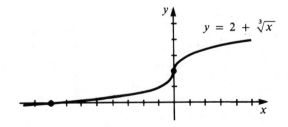

$y = 2 + \sqrt[3]{x}$

■ **Figure 4.24**

■

EXAMPLE 3   ■ **Problem**

Sketch the graph of $y = x + \dfrac{1}{x}$.

■ **Solution**

Note that

$$y' = 1 - x^{-2} = 1 - \frac{1}{x^2} \quad \text{and} \quad y'' = 2x^{-3} = \frac{2}{x^3}$$

1. y is undefined at $x = 0$, so there is no y-intercept.

2. Letting $y = 0$ yields $x + (1/x) = 0$, or $x^2 + 1 = 0$. Since $x^2 + 1$ cannot equal zero, there are no x-intercepts.

3. $y' = 0$ when $x = 1$ and -1. The second derivative test shows that $(1, 2)$ is a relative minimum point, and $(-1, -2)$ is a relative maximum point.

4. y'' is never zero. Since y is undefined at $x = 0$ and y'' is never zero, there are no points of inflection, even though the sense of concavity is different for $x < 0$ and $x > 0$.

5. Neither the function nor the first and second derivatives are defined at $x = 0$, so the line $x = 0$ is a vertical asymptote.

6. As x becomes large, the term $1/x$ becomes very small. Therefore, the graph is approximately that of $y = x$ for large x. Figure 4.25 is a sketch of the curve.

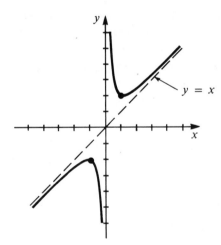

■ **Figure 4.25**   ■

━━━━━━   **EXERCISES FOR SECTION 4.6**

Use all the methods at your disposal to sketch the graphs of the functions indicated in Exercises 1–18.

1. $y = 2x^3 + 3x^2 - 12x$ 2. $y = 4x^3 - 3x^4$

3. $y = 2x^2 - x^4$ 4. $y = 2x^5 - 5x^2$

5. $y = 5 + \dfrac{3}{x}$

6. $y = 7 - \dfrac{2}{x}$

7. $y = 2x + \dfrac{1}{x}$

8. $y = x - \dfrac{2}{x}$

9. $y = x + \dfrac{1}{x^2}$

10. $y = -x + \dfrac{3}{x^2}$

11. $y = x^{1/3} + 2$

12. $y = x^{-1/2}$

13. $y = \dfrac{x^2 - 4}{x^2}$

14. $y = \dfrac{x - x^2}{x^2}$

15. $y = x^3 - 3x + 1$

16. $y = x^2 + \dfrac{1}{x}$

17. $y = \dfrac{2}{x} - \dfrac{1}{x^3}$

18. $y = (x^2 - 1)(x^2 - 5x + 6)$

 19. The current (in amperes) in a switching circuit varies with time (in seconds) according to $i = t^2 - t^3$. Sketch the current-time graph from $t = 0$ to $t = 3$ sec.

 20. The stress S (in pounds) in a cantilever beam varies with the distance x (in feet) from the fixed end. Sketch the stress curve from $x = 0$ to $x = 4$ ft given that $S = 12x - x^3$.

 21. The thrust (in pounds) of a new rocket motor is found to vary with time (in minutes) in accordance with $T = 5 + t^2 - 4t^3$. Sketch the thrust-time curve from $t = 0$ to $t = 1$ min.

 22. The power (in watts) in a system varies with time (in seconds) as $p(t) = 6t^2 - t^3$. Sketch the power-time curve for $t \geq 0$.

▬▬ 4.7 APPLIED MAXIMUM AND MINIMUM PROBLEMS

In applied problems, we are interested primarily in finding the absolute maximum (or minimum) of a function. The technique is illustrated by the following examples.

EXAMPLE 1   ■ **Problem**

The work done (in foot-pounds) by a solenoid varies with time (in seconds) according to $w = 6t^2 - t^4$. What is the greatest power produced by the solenoid for $t \geq 0$?

■ **Solution**

Power is the rate of change of work with respect to time. Therefore,

$$p = \frac{dw}{dt} = 12t - 4t^3$$

We now wish to find the maximum power. The critical values are found by letting $dp/dt = 0$.

$$\frac{dp}{dt} = 12 - 12t^2$$

$$12 - 12t^2 = 0$$

$$t = \pm 1$$

We need not consider the critical value $t = -1$ since negative times are not feasible. Taking d^2p/dt^2, we have $d^2p/dt^2 = -24t$. When $t = 1$, $d^2p/dt^2 = -24$, and therefore, the power is a maximum. The power at $t = 1$ sec is

$$p = 12(1) - 4(1)^3 = 8 \text{ ft-lb/sec} \qquad \blacksquare$$

EXAMPLE 2   ■ **Problem**

A packaging engineer wishes to construct a cylindrical oil can that will have a volume of 1000 in.^{3}. For minimization of the cost of producing the can, as little material as possible should be used in constructing the can. What are the dimensions of the can having minimum surface area for the given volume?

■ **Solution**

To determine the minimum surface area, we must know the relationship between the volume V, the surface area A, and the radius r. See Figure 4.26. The volume V of a cylinder of radius r and height h is

$$V = \pi r^2 h \qquad \qquad \textbf{(4.9)}$$

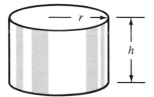

■ **Figure 4.26**

The surface area of the cylinder including the top is

$$A = 2\pi r^2 + 2\pi r h \qquad \qquad \textbf{(4.10)}$$

Equation (4.10) contains both r and h, as does Equation (4.9). Since the volume is given, we can write Equation (4.9) as

$$1000 = \pi r^2 h \qquad \text{or} \qquad h = \frac{1000}{\pi r^2} \qquad \qquad \textbf{(4.11)}$$

Substituting this expression into Equation (4.10) yields A as a function of the radius r only:

$$A = 2\pi r^2 + 2\pi r \left(\frac{1000}{\pi r^2}\right) = 2\pi r^2 + \frac{2000}{r}$$

$$= 2\pi r^2 + 2000r^{-1}$$

(4.12)

The derivative of this function with respect to the radius r is

$$\frac{dA}{dr} = 4\pi r - 2000r^{-2}$$

We find the critical values by setting $dA/dr = 0$:

$$4\pi r - 2000r^{-2} = 0$$

We multiply this equation by r^2 to get

$$4\pi r^3 - 2000 = 0$$
$$r^3 = \frac{2000}{4\pi}$$
$$r \approx 5.4 \text{ in.}$$

We check by using the second derivative test with

$$\frac{d^2A}{dr^2} = 4\pi + 4000r^{-3}$$

The second derivative is positive for all positive values of r; therefore, the area will be a minimum for $r = 5.4$ in. The height can be found by substituting this value of r into Equation (4.11):

$$h = \frac{1000}{\pi(5.4)^2} \approx 10.9 \text{ in.}$$

The minimum area is 553.0 in.^{2}, which we find by substituting $r = 5.4$ in. and $h = 10.9$ in. into Equation (4.10).   ■

The following steps are recommended for solving applied maximum and minimum value problems.

<table>
<tr><td>PROCEDURE</td><td>

Solving Applied Maximum and Minimum Value Problems

1. Study the problem and determine what relationships exist between the variables in the problem.

2. Determine which variable is to be either maximized or minimized. Using the equations found in step 1, write the variable of interest as a function of only one of the other variables.

3. Find the critical values by setting the derivative of the function to be maximized or minimized equal to zero.

4. Test the critical values to establish maximum and minimum values.

</td></tr>
</table>

EXAMPLE 3   ■ **Problem**

Find the maximum area of the rectangle that can be inscribed in the first quadrant under the parabola $y = 9 - x^2$.

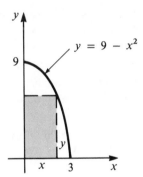

■ **Figure 4.27**

■ **Solution**

1. The area of the rectangle shown in Figure 4.27 is given by $A = xy$. Furthermore, since the corner of the rectangle is on the parabola, we know that $y = 9 - x^2$.

2. Replacing y in the area formula with $9 - x^2$, we get $A = x(9 - x^2)$.

3. We find $dA/dx = 9 - 3x^2$. Setting this expression equal to zero, we get

$$9 - 3x^2 = 0$$

$$x^2 = 3$$

$$x = \pm\sqrt{3}$$

The critical value is $\sqrt{3}$, since the rectangle is in the first quadrant.

4.   The second derivative is negative for positive x, so we conclude that $x = \sqrt{3}$ is a maximum. Therefore, the area is

$$A = \sqrt{3}\,[9 - (\sqrt{3})^2] = 6\sqrt{3} \approx 10.4 \text{ square units}$$ ■

EXERCISES FOR SECTION 4.7

1.   The velocity (in feet per second) of a rocket varies with time (in seconds) according to $v = 4t - t^2$. At what time does the rocket reach its maximum velocity? What is the magnitude of the velocity at this time?

2.   The displacement of a cam-operated pushrod is given by $s = 1 + 3t^2 - t^3$, where t is the elapsed time (in seconds) and s is in inches. What is the maximum displacement of the pushrod?

3.   The current (in amperes) in a circuit is given by the equation $i = 4 + 6t - t^2$, where t is time (in seconds). At what time does the current reach its maximum value? What is the maximum current?

4.   The charge-time equation for a certain electric circuit is $q = 0.24t - t^3$. At what time is the charge transferred a maximum?

5.   Electric charge (in coulombs) varies with time (in seconds) according to $q = 0.6t^2 - t^3$. At what time does the maximum current occur? What is the maximum current?

6.   What is the maximum strength of a new polymer film if the strength (in pounds) varies with temperature (in degrees) according to $s = 5.9 + 7.5T - 0.5T^2$?

7.   The displacement (in inches) of an electron in a magnetic field varies with time (in seconds) according to $s = t + 3t^2 - 4t^3$. What is the maximum velocity that the electron achieves?

8.   The energy (in joules) in an inductance coil varies with time (in seconds) according to $w = 3 + 4t - 3t^2$. What is the maximum energy of the coil?

9.   Find two integers whose sum is 30 and whose product is a maximum.

10. Find two integers whose sum is 12 such that the product of one number by the square of the other is a maximum.

11. A box is to be constructed with a volume of 32 ft³. The box is to have a square base. If the top is not included, what are the dimensions of the box that will have a minimum surface area?

12. A storage bin is to be constructed with a square base and no top. What is the maximum volume of the storage bin if only 432 ft² of material can be used?

13. A box without a top is to be constructed from a 10-in. × 16-in. metal sheet. The box is to be formed by cutting equal squares from each corner of the sheet and then bending up the sides. What size square should be cut to give the maximum volume?

14. Find the dimensions of an open cylindrical oil drum having a minimum surface area that will hold 64 ft^3 of oil.

15. A rectangular field is to be enclosed by 1000 ft of fence. What is the maximum area that can be enclosed?

16. A battery has an electromotive force E and an internal resistance r. When a current I is delivered by the battery, its power output is given by $P = EI - rI^2$. For what current is the power output a maximum?

17. The strength of a beam having a rectangular cross section varies directly as the breadth and the square of the depth. What is the strongest beam that can be cut from a log whose diameter is 10 in.?

18. The Ace Safe Company sells n safes per month at S dollars each. The cost of manufacturing n safes varies in accordance with $C = 600 + 10n + n^2$ (in dollars/per month). The selling price of each safe is $S = \$70$. How many safes should be made each month to maximize the profit? (*Hint:* Profit equals total income minus cost.)

19. A capacitor is to be sealed in a cylindrical container having a volume of 2000 cm^3. Find the dimensions of the container requiring the least amount of material.

20. A rectangular field of given area is to be fenced off along the bank of a river. If no fence is to be built along the river, show that the least amount of fencing is needed when the field is twice as long as it is wide.

21. Find the area of the largest rectangle that can be inscribed in the area bounded by $x = 6$, $y = 0$, and $y^2 = 8x$.

22. Find the area of the largest rectangle that can be inscribed in the area bounded by $x = 0$, $y = 0$, and $y = 8 - x^3$.

23. Find the minimum distance between the point $(5, 0)$ and the graph of $y^2 = 2x$.

24. Find the minimum distance between the point $(0, 5)$ and the graph of $y = 2x$.

◼◼◼◼ 4.8 DIFFERENTIALS

The symbol dx when considered by itself is called the **differential of x**. Similarly, dy is called the **differential of y.** We define the differential dx to be equal to any arbitrarily chosen increment of x, Δx.

We define the differential dy as the product of the derivative of the function times the differential dx. That is,

DEFINITION

> **The Differential of y**   If $y = f(x)$ is a differentiable function, then the *differential of y* is given by
>
> $$dy = f'(x)\,dx \qquad\qquad (4.13)$$
>
> where $f'(x)$ is the derivative of $f(x)$.

The definition in (4.13) can be divided by dx to yield the following familiar statement:

$$\frac{dy}{dx} = f'(x) \qquad\qquad (4.14)$$

Because of the relationship of Equations (4.13) and (4.14), we may interpret dy/dx as a fraction.

EXAMPLE 1   ■ **Problem**

Find the differential of $y = x^5 - 3x$.

■ **Solution**

$$\frac{dy}{dx} = 5x^4 - 3$$

then

$$dy = (5x^4 - 3)\,dx \qquad\qquad ■$$

We use the letter d as a symbol for the **differential operator.** When written before a function, it indicates that we are to find the differential of that function. Specifically,

$$d(f(x)) = f'(x)\,dx \qquad\qquad (4.15)$$

EXAMPLE 2   ■ **Problem**

Evaluate $d(\sqrt[3]{x})$.

■ **Solution**

$$d(\sqrt[3]{x}) = d(x^{1/3}) = \frac{1}{3}x^{-2/3}\,dx \qquad\qquad ■$$

We can best explain the physical meaning of the differentials dx and dy by comparing them graphically with the increments Δx and Δy, respectively. To this end, we have drawn two identical curves in Figure 4.28.

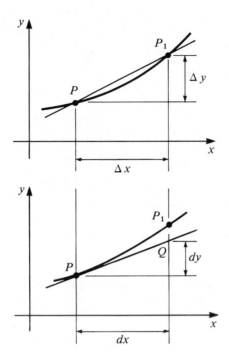

■ **Figure 4.28**

As you can see in the figure, dx and Δx are equal increments on the x-axis. To emphasize the difference between dy and Δy, we draw a line $\overline{PP_1}$ in the upper graph and a line tangent to the graph at P in the lower one. The increment Δy, shown in the upper graph, is the *exact* change in the function y. In the lower graph, we see that *dy represents the corresponding change along the tangent line.* Algebraically, dy and Δy are related by

$$\Delta y = dy + \overline{P_1Q}$$

where $\overline{P_1Q}$ is the vertical distance from the tangent line to the curve at the point P_1. We observe that dy will always differ from Δy by an amount $| \overline{P_1Q} |$. The value of dy is not necessarily less than Δy. Had we drawn the curve between P and P_1 to be concave down, dy would have been greater than Δy.

We use the differential dy to estimate the change in the value of the function Δy caused by a small change in x, since when dx is small, dy and Δy are nearly equal.

EXAMPLE 3   ■ **Problem**

A square sheet of steel has an initial side length of 20 cm. After the sheet is heated, the side length is 20.03 cm. See Figure 4.29. Find the approximate change in area and the exact change in area.

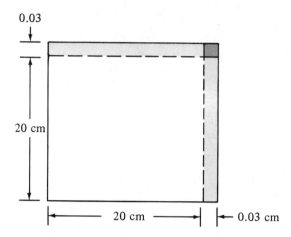

0.03

20 cm

20 cm → |← 0.03 cm

■ **Solution**

The relationship between the area A and the side length s is $A = s^2$. The derivative of A with respect to side length s gives

$$\frac{dA}{ds} = 2s$$

so

$$dA = 2s\, ds$$

Letting $s = 20$ and $ds = 0.03$ gives

$$dA = 2(20)(0.03) = 1.2 \text{ in.}^2$$

which is the approximate change in the area.

The exact change ΔA is found by

$$\Delta A = (20.03)^2 - (20)^2 = 401.2009 - 400.000 = 1.2009 \text{ in.}^2$$

Notice that the approximate value is only 0.0009 in.^{2} less than the actual value. The difference between the actual value and that obtained by using differentials is numerically equal to the small square area in Figure 4.29. ■

EXAMPLE 4 ■ **Problem**

The work done by a particle in a certain electric field is given by $W = s^4$, where s is the displacement (in meters). What is the approximate work done on the particle in moving it from 1.98 to 2.01 m? Work is measured in joules if displacement is given in meters.

▪ Solution

The approximate work done is given by

$$dW = 4s^3\, ds$$

Here, we choose $s = 2.00$ since it is easier to use in computations than either 1.98 or 2.01. Remember that dW is only an approximation to ΔW, and therefore, we assign to s the most convenient value in the interval. Letting $ds = 0.03$, we have

$$dW = 4(2.00)^3(0.03) = 0.96 \text{ J}$$ ▪

EXERCISES FOR SECTION 4.8

Find the differential of the expressions in Exercises 1–10.

1. $y = 4x^2$

2. $p = t^3 + 4t$

3. $z = \dfrac{1}{\sqrt{x}}$

4. $y = \dfrac{x^3 - x}{\sqrt{x}}$

5. $s = \sqrt{3t}$

6. $\theta = (\phi + 4)^2$

7. $a = 4\sqrt[3]{b^2}$

8. $y = t^4 - \dfrac{1}{t^2}$

9. $m = p^{-3} - p^{-2} - 4p$

10. $i = 4\pi t^2$

In Exercises 11–15, using differentials, find the approximate change in the function for the indicated change in the independent variable.

11. $y = 4x^3$, for $x = 5.0$ to $x = 5.1$

12. $z = 3t^{4/3}$, for $t = 8.00$ to $t = 8.04$

13. $s = \sqrt{t}$, for $t = 8.95$ to $t = 9.00$

14. $p = r^4 - 3r^2$, for $r = 3.98$ to $r = 4.01$

15. $y = (x + 2)^2$, for $x = 15.0$ to $x = 14.8$

16. The horsepower of an internal combustion engine is given by $P = nd^2$, where n is the number of cylinders and d is the bore of each cylinder. What is the approximate increase in horsepower for an eight-cylinder engine if the cylinders are rebored from 3.500 to 3.525 in.?

17. The current (in amperes) in a circuit changes according to $i = 0.1t^4 + 5$. What is the approximate change in current from $t = 4.00$ to $t = 4.02$ sec?

18. The acceleration (in feet per second per second) of an object is $a = (t + 2)/\sqrt{t}$. What is the approximate change in velocity from $t = 4.00$ to $t = 4.08$ sec?

19. The current (in amperes) in a circuit varies according to $i = (t + 1)^{1/3}$. What is the approximate charge transferred between $t = 0.500$ and $t = 0.501$ sec?

20. The power (in watts) in a resistor is given by $p = ri^2$, where r is the resistance (in ohms) and i is the current (in amperes). The current in a 750-Ω resistor changes from 3.000 to 3.005 A. What is the approximate change in power?

21. The heat radiation R from an incandescent lamp is $R = \sigma T^4$, where σ is a constant and T is the operating temperature. What is the approximate change in radiation if the temperature changes from $T = 2500°$ to $T = 2510°$. Assume that $\sigma = 5 \times 10^{-10}$.

22. A circular disk is supposed to have a radius of 2.00 in. During the machining operation, the radius is cut to 1.98 in. What is the approximate loss in area?

23. What is the approximate error in the volume of a cubical box if the sides are made 25.10 cm long instead of 25.00 cm?

24. A 15-in. square plate is to be constructed so that its area will not be in error by more than 0.03 in.^{2}. What is the approximate tolerance that must be placed on the side length?

25. A spherical oil tank is to be constructed with a radius of 25 ft. What is the approximate error in the volume of the tank if it is constructed with a radius of 24 ft?

26. The force of attraction between two unlike magnetic poles is given by $F = 10/s^2$, where F is the force (in dynes) and s is the distance (in centimeters) separating the two poles. The poles are initially 20 cm apart. What is the approximate change in the force of attraction when $ds = 0.5$ cm?

27. A circular ring has an inside radius of 5 in. and an outside radius of 10 in. What change in the inside radius will decrease the area of the ring by 1 in.^{2}?

28. In Exercise 27, what change in outside radius will decrease the area by 1 in.^{2}?

29. Refer to Exercise 27. Assume that the inside radius and the outside radius change simultaneously by equal amounts. What equal changes in the radii will decrease the area by 1 in.^{2}?

REVIEW EXERCISES FOR CHAPTER 4

1. The velocity (in centimeters per second) of a particle is given by $v = 4t - 5t^2$, where t is the elapsed time (in seconds). Find the equation for the acceleration.

2. The displacement (in feet) of an object varies with the elapsed time (in seconds) according to $s = \sqrt{t}$. Find the expressions for the velocity and the acceleration.

3. What are the displacement, velocity, and acceleration of an object at $t = 0.5$ sec if the displacement equation is given by $s = 0.2t^{2.7}$. Displacement is in feet.

4. What are the displacement, velocity, and acceleration of an object at $t = 3$ sec if the displacement equation is $s = 0.05t^4 + 2t$? Displacement is in meters.

Exercises 5–10

5. What is the current in a 1-millifarad (0.001 F) capacitor at $t = 8$ sec if the applied voltage (in volts) is given by $v = 6 + \sqrt[3]{t}$?

6. The current (in amperes) in a circuit is $i = \sqrt{\frac{1}{2}t + 2}$ where t is in seconds. The circuit contains an inductance of 5 H. What is the induced voltage when $t = 0.1$ sec?

7. What is the equation for the induced voltage in a coil of wire having an inductance of 2 H if the current (in amperes) in the coil varies according to $i = (t^2 + t)^3$?

8. What is the current in a circuit at $t = 0.5$ sec if the electric charge (in coulombs) varies with time (in seconds) according to $q = t^2 + 3t + 1$?

9. The current (in amperes) through an inductance coil of 5 millihenrys (0.005 H) is given by $i = (t^2 + 3)^4$, where t is in seconds. Find the voltage across the coil when $t = 1$ sec.

10. The charge (in coulombs) that is transferred in a circuit varies with time as $q = \sqrt{(t^2 + 1)}/(t^3 + 4)$. Find the expression for the current.

11. Each side of a square is increasing at a constant rate of 3 cm/sec. How fast is the area increasing when $x = 15$ cm?

12. A spherical balloon is being filled with gas at a constant rate of 5 ft³/sec. How fast is the surface of the balloon increasing when the radius is 4 ft?

13. A boat is pulled in by a rope from a dock that is 6 ft above the surface of the water. The rope is pulled in at the rate of 3 ft/sec. How fast is the boat approaching the dock when it is 50 ft away?

14. A balloon released 200 ft east of an observer rises vertically at a constant rate of 16 ft/sec. How fast is the distance from the observer to the balloon changing 0.5 min after the balloon is released?

15. Locate any relative extremes and points of inflection of $y = x^3 - 24x^2 - 5$. Sketch the graph.

16. Sketch the graph of $y = \frac{1}{4}x^4 - x^3 - 2$. Indicate relative extremes, points of inflection, and intercepts.

17. Sketch the graph of $y = x(x^2 - 4)$. Locate relative extremes, points of inflection, and intercepts.

18. The work being done (in foot-pounds) by a machine varies with time (in seconds) according to

$$w = 18t^2 + 4t^3 - t^4$$

What is the maximum power output? Assume $t \geq 0$.

19. The velocity (in feet per second) of an object varies with time according to $v = 3t^2 - t^3$. Find the maximum acceleration of the object. At what time does it occur?

20. A cylindrical can, open at one end, is to have a surface area of 50 in.^{2}. Determine the dimensions that will give the maximum volume.

21. A wire of length L is to be divided into two pieces. One piece is to be used to construct the perimeter of a square, the other to make a circle. How should the division be made to yield the maximum total area?

Evaluate the following differentials.

22. $d\left(3x^5 + \dfrac{2}{x}\right)$

23. $d(\sqrt{x} + 3\sqrt[4]{x})$

24. $d(\sqrt{2 - 3x})$

25. $d(\sqrt[3]{1 - 5x})$

26. The current (in amperes) in a circuit varies with time according to $i = 0.5t^3 + 2t^2$. What is the approximate change in current from 1.99 to 2.00 sec?

27. A ball bearing is to have a diameter of 0.1 in. Estimate the error in the volume given that the actual diameter is 0.096 in.

28. The graph of a polynomial function is shown in the accompanying figure. Answer the following questions.

(a) Assuming the polynomial has only real roots, what is the degree of $f(x)$?
(b) For which values of x is $f(x) = 0$?
(c) For which values of x is $f'(x) = 0$?
(d) Where is $f''(x) = 0$?

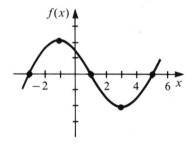

CHAPTER 5

The Antiderivative and Its Applications

In differential calculus, you learn how to find the derivative of a function. Perhaps you have wondered if it is possible to reconstruct a function when you know its derivative. The answer to this question is yes. The process is called **antidifferentiation** or **integration.**

5.1 THE ANTIDERIVATIVE OR INTEGRAL OF A FUNCTION

Antidifferentiation is quite simply the reverse of differentiation. Thus, antidifferentiation problems are essentially equivalent to the following problem: Given $f(x)$, find a function F such that $F'(x) = f(x)$. The function F is called the **antiderivative** or **integral** of f. There are two commonly used notations for the antiderivative of $f(x)$, namely,

$$D_x^{-1} f(x) \tag{5.1}$$

and

$$\int f(x)\, dx \tag{5.2}$$

Comment: Because of its wide acceptance in the field of engineering, we will use $\int f(x)\, dx$ as the symbol the antiderivative of $f(x)$. Furthermore, we will use the terminology *antiderivative* and *integral* interchangeably.

We need a brief discussion on the notation here. The symbol $\int$ is called an **integral sign.** The function $f(x)$ is called the **integrand**, and the dx indicates that the variable of integration is x. The reason for such a seemingly artificial notation for the antiderivative will become clearer in the next chapter.

From Chapter 3 and the definition of the antiderivative it follows that

$$\text{since} \quad \frac{d}{dx}(x^3) = 3x^2 \quad \text{we have} \quad \int 3x^2\, dx = x^3$$

$$\text{since} \quad \frac{d}{dx}(x^3 + 2) = 3x^2 \quad \text{we have} \quad \int 3x^2 \, dx = x^3 + 2$$

$$\text{since} \quad \frac{d}{dx}(x^3 - \sqrt{8}) = 3x^2 \quad \text{we have} \quad \int 3x^2 \, dx = x^3 - \sqrt{8}$$

Notice that $\int 3x^2 \, dx$ has three "correct" answers depending on which constant term appears in the original function. It has three answers because the derivative of a constant is zero. Since, in general, we will have no prior knowledge of the constant term, we write the integral of $3x^2$ as

$$\int 3x^2 \, dx = x^3 + C$$

where C may be *any* arbitrary constant. In this way, we account for the existence of a constant term in the function we are trying to reconstruct. A constant arising in this way is called the **constant of integration.**

Geometrically, the constant of integration has the effect of raising or lowering the graph of $y = f(x)$; the shape of the graph remains the same. Therefore, $y = x^3 + C$ represents a family of curves differing only in their vertical location in the plane. Several members of the family $y = x^3 + C$ are shown in Figure 5.1.

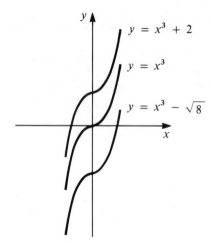

$y = x^3 + 2$

$y = x^3$

$y = x^3 - \sqrt{8}$

■ **Figure 5.1**

Just as we have formulas for differentiation, we have formulas for integration. At present, our best source for such formulas is a differentiation formula. Thus, since

$$\frac{d}{dx}\left(\frac{x^{n+1}}{n + 1}\right) = \frac{(n + 1)x^{(n+1)-1}}{n + 1} = x^n$$

for all n except $n = -1$, we have the following formula.

FORMULA	**Integral of x^n**
	$$\int x^n \, dx = \frac{x^{n+1}}{n+1} + C, \qquad n \neq -1 \qquad\qquad (5.3)$$

The exceptional case, $n = -1$, is quite important but must be studied at a later time.

EXAMPLE 1   ■ **Problem**

Find $\int x^5 \, dx$.

■ **Solution**

Using Formula (5.3), we find

$$\int x^5 \, dx = \frac{x^6}{6} + C$$

■

EXAMPLE 2   ■ **Problem**

Find $\int \frac{1}{x^3} \, dx$.

■ **Solution**

We use formula (5.3) to find

$$\int \frac{1}{x^3} \, dx = \int x^{-3} \, dx = \frac{x^{-2}}{-2} + C$$

$$= \frac{-1}{2x^2} + C$$

■

EXAMPLE 3   ■ **Problem**

Find $\int \sqrt{x} \, dx$.

■ **Solution**

Again, we use Formula (5.3).

$$\int \sqrt{x} \, dx = \int x^{1/2} \, dx = \frac{x^{3/2}}{3/2} + C$$

$$= \frac{2}{3} x^{3/2} + C$$

■

The following linearity properties are useful in taking integrals of sums.

PROPERTIES

> **Linearity Properties**
>
> 1. If c is a constant, then $\int cf(x)\,dx = c\int f(x)\,dx$.
> 2. $\int [f(x) + g(x)]\,dx = \int f(x)\,dx + \int g(x)\,dx$.

EXAMPLE 4   ■ **Problem**

Find $\int (4x^2 + 5x + 6)\,dx$.

■ **Solution**

Here, we apply both linearity properties to obtain

$$\int (4x^2 + 5x + 6)\,dx = 4\int x^2\,dx + 5\int x\,dx + 6\int dx$$

$$= \frac{4}{3}x^3 + \frac{5}{2}x^2 + 6x + C$$

Only one constant is written in this solution because the sum of three arbitrary constants is, in turn, an arbitrary constant.   ■

EXAMPLE 5   ■ **Problem**

Find $\int \dfrac{5\,dt}{t^\pi}$.

■ **Solution**

Use Formula (5.3) and Property 1.

$$\int \frac{5\,dt}{t^\pi} = 5\int t^{-\pi}\,dt = \frac{5t^{-\pi+1}}{-\pi + 1} + C$$

Beginning students often forget that when they apply Property 1, the C must be a constant. That is, only constants may be moved across the integration symbol.   ■

EXAMPLE 6   ■ **Problem**

Find $\int x(x - 1)\,dx$.

■ **Solution**

Note that

$$\int x(x - 1)\,dx \qquad \text{is } not \text{ equal to} \qquad x\int (x - 1)\,dx$$

since the *variable x* may not be moved across the integral sign. Furthermore, we note that

$$\int x(x - 1)\, dx \quad \text{is } not \text{ equal to} \quad (\int x\, dx) \cdot (\int (x - 1)\, dx)$$

since the integral of a product is not the product of integrals.

To find this integral, we first expand the given product and then use the linearity properties. Thus,

$$\int x(x - 1)\, dx = \int (x^2 - x)\, dx = \tfrac{1}{3} x^3 - \tfrac{1}{2} x^2 + C \qquad \blacksquare$$

EXAMPLE 7 ■ **Problem**

Find $\int (2x - 3)^2\, dx$.

■ **Solution**

First, expand the binomial. Then, evaluate the integrals. Thus,

$$\int (2x - 3)^2\, dx = \int (4x^2 - 12x + 9)\, dx$$
$$= 4 \int x^2\, dx - 12 \int x\, dx + 9 \int dx$$
$$= \tfrac{4}{3} x^3 - 6x^2 + 9x + C \qquad \blacksquare$$

EXAMPLE 8 ■ **Problem**

Find $\int x^{1/3} (x^2 - x^{-1})\, dx$.

■ **Solution**

This integral may be determined if we first perform the indicated multiplication of $(x^2 - x^{-1})$ by $x^{1/3}$. Then,

$$\int x^{1/3} (x^2 - x^{-1})\, dx = \int (x^{7/3} - x^{-2/3})\, dx = \int x^{7/3}\, dx - \int x^{-2/3}\, dx$$

$$= \frac{x^{10/3}}{10/3} - \frac{x^{1/3}}{1/3} + C = \frac{3}{10} x^{10/3} - 3x^{1/3} + C \qquad \blacksquare$$

To determine a specific member of a family, we must have additional information to find a particular value of C. Typically, we are given the coordinates of a point on the graph of a member of the family.

EXAMPLE 9 ■ **Problem**

Find the equation of the graph passing through the point $(2, 5)$ if the slope of the curve is given by $3x$.

■ **Solution**

The slope of a curve is given by its derivative, so we know that

$$\frac{dy}{dx} = 3x$$

The family of functions having this derivative is

$$y = \int 3x \, dx = \tfrac{3}{2} x^2 + C \tag{5.4}$$

Since the curve passes through (2, 5), the coordinates of this point must satisfy (5.4). Substituting $x = 2$ and $f(x) = 5$, we get

$$5 = \tfrac{3}{2} (2)^2 + C$$

or

$$C = -1$$

The desired equation is then $y = \tfrac{3}{2} x^2 - 1$; its graph is shown in Figure 5.2.

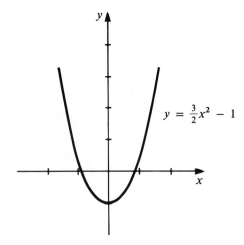

■ **Figure 5.2**

■

![y = 3/2 x² − 1 label inside figure] $y = \frac{3}{2}x^2 - 1$

EXERCISES FOR SECTION 5.1

Find the following integrals.

1. $\displaystyle\int 3 \, dx$

2. $\displaystyle\int y \, dy$

3. $\displaystyle\int 3z \, dz$

4. $\displaystyle\int (z^2 - 3) \, dz$

5. $\displaystyle\int (x^3 + x^2) \, dx$

6. $\displaystyle\int 5x^4 \, dx$

7. $\int (3x^7 + 6x^9) \, dx$

8. $\int \frac{x^3}{2} \, dx$

9. $\int p^{1/3} \, dp$

10. $\int 4r^{4/5} \, dr$

11. $\int \sqrt{3\phi} \, d\phi$

12. $\int q^{-3} \, dq$

13. $\int \frac{dx}{x^5}$

14. $\int \frac{dy}{\sqrt{y}}$

15. $\int x(x - x^3) \, dx$

16. $\int x \sqrt{x} \, dx$

17. $\int \frac{\theta^2 - 2\theta^3}{\theta^5} \, d\theta$

18. $\int \frac{dt}{2 \sqrt{t^3}}$

19. $\int (x - 3)^2 \, dx$

20. $\int x(3x + 2)^2 \, dx$

21. $\int \frac{(x^2 + 1)^2}{x^2} \, dx$

22. $\int (\sqrt{x} - 3)^2 \, dx$

23. $\int \frac{2x^3 - 3x^2 + x}{\sqrt[3]{x}} \, dx$

24. $\int \frac{3 \, dx}{\sqrt[3]{x^2}}$

25. $\int \left(x^3 + \frac{1}{x^3} \right) dx$

26. $\int (\sqrt[3]{x} - x) \sqrt{x} \, dx$

27. $\int (t^3)^4 \, dt$

28. $\int (\sqrt{t^5})^{1/3} \, dt$

29. $\int (\sqrt[3]{\phi} + \sqrt[4]{\phi})^2 \, d\phi$

30. $\int (x^3 - \sqrt{x})(x - 4) \, dx$

31. $\int p^{1/8} (1 - p^{-1/3}) \, dp$

32. $\int (x - 2)(x + 1)^2 \, dx$

33. $\int x^{0.35} \, dx$

34. $\int 0.23t^{1.4} \, dt$

35. $\int (17 + x^{-0.25}) \, dx$

36. $\int \frac{ds}{0.2s^{0.9}}$

37. $\int x^e \, dx$

38. $\int t^\pi \, dt$

In Exercises 39–43, determine the equation of the graph having the given slope and passing through the indicated point.

39. $\dfrac{dy}{dx} = 3x$; (1, 2)

40. $\dfrac{dq}{dt} = 4t^2 + 5$; (0, 3)

41. $\dfrac{ds}{dr} = \sqrt[3]{r}$; (8, −2)

42. $\dfrac{dy}{dx} = x^3 - 4x$; (2, 0)

43. $\dfrac{dp}{dt} = \sqrt{t} + \dfrac{1}{\sqrt{t}}$; (4, 1)

■ 5.2 THE INTEGRAL OF $u^n\, du$

In this section, we discuss the integration formula for integrand functions of the form $u^n\, du$, where u is a function of x. We obtain this formula by differentiating the function $u^{n+1}/(n + 1)$. Thus,

$$\frac{d}{dx}\left(\frac{u^{n+1}}{n + 1}\right) = u^n \frac{du}{dx}$$

Therefore, the integral of $u^n\, du/dx$ is

$$\int u^n \frac{du}{dx}\, dx = \frac{u^{n+1}}{n + 1} + C, \qquad n \neq -1 \tag{5.5}$$

This expression is commonly shortened to the following more compact form called the power rule.

FORMULA

> **The Power Rule**
>
> $$\int u^n\, du = \frac{u^{n+1}}{n + 1} + C, \qquad n \neq -1 \tag{5.6}$$

Keep in mind that Formula (5.6) is merely a compact form of (5.5). Consequently, the du in (5.6) must be interpreted to be the derivative of u with respect to x times dx; that is, $du = (du/dx)\, dx$. Thus, to use (5.6) for functions taken to a power, we must have the derivative of u as a factor in the integrand. For example, note carefully the difference between the following two integrals:

$$\int (x^2 + 6)^3(2x)\, dx \qquad \text{and} \qquad \int (x^2 + 6)^3\, dx \tag{5.7}$$

The first integral in (5.7) is precisely the form of (5.6), since $(d/dx)(x^2 + 6) = 2x$ is a part of the integrand. Thus,

$$\int (x^2 + 6)^3(2x)\,dx = \tfrac{1}{4}(x^2 + 6)^4 + C$$

a result that can be verified by taking the derivative of $\tfrac{1}{4}(x^2 + 6)^4$. However,

$$\int (x^2 + 6)^3\,dx \quad \text{is } not \text{ equal to} \quad \tfrac{1}{4}(x^2 + 6)^4 + C$$

since the power rule may not be applied unless the derivative of $(x^2 + 6)$ is present in the integrand. We are not saying that the second integral of Equation (5.7) cannot be found but that you may not use (5.6) immediately. In fact, to integrate $(x^2 + 6)^3$, we first expand as follows:

$$\int (x^2 + 6)^3\,dx = \int (x^6 + 18x^4 + 108x^2 + 216)\,dx$$
$$= \tfrac{1}{7}x^7 + \tfrac{18}{5}x^5 + 36x^3 + 216x + C$$

A little ingenuity is often required to obtain the form required by Formula (5.6). When manipulating the integrand, keep in mind that only constants may be brought inside or outside the integral sign without affecting the value. For example, we can put the integral

$$\int (2x + 5)^7\,dx$$

in the form of (5.6) by multiplying inside the integral sign by $\tfrac{2}{2}$ and using Property 1 to write $\tfrac{1}{2}$ outside the integral. Thus,

$$\int (2x + 5)^7\,dx = \frac{1}{2}\int (2x + 5)^7(2\,dx) = \frac{1}{2}\frac{(2x + 5)^8}{8} + C$$
$$= \frac{(2x + 5)^8}{16} + C$$

On the other hand, the integral

$$\int (x^2 + 5)^7\,dx$$

cannot be put into form of (5.6), since you may not take $1/2x$ outside the integral sign while leaving the needed $2x$ inside the integral sign.

EXAMPLE 1   ■ **Problem**

Find $\int x\,(3x^2 - 1)^{1/2}\,dx$.

■ **Solution**

Here, we let $u = 3x^2 - 1$, then $du/dx = 6x$. Thus, to apply Formula (5.6), we need a 6 inside the integral sign. Multiplying inside by 6 and outside by $\tfrac{1}{6}$, we obtain

$$\tfrac{1}{6}\int 6x\,(3x^2 - 1)^{1/2}\,dx$$

Formula (5.6) may now be applied with $n = \tfrac{1}{2}$. The desired integral is given by

$$\frac{1}{6} \int (3x^2 - 1)^{1/2}(6x\,dx) = \frac{1}{6}\frac{(3x^2 - 1)^{3/2}}{3/2} + C = \frac{1}{9}(3x^2 - 1)^{3/2} + C$$

$$\underbrace{\hphantom{(3x^2-1)^{1/2}}}_{u^n} \quad \underbrace{\hphantom{(6x\,dx)}}_{du} \qquad \underbrace{\hphantom{\frac{(3x^2-1)^{3/2}}{3/2}}}_{\dfrac{u^{n+1}}{n+1}}$$

EXAMPLE 2   ■ **Problem**

Find $\displaystyle\int \frac{x^2\,dx}{(x^3 + 1)^2}$.

■ **Solution**

To find this integral, we first rewrite it in the form of $\int x^2(x^3 + 1)^{-2}\,dx$. Then, if we let $u = (x^3 + 1)$, we have $n = -2$ and $du/dx = 3x^2$. Thus, before applying Formula (5.6), we must multiply inside the integral sign by 3 and outside by $\frac{1}{3}$. Remember that Formula (5.6) cannot be applied until the required term (du/dx) is present.

$$\int x^2(x^3 + 1)^{-2}\,dx = \frac{1}{3}\int (x^3 + 1)^{-2}(3x^2\,dx) = \frac{1}{3}\frac{(x^3 + 1)^{-1}}{-1} + C$$

$$= -\frac{1}{3}(x^3 + 1)^{-1} + C \qquad\qquad ■$$

EXAMPLE 3   ■ **Problem**

Show that $\int (x^3 - 3)^2 x^5\,dx$ is *not* of the form $\int u^n\,du$.

■ **Solution**

Letting $u = x^3 - 3$, we have $n = 2$ and $du/dx = 3x^2$. The term x^5 and not $3x^2$ appears in the integrand. *Since a variable cannot be moved across an integral sign,* there is no way that x^5 can be rearranged to give $3x^2$. We must, therefore, conclude that the given integral is not the form $\int u^n\,du$. ■

EXAMPLE 4   ■ **Problem**

Find $\int (x^3 - 3)^2 x^5\,dx$.

■ **Solution**

In the preceding example, we showed that this integral cannot be found by using Formula (5.6). However, it can be found by first expanding the integrand. Thus,

$$\int (x^3 - 3)^2 x^5\,dx = \int (x^{11} - 6x^8 + 9x^5)\,dx = \frac{x^{12}}{12} - \frac{2x^9}{3} + \frac{3x^6}{2} + C \qquad ■$$

EXAMPLE 5 ▪ **Problem**

Find $\int (x^2 + 1)(x^3 + 3x)^{1/2} \, dx$.

▪ **Solution**

We let $u = x^3 + 3x$; then $du/dx = 3x^2 + 3 = 3(x^2 + 1)$. Multiplying by 3 and $\frac{1}{3}$, we write

$$\int (x^2 + 1)(x^3 + 3x)^{1/2} \, dx = \tfrac{1}{3} \int (x^3 + 3x)^{1/2} (3)(x^2 + 1) \, dx$$

Now, applying Formula (5.6), we get

$$\int (x^2 + 1)(x^3 + 3x)^{1/2} \, dx = \frac{1}{3} \frac{(x^3 + 3x)^{3/2}}{3/2} + C$$

$$= \frac{2}{9} (x^3 + 3x)^{3/2} + C \qquad \blacksquare$$

EXAMPLE 6 ▪ **Problem**

Given $y' = \sqrt{2x + 1}$, find the function whose graph passes through $(0, 2)$.

▪ **Solution**

The family of functions is given by

$$y = \int \sqrt{2x + 1} \, dx = \int (2x + 1)^{1/2} \, dx = \tfrac{1}{2} \int (2x + 1)^{1/2}(2 \, dx)$$

$$= \tfrac{1}{3} (2x + 1)^{3/2} + C$$

Letting $x = 0$ and $y = 2$, we get

$$2 = \tfrac{1}{3} (1)^{3/2} + C \qquad \text{or} \qquad C = \tfrac{5}{3}$$

The desired function is $y = \tfrac{1}{3} (2x + 1)^{3/2} + \tfrac{5}{3}$. $\qquad \blacksquare$

▬▬▬▬ EXERCISES FOR SECTION 5.2

In Exercises 1–26, find the indicated integral.

1. $\displaystyle \int (2x + 1)^3 \, dx$

2. $\displaystyle \int (2 + 3t)^4 \, dt$

3. $\displaystyle \int (1 - 5x)^3 \, dx$

4. $\displaystyle \int (a - bx)^2 \, dx$

5. $\displaystyle \int \sqrt{3 + 4p} \, dp$

6. $\displaystyle \int (5x + 1)^{1/3} \, dx$

7. $\displaystyle \int x(x^2 + 1)^4 \, dx$

8. $\displaystyle \int x^3(4 + x^4)^{1/4} \, dx$

9. $\int (z + 1)\sqrt[3]{z^2 + 2z} \, dz$

10. $\int (3y + 5)^{-2} \, dy$

11. $\int \dfrac{ds}{(1 - s)^2}$

12. $\int \dfrac{s \, ds}{\sqrt{16 + s^2}}$

13. $\int \sqrt[3]{4 + 2x} \, dx$

14. $\int (\sqrt{x} + 1)^2 \, dx$

15. $\int \dfrac{dx}{(2 + 4x)^4}$

16. $\int \dfrac{3t \, dt}{(t^2 + 4)^2}$

17. $\int \dfrac{4r^2 \, dr}{\sqrt{r^3 + 1}}$

18. $\int \dfrac{(x^2 - 1) \, dx}{\sqrt[3]{x^3 - 3x}}$

19. $\int \dfrac{(4 + \sqrt{x})^2 \, dx}{\sqrt{x}}$

20. $\int \dfrac{\sqrt{x^{1/3} + 4} \, dx}{x^{2/3}}$

21. $\int (x^2 + 4)^2 \, dx$

22. $\int \sqrt{x} \, (1 + \sqrt{x})^2 \, dx$

23. $\int x(x^2 - 3)^2 \, dx$

24. $\int x^2 (3 - x^2)^2 \, dx$

25. $\int (x^2 + 2)^3 \, dx$

26. $\int (x^2 - 3)^3 \, dx$

In Exercises 27–30, determine the equation of the graph having the given slope and passing through the indicated point.

27. $y' = (3x + 2)^3$; $(0, 5)$

28. $\dfrac{dy}{dx} = (1 - x)^4$; $(1, -2)$

29. $\dfrac{ds}{dt} = \sqrt{\dfrac{1}{2}t + 1}$; $(6, 0)$

30. $\dfrac{di}{dt} = \dfrac{2t}{\sqrt{t^2 + 5}}$; $(2, 3)$

31. Explain why $\int x \sqrt{2x + 3} \, dx$ cannot be found by using $\int u^n \, du$.

32. Explain why $\int (2 - x^3)^{-2} \, dx$ cannot be found by using $\int u^n \, du$.

33. Integrals of the type $\int dt/(5 - t)^2$ occur in the analysis of certain variable inductance electric circuits. Find the indicated integral.

34. Integrals of the type $\int \sqrt{4t - 3} \, dt$ occur when finding the distance traveled by certain charged particles. Find the indicated integral.

35. Find the family of functions whose derivative is $dy/dx = x/\sqrt{x^2 + 1}$.

36. Find the family of functions whose derivative is $y' = x^3(3x^4 - 2)^2$.

◼◼◼◼ 5.3 LINEAR MOTION

The study of the motion of an object without regard to the forces causing the motion is called *kinematics*. That is, kinematics is the study of the relationship between displacement, velocity, and acceleration of an object. We have already seen how we can find the velocity and the acceleration from the distance-time function.

Integration allows us to reverse this process and find the velocity from the acceleration and the displacement if we know the velocity. Stating this process symbolically, since $v = ds/dt$, we have the following formula.

FORMULA

> **Displacement Formula**
>
> $$s = \int v\, dt \qquad\qquad\qquad\qquad (5.8)$$

Equation (5.8) says that the displacement-time function is the antiderivative of the velocity-time function. Similarly, the velocity-time function is the antiderivative of the acceleration-time function. That is, since $a = dv/dt$, we have the following formula.

FORMULA

> **Velocity Formula**
>
> $$v = \int a\, dt \qquad\qquad\qquad\qquad (5.9)$$

The following examples will help to explain the use of Equations (5.8) and (5.9) in solving kinematic problems.

EXAMPLE 1 ◼ **Problem**

The velocity of a car changes with time according to $v = 180\sqrt{t}$, where v is in feet per second when t is in seconds. Assuming the initial displacement of the car is zero, find the expression for its displacement as a function of time.

◼ **Solution**

Writing the velocity function in the form $v = 180t^{1/2}$, we have

$$s = \int 180t^{1/2}\, dt = 180\left(\frac{t^{3/2}}{3/2}\right) + C = 120t^{3/2} + C$$

Since $s = 0$ when $t = 0$, we get

$$s = 120t^{3/2}$$

as the desired function.   ■

EXAMPLE 2   ■ **Problem**

The acceleration (in centimeters per second squared) of a particle is given by $a = 6t + 10$. The initial velocity of the particle is 3 cm/sec, and the initial displacement is zero. How far has the particle moved after 2 sec, and what is its velocity at this time?

■ **Solution**

The equation for the velocity is

$$v = \int a\, dt = \int (6t + 10)\, dt = 3t^2 + 10t + K$$

The constant K is evaluated by using the condition $v = 3$ when $t = 0$. Thus,

$$3 = 3(0)^2 + 10(0) + K$$

or

$$K = 3$$

and, therefore, the velocity (in cm/sec) is given by the function

$$v = 3t^2 + 10t + 3$$

We now use the velocity-time function to find the distance-time function. That is,

$$s = \int v\, dt = \int (3t^2 + 10t + 3)\, dt = t^3 + 5t^2 + 3t + K_1$$

Substituting the condition $s = 0$ when $t = 0$, we have

$$K_1 = 0$$

and therefore,

$$s = t^3 + 5t^2 + 3t$$

The distance traveled by the particle in 2 sec is then

$$s = (2)^3 + 5(2)^2 + 3(2) = 34 \text{ cm}$$

And its velocity at that time is

$$v = 3(2)^2 + 10(2) + 3 = 35 \text{ cm/sec}$$ ■

Before we give another example, we consider some facts about objects moving through the atmosphere without a propelling force, such as a thrown ball or a bullet after it leaves the gun barrel. Objects of this type are called **projectiles**. First, note that all projectiles have a constant acceleration toward the earth of approximately 32 ft/sec^2. Second, if a projectile is thrown upward, it will be decelerated 32 ft/sec^2 until its velocity is zero — at that instant it starts to fall. It is then accelerated 32 ft/sec^2 until it strikes the ground. When it strikes the ground, its speed is the same as it was when it was launched upward.

EXAMPLE 3   ■ **Problem**

A bullet is fired vertically upward from the roof of a building that is 75 ft high. The bullet has an initial velocity of 500 ft/sec.

a.   What is the velocity of the bullet after 10 sec?

b.   How long does it take the bullet to reach its maximum altitude?

c.   How high is the bullet above the ground when it reaches maximum altitude?

■ **Solution**

a.   The acceleration function in this case is

$$a = -32$$

The negative sign indicates that the bullet will be decelerated. The equation for the velocity is then given by

$$v = \int a \, dt = \int (-32) \, dt = -32t + C_1$$

We can evaluate the constant of integration by using $t = 0$ and $v = 500$. Thus,

$$500 = -32(0) + C_1$$
$$C_1 = 500$$

and, therefore, the velocity (in ft/sec) is given by

$$v = -32t + 500 \tag{5.10}$$

The velocity when $t = 10$ sec is then

$$v = -32(10) + 500 = 180 \text{ ft/sec}$$

b.   The time required for the bullet to reach maximum altitude is found by letting $v = 0$ in Equation (5.10) and solving for t. We let $v = 0$ because the instantaneous velocity of the bullet at its maximum altitude is zero. Thus,

$$0 = -32t + 500$$

and

$$t = \frac{-500}{-32} \approx 15.6 \text{ sec}$$

c.   To find the maximum altitude above the ground, we need to know the displacement function. This function is found by antidifferentiating the velocity function; that is,

$$s = \int v \, dt = \int (-32t + 500) \, dt = -16t^2 + 500t + C_2$$

We evaluate C_2 by noting that the altitude above the ground when $t = 0$ is $s = 75$. (The bullet is already 75 ft above the ground when the gun is fired.) So,

$$75 = -16(0)^2 + 500(0) + C_2$$
$$C_2 = 75$$

Therefore,

$$s = -16t^2 + 500t + 75$$

Substituting $t = 15.6$ sec into this equation, we get

$$s = -16(15.6)^2 + 500(15.6) + 75 \approx 3980 \text{ ft} \qquad \blacksquare$$

Comment: Compare this example with Example 3, Section 4.1. In that example, we began with the distance function and obtained the acceleration function. Here, we reversed the process.

EXERCISES FOR SECTION 5.3

Exercises 1–22.

1. The acceleration of a rocket sled starting from rest is $a = \sqrt{t}$, where t is the time in seconds. What is the equation for the velocity of the sled?

2. Find the equation for the displacement of the sled in Exercise 1.

3. A car starts from rest and accelerates 8 ft/sec^2. What is the elapsed time for the car to reach a velocity of 88 ft/sec? (Note: 88 ft/sec = 60 mi/hr.)

4. In Exercise 3, how far has the car traveled from the starting line by the time it reaches a velocity of 100 ft/sec?

5. An airplane lands with a velocity of 200 ft/sec. At the instant of touchdown the airplane decelerates 10 ft/sec^2 until it comes to rest. How far does the plane taxi before coming to rest?

6. A new braking system has been designed to decelerate an airplane 4 ft/sec^2. What is the minimum runway necessary for planes landing with a velocity of 200 ft/sec?

7. The velocity of a particle is equal to the square of the elapsed time. What is the equation for the displacement of the particle if $t = 2, s = 3$ is a boundary condition?

8. A stone is dropped from the top of a building. The building is 50 ft high. How long will it take the stone to reach the ground?

9. How long will it take the stone in Exercise 8 to reach the ground if it is thrown downward with a velocity of 5 ft/sec?

10. The velocity of a particle is $v = t^{4/3}$, where v is in feet per second and t is in seconds. Find the acceleration of the particle when $t = 27$ sec.

11. A baseball is thrown vertically upward with a velocity of 96 ft/sec. Find the velocity of the ball after 2 sec.

12. In Exercise 11, determine the maximum altitude reached by the ball. How long does it take to reach this height?

13. The kinetic energy of a moving object is the capacity of the object to do work by virtue of its motion. The formula for kinetic energy is $KE = wv^2/2g$, where w is the weight, v is the velocity, and g is the acceleration of gravity. A 2000-lb car starts from rest and accelerates according to $a = \frac{1}{4}t^{-1/2}$. What is its kinetic energy when $t = 2$ sec? Assume $g = 32$ ft/sec^2.

14. What is the kinetic energy of a 2-lb rocket 3 sec after it is launched if its acceleration is given by $a = 3t^3$. Assume a is in feet per second squared and t is time in seconds.

15. The Mach number N of an object moving through the atmosphere is defined by $N = v/v_s$, where v is the velocity of the object and v_s is the speed of sound. Under standard conditions, $v_s = 1110$ ft/sec. A meteorological probe is dropped from an altitude of 10,000 ft. What is the Mach number of the probe 20 sec after it is released if its acceleration (in feet per second squared) is $a = 32 - 0.1t$?

16. A missile accelerates from rest according to $a = 20t + 100$ (in feet per second squared). How long does it take the missile to reach Mach 1?

17. A spaceship moving through a force field experiences an acceleration (in meters per second squared) of $a = 300 - 2t$. What is the velocity equation for the spaceship if its initial velocity is 3000 m/sec? Assume t is in seconds.

18. What is the maximum velocity of the spaceship in Exercise 17?

19. The velocity (in feet per second) of an object is given by the equation $v = (5.1t + 1)^{0.7}$. If $s = 25$ ft when $t = 2.8$ sec, what is the equation for the displacement of the object?

20. The acceleration (in centimeters per second squared) of a particle is given by $a = (0.3t + 0.2)^{0.3}$. What is the velocity equation if $v = 51.8$ cm/sec when $t = 1.5$ sec? What are the velocity and the acceleration of the particle when $t = 1.8$ sec?

21. Power is the rate of change of work, so it follows that work is the integral of power. What is the work equation of a particle if its power equation is $p = \sqrt{0.01t + 1}$, where p is in foot-pounds per second and t is in seconds? Assume $W = 0$ when $t = 0$.

22. The power (in foot-pounds per second) generated by a mechanical system is given by $p = 5(2t + 1)^{0.2}$. What is the work equation of the system if $W = 0$ when $t = 0$?

■■■■ 5.4 SERIES AND PARALLEL CIRCUITS

The many electronic devices on the market today are made up of varies series and parallel combinations of resistors, inductors, and capacitors. The relationship that

exists between current and voltage in these RLC* combinations is of primary importance in the design of electronic equipment. To determine the current in either a series or a parallel RLC circuit when a voltage is applied to the combination, we use
two laws basic to the study of series and parallel circuits: Kirchhoff's laws. In this
section, we will describe how to use Kirchhoff's laws, but first we extend the electrical concepts introduced in Section 4.2.

The electric charge transferred past to a point can be found from the definition of current. Recall that current is defined by $i = dq/dt$. The equation for the
charge is then found by taking the antiderivative of the current-time function. Symbolically, this formula is as follows:

FORMULA

> **Electric Charge Formula**
>
> $$q = \int i \, dt \qquad\qquad\qquad (5.11)$$

EXAMPLE 1   ■ **Problem**

The current (in amperes) in a resistor is given by $i = \sqrt{2t + 9}$. What is the equation for the charge if $q = 0$ when $t = 0$?

■ **Solution**

Using $i = (2t + 9)^{1/2}$ in Equation (5.11), we have

$$q = \int i \, dt = \int (2t + 9)^{1/2} \, dt = \frac{1}{2} \int 2(2t + 9)^{1/2} \, dt$$
$$= \frac{1}{2} \frac{(2t + 9)^{3/2}}{3/2} + K = \frac{1}{3} (2t + 9)^{3/2} + K$$

where K is the arbitrary constant. Letting q and t equal zero, we have

$$0 = \tfrac{1}{3} (9)^{3/2} + K \qquad \text{or} \qquad K = -9$$

Therefore, the equation for the charge is

$$q = \tfrac{1}{3} (2t + 9)^{3/2} - 9 \qquad\qquad\qquad ■$$

Another relationship that we recall from Section 4.2 is

$$v = L \frac{di}{dt}$$

*The letters RLC are used to denote the combination of a resistor, an inductor, and a capaci-
tor.

which gives the voltage induced in a coil of inductance L by a given current.

$$\frac{di}{dt} = \frac{1}{L} v$$

So the current required to cause a voltage drop v is given by the following formula.

FORMULA

Current in an Inductor

$$i = \frac{1}{L} \int v \, dt \qquad\qquad (5.12)$$

Similarly, we have the current in a capacitor given by $i = C \, (dv/dt)$, and therefore, the equation for the voltage across the plates of the capacitor can be written as follows.

FORMULA

Voltage across the Plates of a Capacitor

$$v = \frac{1}{C} \int i \, dt \qquad\qquad (5.13)$$

Now, we are ready to discuss Kirchhoff's two laws. *Kirchhoff's first law* states that at any instant, the algebraic sum of the currents flowing toward any point in a circuit is equal to zero. This law is often written in the form

$$\Sigma \text{ currents} = 0$$

where the Greek letter sigma, Σ, is used to mean "the sum of." This symbol will be explained in more detail in the next chapter.

To illustrate the use of Kirchhoff's first law, we consider the circuit in Figure 5.3. This figure shows a generator connected to a parallel *RLC* circuit. If the generator supplies a voltage v to the parallel circuit, a current will flow in the exterior circuit and in each branch of the parallel arrangement. Kirchhoff's first law indicates that the algebraic sum of currents at point P will equal zero. For convenience, we let currents moving toward the point be positive and currents moving away from the point be negative. Using this convention in Figure 5.3, we get

$$\Sigma i = i - i_R - i_L - i_C = 0 \qquad\qquad (5.14)$$

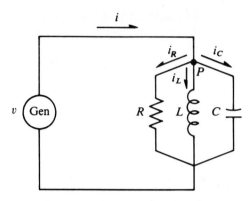

■ Figure 5.3

In addition, the voltage drop across each branch of the parallel circuit will be equal to the voltage drop across the terminals of the generator. Thus, the current in each branch can be expressed in terms of the applied voltage. Therefore, the current in the resistor is given by *Ohm's law* to be $i_R = v/R$; the current in the inductance coil is given by $i_L = 1/L \int v \, dt$; and the current in the capacitor is given by $i_C = C \, dv/dt$. Making these substitutions in Equation (5.14), we have the following formula.

FORMULA

> **Current in a Parallel *RLC* Circuit**
>
> $$i = \frac{v}{R} + \frac{1}{L} \int v \, dt + C \frac{dv}{dt} \qquad\qquad (5.15)$$

The total current can then be found if we know the equation for the voltage being applied and the numerical values of R, L, and C.

EXAMPLE 2   ■ **Problem**

The voltage being applied across a parallel *RLC* circuit is given by $v = t^4 + 10$. Find the equation for the total current if $R = 2$ ohms (Ω), $L = 5$ henrys (H), and $C = 10$ farads (F)*. The current is initially zero.

■ **Solution**

This solution is obtained by a direct substitution of the given values into Equation (5.15):

*In practice, a capacitance of 10 F is extremely large. Capacitors are generally of the order of 10^{-6} F. We have ignored this practical consideration in this section so that the numerical calculations will be simpler.

$$i = \frac{v}{R} + C\frac{dv}{dt} + \frac{1}{L}\int v\,dt$$

$$= \frac{(t^4 + 10)}{2} + 10\frac{d}{dt}(t^4 + 10) + \frac{1}{5}\int (t^4 + 10)\,dt$$

$$= \frac{t^4}{2} + 5 + 40t^3 + \frac{t^5}{25} + 2t + K$$

$$= 0.04t^5 + 0.5t^4 + 40t^3 + 2t + 5 + K$$

The value of K can be found by letting $i = 0$ and $t = 0$ in this equation. Thus,

$$K = -5$$

Consequently, the equation for the total current is

$$i = 0.04t^5 + 0.5t^4 + 40t^3 + 2t \qquad ■$$

Kirchhoff's second law is concerned with voltage drops around a closed loop. It states that at any instant, the algebraic sum of voltage drops around the circuit loop is zero. This statement is denoted by

$$\Sigma \text{ voltage drops} = 0 \qquad\qquad (5.16)$$

Kirchhoff's second law can be used to find the voltage being supplied to a series *RLC* circuit like the one shown in Figure 5.4 if we know the equation for the current.

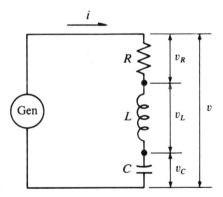

■ **Figure 5.4**

Here, we consider the total voltage drop to be positive and the voltage drops across the individual components to be negative. Using this convention and applying Kirchhoff's second law, we have

$$\Sigma v = v - v_R - v_L - v_C = 0 \qquad\qquad (5.17)$$

Since the components are connected in series, the current in each component is the same as the total current. This fact allows us to write Equation (5.17) as follows:

FORMULA

> **Voltage across a Series RLC Circuit**
>
> $$v = iR + L\frac{di}{dt} + \frac{1}{C}\int i\,dt \qquad\qquad (5.18)$$

EXAMPLE 3   ■ **Problem**

A series RLC circuit has $R = 25\ \Omega$, $L = 3$ H, and $C = 0.5$ F. The current in the circuit is given by $i = 3t$. Find the voltage being applied when $t = 0.2$ sec. Assume the capacitor is initially discharged.

■ **Solution**

The equation for the voltage is found by substituting into Equation (5.18) and solving it for v.

$$v = 25(3t) + 3\frac{d}{dt}(3t) + \frac{1}{0.5}\int 3t\,dt = 75t + 9 + 3t^2 + K$$

$$= 75t + 3t^2 + K_1$$

where $K_1 = 9 + K$. The fact that the capacitor was initially discharged means that the initial voltage is zero. Hence, the value of K_1 is

$$K_1 = 0$$

The equation for the applied voltage is then

$$v = 75t + 3t^2$$

Substituting $t = 0.2$, we have

$$v = 75(0.2) + 3(0.2)^2 = 15.1\text{ V}$$ ■

EXERCISES FOR SECTION 5.4

Exercises 1–20

1. The current (in amperes) in a resistor varies with time (in seconds) according to $i = 0.1t$. What is the charge transferred after 1 sec if the initial charge is 0.25 C?

2. A capacitor contains an initial charge of 0.5 C. The terminals of the capacitor are then connected to a circuit that delivers a current (in amperes) of $i = 0.5t^4 - 0.4t$. What is the charge on the plates of the capacitor after 2 sec?

3. The voltage (in volts) induced in a coil of wire is $v = (3t + 1)^{1/2}$. The current in the coil

is initially $\frac{1}{2}$ A, and the inductance is 4 H. What is the equation for the current in the coil?

4. What is the current in the coil of Exercise 3 after 0.1 sec?

5. The voltage across the plates of a capacitor is initially 6 V. What is the equation for the voltage across the plates of the capacitor if the current (in amperes) in the capacitor is $i = (t + 3)^2$? Assume that the capacitance of the capacitor is 0.5 F.

6. Find the voltage across the plates of the capacitor in Exercise 5 when $t = 0.2$ sec.

7. The power in a resistor is given by $p = ri^2$. Find the energy dissipated by a 100-Ω resistor after 8 sec, given that the current (in amperes) being supplied to the resistor is $i = \sqrt[3]{t}$. Assume that no current flows in the resistor initially.

8. Another expression for the power in an electric circuit is $p = vi$, where v is the voltage in volts and i is the current in amperes. The voltage is $v = 3t + 2$, and the current is $i = \sqrt{t}$. What is the energy W delivered to the circuit in 10 sec? (Let $W = 0$ when $t = 0$.)

9. A 25-Ω resistor and a 5-H inductance coil are connected in parallel, and a voltage (in volts) of $v = 7t^3 + 5$ is applied to the combination. What is the equation of the resulting current if $i = 2$ A when $t = 0$?

10. The current (in amperes) in a series RL circuit is $i = 15\sqrt{t}$. What is the voltage across the combination when $t = 9$ sec if $R = 12\ \Omega$ and $L = 2$ H?

11. Find the formula for the voltage across a series RC circuit required to deliver a current (in amperes) of $i = 5t^3 - 4t + 3$, given that $R = 50\ \Omega$ and $C = 10$ F. Assume that the initial voltage across the circuit is 1 V.

12. Find the voltage in the circuit in Exercise 11 when $t = 1$ sec.

13. The voltage (in volts) across a 200-Ω resistor and a 0.1-F capacitor connected in parallel is $v = 1 - t$. What is the current in the combination after 2 sec?

14. Consider a parallel RLC circuit with $R = 20\ \Omega$, $L = 0.5$ H, and $C = 5$ F. Let the applied voltage (in volts) be $v = t^2 + 4$. Find the formula for the current if $i = 1.5$ A when $t = 0$. (See the accompanying figure.)

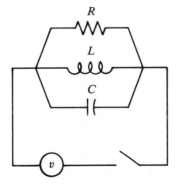

15. The current (in amperes) in a series RLC circuit is to vary according to $i = t^2$. In this

circuit, $R = 20\ \Omega$, $L = 500$ H, and $C = 10^{-2}$ F. Assuming that $v = 0$ when $t = 0$, find the formula for the applied voltage. (See the accompanying figure.)

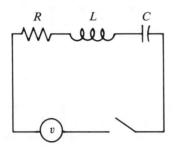

16. Use the expression for the voltage obtained in the Exercise 15 to calculate the voltage drop across the circuit when $t = 0.1$ sec.

17. If $R = 10\ \Omega$, $L = 100$ H, and $C = 10^{-2}$ F, what voltage (in volts) must be applied to a series RLC circuit to yield a current of $i = 3t - 2$? Assume $v = 5$ V when $t = 1$ sec. What is the voltage when $t = 0.1$ sec?

18. In a parallel RLC circuit, $R = 10\ \Omega$, $L = 0.25$ H, and $C = 2$ F. Find the formula for the total current if the applied voltage (in volts) is $v = t^{5/3}$. Assume $i = 0$ when $t = 0$.

19. Consider a parallel RLC circuit with $R = 5\ \Omega$, $L = 0.1$ H, and $C = 3$ F. If $v = t + 4t^2$ is applied to the circuit, what is the rate of change of current with respect to time?

20. What is the time rate of change of current in a parallel RLC circuit with $R = 25\ \Omega$, $L = 0.5$ H, and $C = 0.2$ F if the applied voltage varies with time according to $v = 6 + 1.5t$?

REVIEW EXERCISES FOR CHAPTER 5

Find the integrals in Exercises 1–20.

1. $\displaystyle \int (x - 1)\, dx$

2. $\displaystyle \int x(x^2 + 1)\, dx$

3. $\displaystyle \int (5x - x^3)\, dx$

4. $\displaystyle \int x^{\pi}\, dx$

5. $\displaystyle \int \frac{dx}{4\sqrt{x}}$

6. $\displaystyle \int \frac{x^2 - x - 5}{\sqrt{x}}\, dx$

7. $\displaystyle \int (r^{-1/3} + 5r^{2/3})\, dr$

8. $\displaystyle \int \left(3t + \frac{1}{3t^2}\right) dt$

9. $\displaystyle \int (\sqrt[4]{x} - 9)^2\, dx$

10. $\displaystyle \int \frac{i^{1/2} - i^2}{i}\, di$

11. $\displaystyle \int x^2(x^3 + 1)^2\, dx$

12. $\displaystyle \int x^2\sqrt{x^3 + 1}\, dx$

13. $\int (x^2 + 1)^4 x \, dx$

14. $\int x(x^2 + 1)^{1/4} \, dx$

15. $\int \sqrt[5]{16x + 3} \, dx$

16. $\int \dfrac{dx}{\sqrt{2x + 3}}$

17. $\int \dfrac{4t^2 \, dt}{(5t^3 - 7)^{13}}$

18. $\int (x + 1)(x^2 + 2x)^{10} \, dx$

19. $\int \dfrac{dx}{\sqrt{5x - 7}}$

20. $\int \dfrac{(2x - 1) \, dx}{(x^2 - x - 14)^5}$

21. Find the equation of the curve that passes through $(-1, 2)$ and whose slope is given by $y' = 6x - 5$.

22. Find the equation of the curve passing through $(2, 1)$ whose slope is given by $y' = -x^2$.

 23. The velocity (in feet per second) of a car varies with time according to $v = 3t^2 + 5t$. How far does the car move in the first 3 sec? What are its velocity and acceleration at that time?

 24. Find the expression for the charge (in coulombs) transferred in a circuit, if the current (in amperes) is $i = t^{3/2} + 2$ and $q(1) = 3$.

 25. The acceleration (in ft/sec²) of a rocket sled is $a = \sqrt{t}$. If the sled starts with a velocity of 10 ft/sec, what is the equation for the displacement of the sled?

 26. A bullet is fired vertically upward from the roof of a building 100 ft high with an initial velocity of 750 ft/sec. What is the velocity of the bullet after 5 sec? How long does it take the bullet to reach its maximum altitude, and how high is the bullet above the ground at that time?

 27. The acceleration (in feet per second squared) of an object traveling in a straight line is given by $a = -4t + 1$. What are the velocity and displacement functions if the object starts from rest and the initial displacement is 5 ft?

 28. Find the current in the circuit shown in the accompanying figure. Assume $i = 0$ when $t = 0$.

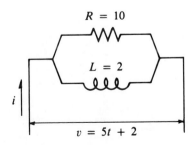

29. The voltage (in volts) being applied across a parallel *RLC* circuit is given by $v = t^3 + 3t$. Find the equation for the total current given that $R = 5\ \Omega$, $L = 3$ H, and $C = 2$ F. The current is initially zero.

30. Find the voltage that must be applied to the circuit to produce the current of $i = 3t^2$. Assume the initial voltage is 2 V. See the accompanying figure.

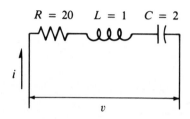

31. Repeat Exercise 30 with $i = t^{3/2}$ and an initial voltage of 5 V.

CHAPTER 6

The Definite Integral

The study of calculus is divided into two main branches, namely, differential calculus and integral calculus. The derivative, which is the basic concept in differential calculus, was introduced in Chapter 2. In this chapter, we initiate the study of *integral calculus*, the origin of which can be traced to the problem of calculating the area bounded by a curve. The basic concept in integral calculus is the *definite integral*.

6.1 SUMMATION NOTATION

In integral calculus, we are frequently concerned with writing sums of numbers, such as the sum of the first five integers $(1 + 2 + 3 + 4 + 5)$ or the first eight squared integers $(1^2 + 2^2 + 3^2 + 4^2 + 5^2 + 6^2 + 7^2 + 8^2)$. A notation often used to express the terms of a sum is called **sigma notation.** The Greek letter Σ (capital sigma) means to form the sum of the indicated terms and is defined as follows:

DEFINITION

> **Sigma Notation**
>
> $$\sum_{k=1}^{n} a_k = a_1 + a_2 + a_3 + \cdots + a_n$$
>
> This expression is read, "the summation of a sub k from $k = 1$ to $k = n$."

The subscript k, called the **index of summation,** is often called a *dummy variable.* Its only purpose is to indicate the steps in the summation process. The choice of the letter k for the index is arbitrary; any other symbol would serve as well. The 1 and the n indicate the range of the index. In the definition, k begins at 1 and increases in steps of 1 until it becomes equal to n.

EXAMPLE 1   **a.** The sum of the first 100 integers may be written as $\sum\limits_{k=1}^{100} k$.

 b. The sum of the first eight squared integers may be written as $\sum\limits_{k=1}^{8} k^2$. ■

Writing a sum such as $\sum\limits_{k=2}^{5} k^2$ in the form $4 + 9 + 16 + 25$ is called *expanding* the summation or writing it in *expanded form*.

EXAMPLE 2   The sum $\sum\limits_{k=2}^{4} x^k$ in expanded form is

$$\sum_{k=2}^{4} x^k = x^2 + x^3 + x^4$$

■

EXAMPLE 3   The sum $\sum\limits_{i=1}^{10} (-1)^{i-1}$ in expanded form and simplified is

$$\sum_{i=1}^{10} (-1)^{i-1} = (-1)^0 + (-1)^1 + (-1)^2 + \cdots + (-1)^9 = 0$$

■

Two basic rules govern the use of the summation symbol. The first allows us to factor out a constant coefficient, and the second allows us to compute the sum of sums on a term-by-term basis. Together these two properties are called the *linearity properties*.

PROPERTIES

> **Linearity Properties for Σ**   If c is a constant, then
>
> **1.** $\displaystyle\sum_{k=1}^{n} ca_k = c \sum_{k=1}^{n} a_k$
>
> **2.** $\displaystyle\sum_{k=1}^{n} (a_k + b_k) = \sum_{k=1}^{n} a_k + \sum_{k=1}^{n} b_k.$

As an example of the use of these properties, the sum

$$\sum_{k=1}^{3} k(5k + 1) = 1(6) + 2(11) + 3(16) = 76$$

can also be written as

$$\sum_{k=1}^{3} (5k^2 + k) = 5 \sum_{k=1}^{3} k^2 + \sum_{k=1}^{3} k$$

$$= 5(1 + 4 + 9) + (1 + 2 + 3) = 76$$

Either way, the value is the same.

You should also be aware of the fact that a sum may be represented by more than one summation expression. One such instance is given in the following example.

EXAMPLE 4   Showing that

$$\sum_{k=1}^{n} \frac{1}{k} = \sum_{k=0}^{n-1} \frac{1}{k + 1}$$

is merely a matter of writing them both in expanded form. The index on the left summation ranges from 1 to n. The index on the right ranges from 0 to $n - 1$. The second sum is said to be obtained from the first by a **shift of index**. ∎

EXAMPLE 5   ■ **Problem**

Write the summation $\sum_{k=0}^{n} a_k (k - 1)x^k$ as a summation whose index begins at 2.

■ **Solution**

Let $k = t - 2$. Then, when $k = 0$, $t = 2$. When $k = n$, $t = n + 2$. Thus, the summation becomes

$$\sum_{t=2}^{n+2} a_{t-2} (t - 3)x^{t-2}$$

Letting $k = t$, this sum becomes

$$\sum_{k=2}^{n+2} a_{k-2} (k - 3)x^{k-2}$$ ∎

EXAMPLE 6   The sum of the first six terms of the *harmonic sequence* is $1 + \frac{1}{2} + \frac{1}{3} + \frac{1}{4} + \frac{1}{5} + \frac{1}{6}$. The first term can be written as $\frac{1}{1}$. Since the denominator varies from 1 to 6 in steps of 1, the sum can be written in summation notation as

$$\sum_{n=1}^{6} \frac{1}{n}$$ ∎

EXAMPLE 7   The sum of the first n squared integers can be found by expanding

$$\sum_{k=1}^{n} k^2$$

For instance, the sum of the first five squared integers is

$$\sum_{k=1}^{5} k^2 = 1^2 + 2^2 + 3^2 + 4^2 + 5^2 = 55$$

Mathematicians have shown that the sum of the first n squared integers can also be calculated by the formula

$$\sum_{k=1}^{n} k^2 = \frac{1}{6}n(n + 1)(2n + 1)$$

We are not interested in proving this fact, but we will verify it for $n = 5$:

$$\sum_{k=1}^{5} k^2 = \frac{1}{6}(5)(6)(11) = 55$$

This result agrees with the previous result. A variation of this formula will be used in Section 6.3. ∎

EXERCISES FOR SECTION 6.1

Write Exercises 1–10 in expanded form and evaluate.

1. $\displaystyle\sum_{m=1}^{5} \frac{1}{m + 2}$

2. $\displaystyle\sum_{k=2}^{3} \frac{1 - k}{1 + k}$

3. $\displaystyle\sum_{n=0}^{5} n^2$

4. $\displaystyle\sum_{k=0}^{3} \left(-\frac{1}{2}\right)^k$

5. $\displaystyle\sum_{j=1}^{5} 3j$

6. $\displaystyle\sum_{k=0}^{5} (-1)^k$

7. $\displaystyle\sum_{k=1}^{5} (1 + k)$

8. $\displaystyle\sum_{k=0}^{5} [2 + 3(k - 1)]$

9. $\displaystyle\sum_{n=-2}^{2} n^2$

10. $\displaystyle\sum_{k=-1}^{2} (3k - 1)$

Write the terms of the summation in Exercises 11 and 12.

11. $\displaystyle\sum_{k=1}^{5} x_k$

12. $\displaystyle\sum_{k=1}^{5} x^k$

Represent the summations in Exercises 13–17 by using sigma notation.

13. $1 + 2 + 3 + 4$

14. $\frac{1}{2} + \frac{1}{4} + \frac{1}{8}$

15. $1 + 4 + 9 + 16$

16. $x_0 + x_1 + x_2 + x_3 + x_4$

17. $2x + 4x + 8x + 16x + 32x$

Which of Exercises 18–23 are true?

18. $\displaystyle\sum_{k=1}^{n} ax_k = a \sum_{k=1}^{n} x_k$

19. $\displaystyle\sum_{i=1}^{n} (x_i + d) = nd + \sum_{i=1}^{n} x_i$

20. $\displaystyle\sum_{k=1}^{n} d = nd$

21. $\displaystyle\sum_{k=2}^{5} a_{k-2} = \sum_{k=0}^{3} a_k$

22. $\displaystyle\sum_{n=1}^{4} n^2 = \sum_{n=2}^{5} n^2$

23. $\displaystyle\sum_{n=1}^{3} x^n = \sum_{t=2}^{4} x^{t-1}$

24. If a formula for S_n can be found without the summation sign, the sequence is said to be expressed in *closed form*. For example, the sequence of sums corresponding to 1, 4, 9, 16, ..., n^2, ... is given by $S_n = \sum_{k=1}^{n} k^2$. Mathematicians have shown that S_n is given by the formula $S_n = \frac{1}{6}n(n + 1)(2n + 1)$. Directly verify this formula for $n = 1, 2,$ and 3. Then, find the sum of the first one hundred squared integers.

25. Verify the formula for the sum of the first n cubes,

$$\sum_{k=1}^{n} k^3 = \frac{1}{4}n^2(n + 1)^2$$

for $n = 1, 2,$ and 3. Then, find the sum of the first fifty cubed integers.

6.2 TWO PROBLEMS INVOLVING THE SUMMATION OF ELEMENTS

An Area Problem

We introduce the definite integral in much the same way that we introduced the derivative, that is, by identifying the common thread in the solutions of seemingly unrelated problems. The first problem involves finding the area bounded by a curve $y = g(x)$ and the x-axis over an interval $a \leq x \leq b$ as shown in Figure 6.1. The shaded area is called the **area under the curve**.

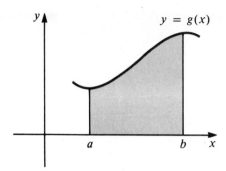

■ **Figure 6.1**

We begin our discussion of the area problem by describing a means for estimating the area under the curve in Figure 6.2, using a series of rectangles.

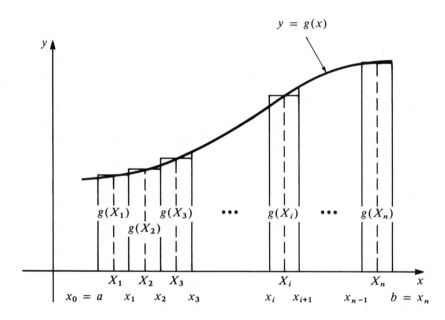

■ **Figure 6.2**

PROCEDURE	**Estimating the Area under a Curve**

1. Subdivide the interval $[a, b]$ into n subintervals of width $\Delta x = (b - a)/n$.

2. Pick a point in each of the n subintervals to evaluate the function. Designate the point chosen in the first subinterval by X_1; in the second, by X_2; and so on, until you reach the nth subinterval. In a typical subinterval, the point is labeled X_i, where i is an integer from 1 to n.

3. Draw a rectangle above each subinterval so that its height is $g(X_i)$. The area of a typical rectangle is then height $\cdot$ width $= g(X_i)\, \Delta x$.

4. Form the sum

$$A_n = g(X_1)\,\Delta x + g(X_2)\,\Delta x + \cdots + g(X_n)\,\Delta x = \sum_{i=1}^{n} g(X_i)\,\Delta x$$

which gives the estimate. When the value of n changes, the estimate changes. Usually, the more subdivisions used, the better the estimate is.

EXAMPLE 1   ■ **Problem**

Estimate the area bounded by $y = x^2$ and the x-axis over the interval $0 \le x \le 2$. Use four subintervals, and select X_i to be the midpoint of each subinterval. See Figure 6.3.

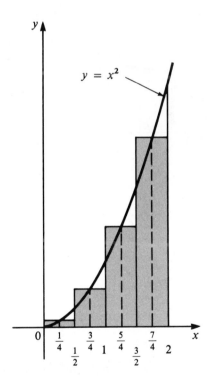

■ **Figure 6.3**

■ **Solution**

The endpoints of the subintervals are $x_0 = 0$, $x_1 = \frac{1}{2}$, $x_2 = 1$, $x_3 = \frac{3}{2}$, $x_4 = 2$. One-half of a subinterval is $\frac{1}{2}(\frac{1}{2}) = \frac{1}{4}$, so $X_1 = 0 + \frac{1}{4} = \frac{1}{4}$. Likewise, X_2 is $\frac{1}{4}$ to the right of x_1, or $X_2 = \frac{3}{4}$; X_3 is $\frac{1}{4}$ to the right of x_2, or $X_3 = \frac{5}{4}$; X_4 is $\frac{1}{4}$ to the right of x_3 or $X_4 = \frac{7}{4}$. The approximating sum is then

$$A_{est} = \sum_{i=1}^{4} g(X_i)\, \Delta x = \sum_{i=1}^{4} X_i^2\, \Delta x$$

$$= \left(\frac{1}{4}\right)^2 \left(\frac{1}{2}\right) + \left(\frac{3}{4}\right)^2 \left(\frac{1}{2}\right) + \left(\frac{5}{4}\right)^2 \left(\frac{1}{2}\right) + \left(\frac{7}{4}\right)^2 \left(\frac{1}{2}\right)$$

$$= \frac{1}{2}\left(\frac{1}{16} + \frac{9}{16} + \frac{25}{16} + \frac{49}{16}\right) = \frac{1}{2}\left(\frac{84}{16}\right) = \frac{21}{8}$$

$$= 2.625 \text{ square units}$$

■

EXAMPLE 2   ■ **Problem**

Estimate the area that is bounded by $y = \sqrt{x + 1}$ and the x-axis over the interval $1 \le x \le 4$ by using three subintervals and selecting X_i to be the midpoint of each subinterval. See Figure 6.4.

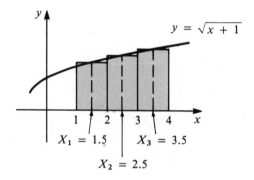

■ **Figure 6.4**

■ **Solution**

The width of each subinterval is $\Delta x = (4 - 1)/3 = 1$. From the figure, $X_1 = 1.5$, $X_2 = 2.5$, and $X_3 = 3.5$. Therefore,

$$A_{est} = \sum_{i=1}^{3} \sqrt{X_i + 1}\, \Delta x$$

$$= \sqrt{1.5 + 1}\,(1) + \sqrt{2.5 + 1}\,(1) + \sqrt{3.5 + 1}\,(1)$$

$$= 1.58 + 1.87 + 2.12 = 5.57 \text{ square units}$$

■

We expect the accuracy of our estimates to improve as we take more and more rectangles, but the important thing is that you understand how the estimates are made. This process is of immense practical value as well as a prelude to finding the precise area under a curve.

A Displacement Problem

The distance traveled by an object moving with a constant velocity for a specified time can be calculated by the formula

$$S = V \cdot T \quad \text{or} \quad \textit{displacement} = \textit{velocity} \cdot \textit{time}$$

For example, a car moving at a constant speed of 40 mi/hr will travel 80 mi in 2 hr.

Suppose the velocity (in feet per second) of a rocket sled varies with time according to $v = 25t^2$ during the first 2 sec of motion. Figure 6.5 depicts the variation in velocity for this interval. Notice that $v = 0$ when $t = 0$; $v = 25$ ft/sec when $t = 1$ sec; and $v = 100$ ft/sec when $t = 2$ sec. The formula $S = V \cdot T$ cannot be used to calculate the distance traveled by the sled, because the velocity is not constant. However, by subdividing the given time interval into *small* increments Δt, and assuming the velocity is constant over each Δt, we can use this formula to estimate the distance traveled by the sled during each subinterval. The sum of these estimates is then an estimate of the total distance traveled.

■ **Figure 6.5**

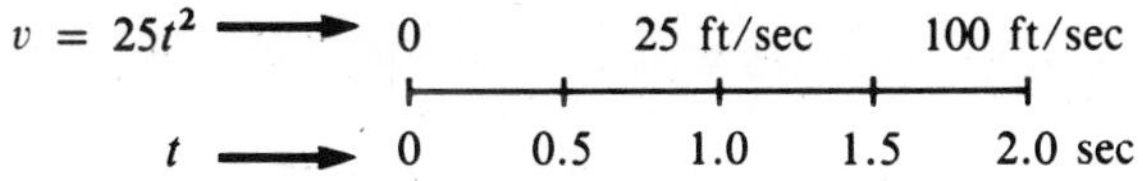

$v = 25t^2 \longrightarrow$ 0   25 ft/sec   100 ft/sec

$t \longrightarrow$ 0   0.5   1.0   1.5   2.0 sec

EXAMPLE 3   ■ **Problem**

Use four subintervals to estimate the total distance traveled by a rocket sled in the first 2 sec, given a velocity (in feet per second) of $v = 25t^2$.

■ **Solution**

We divide the 2 sec into 0.5-sec increments. We can make a reasonable estimate of the distance traveled on each of these subintervals of time by choosing v at the midpoint. Thus,

$$v_1 = 25(0.25)^2 = 1.56 \text{ ft/sec} \qquad s_1 = v_1 \cdot \Delta t = 1.56(0.5) = 0.78 \text{ ft}$$
$$v_2 = 25(0.75)^2 = 14.06 \text{ ft/sec} \qquad s_2 = v_2 \cdot \Delta t = 14.06(0.5) = 7.03 \text{ ft}$$
$$v_3 = 25(1.25)^2 = 39.06 \text{ ft/sec} \qquad s_3 = v_3 \cdot \Delta t = 39.06(0.5) = 19.53 \text{ ft}$$
$$v_4 = 25(1.75)^2 = 76.56 \text{ ft/sec} \qquad s_4 = v_4 \cdot \Delta t = 76.56(0.5) = 38.28 \text{ ft}$$

The estimate of the total distance traveled by the sled in 2 sec is then given by

$$S_{\text{est}} = \sum_{i=i}^{4} v_i \cdot \Delta t = 0.78 + 7.03 + 19.53 + 38.28 = 65.62 \text{ ft}$$

We have no indication at this time of the accuracy of this estimate. However, we suspect that we can improve our estimate by increasing the number of subintervals or changing the velocity estimate within each subinterval.   ■

The method used in Example 3 is generalized as follows: If the velocity of an object is given by $v = V(t)$ over the interval $a \leq t \leq b$, partition the interval into n equal subintervals of duration $\Delta t = (b - a)/n$. The estimate of the distance traveled during a typical Δt is $\Delta s_i = V(T_i) \Delta t$, where T_i denotes a time within the subinterval. See Figure 6.6. An estimate of the total distance traveled by the object is then given by the following formula.

FORMULA

Estimate of Total Distance

$$S_{est} = \sum_{i=1}^{n} V(T_i) \Delta t$$

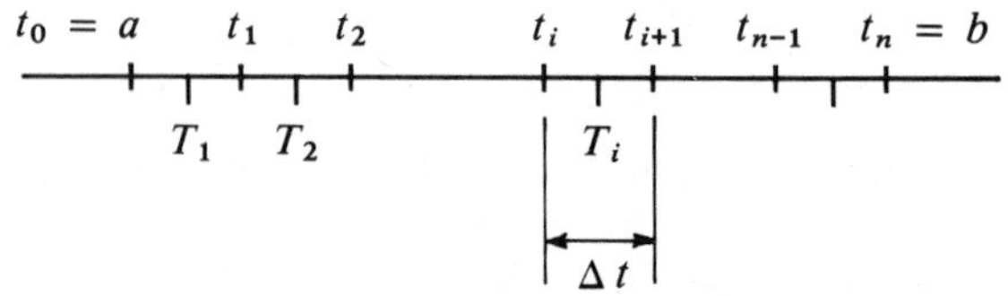

■ **Figure 6.6**

EXAMPLE 4   ■ **Problem**

The velocity (in centimeters per second) of a charged particle is given by $v = 1/(t + 1)$ during the first 6 sec of movement. Estimate the distance traveled by the particle by using three subintervals and selecting T_i to be the midpoint of each subinterval. See Figure 6.7.

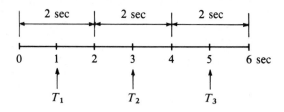

■ **Figure 6.7**

■ **Solution**

Here, the duration of each subinterval is

$$\Delta t = \frac{6 - 0}{3} = 2 \text{ sec}$$

From Figure 6.7, the values of T_i are $T_1 = 1$, $T_2 = 3$, and $T_3 = 5$. Thus,

$$S_{est} = \sum_{n=1}^{3} \frac{1}{T_i + 1} \Delta t = \left(\frac{1}{1 + 1}\right)(2) + \left(\frac{1}{3 + 1}\right)(2) + \left(\frac{1}{5 + 1}\right)(2)$$

$$= 1.0 + 0.5 + 0.3 = 1.8 \text{ cm}$$ ∎

EXERCISES FOR SECTION 6.2

In Exercises 1–10, estimate the area under the indicated curve, choosing X_i to be the midpoint of each subinterval. Sketch each curve, and show the approximating rectangles to scale.

1. $y = 0.6x^2$ from $x = 0$ to $x = 3$; use three subintervals.

2. $y = 2 + x^2$ from $x = 0$ to $x = 4$; use four subintervals.

3. $y = 1 + 3x^2$ from $x = 0$ to $x = 2$; use two subintervals.

4. $y = 2.4 + 0.2x^2$ from $x = 1$ to $x = 4$; use three subintervals.

5. $y = 0.1x^2 + 1.3$ from $x = 2$ to $x = 3$; use four intervals.

6. $y = x^2$ from $x = 1$ to $x = 2$; use five subintervals.

7. $y = x^2$ from $x = 1$ to $x = 1.5$; use five subintervals.

8. $y = 4 - x^2$ from $x = 0$ to $x = 2$; use five subintervals.

9. $y = \sqrt{x}$ from $x = 0$ to $x = 4$; use four subintervals.

10. $y = \sqrt{x + 1}$ from $x = 0$ to $x = 1$; use four subintervals.

11. Estimate the area under the curve $y = \frac{1}{4}x^2$ from 0 to 2, using four subintervals. Select X_i to be (a) the left-hand endpoint and (b) the right-hand endpoint. Sketch each case.

12. From Exercise 11, we know that the area under the curve $y = \frac{1}{4}x^2$ from 0 to 2 is greater than _____ and less than _____ .

13. Estimate the area under the curve $y = x^2$ from 1 to 3, using four subintervals. Select X_i to be (a) the left-hand endpoint and (b) the right-hand endpoint. Sketch each area, and show the indicated rectangles.

14. From Exercise 13, we know that the area under the curve $y = x^2$ from 1 to 3 is greater than _____ and less than _____ .

In Exercises 15–20, assume the velocity (in feet per second) of a car is given by $v = 10t^2$ from $t = 0$ to $t = 4$ sec.

15. Estimate the distance traveled, using four subintervals. Choose T_i to be the midpoint of each subinterval.

16. What is the velocity of the car at T_1, T_2, T_3, and T_4 in Exercise 15?

17. Estimate the distance traveled by the car, using four subintervals, if (a) T_i is the left-hand endpoint of each subinterval and (b) T_i is the right-hand endpoint of each subinterval.

18. From Exercise 17, we can say that the distance traveled by the car from time $t = 0$ to $t = 4$ sec is greater than _____ and less than _____ .

19. Estimate the distance traveled by the car, using eight subintervals, if (a) T_i is the left-hand endpoint of each subinterval and (b) T_i is the right-hand endpoint of each subinterval.

20. From Exercise 19, we can say that the distance traveled by the car from time $t = 0$ to $t = 4$ sec is greater than _____ and less than _____ .

In Exercises 21–24, assume the velocity (in centimeters per second) of a particle in a magnetic field is given by $v = 9 - t^2$ from $t = 0$ to $t = 3$ sec.

21. Estimate the distance traveled by the particle, using three subintervals. Choose T_i to be the midpoint of each subinterval.

22. What is the velocity of the particle in Exercise 21 at T_1, T_2, and T_3?

23. Estimate the distance traveled by the particle, using six subintervals, given that (a) T_i is the left-hand endpoint of each subinterval and (b) T_i is the right-hand endpoint of each subinterval.

24. From Exercise 23, we can say that the distance traveled by the particle is greater than _____ and less than _____ .

In Exercises 25 and 26, assume the velocity (in meters per second) of a rocket is given by $v = \sqrt{t}$ for the first 20 sec of flight.

25. Estimate the distance travled by the rocket, using five subintervals. Choose T_i to be the midpoint of each subinterval.

26. Estimate the distance traveled by the rocket, using four subintervals. Choose T_i to be the midpoint of each subinterval.

Use the following BASIC program to estimate the areas in Exercises 1–10.

```
10   REM   USING RECTANGLES TO FIND AREA
20   HOME
30   PRINT "THIS PROGRAM WILL FIND THE VALUE OF THE"
40   PRINT "INTEGRAL USING THE LEFT OR RIGHT"
50   PRINT "ENDPOINTS, WITH AN INTERVAL AND A VALUE"
60   PRINT "OF N SUPPLIED BY THE USER."
70   PRINT : PRINT "THE FUNCTION IN THIS PROGRAM"
80   PRINT "IS    X^2 - 4*X + 5"
90   PRINT "IF THE USER CHANGES THE FUNCTION,"
100   PRINT "THE NEW FUNCTION WILL REMAIN UNTIL"
110   PRINT "THE USER CHANGES IT AGAIN, OR"
120   PRINT "RELOADS THE PROGRAM"
130   PRINT : PRINT "DO YOU WISH TO CHANGE THE FUNCTION? Y/N"
140   INPUT A$
150   IF A$ = "Y" THEN 400
160   PRINT "ENTER THE ENDPOINTS OF THE INTERVAL"
```

```
170  INPUT A,B
180  PRINT "HOW MANY SUBINTERVALS DO YOU WISH TO USE"
190  INPUT N
200 H = (B - A) / N
210  PRINT "DO YOU WISH TO USE THE LEFT OR RIGHT ENDPOINTS? L/R"
220  INPUT R$
230 K = 0
240  IF R$ = "L" THEN K =  - 1
250  HTAB 12: FLASH : PRINT "PATIENCE PLEASE"
260  HTAB 12: PRINT "CALCULATING!!!": NORMAL
270  DEF  FN F(X) = X ^ 2 - 4 * X + 5
280 S = 0
290  FOR I = 1 TO N
300 S = S +  FN F(A + (I + K) * H)
310  NEXT I
320 S = S * H
330  PRINT "THE VALUE OF THE INTEGRAL IS"
335 S =  INT (10000 * S) / 10000
340  PRINT "APPROXIMATELY ";S
350  PRINT : PRINT "DO YOU WISH TO RUN THIS PROGRAM AGAIN? Y/N"
360  INPUT B$
370  HOME
380  IF B$ = "Y" THEN 70
390  GOTO 430
400  PRINT "TYPE"
410  INVERSE : PRINT "270 DEF FN F(X) = (YOUR FUNCTION) <RET>"
420  PRINT "RUN 160 <RET>": NORMAL
430  END
```

■■■■■ 6.3 THE DEFINITE INTEGRAL

The estimates of the area under a curve and the displacement of an object can be made precise by using the concept of a limit.

Area under a Curve

Consider the area bounded by the curve $y = x^2$ and the x-axis over the interval $0 \le x \le 2$, as shown in Figure 6.8(a). In the previous section, we used the sum

$$A_n = \sum_{i=1}^{n} X_i^2 \Delta x$$

to approximate this area. See Figure 6.8(b).

To find the precise area bounded by the curve $y = x^2$ and the x-axis over the interval $0 \le x \le 2$, we note that we expect the approximating sum

$$A_n = \sum_{i=1}^{n} X_i^2 \Delta x$$

to approach the area under the curve as the number of subdivisions is increased.

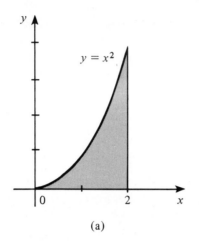

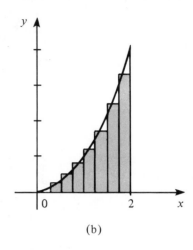

■ Figure 6.8   (a)   (b)

This remark is the key to the solution of the area problem and leads us to conclude that the area under a curve $y = x^2$ from $x = 0$ to $x = 2$ is the limit of the finite sum of rectangular areas as the number of subintervals increases without bound. That is,

$$A = \lim_{n \to \infty} \sum_{i=1}^{n} X_i^2 \, \Delta x \qquad\qquad (6.1)$$

The limiting value must be the same regardless of the choice of X_i in each subinterval. This definition can be used to compute the exact area under the curve $y = x^2$ from $x = 0$ to $x = 2$.

EXAMPLE 1   **■ Problem**

Use Equation (6.1) to show that the precise area under the curve $y = x^2$ on the interval $[0, 2]$ is equal to $\frac{8}{3}$.

■ Solution

We choose X_i to be the left endpoint of each subinterval. See Figure 6.9. To find the area under the curve we proceed by first finding the area of the *rectangles* below the curve $y = x^2$ shown in Figure 6.9. Since the interval $0 \le x \le 2$ is divided into n equal parts, each of width Δx, we can write the successive subdivision points as follows:

$$x_0 = 0$$
$$x_1 = \Delta x$$
$$x_2 = 2\,\Delta x$$
$$x_3 = 3\,\Delta x$$
$$\cdot$$
$$\cdot$$
$$\cdot$$
$$x_{n-1} = (n - 1)\,\Delta x$$
$$x_n = n\,\Delta x$$

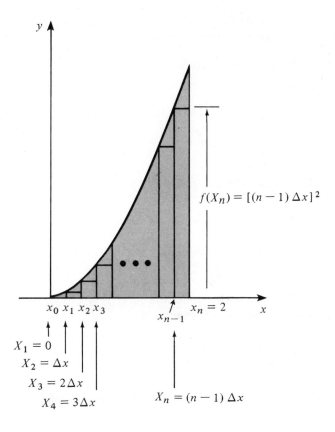

$$f(X_n) = [(n-1)\,\Delta x]^2$$

$$x_n = 2$$

$$x_{n-1}$$

$$X_1 = 0$$
$$X_2 = \Delta x$$
$$X_3 = 2\Delta x$$
$$X_4 = 3\Delta x$$

$$X_n = (n-1)\,\Delta x$$

■ Figure 6.9

When X_i is the left endpoint of each subinterval, the area of the successive rectangles is given by

$$f(X_1)\,\Delta x = (0)^2\Delta x = 0$$
$$f(X_2)\,\Delta x = (\Delta x)^2\Delta x = 1^2(\Delta x)^3$$
$$f(X_3)\,\Delta x = (2\,\Delta x)^2\Delta x = 2^2(\Delta x)^3$$
$$f(X_4)\,\Delta x = (3\,\Delta x)^2\Delta x = 3^2(\Delta x)^3$$

$$\cdot$$
$$\cdot$$
$$\cdot$$

$$f(X_n)\,\Delta x = [(n-1)\,\Delta x]^2\Delta x = (n-1)^2(\Delta x)^3$$

Summing these areas, we obtain

$$A_n = 0 + 1^2(\Delta x)^3 + 2^2(\Delta x)^3 + 3^2(\Delta x)^3 + \cdots + (n-1)^2(\Delta x)^3$$
$$= [1^2 + 2^2 + 3^2 + \cdots + (n-1)^2](\Delta x)^3$$
$$= \sum_{k=1}^{n-1} k^2(\Delta x)^3 = (\Delta x)^3 \sum_{k=1}^{n-1} k^2$$

A formula for the sum of the squares of the first $(n - 1)$ integers (see Example 7, Section 6.1) allows us to write

$$\sum_{k=1}^{n-1} k^2 = \frac{n(n-1)(2n-1)}{6}$$

In addition, we note that Δx can be written as

$$\Delta x = \frac{2-0}{n} = \frac{2}{n}$$

Therefore, A_n reduces to

$$A_n = (\Delta x)^3 \sum_{k=1}^{n-1} k^2 = \left(\frac{2}{n}\right)^3 \left[\frac{n(n-1)(2n-1)}{6}\right]$$

$$= \frac{8(2n^3 - 3n^2 + n)}{6n^3} = \frac{8}{6}\left(2 - \frac{3}{n} + \frac{1}{n^2}\right)$$

The precise area under the curve $y = x^2$ from $x = 0$ to $x = 2$ is then the limit of A_n as n increases without bound.

$$A = \lim_{n \to \infty} \sum_{k=1}^{n-1} k^2(\Delta x)^3 = \lim_{n \to \infty} \frac{8}{6}\left(2 - \frac{3}{n} + \frac{1}{n^2}\right) = \frac{8}{3} \qquad \blacksquare$$

Comment: The method used in Example 1 to evaluate the area under the curve $y = x^2$ is tedious at best. In the next section, we will show you a much easier way to solve this problem.

Displacement

In Example 3 of Section 6.2, the displacement of the rocket sled, whose velocity during the first 2 sec is given by $v = 25t^2$, was estimated by

$$S_{\text{est}} = \sum_{i=1}^{n} 25T_i^2 \Delta t$$

This sum approaches the precise displacement of the sled as the number of subdivisions is increased. So, we conclude that the precise displacement is given by

$$s = \lim_{n \to \infty} \sum_{i=1}^{n} 25T_i^2 \Delta t = 25\left(\lim_{n \to \infty} \sum_{i=1}^{n} T_i^2 \Delta t\right)$$

Notice that the quantity in the parentheses is similar to the limit encountered in the solution to the area under the curve $y = x^2$ from $x = 0$ to $x = 2$. Therefore, the results obtained for the area problem can be used here. Since the area under the curve $y = x^2$ from $x = 0$ to $x = 2$ is $\frac{8}{3}$ square units, we conclude that the precise displacement of the indicated rocket sled from $t = 0$ to $t = 2$ sec is

$$s = 25\left(\frac{8}{3}\right) = \frac{200}{3} \text{ ft}$$

Definition and Basic Properties of the Definite Integral

The fact that the area and the displacement problems just considered are mathematically identical suggests that we generalize the discussion of the limit of such sums. Hence, to avoid reference to specific applications, we denote the functional values by $f(x)$ and study the limit

$$\lim_{n \to \infty} \sum_{i=1}^{n} f(X_i)\Delta x$$

This limit, which is the focal point of integral calculus, leads to the following definition.

DEFINITION

The Definite Integral If f is a function defined on an interval $a \leq x \leq b$ and

$$\lim_{n \to \infty} \sum_{i=1}^{n} f(X_i)\Delta x$$

exists regardless of the way in which X_i is chosen, we call that number the *definite integral* of the function f from a to b and denote it by

$$\int_a^b f(x)\ dx$$

The symbol $\int_a^b f(x)\ dx$ consists of four parts to which we have occasion to refer. The $\int$ is called the **integral sign**. It comes from the letter S, which was originally used to denote summation. The letters a and b stand for the extreme points of the interval $a \leq x \leq b$ and are called the **limits of integration**. There is no stipulation on the magnitudes of a and b, but we do insist that they be real numbers. The function f is known as the **integrand**. And the dx—which, as we indicated in Chapter 4, suggests a small interval on the x-axis—is known as the **differential**.

Note that $\int f(x)\ dx$ and $\int_a^b f(x)\ dx$ are conceptually quite different. The first quantity is the antiderivative of $f(x)$ and represents an entire family of *functions*. The second quantity is a *real number* called the definite integral of $f(x)$. Any process by which the quantity $\int_a^b f(x)\ dx$ is evaluated is called **integration**.

Area and displacement are special applications of the definite integral. The following table emphasizes this point.

If $f(x)$ represents	$\sum\limits_{i=1}^{n} f(X_i)\, \Delta x$ represents	$\int_a^b f(x)\, dx$ represents
Curve in the plane	Approximate area under curve	Exact area
Velocity	Approximate displacement	Exact displacement

The answers to the two special problems can now be written in terms of the definite integral. Thus, the area under the curve $y = x^2$ from 0 to 2 is given by

$$A = \int_0^2 x^2\, dx = \frac{8}{3} \text{ square units}$$

The displacement of the rocket sled from $t = 0$ to $t = 2$ sec is given by

$$S = \int_0^2 25t^2\, dt = \frac{200}{3} \approx 66.67 \text{ ft}$$

The definite integral has the following useful properties.

Linearity Properties of the Definite Integral

1. If c is a constant, then

$$\int_a^b cf(x)\, dx = c \int_a^b f(x)\, dx$$

2. $\displaystyle\int_a^b [f(x) + g(x)]\, dx = \int_a^b f(x)\, dx + \int_a^b g(x)\, dx$

As with antiderivatives, you must remember that Property 1 may be used only when c is a constant. By using Property 1 and the value of $\int_0^2 x^2\, dx$, we can solve many other definite integrals.

EXAMPLE 2   ■ **Problem**

Use Property 1 of definite integrals and $\int_0^2 x^2\, dx = \frac{8}{3}$ to evaluate $\int_0^2 3x^2\, dx$.

■ **Solution**

The definite integral can be written as

$$\int_0^2 3x^2\, dx = 3 \int_0^2 x^2\, dx = 3\left(\frac{8}{3}\right) = 8$$ ■

Comment: We use the definite integral $\int_a^b f(x)\, dx$ to calculate the area under the graph of the function f from $x = a$ to $x = b$. Conversely, we can use the area under the graph of a function f from $x = a$ to $x = b$ to obtain the value of the definite integral $\int_a^b f(x)\, dx$. In the following examples and exercises, you are to use the area under a curve to evaluate the given definite integral. In the next section, we will use the definite integral to calculate areas.

EXAMPLE 3   ■ **Problem**

Evaluate $\int_0^2 x\, dx$ by using a geometric formula.

■ **Solution**

The graph of the integrand function is the straight line $y = x$, which is shown in Figure 6.10. The area under this curve from $x = 0$ to $x = 2$ is triangular. Since the area of a triangle is $A = \frac{1}{2}bh$, the value of $\int_0^2 x\, dx$ is

$$\int_0^2 x\, dx = \frac{1}{2}(2)(2) = 2$$

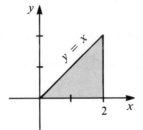

■ **Figure 6.10**

EXAMPLE 4   ■ **Problem**

Use the properties of a definite integral and results of previous examples to evaluate $\int_0^2 (4x + 9x^2)\, dx$.

■ **Solution**

Using Properties 1 and 2, we get

$$\int_0^2 (4x + 9x^2)\, dx = 4 \int_0^2 x\, dx + 9 \int_0^2 x^2\, dx = 4(2) + 9\left(\frac{8}{3}\right) = 32$$

EXAMPLE 5   ■ **Problem**

Evaluate $\int_0^2 \sqrt{4 - x^2}\, dx$ by using a geometric formula.

■ **Solution**

To evaluate this definite integral, we observe that the integrand function $y = \sqrt{4 - x^2}$ can be written $y^2 = 4 - x^2$, or equivalently, $x^2 + y^2 = 4$. In this form we recognize the equation of a circle with center at the origin and radius 2. See Figure 6.11. Therefore,

$$\int_0^2 \sqrt{4 - x^2}\, dx = \frac{1}{4} \text{ (area of circle)} = \frac{1}{4}(\pi \cdot 2^2) = \pi$$

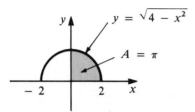

■ **Figure 6.11**   ■

![section marker] **EXERCISES FOR SECTION 6.3**

All definite integrals in this exercise set are to be evaluated by using the results of previous examples or purely geometric considerations. In Exercises 1–10, use Properties 1 and 2 and the results from previous examples to evaluate the given definite integrals.

1. $\int_0^2 8x^2\, dx$

2. $\int_0^2 \pi x^2\, dx$

3. $\int_0^2 \frac{x^2}{2}\, dx$

4. $\int_0^2 0.05t^2\, dt$

5. $\int_0^2 3x\, dx$

6. $\int_0^2 \frac{1}{2}t\, dt$

7. $\int_0^2 100z\, dz$

8. $\int_0^2 25x\, dx$

9. $\int_0^2 3\sqrt{4 - x^2}\, dx$

10. $\int_0^2 \frac{\sqrt{4 - x^2}}{3}\, dx$

In Exercises 11–14, evaluate the indicated definite integrals. (*Hint:* Find the area under the curve $y = 1$ from $x = 0$ to $x = 2$.)

11. $\int_0^2 4\, dx$

12. $\int_0^2 3\, dx$

13. $\int_0^2 \frac{1}{3}\, dt$

14. $\int_0^2 0.8\, dz$

Use the results from the foregoing exercises and examples to evaluate the integrals in Exercises 15–18.

15. $\int_0^2 (2x^2 + 5x) \, dx$

16. $\int_0^2 (s^2 + 7s) \, ds$

17. $\int_0^2 (4x^2 + 2x + 3) \, dx$

18. $\int_0^2 (7 + 3x - x^2) \, dx$

19. The velocity (in centimeters per second) of a solenoid-actuated plunger is given by $v = 3t - t^2$. Using *only* the results from the foregoing exercises, find the distance traveled by the plunger from 0 to 2 sec.

20. The velocity (in centimeters per second) of a charged particle in a magnetic field is given by $v = 10 + \frac{1}{2}t^2$. Using available results, compute the distance traveled by the particle from $t = 0$ to $t = 2$ sec.

21. Find the area under the curve $y = 2x - x^2$ from 0 to 2 by using available information. Make a sketch of the area.

22. Find the area under the curve $y = 4 - x^2$ from 0 to 2 by using the available information. Make a sketch of the area.

6.4 THE FUNDAMENTAL THEOREM OF CALCULUS

From the foregoing discussion, we see that the value of the definite integral of a function depends on the endpoints of the interval of integration. Thus, if the upper limit of integration is changed, we expect the value of the definite integral to change. By considering the upper limit of the interval of integration as a variable, we may define a function F on the interval $a \le t \le b$ such that

$$F(x) = \int_a^x f(t) \, dt$$

where x is an arbitrary point in the interval. See Figure 6.12. We call $F(x)$ the **indefinite integral** of f. Note that we now have used the word *integral* in three ways: the first in connection with the antiderivative, the second in connection with the definite integral, and the third in connection with the indefinite integral. Frequently, each concept is simply referred to as an *integral*, with only the context of the problem to tell you how the word is used.

Our interest in the indefinite integral is motivated by the fact that it leads to one of the more remarkable results in mathematics. This result shows the close relationship between the derivative and the indefinite integral and is called the **fundamental theorem of calculus**.

Figure 6.13 is a graph of a continuous function f over an interval $a \le x \le b$.

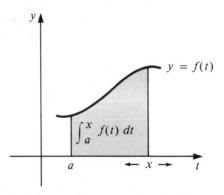

We want to investigate the manner in which the area A changes as x changes. Assume that if the value of the independent variable changes from x to x_1, then the area under the curve will change by an amount ΔA. (This change in area is represented by the darker shading in Figure 6.13.) To find the relationship between Δx and ΔA, we construct two rectangles, as shown in Figure 6.13. One rectangle is formed by drawing a line parallel to the x-axis from A to D. The other is formed by extending line AF and then drawing a line parallel to the x-axis from C to B. The smaller rectangle $ADEF$ has an area of $\Delta A_s = f(x)\,\Delta x$, and the area of the larger rectangle $BCEF$ is $\Delta A_l = f(x_1)\,\Delta x$. Obviously, ΔA is greater than ΔA_s and is less than ΔA_l, a fact that may be written as

$$f(x)\,\Delta x < \Delta A < f(x_1)\,\Delta x$$

Dividing the inequality by Δx yields

$$f(x) < \frac{\Delta A}{\Delta x} < f(x_1)$$

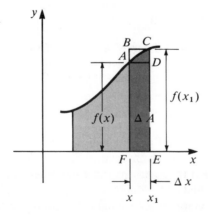

What happens to this inequality if we let $\Delta x \to 0$? As $\Delta x \to 0$, the middle term becomes

$$\lim_{\Delta x \to 0} \frac{\Delta A}{\Delta x} = \frac{dA}{dx}$$

and simultaneously, $f(x_1) \to f(x)$. But if $f(x_1)$ approaches $f(x)$ as a limit and $\Delta A/\Delta x$ lies between $f(x_1)$ and $f(x)$, then $\Delta A/\Delta x$ must also approach $f(x)$. Hence,

$$\frac{dA}{dx} = f(x)$$

Recalling that $A = \int_a^x f(t)\, dt$, we have proved the following theorem.

THEOREM 1

> **The Fundamental Theorem of Calculus**   If f is a continuous function over the interval $a \leq x \leq b$, then
>
> $$\frac{d}{dx}\left[\int_a^x f(t)\, dt \right] = f(x)$$
>
> Expressed in words, the theorem states that the derivative of an indefinite integral is the integrand function evaluated at the upper limit. Thus, if
>
> $$F(x) = \int_a^x f(t)\, dt$$
>
> then
>
> $$F'(x) = f(x)$$

EXAMPLE 1   ■ **Problem**

Let $F(x) = \int_1^x \frac{1}{t^2}\, dt$. Find $F'(2)$.

■ **Solution**

We have $dF/dx = 1/x^2$, and hence,

$$F'(2) = \frac{1}{4}$$

■

EXAMPLE 2   For $G(x) = \int_2^x \sqrt{t}\, dt$, $G'(x)$ is

$$\frac{dG}{dx} = \sqrt{x}$$

■

The fundamental theorem of calculus demonstrates the essential unity of the subject of calculus, and as a most profitable by-product, we are provided with a simple method for evaluating definite integrals. Consider the indefinite integral

$$F(x) = \int_a^x f(t)\, dt \tag{6.2}$$

Then by Theorem 1, we have $F'(x) = f(x)$. Suppose G is any antiderivative of f. Since $G(x)$ and $F(x)$ have the same derivative, they differ at most by a constant; that is,

$$F(x) = G(x) + C \qquad\qquad (6.3)$$

From (6.2) and (6.3), it follows that

$$F(b) = \int_a^b f(t)\, dt = G(b) + C$$

The value of C can be found by observing that

$$F(a) = \int_a^a f(t)\, dt = 0 = G(a) + C$$

or

$$C = -G(a)$$

Therefore,

$$\int_a^b f(t)\, dt = G(b) - G(a)$$

This result justifies the following theorem.

THEOREM 2

Evaluation of a Definite Integral   Let G be any antiderivative of f. Then

$$\int_a^b f(t)\, dt = G(b) - G(a)$$

Expressed in words, the theorem states that the value of a definite integral is equal to the value of the antiderivative of the intergrand function at the upper limit minus its value at the lower limit.

When we evaluate a definite integral, it is customary to use the notation

$$\int_a^b f(x)\, dx = \left[G(x) \right]_a^b = G(b) - G(a)$$

where $G(x)$ represents *any* antiderivative of $f(x)$. The intermediate step shows the antiderivative of the integrand function, except for the arbitrary constant, with the limits of integration written to the right of the bracket.

Comment: Here, you see another reason for developing your skill in finding antiderivatives. If you can obtain *any* antiderivative for f, the problem of evaluating $\int_a^b f(x)\, dx$ is essentially finished. For this reason, some people equate the process of antidifferentiation and integration, which is why the integral sign without any limits is widely used to indicate an antiderivative.

EXAMPLE 3 ▪ **Problem**

Evaluate $\int_0^2 x^2 \, dx$.

▪ **Solution**

We find

$$\int_0^2 x^2 \, dx = \left[\frac{x^3}{3}\right]_0^2 = \frac{2^3}{3} - \frac{0^3}{3} = \frac{8}{3}$$

Note that $\frac{8}{3}$ is the value obtained previously for the area under the curve $y = x^2$ from $x = 0$ to $x = 2$. The method is certainly quicker than the method used in Section 6.3 to evaluate the limit of a finite sum. ▪

EXAMPLE 4 ▪ **Problem**

Evaluate $\int_1^8 x^{-1/3} \, dx$.

▪ **Solution**

$$\int_1^8 x^{-1/3} \, dx = \left[\frac{3}{2} x^{2/3}\right]_1^8 = \frac{3}{2} (8)^{2/3} - \frac{3}{2}(1)^{2/3} = \frac{9}{2}$$ ▪

EXAMPLE 5 ▪ **Problem**

Evaluate $\int_0^2 (4x^3 + 1) \, dx$.

▪ **Solution**

$$\int_0^2 (4x^3 + 1) \, dx = \left[x^4 + x\right]_0^2 = (16 + 2) - (0) = 18$$ ▪

EXAMPLE 6 ▪ **Problem**

Evaluate $\int_2^3 \frac{dy}{y^3}$.

▪ **Solution**

We find

$$\int_2^3 \frac{dy}{y^3} = \int_2^3 y^{-3} \, dy = \left[\frac{y^{-2}}{-2}\right]_2^3 = \left[-\frac{1}{2y^2}\right]_2^3 = -\frac{1}{18} + \frac{1}{8} = \frac{5}{72}$$ ▪

EXAMPLE 7 ▪ **Problem**

Evaluate $\int_0^4 \sqrt{3x + 4} \, dx$.

■ **Solution**

We have

$$\int_0^4 \sqrt{3x + 4} \, dx = \frac{1}{3} \int_0^4 3(3x + 4)^{1/2} \, dx = \left[\frac{2}{9} (3x + 4)^{3/2} \right]_0^4$$

$$= \frac{2}{9}(16)^{3/2} - \frac{2}{9}(4)^{3/2} = \frac{128}{9} - \frac{16}{9} = \frac{112}{9}$$

■

EXERCISES FOR SECTION 6.4

In Exercises 1–8, find the indicated derivative.

1. $F(x) = \int_1^x 4t^2 \, dt$; find $F'(x)$.

2. $G(x) = \int_2^x (t^2 - 1) \, dt$; find $G'(x)$.

3. $F(t) = \int_3^t \sqrt{4x} \, dx$; find $F'(t)$.

4. $H(z) = \int_4^z \cos x \, dx$; find $H'(z)$.

5. $F(t) = \int_1^t \sin x \, dx$; find $F'(t)$.

6. $g(t) = \int_3^t \sin x \, dx$; find $g'(t)$.

7. $H(x) = \int_0^x (3 - t)^{-1} \, dt$; find $H'(x)$.

8. $G(x) = \int_2^x \sqrt{t + 2} \, dt$; find $G'(x)$.

In Exercises 9–32, evaluate the given definite integrals.

9. $\int_0^1 5x^4 \, dx$

10. $\int_0^3 \theta^3 \, d\theta$

11. $\int_1^2 3t \, dt$

12. $\int_1^2 t^4 \, dt$

13. $\int_2^3 12x^{-3} \, dx$

14. $\int_1^3 x^{-3} \, dx$

15. $\int_0^4 \sqrt{t} \, dt$

16. $\int_1^8 \sqrt[3]{m} \, dm$

17. $\int_2^4 (3t^2 - 5) \, dt$

18. $\int_1^4 z^2 \sqrt{z} \, dz$

19. $\int_{-2}^1 (q - q^3) \, dq$

20. $\int_{-1}^3 (v^2 + 2)^2 \, dv$

21. $\int_{-8}^0 (x^{1/3} + 2)x \, dx$

22. $\int_5^{10} \frac{dw}{w^2}$

23. $\int_{1/4}^{1/2} \frac{dx}{2x^2}$

24. $\int_{-3}^{-1} p^2 \left(1 - \frac{1}{p^2} \right) dp$

25. $\int_2^4 \sqrt{8x} \, dx$

26. $\int_1^4 x^{1/2}(3x^2 - 5x) \, dx$

27. $\int_0^1 (2x - 1)^3 \, dx$

28. $\int_0^2 (2x - 2)^4 \, dx$

29. $\int_0^7 \sqrt{t + 1} \, dt$

30. $\int_0^5 3\sqrt{x + 4} \, dx$

31. $\int_1^2 \dfrac{x \, dx}{(x^2 + 1)^2}$

32. $\int_{-1}^1 \dfrac{x \, dx}{\sqrt{x^2 + 3}}$

In Exercises 33–40, make a sketch of the graph of the given function, and compute the area under the curve for the indicated interval.

33. $y = x^2; 1 \le x \le 3$

34. $y = x^2; 1 \le x \le 4$

35. $y = 2 + x^3; 0 \le x \le 2$

36. $y = 1 + x^2; 0 \le x \le 3$

37. $y = \sqrt{x}; 0 \le x \le 4$

38. $y = \sqrt{x}; 1 \le x \le 9$

39. $y = 1/x^2; 1 \le x \le 5$

40. $y = 1/\sqrt{x}; 1 \le x \le 4$

41. The velocity (in feet per second) of a car varies with time according to $v = 2t + \sqrt{t}$. Compute the distance traveled by the car from $t = 0$ to $t = 9$ sec.

42. In Exercise 41, how far does the car move from $t = 9$ to $t = 16$ sec?

▮▮▮▮ 6.5 PROPERTIES OF THE DEFINITE INTEGRAL

The definite integral has several important properties. The validity of these properties can be proven, but we will be content to simply explain them.

PROPERTY 3 | **Interchange of the Limits of Integration**

$$\int_a^b f(x) \, dx = -\int_b^a f(x) \, dx$$

Expressed in words, when the upper and lower limits of integration are interchanged, the sign of the definite integral changes.

EXAMPLE 1   ▪ **Problem**

Show that $\int_1^2 3y^2 \, dy = -\int_2^1 3y^2 \, dy.$

■ **Solution**

The integral on the left has the value

$$\int_1^2 3y^2 \, dy = \left[y^3 \right]_1^2 = 8 - 1 = 7$$

The value of the integral on the right is

$$-\int_2^1 3y^2 \, dy = \left[-y^3 \right]_2^1 = -1 + 8 = 7 \qquad ■$$

PROPERTY 4

Integration over the Union of Two Intervals   If $a \le X \le b$, then

$$\int_a^b f(x) \, dx = \int_a^X f(x) \, dx + \int_X^b f(x) \, dx$$

The plausibility of Property 4 is established by using the area interpretation of the definite integral. Referring to Figure 6.14, we see that the area from a to b is equal to the area from a to X plus the area from X to b.

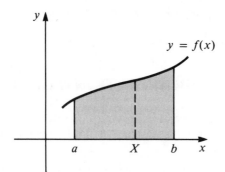

■ **Figure 6.14**

PROPERTY 5

Mean Value Theorem for Integrals   If f is continuous on $[a, b]$, then there is a number X on the interval such that

$$\int_a^b f(x) \, dx = f(X)(b - a)$$

Property 5 has an interesting and useful geometric interpretation. The mean value theorem says that there is a rectangle of width $b - a$ and height $f(X)$ that has the same area as the area under the curve $y = f(x)$. The area of the rectangle in Figure 6.15 is the same as the area under the curve.

The height $f(X)$ is sometimes called the **average height** of the curve on the

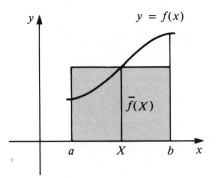

■ **Figure 6.15**

interval $[a, b]$ and is denoted by $\overline{f}(X)$. The average height is given by

$$\overline{f}(X) = \frac{1}{b - a} \int_a^b f(x) \, dx$$

EXAMPLE 2   ■ **Problem**

Find the dimensions of the rectangle that has the same area as the region under the curve $y = x^2$ from $x = 0$ to $x = 2$.

■ **Solution**

From our previous work, we know that the area under the curve $y = x^2$ from 0 to 2 is $\frac{8}{3}$ square units. By Property 5, we have

$$\frac{8}{3} = \overline{f}(X)(2 - 0)$$

Solving for $\overline{f}(X)$, we get

$$\overline{f}(X) = \frac{4}{3}$$

The dimensions of the desired rectangle are then $\frac{4}{3} \cdot 2$. This result is shown geometrically in Figure 6.16.

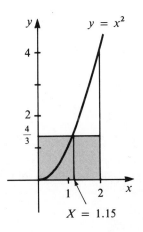

■ **Figure 6.16**

EXAMPLE 3   ▪ **Problem**

Find the point X at which the height of the curve is equal to the average height of $y = x^2$ from 0 to 2.

▪ **Solution**

From Example 2, $\bar{f}(X) = \frac{4}{3}$. Therefore, X must be the value such that $\frac{4}{3} = X^2$, or $X = \sqrt{\frac{4}{3}} \approx 1.15$. Figure 6.16 shows this value.   ▪

EXAMPLE 4   ▪ **Problem**

Find the average velocity of a particle whose velocity (in millimeters per second) varies with time according to $v = 2t^2 + 5t$ over the time interval $[1, 3]$.

▪ **Solution**

The average velocity can be written as

$$\bar{v}(T) = \frac{1}{3-1} \int_1^3 (2t^2 + 5t)\, dt = \frac{1}{2}\left[\frac{2t^3}{3} + \frac{5t^2}{2}\right]_1^3$$

$$= \frac{1}{2}\left[18 + \frac{45}{2}\right] - \frac{1}{2}\left[\frac{2}{3} + \frac{5}{2}\right] = \frac{81}{4} - \frac{19}{12} = \frac{56}{3} \approx 18.7 \text{ mm/sec} \quad ▪$$

EXERCISES FOR SECTION 6.5

In Exercises 1–8, show that the left side of the equation equals the right side.

1. $\int_1^3 x^3\, dx = -\int_3^1 x^3\, dx$

2. $\int_0^3 x^5\, dx = -\int_3^0 x^5\, dx$

3. $\int_0^2 2x\, dx = -\int_2^0 2x\, dx$

4. $\int_{-1}^1 x^2\, dx = -\int_1^{-1} x^2\, dx$

5. $\int_{-1}^3 (x^3 - 5x)\, dx = \int_{-1}^2 (x^3 - 5x)\, dx + \int_2^3 (x^3 - 5x)\, dx$

6. $\int_0^4 s^{3/2}\, ds = \int_0^1 s^{3/2}\, ds + \int_1^4 s^{3/2}\, ds$

7. $\int_0^4 s^{3/2}\, ds = \int_0^2 s^{3/2}\, ds + \int_2^4 s^{3/2}\, ds$

8. $\int_{-1}^2 3x\, dx = \int_{-1}^0 3x\, dx + \int_0^2 3x\, dx$

9. Find the average height of the curve $y = x + 2$ from $x = 0$ to $x = 2$. Draw the curve and indicate $\bar{f}(X)$.

10. How does the answer obtained in Exercise 9 compare with your intuitive notion of the average height of the indicated straight line?

11. Find the average height of the curve $y = 3x^2$ over the interval $[1, 2]$. At what point in $[1, 2]$ does the height of the curve equal the average height?

12. Find the average height of the curve $y = \sqrt{x}$ from $x = 0$ to $x = 3$. At what point X in $[0, 3]$ does the average height occur?

In Exercises 13–20, compute the average height of the curve over the indicated interval. Draw a sketch of the curve showing the average value.

13. $y = 1 + x^2; 0 \leq x \leq 3$

14. $y = 4x^3; 0 \leq x \leq 2$

15. $y = 4x - x^2; 0 \leq x \leq 3$

16. $y = 5 - 2x; 0 \leq x \leq 2$

17. $y = 1/x^2; 1 \leq x \leq 4$

18. $y = \frac{1}{2}x^2; 2 \leq x \leq 5$

19. $y = \sqrt{x}; 0.5 \leq x \leq 1$

20. $y = 1/x^3; 1 \leq x \leq 2$

21. During the first 2 sec of motion, the velocity (in feet per second) of a rocket is given by $v = t^2 + 3t$. What is the rocket's (a) minimum velocity, (b) maximum velocity, and (c) average velocity during this time period?

22. Is the average velocity obtained in Exercise 21(c) equal to the arithmetic average of the two extreme velocities?

23. How far did the rocket in Exercise 21 travel during the first 2 sec?

24. The velocity (in feet per second) of a car is given by $v = 25\sqrt{t}$.
    (a) Compute the average velocity over $[0, 2]$ sec.
    (b) Compute the average velocity over $[2, 4]$ sec.

25. The electric current (in amperes) in a circuit varies with time according to $i = 0.4t^2$. Find the average current in the circuit from $t = 0$ to $t = 5$ sec.

26. The force (in pounds) applied to a brake increases according to $F = 5 + 0.1t^2$, where t is the elapsed time (in seconds). Find the average force applied to the brake from $t = 0$ to $t = 10$ sec.

6.6 APPROXIMATE INTEGRATION

In the previous sections of this chapter, we showed that every definite integral can be interpreted as the algebraic sum of the areas between the graph of the integrand function and the x-axis. The practice of determining this area by finding an anti-derivative and then using the fundamental theorem is limited to those cases where finding such an antiderivative is convenient. In practice, we meet functions that either are not defined by a rule for all values of x or are very difficult or impossible to antidifferentiate.

To integrate any of this broad set of functions, we resort to our area-under-the-curve interpretation. If the area can be divided into parts, the area of each of which is well known, we can often do the integration — that is, find the value of $\int_a^b f(x)\,dx$ — even when we do not know the exact formula for the function.

EXAMPLE 1   ▪ **Problem**

The graph of $f(x)$ is shown in Figure 6.17. Find $\int_1^6 f(x)\,dx$.

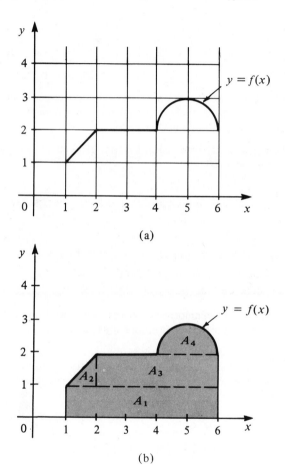

(a)

(b)

▪ **Figure 6.17**

▪ **Solution**

Here, the total area under the curve is given by $A_1 + A_2 + A_3 + A_4$. Each area is a well-known geometric figure.

$$A_1 = 1 \cdot 5 = 5 \qquad A_3 = 1 \cdot 4 = 4$$
$$A_2 = \frac{1}{2} \cdot 1 \cdot 1 = \frac{1}{2} \qquad A_4 = \frac{1}{2}\pi(1)^2 = \frac{1}{2}\pi$$

Using the area interpretation of the definite integral, we get

$$\int_1^6 f(x)\, dx = 5 + \frac{1}{2} + 4 + \frac{1}{2}\pi = 9\frac{1}{2} + \frac{1}{2}\pi \approx 9.5 + 1.57 = 11.07 \qquad \blacksquare$$

In most practical cases, the graph of the integrand function will not permit such a precise division as in the previous example. We usually try to approximate the area by using geometric figures *whose area is well known*. For example, if we were to take a rather general situation, as shown in Figure 6.18, we could approximate $\int_1^4 f(x)\, dx$ by evaluating the areas of the "outer rectangles" in Figure 6.18(a), the "inner rectangles" in Figure 6.18(b), or the trapezoids in Figure 6.18(c). Note further that to find the areas of the rectangles or the trapezoids, the functional values need be known only at discrete points on the interval — in this case, at $x = 1, 2, 3,$ and 4.

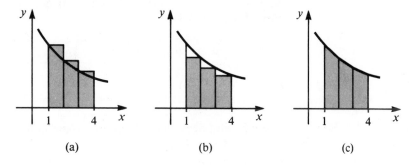

■ **Figure 6.18**   (a)   (b)   (c)

The method of approximating $\int_a^b f(x)\, dx$ by using rectangles is exactly the same technique as was shown in Section 6.2 for defining the definite integral. We exhibit the technique of *approximating trapezoids* in Example 2.

EXAMPLE 2   ■ **Problem**

Using three trapezoids, approximate $\int_2^8 \dfrac{dx}{x}$.

■ **Solution**

The graph of the integrand is shown in Figure 6.19 with the approximating trapezoids. Here, $\int_2^8 dx/x \approx A_1 + A_2 + A_3$.

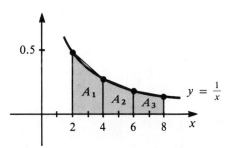

■ **Figure 6.19**

Recalling that *the area of a trapezoid is the height times the average of the bases,* we have

$$A_1 = \frac{2}{2}\left(\frac{1}{2} + \frac{1}{4}\right) = \frac{3}{4} \quad A_2 = \frac{2}{2}\left(\frac{1}{4} + \frac{1}{12}\right) = \frac{5}{12} \quad A_3 = \frac{2}{2}\left(\frac{1}{6} + \frac{1}{8}\right) = \frac{7}{24}$$

The approximate value of $\int_2^8 dx/x$ is then

$$\int_2^8 \frac{dx}{x} \approx \frac{3}{4} + \frac{5}{12} + \frac{7}{24} = \frac{35}{24} = 1.46$$

■

EXAMPLE 3 ■ **Problem**

In an experiment to determine the charge transferred past a point, the current was recorded every 0.1 sec. Determine the average current during the first $\frac{1}{2}$ sec if the following current readings were taken.

t (sec)	0	0.1	0.2	0.3	0.4	0.5	0.6	0.7	0.8
i (A)	0	2.3	2.9	2.4	1.9	1.5	1.2	1.0	0.9

■ **Solution**

We plot the points and draw lines connecting these points. The resulting area consists of five trapezoids. See Figure 6.20.

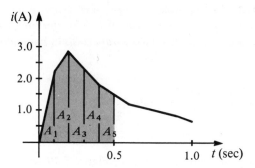

■ **Figure 6.20**

The average current during the first $\frac{1}{2}$ sec is given by

$$i_{avg} = \frac{1}{0.5 - 0}\int_0^{0.5} i\, dt \approx 2(A_1 + A_2 + A_3 + A_4 + A_5)$$

$$= 2\left[\frac{1}{2}(0 + 2.3)0.1 + \frac{1}{2}(2.3 + 2.9)0.1 + \frac{1}{2}(2.9 + 2.4)0.1\right.$$

$$\left. + \frac{1}{2}(2.4 + 1.9)0.1 + \frac{1}{2}(1.9 + 1.5)0.1\right] = 2.05 \text{ A}$$

■

The technique of using trapezoids to estimate areas under curves is known as the **trapezoidal rule.** Although this approach is sometimes modified into a single rule, we encourage you to work with sums of individual trapezoids.

EXERCISES FOR SECTION 6.6

Use trapezoids to estimate the values of the definite integrals given in Exercises 1–12. The number of trapezoids to be used is given to the right of the integrals.

1. $\int_0^3 (4x + 3) \, dx$; 1

2. $\int_1^2 z^2 \, dz$; 4

3. $\int_2^8 \frac{dx}{x^2}$; 3

4. $\int_0^2 \sqrt{4 - x^2} \, dx$; 2

5. $\int_1^4 \sqrt{9 + y} \, dy$; 3

6. $\int_0^2 (2x - 3x^2) \, dx$; 2

7. $\int_1^4 \sqrt{t^2 + t} \, dt$; 3

8. $\int_0^4 \sqrt[3]{3t + 5} \, dt$; 2

9. $\int_1^5 \sqrt[3]{3x^2 - 1} \, dx$; 2

10. $\int_1^2 \frac{dx}{3x^2 + 2}$; 4

11. $\int_0^3 \sqrt{y^2 + 3} \, dy$; 4

12. $\int_1^3 \frac{dx}{x(x + 3)}$; 5

13. The velocity (in centimeters per second) of an electrically charged particle is found to be $v = 1/(1 + t)$. Use three trapezoids to approximate the displacement of the particle from $t = 0$ to $t = 9$ sec.

14. The angular velocity ω (in radians per second) of a radar antenna while tracking a satellite is $\omega = 0.1(1 + t)^{1/2}$. Use two trapezoids to approximate the angular displacement θ from $t = 3$ to $t = 5$ sec.

15. An accelerometer is used to record the vertical acceleration of an airplane due to wind gusts. During a brief period of turbulence the accompanying data was recorded.

t (sec)	0	0.01	0.02	0.03	0.04	0.05	0.06
a (ft/sec^2)	0	0.05	0.12	0.25	0.40	0.45	0.47

Find the average acceleration during this interval.

16. In an experiment to determine the electric charge transferred to a copper plate immersed in a silver solution, the electric current is recorded every 0.1 sec. Find the charge q transferred given that the accompanying table represents the current and that charge is the integral of current.

t (sec)	0	0.1	0.2	0.3	0.4	0.5
i (A)	0	0.1	0.5	0.7	0.8	0.7

17. The rate R of fuel consumption of an experimental engine is monitored every 2 min by a flowmeter, and the data is recorded in the accompanying table. Determine how much fuel F is used in the experiment if

$$F = \int_{t_1}^{t_2} R \, dt$$

t (min)	0	2	4	6	8	10
R (gal/min)	0.2	0.8	1.9	3.0	3.4	3.6

18. The force F delivered by a piston was recorded as a function of the displacement S of the piston from top-dead center. What was the average force delivered by the piston if the relation between F and S is described by the accompanying table?

S (in.)	0	1	2	3	4	5
F (lb)	140	90	50	30	15	10

19. The work done by an expanding gas is

$$\text{work} = \int_{v_0}^{v_1} p \, dv$$

where v is volume (in cubic feet) and p is pressure (in pounds per square foot). What is the work done by a gas if the pressure-volume relationship is as given in the accompanying table?

p (lb/ft^2)	35.2	30.5	26.9	24.3
v (ft^3)	5.7	6.6	7.4	8.3

20. What is the work done on a gas if the pressure-volume relationship is as given in the accompanying table?

p (lb/ft^2)	14.7	16.7	18.7	20.7	22.7
v (ft^3)	3.4	3.0	2.7	2.5	2.4

Use the following BASIC program to estimate the definite integrals in Exercises 1–12 by trapezoidal areas.

```
10   REM   TRAPEZOIDAL RULE
20   HOME
30   PRINT "THIS PROGRAM WILL FIND THE VALUE OF THE"
40   PRINT "INTEGRAL USING THE TRAPEZOIDAL RULE,"
50   PRINT "WITH AN INTERVAL AND A VALUE OF N"
60   PRINT "SUPPLIED BY THE USER."
70   PRINT : PRINT "THE FUNCTION IN THIS PROGRAM"
80   PRINT "IS   X^2 - 4*X + 5"
90   PRINT "IF THE USER CHANGES THE FUNCTION,"
100  PRINT "THE NEW FUNCTION WILL REMAIN UNTIL"
110  PRINT "THE USER CHANGES IT AGAIN, OR"
120  PRINT "RELOADS THE PROGRAM"
```

```
130   PRINT : PRINT "DO YOU WISH TO CHANGE THE FUNCTION? Y/N"
140   INPUT A$
150   IF A$ = "Y" THEN 370
160   PRINT "ENTER THE ENDPOINTS OF THE INTERVAL"
170   INPUT A,B
180   PRINT "HOW MANY SUBINTERVALS DO YOU WISH TO USE?"
190   INPUT N
200   H = (B - A) / N
210   N1 = N - 1
220   HTAB 12: FLASH : PRINT "PATIENCE PLEASE"
230   HTAB 12: PRINT "CALCULATING!!!": NORMAL
240   DEF  FN F(X) = X ^ 2 - 4 * X + 5
250   S = 0
260   FOR I = 1 TO N1
270   S = S +  FN F(A + I * H)
280   NEXT I
290   S = (S + ( FN F(A) +  FN F(B)) / 2) * H
300   PRINT "THE TRAPEZOIDAL APPROXIMATION, USING ";N;" SUBINTERVALS"
305   S =  INT (10000 * S) / 10000
310   PRINT "FROM ";A;" TO ";B;" IS APPROXIMATELY ";S
320   PRINT : PRINT "DO YOU WISH TO RUN THIS PROGRAM AGAIN? Y/N"
330   INPUT B$
340   HOME
350   IF B$ = "Y" THEN 70
360   GOTO 400
370   PRINT "TYPE"
380   INVERSE : PRINT "240 DEF FN F(X) = (YOUR FUNCTION) <RET>"
390   PRINT "RUN 160 <RET>": NORMAL
400   END
```

REVIEW EXERCISES FOR CHAPTER 6

1. Estimate the area under the curve $y = 0.5x^2$ from 0 to 4 by using four subintervals and selecting X_i to be the midpoint of each subinterval. Sketch the graph, and show each rectangle to scale.

2. Estimate the area under the curve $y = 4 - x^2$ from 0 to 2 by using four subintervals and selecting X_i to be the midpoint of each subinterval. Sketch the graph, and show each rectangle to scale.

3. Estimate the area under the curve $y = 1/x$ from 1 to 5. Use two equal subintervals, and choose X_i as the midpoint of each subinterval.

4. Estimate the average height of the curve $y = x^3$ from $x = 1$ to $x = 4$. Use three subintervals, and choose X_i as the midpoint of each subinterval.

5. Estimate the displacement of a car from $t = 0$ to $t = 2$ sec given that the velocity (in feet per second) of the car is expressed as $v = 5t^2 + 3$. Use four subintervals, and choose T_i to be the midpoint of each subinterval.

6. Find the value of $\int_0^3 \sqrt{9 - x^2}\, dx$ by using a geometric formula.

In Exercises 7–10, use the fact that $\int_1^5 f(x)\,dx$ 12 and $\int_1^5 g(x)\,dx = 9$ to evaluate the given definite integrals.

7. $\displaystyle\int_1^5 3f(t)\,dt$

8. $\displaystyle\int_1^5 8g(x)\,dx$

9. $\displaystyle\int_1^5 [f(x) + 5g(x)]\,dx$

10. $\displaystyle\int_1^5 [2f(t) - g(t)]\,dt$

In Exercises 11–24, evaluate the given definite integrals.

11. $\displaystyle\int_1^3 2x\,dx$

12. $\displaystyle\int_0^4 \tfrac{1}{2}t\,dt$

13. $\displaystyle\int_2^3 (3x + 1)\,dx$

14. $\displaystyle\int_1^5 (3 - x)\,dx$

15. $\displaystyle\int_{-2}^0 x^2\,dx$

16. $\displaystyle\int_{-1}^1 (3x^2 - 2)\,dx$

17. $\displaystyle\int_0^8 (1 + \sqrt[3]{t})\,dt$

18. $\displaystyle\int_1^4 x^{3/2}\,dx$

19. $\displaystyle\int_0^2 (2x - 4)^3\,dx$

20. $\displaystyle\int_3^4 \left(\frac{3}{t^4}\right)\,dt$

21. $\displaystyle\int_0^3 x(2x^2 - 25)^{-2}\,dx$

22. $\displaystyle\int_0^1 \sqrt{x + 3}\,dx$

23. $\displaystyle\int_0^1 (\sqrt{x} + 2x)\sqrt{x}\,dx$

24. $\displaystyle\int_{-1}^2 (x^2 + 3)^2\,dx$

In Exercises 25–30, make a sketch of the graph of the given function, and compute the area between the curve and the x-axis for the indicated interval.

25. $y = 5 - x^2$; $-1 \le x \le 0$

26. $y = x(2x + 7)$; $-2 \le x \le 0$

27. $y = 1 + \sqrt{x}$; $0 \le x \le 4$

28. $y = \sqrt[3]{x}$; $1 \le x \le 8$

29. $y = (x - 3)^2$; $3 \le x \le 6$

30. $y = (x + 1)^2$; $-1 \le x \le 1$

31. Find the average height of the curve $y = 1 + x^3$ from $x = 0$ to $x = 5$.

32. Find the average height of the curve $y = 3x - 2x^2$ from $x = 0$ to $x = 3$.

33. What is the displacement of a charged particle from $t = 0$ to $t = 3$ sec if the velocity (in centimeters per second) of the particle is given by $v = 16t + \tfrac{1}{2}t^2$?

34. What is the average velocity of the particle in Exercise 33?

35. Find the distance traveled on $0 \le t \le 2$ sec by a particle whose velocity (in feet per second) is given by $v = t(t + 1)$. Find the average velocity over this interval.

36. What is the average force applied over the 2-sec interval $0 \le t \le 2$ if the force varies with t according to $F(t) = 2t^2 + 3t - 2$?

37. Use four trapezoids to estimate the value of $\int_0^8 \sqrt{t^2 + 4} \, dt$.

38. Use three trapezoids to estimate the value of $\int_0^{0.6} \sqrt{2x + x^2} \, dx$.

CHAPTER 7

Applications of the Definite Integral

The definite integral of a function is the limit of a sum of terms of the form $f(x_i)\,\Delta x_i$. Once the form of this term is determined, we can perform the integration. Each individual term is called an **element of integration.** As you will see, in many of the applications, there will be some geometric clue about the nature of this element of integration.

For convenience, we denote an element of integration by $f(x)\,dx$ or $y\,dx$ (or something like this with other letters), instead of the more formal $f(x_i)\,\Delta x_i$.

In Chapter 6, we introduced the definite integral by identifying the common thread in the solution to an area problem and a displacement problem. In this chapter, we will look at some other applications of the definite integral. These applications vary from the computation of the centroid of an area between two curves to that of the force on a dam exerted by the stored water. The integration problems covered in this chapter typify the use of the definite integral in engineering applications.

7.1 AREA

In the area problems that we have considered so far, the curve has been *above* the x-axis. However, as the first example shows, this restriction is not necessary.

EXAMPLE 1 ■ **Problem**

Find the area bounded by the curve $y = x^2 - 5x$ and the x-axis from $x = 1$ to $x = 4$. See Figure 7.1.

■ **Solution**

The area between the curve and the x-axis with a typical rectangular strip is shown in Figure 7.1. Since the graph of $y = x^2 - 5x$ lies below the x-axis on the interval $(1, 4)$, $y(x)$ is negative. Hence, the area of a typical rectangular element is given by

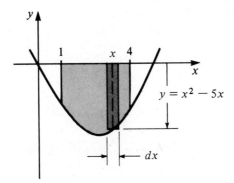

■ Figure 7.1

$$dA = -y(x)\, dx = -(x^2 - 5x)\, dx$$

The desired area is then given by

$$A = -\int_1^4 (x^2 - 5x)\, dx$$

Using Theorem 2, Chapter 6 we have

$$A = -\int_1^4 (x^2 - 5x)\, dx = -\left[\frac{x^3}{3} - \frac{5x^2}{2}\right]_1^4$$

$$= -\left[\left(\frac{64}{3} - \frac{80}{2}\right) - \left(\frac{1}{3} - \frac{5}{2}\right)\right] = \frac{33}{2}$$

■

We frequently encounter functions whose graphs lie alternately above and below the x-axis over the interval of integration, as shown in Figure 7.2. To find the area bounded by this curve, the x-axis, and the lines $x = a$ and $x = b$, we must realize that $f(x)$ is positive between a and c and negative between c and b. The area A_1 is given by

$$A_1 = \int_a^c f(x)\, dx$$

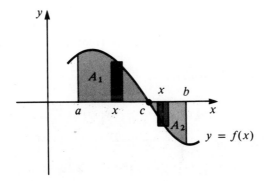

■ Figure 7.2

To obtain a positive area from $x = c$ to $x = b$, we write the area of the typical element as $dA_2 = -f(x)\,dx$. Hence

$$A_2 = -\int_c^b f(x)\,dx$$

The desired area is then $A = A_1 + A_2$.

EXAMPLE 2   ■ **Problem**

Consider the curve $y = x^2 - 4$ from $x = 0$ to $x = 5$. Find the area bounded by the curve and the x-axis. See Figure 7.3.

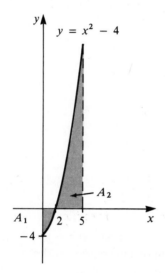

■ **Figure 7.3**

■ **Solution**

We locate the points at which the curve crosses the x-axis. By letting $y = 0$ and solving for x, we obtain

$$x^2 - 4 = 0 \quad\text{or}\quad x = \pm\, 2$$

Thus, the curve has intercepts at both 2 and -2. Only $x = 2$ is of interest since it is the value within the interval of integration. Hence,

$$A_1 = -\int_0^2 (x^2 - 4)\,dx = -\left[\frac{x^3}{3} - 4x\right]_0^2 = \frac{16}{3}$$

$$A_2 = \int_2^5 (x^2 - 4)\,dx = \left[\frac{x^3}{3} - 4x\right]_2^5 = \left[\left(\frac{125}{3} - 20\right) - \left(\frac{8}{3} - 8\right)\right] = 27$$

$$A = A_1 + A_2 = \frac{16}{3} + 27 = \frac{97}{3}$$

■

EXAMPLE 3   ▪ **Problem**

Find the average height of the curve in Example 2. See Figure 7.4.

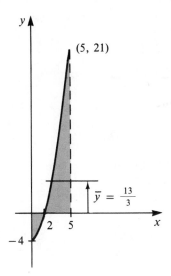

▪ **Figure 7.4**

▪ **Solution**

The average height of a curve was given in Section 6.4 to be

$$\bar{y} = \frac{1}{b-a} \int_a^b f(x)\, dx$$

Therefore,

$$\bar{y} = \frac{1}{5-0} \int_0^5 (x^2 - 4)\, dx = \frac{1}{5}\left[\frac{x^3}{3} - 4x\right]_0^5 = \frac{1}{5}\left(\frac{65}{3}\right) = \frac{13}{3}$$

Notice that the fact that the curve crosses the x-axis between 0 and 5 does *not* affect the use of the formula for average height.   ▪

The area bounded by a curve and the y-axis can also be stated as a definite integral. The next example shows how.

EXAMPLE 4   ▪ **Problem**

Find the area bounded by the curve $y = x^{1/2}$, the y-axis, and the lines $y = 1$ and $y = 3$. See Figure 7.5.

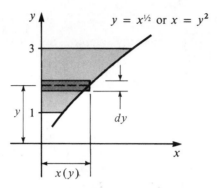

$y = x^{1/2}$ or $x = y^2$

■ Figure 7.5

■ Solution

In this problem, a typical element of integration is a horizontal strip dy units wide. Denoting the length of a typical rectangle by $x(y)$, we have

$$dA = x(y)\,dy$$

The desired area may then be written as

$$A = \int_1^3 x(y)\,dy$$

Before we can evaluate this definite integral, we must replace $x(y)$ with its functional value in terms of y. Since the given equation can be expressed in the form $x = y^2$, we have

$$A = \int_1^3 x(y)\,dy = \int_1^3 y^2\,dy = \left[\frac{y^3}{3}\right]_1^3 = 9 - \frac{1}{3} = \frac{26}{3} \qquad ■$$

EXERCISES FOR SECTION 7.1

In Exercises 1–10, find the *total area* bounded by the curve and the *x*-axis for the indicated interval on the *x*-axis. Sketch each figure.

1. $y = 2x - 6; x = 2$ to $x = 5$

2. $y = 1 - x^2; x = 0$ to $x = 3$

3. $y = x^3; [-1, 2]$

4. $y = x^2 - 2x; [1, 4]$

5. $y = x^3 + 1; -3 \le x \le 0$

6. $y = x^2 - 5x + 6; x = 0$ to $x = 4$

7. $y = x^2 + x - 2; x = -1$ to $x = 3$

8. $y = x^3 - 3x^2 + 2x; 0 \le x \le 2$

9. $y = 4 - x^2; -1 \le x \le 3$

10. $y = x - x^2; -1 \le x \le 3$

In Exercises 11–14, find the average height of the curve over the indicated interval on the *x*-axis.

11. $y = \sqrt[3]{x}; [0, 8]$

12. $y = x^2 - 4x; [0, 2]$

13. $y = x^{-1/2}; 4 \le x \le 9$

14. $y = \frac{1}{x^2}; \frac{1}{2} \le x \le 2$

In Exercises 15–20, find the area bounded by the curve, the *y*-axis, and the indicated interval on the *y*-axis. Sketch each figure.

15. $y = x; 2 \le y \le 5$

16. $y^2 = x^3; 1 \le y \le 8$

17. $y = 6x - 6; y = 3$ to $y = 6$

18. $y = x^{1/2} - 1; y = 0$ to $y = 3$

19. $y^3 = x^2; [1, 4]$

20. $y = x^3; [0, 8]$

21. What is the average height of a curve that bounds equal areas above and below the *x*-axis on the interval of integration?

7.2 AREA BETWEEN TWO CURVES

In this section, we consider areas bounded by two curves in the *xy*-plane. A special case occurs when one of the curves is the *x*- or the *y*-axis, a problem discussed in the previous section.

EXAMPLE 1   ▪ **Problem**

Find the area bounded by $y = 7x$, $y = x^2$, $x = 2$, and $x = 5$. This area is shown in Figure 7.6.

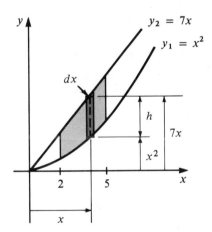

▪ **Figure 7.6**

▪ **Solution**

In Figure 7.6, we see that a typical rectangle is bounded above by the curve $y_2 = 7x$ and below by the curve $y_1 = x^2$. The height of a typical element is then of the form

$$h = 7x - x^2$$

The area of a typical rectangle is

$$dA = (7x - x^2)\, dx$$

The area bounded by the two curves is then given by

$$A = \int_2^5 (7x - x^2)\, dx = \left[\frac{7x^2}{2} - \frac{x^3}{3}\right]_2^5 = \left[\frac{175}{2} - \frac{125}{3}\right] - \left[\frac{28}{2} - \frac{8}{3}\right]$$

$$= \frac{207}{6} = \frac{69}{2} \qquad\blacksquare$$

In using the definite integral to find the area between curves, give careful consideration to the choice of elements. That is, decide whether to use elements parallel to the y-axis or parallel to the x-axis. In most cases, a sketch of the area will suggest the type of element.

EXAMPLE 2   ■ **Problem**

Find the area formed by the intersection of the curves $y = x - 4$ and $y^2 = 2x$.

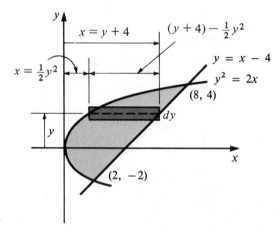

■ **Figure 7.7**

■ **Solution**

Figure 7.7 shows a horizontal element with its right endpoint on the line $y = x - 4$ and the left endpoint on $y^2 = 2x$. Thus, the area of a typical rectangular element is

$$dA = [(y + 4) - \tfrac{1}{2} y^2]\, dy$$

The limits of integration are obtained from the simultaneous solution of the two equations. Substituting $y = x - 4$ into $y^2 = 2x$, we have

$$(x - 4)^2 = 2x$$
$$x^2 - 8x + 16 = 2x$$
$$(x - 2)(x - 8) = 0$$
$$x = 2 \qquad y = -2$$
$$x = 8 \qquad y = 4$$

Since the elements are horizontal, we choose $y = -2$ and $y = 4$ for the limits of integration. Thus, the desired area is

$$A = \int_{-2}^{4} \left[(y + 4) - \frac{y^2}{2} \right] dy = \left[\frac{y^2}{2} + 4y - \frac{y^3}{6} \right]_{-2}^{4}$$

$$= \left(8 + 16 - \frac{64}{6} \right) - \left(2 - 8 + \frac{8}{6} \right) = 18 \qquad \blacksquare$$

EXAMPLE 3 ▪ **Problem**

Find the area for Example 2 by using vertical elements instead of horizontal elements. See Figure 7.8.

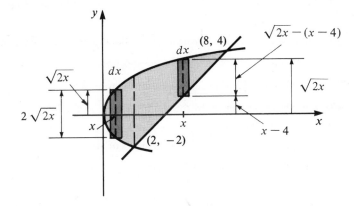

▪ **Figure 7.8**

▪ **Solution**

Figure 7.8 shows that two different types of vertical elements must be used.

1. The elements between $x = 0$ and $x = 2$ have *both* ends on the curve $y^2 = 2x$, so the height of a typical element is $2\sqrt{2x}$.

2. The elements between $x = 2$ and $x = 8$ have their upper end on the parabola and their lower end on the straight line, so the height of each element is given by $\sqrt{2x} - (x - 4)$. Thus, the area between the curves is

$$A = \int_{0}^{2} 2\sqrt{2x}\, dx + \int_{2}^{8} (\sqrt{2x} - x + 4)\, dx$$

$$= \left[\frac{4\sqrt{2}}{3} x^{3/2} \right]_{0}^{2} + \left[\frac{2\sqrt{2}}{3} x^{3/2} - \frac{1}{2} x^2 + 4x \right]_{2}^{8}$$

$$= \left[\frac{16}{3} - 0 \right] + \left[\left(\frac{64}{3} - 32 + 32 \right) - \left(\frac{8}{3} - 2 + 8 \right) \right] = 18 \qquad \blacksquare$$

Comment: The solution can be obtained by either choice of elements. However, the use of horizontal elements is easier in the case just discussed.

EXERCISES FOR SECTION 7.2

Find the area bounded by the given curves. Sketch the figure. Choose the most convenient element of area in solving each problem.

1. $y = 2x; y = x^2$

2. $y = 4x - x^2; y = x$

3. $y = x^4; y = x^2$

4. $y = 9x; y = x^3$, first quadrant only

5. $y = x^2; y = 4$

6. $y^2 = x; x = 3$

7. $y = 2x - x^2; y = x^2$

8. $y^2 = 4x; y^2 = x^3$

9. $y = -x; y = x^2 - 5x$

10. $y = x^2 - 4; 3x + y = 0$

11. $3y + x - 4 = 0; y^2 = x$

12. $x = 4y - y^2; y^2 = 3x$

13. Find the area bounded by the curve $y = x^2 - 5$ and the straight line passing through the points $(-2, -1)$ and $(3, 4)$.

14. Find the area bounded by $y^2 = x^3$ and the straight line passing through $(1, 1)$ and $(4, 8)$.

15. Using horizontal elements, find the area bounded by $y^2 = 4x$ and $y = 2x - 4$.

16. Solve Exercise 15 by using vertical elements of area.

17. Find the area bounded by the straight lines $y = x$, $y = -3x + 4$, and $x = 0$. Use vertical elements of area.

18. Solve Exercise 17 by horizontal elements of area.

19. Find the area bounded by the curve $y = x^3 - 6x^2$ and the chord connecting the points $(1, -5)$ and $(6, 0)$.

20. Find the area bounded by the curves $y = 5 + x^2$, $y = 2x + x^2$, $x = -1$, and $x = 1$. Sketch the figure.

21. Find the area bounded by the curves $y = 4 + x^2$, $y = x - x^2$, $x = -2$, and $x = 2$. Sketch the figure.

22. Find the area bounded by the curves $y = x^3$, $y = -x - 3$, $y = -1$, and $y = 2$. Sketch the figure.

23. Find the area bounded by the curves $y^2 = x$, $y = 2x + 1$, $y = 0$, and $y = 3$. Sketch the figure.

7.3 CENTROIDS

Figure 7.9 shows a rod that is pivoted at point P with a force F applied to the rod. This applied force will cause the rod to rotate about P. By our knowledge of levers, we know that the tendency of a force to produce rotation is dependent on the distance between the force and the axis of rotation. That is, the farther the line of action of F is from P, the greater will be the tendency of F to produce rotation. In mechanics, the tendency of a force to produce rotation is called torque, or **moment**.

The quantitative measure of the moment of a force is defined to be the product of the force times the perpendicular distance between the force and the axis of rotation. The perpendicular distance L between the force and the axis of rotation is called the *moment arm of the force*.

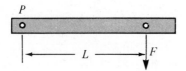

■ **Figure 7.9**

DEFINITION	**Moment** The *moment* of a force is defined as $$M = F \cdot L$$ *where F is the force and L is the moment arm.*

The units of moment are the product of the units used for force and moment arm.

EXAMPLE 1 ■ **Problem**

Compute the moment of a 10-lb force if the moment arm is 5 ft, 12 ft, and 3 in.

■ **Solution**

The moment for each case is

$$M = 10 \cdot 5 = 50 \text{ ft-lb}$$
$$M = 10 \cdot 12 = 120 \text{ ft-lb}$$
$$M = 10 \cdot 3 = 30 \text{ in.-lb}$$

If we wish to express the third answer in foot-pounds, we write

$$M = 10 \left(\frac{3}{12}\right) = 2.5 \text{ ft-lb}$$

■

The idea of a moment of force can be generalized to a system of weights. Figure 7.10 shows a system of three weights on a coordinate line. We can think of the line as a seesaw with a fulcrum at the origin. Then, each weight has an associated

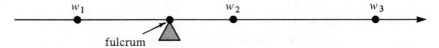

■ **Figure 7.10**

fulcrum

moment, which represents the tendency of the weight to cause the line to rotate, clockwise for a positive moment and counterclockwise for a negative moment. The total moment is the sum of the individual moments:

$$M = w_1x_1 + w_2x_2 + w_3x_3$$

EXAMPLE 2   ■ **Problem**

Weights of 10, 15, and 25 lb are located on the x-axis at -3, 1, and 4 ft, respectively. Find the moment with respect to the origin.

■ **Solution**

The moment is

$$M = 10(-3) + 15(1) + 25(4) = 85 \text{ ft-lb}$$

That is, this system of weights will cause the line to rotate clockwise.   ■

A system is said to be in **rotational equilibrium** if the total moment is zero; in this case, the system has no tendency to rotate. A system that is not in equilibrium can be put into equilibrium by moving the fulcrum to point $\overline{x}$, called the **centroid**, for which the sum of the moments about this point is zero. For the system in Figure 7.10, we have

$$w_1(x_1 - \overline{x}) + w_2(x_2 - \overline{x}) + w_3(x_3 - \overline{x}) = 0$$

Solving for $\overline{x}$ (which takes a little work) gives the following formula:

$$\overline{x} = \frac{w_1x_1 + w_2x_2 + w_3x_3}{w_1 + w_2 + w_3}$$

The numerator of the expression for $\overline{x}$ is the total moment of the system with respect to the origin. The denominator is the total weight of the system.

FORMULA

Centroid of a System

$$\overline{x} = \frac{\text{sum of the moments}}{\text{sum of the weights}}$$

EXAMPLE 3   ■ **Problem**

Locate the centroid of the system of Example 2.

■ **Solution**

Since the total weight is 50 lb,

$$\bar{x} = \frac{85 \text{ ft-lb}}{50 \text{ lb}} = 1.7 \text{ ft}$$

■

The concept of moment can be extended to plane regions by simply making a correspondence between area and weight. Then, if we think of the centroid of a region as the point at which all of the area is concentrated, the moment of the region is the product of the area times the perpendicular distance from the centroid to the axis of rotation. The centroid of a plane region corresponds to its geometric center. For a region in the xy-plane, the coordinates are designated by $(\bar{x}, \bar{y})$. The coordinate $\bar{x}$ is called the *moment arm of the area about the y-axis,* and $\bar{y}$ is called *the moment arm of the area about the x-axis.*

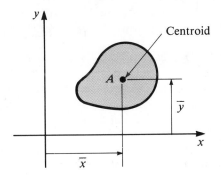

■ **Figure 7.11**

Referring to Figure 7.11, we have the moment M_y of a given area A about the y-axis as

$$M_y = A\bar{x} \qquad\qquad (7.1)$$

The moment M_x of A about the x-axis is

$$M_x = A\bar{y} \qquad\qquad (7.2)$$

where $\bar{x}$ and $\bar{y}$ are the coordinates of the centroid of A.

The following facts are basic to the computation of centroids and moments.

1. The centroid of a rectangle is at its geometric center.
2. The moment of an area about a given axis is equal to the sum of the moments of its component areas about the same axis.

EXAMPLE 4 ■ **Problem**

Compute M_x and M_y for the area shown in Figure 7.12(a).

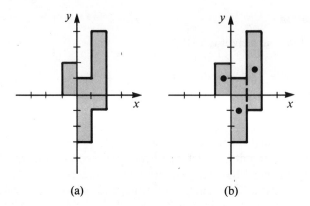

■ **Figure 7.12**   (a)   (b)

■ **Solution**

Figure 7.12(b) shows the given area divided into three rectangles, with the centroid of each rectangle indicated by a dot. We use the coordinates from the figure to determine the various areas and moment arms. The moment M_x is

$$M_x = 2(1) + 4(-1) + 5(1.5) = 5.5$$

Similarly, the moment M_y is

$$M_y = 2(-\tfrac{1}{2}) + 4(\tfrac{1}{2}) + 5(1.5) = 8.5$$ ■

To determine the coordinates of the centroid of a given area, we solve Equation (7.1) and (7.2) for $\overline{x}$ and $\overline{y}$, respectively.

FORMULA

Centroid Coordinates of an Area

$$\overline{x} = \frac{M_y}{A} \tag{7.3}$$

$$\overline{y} = \frac{M_x}{A} \tag{7.4}$$

EXAMPLE 5   ■ **Problem**

Find the centroid of the area shown in Figure 7.12(a).

■ **Solution**

From Example 4, $M_x = 5.5$ and $M_y = 8.5$. The total area is

$$A = (1)(2) + (1)(4) + (1)(5) = 11$$

Hence,

$$\bar{x} = \frac{8.5}{11} \approx 0.77 \quad \text{and} \quad \bar{y} = \frac{5.5}{11} = 0.5$$ ■

In cases where A, M_x, and M_y must be found by integration, the following outline should be followed.

PROCEDURE

Finding Moments of Areas

1. The given area is divided into n rectangular elements.
2. The centroid of each of these rectangles is located at its geometric center.
3. The moment of each element is the product of the elemental area times the directed distance from the centroid of the element to the axis.
4. The moment of the total area is the sum of these elemental moments, which is found by integration.

EXAMPLE 6 ■ **Problem**

Find the centroid of the area bounded by the curve $y = x^2$ and the lines $x = 0$ and $x = 3$. See Figure 7.3.

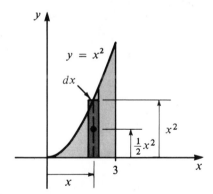

■ **Figure 7.13**

■ **Solution**

A typical rectangular element is shown in Figure 7.13. The area of this element is given by

$$dA = x^2 \, dx$$

So, the area under the curve from $x = 0$ to $x = 3$ is

$$A = \int_0^3 x^2 \, dx = \left[\frac{x^3}{3} \right]_0^3 = 9$$

To compute M_y, we first compute the elemental moments as follows:

$$dM_y = (\text{distance of centroid from } y\text{-axis}) \cdot dA$$

The distance of the centroid of a typical element of area to the y-axis is just x. Consequently,

$$dM_y = x \, dA = x(x^2 \, dx) = x^3 \, dx$$

Therefore, the total moment of the given area about the y-axis is the sum of these elemental moments. Note that the limits of integration are the same as those used in finding the area. Thus,

$$M_y = \int_0^3 x^3 \, dx = \left[\frac{x^4}{4} \right]_0^3 = \frac{81}{4}$$

The elemental moment about the x-axis is given by

$$dM_x = (\text{distance of centroid to } x\text{-axis}) \cdot dA$$

Since the centroid is at the geometric center, its distance from the x-axis is exactly one-half the height of each rectangle. Hence,

$$dM_x = \tfrac{1}{2}x^2 \, dA = \tfrac{1}{2}x^2(x^2 \, dx) = \tfrac{1}{2}x^4 \, dx$$

And the moment of the total area is

$$M_x = \frac{1}{2} \int_0^3 x^4 \, dx = \frac{1}{2} \left[\frac{x^5}{5} \right]_0^3 = \frac{243}{10}$$

Finally (see Figure 7.14), we substitute A, M_x, and M_y into Equations (7.3) and (7.4) to obtain

$$\bar{x} = \frac{M_y}{A} = \frac{81/4}{9} = \frac{9}{4} = 2.25$$

$$\bar{y} = \frac{M_x}{A} = \frac{243/10}{9}$$

$$= \frac{27}{10} = 2.70$$

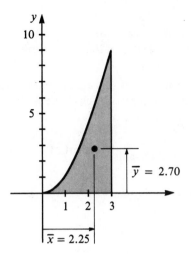

■ Figure 7.14

$\bar{y} = 2.70$

$\bar{x} = 2.25$

EXAMPLE 7 **■ Problem**

Find the centroid of the area bounded by the curve $y = x^3$, the line $y = 10 - x$, and the y-axis.

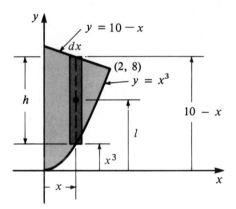

■ Figure 7.15

■ Solution

Choose a vertical element of area, as shown in Figure 7.15. The height of this element is given by

$$h = (10 - x) - x^3$$

since its upper end lies on the line $y = 10 - x$ and its lower end lies on the curve $y = x^3$. The area of the element is then

$$dA = (10 - x - x^3)\, dx$$

And the desired area is

$$A = \int_0^2 [(10 - x) - x^3] \, dx = \left[10x - \frac{x^2}{2} - \frac{x^4}{4} \right]_0^2 = 14$$

Next, we find the moments of this area about the respective axes. Referring to Figure 7.15, we see that the moment of each element about the y-axis is

$$dM_y = x \, dA = x(10 - x - x^3) \, dx$$

Thus, the total moment about the y-axis is

$$M_y = \int_0^2 x[(10 - x) - x^3] \, dx = \int_0^2 (10x - x^2 - x^4) \, dx$$

$$= \left[5x^2 - \frac{x^3}{3} - \frac{x^5}{5} \right]_0^2 = \frac{164}{15}$$

Since the centroid lies at the center of the rectangular elements, its ordinate value l is the average of the upper and lower y values of each rectangle. Thus,

$$l = \tfrac{1}{2}(y_{upper} + y_{lower}) = \tfrac{1}{2}[(10 - x) + x^3]$$

Therefore, the moment of a typical element about the x-axis is

$$M_x = l \, dA = \tfrac{1}{2}(10 - x + x^3)(10 - x - x^3) \, dx = \tfrac{1}{2}[(10 - x)^2 - (x^3)^2]$$

And the total moment about this axis is

$$M_x = \frac{1}{2} \int_0^2 [(10 - x)^2 - (x^3)^2] \, dx = \frac{1}{2} \int_0^2 [x^2 - 20x + 100 - x^6] \, dx$$

$$= \frac{1}{2} \left[\frac{x^3}{3} - 10x^2 + 100x - \frac{x^7}{7} \right]_0^2 = \frac{1516}{21}$$

The coordinates of the centroid are then

$$\bar{x} = \frac{M_y}{A} = \frac{164/15}{14} = \frac{82}{105} \approx 0.78$$

$$\bar{y} = \frac{M_x}{A} = \frac{1516/21}{14} = \frac{758}{147} \approx 5.16$$

∎

EXAMPLE 8   ▪ **Problem**

Find the centroid of the area bounded by the curve $y^2 = 4 - x$ and the y-axis. See Figure 7.16.

▪ **Solution**

By observing the nature of the area, we note that the best approach is to use horizontal elements of area. The horizontal elements are being summed from $y = -2$ to $y = 2$, so the area is

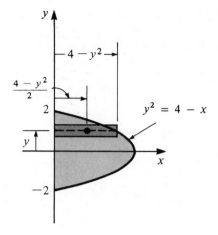

■ **Figure 7.16**

$$A = \int_{-2}^{2} (4 - y^2) \, dy$$

The area is symmetric about the x-axis, so this integral may be simplified by integrating from 0 to 2 and multiplying the integral by 2. Hence,

$$A = 2 \int_{0}^{2} (4 - y^2) \, dy = 2 \left[4y - \frac{y^3}{3} \right]_{0}^{2} = \frac{32}{3}$$

Furthermore, since the centroid must lie on the axis of symmetry, we immediately conclude that $\overline{y} = 0$.

The value of $\overline{x}$ is now computed. The moment arm of each element about the y-axis is $\frac{1}{2}(4 - y^2)$. Therefore, the moment of each element about the y-axis is of the form

$$dM_y = \frac{1}{2}(4 - y^2) \, dA = \frac{1}{2}(4 - y^2)(4 - y^2) \, dy = \frac{1}{2}(4 - y^2)^2 \, dy$$

And M_y is given by

$$M_y = \frac{1}{2} \int_{-2}^{2} (4 - y^2)^2 \, dy = \int_{0}^{2} (4 - y^2)^2 \, dy$$

$$= \int_{0}^{2} (16 - 8y^2 + y^4) \, dy$$

$$= \left[16y - \frac{8y^3}{3} + \frac{y^5}{5} \right]_{0}^{2} = \frac{256}{15}$$

The value of $\overline{x}$ is then

$$\overline{x} = \frac{M_y}{A} = \frac{256/15}{32/3} = \frac{8}{5} = 1.6$$

■

EXERCISES FOR SECTION 7.3

In Exercises 1–4, compute M_x and M_y for the indicated areas. In each exercise, divide the area into three rectangles.

1.

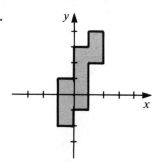

2.

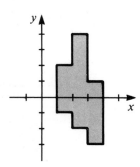

3.

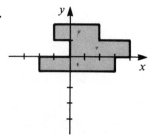

4.

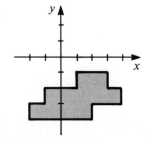

5. Determine the location of the centroid of the area in Exercise 1.

6. Determine the location of the centroid of the area in Exercise 3.

Determine the location of the centroid of the given areas in Exercises 7–23 by using the most convenient elemental moment. Sketch each figure.

7. $y = 2x, x = 3, y = 0$ **8.** $y = 5x, x = 2, y = 0$

9. $y - 3x = 0, x = 0, y = 4$ **10.** $2y + x = 6, x = 0, y = 0$

11. $y - x = 3, x = 0, x = 4, y = 0$ **12.** $2y + x - 4 = 0, x = 0, x = 3, y = 0$

13. $y = x^2, x = 4, y = 0$ **14.** $y = x^3, x = 1, y = 0$

15. $y = x^2, x = 0, y = 9$ **16.** $y = x^3, x = 0, y = 8$

17. $y = x^2 + 2x, y = -2x$ **18.** $y = x^2 - 8x, y = x$

19. $y = 4 - x^2, x = 2, y = 4$ **20.** $y^2 = 4x, y = x - 3$

21. $y - x = 0, 3x + y - 4 = 0, x = 0$ **22.** $y^2 = 4x, y = 2x - 4$

23. $y = -x^2 + 6x - 3, y = (x - 3)(x + 1)$

The following BASIC program computes the location of the centroid of a region bounded by $y = f(x)$, the x-axis, $x = a$, and $x = b$. Use this program to locate the centroid in Exercises 7, 8, 10, 11, 12, 13, and 14.

```
10   REM   PROGRAM TO FIND THE CENTROID
20   REM   USING RIGHT ENDPOINTS OF SUBINTERVALS
30   HOME
40   PRINT "THIS PROGRAM WILL FIND THE CENTROID"
50   PRINT "OF A REGION BOUNDED BY Y=F(X)"
60   PRINT "THE X-AXIS, X=A, AND X=B."
70   PRINT "THE USER WILL SUPPLY THE ENDPOINTS"
80   PRINT "AND THE NUMBER OF SUBINTERVALS."
90   PRINT : PRINT "THE FUNCTION IN THIS PROGRAM"
100  PRINT "IS  X^2 - 4*X + 5"
110  PRINT "IF THE USER CHANGES THE FUNCTION,"
120  PRINT "THE NEW FUNCTION WILL REMAIN UNTIL"
130  PRINT "THE USER CHANGES IT AGAIN, OR"
140  PRINT "RELOADS THE PROGRAM."
150  PRINT : PRINT "DO YOU WISH TO CHANGE THE FUNCTION? Y/N"
160  INPUT A$
170  IF A$ = "Y" THEN 470
180  PRINT "ENTER THE ENDPOINTS OF THE INTERVAL"
190  INPUT A,B
200  PRINT "HOW MANY SUBINTERVALS DO YOU WISH TO USE"
210  INPUT N
220  H = (B - A) / N
230  HTAB 12: FLASH : PRINT "PATIENCE PLEASE"
240  HTAB 12: PRINT "CALCULATING!!!": NORMAL
250  DEF  FN F(X) = X ^ 2 - 4 * X + 5
260  S = 0
270  SX = 0
280  SY = 0
290  FOR I = 1 TO N
300  X = A + I * H
310  Y =  FN F(X)
320  S = S + Y * H
330  SX = SX + (H * Y ^ 2) / 2
340  SY = SY + H * X * Y
350  NEXT I
353  S =  INT (1000 * S) / 1000
355  SX =  INT (1000 * SX) / 1000
357  SY =  INT (1000 * SY) / 1000
360  PRINT "AREA                = ";S
370  PRINT "MOMENT ABOUT X-AXIS = ";SX
380  PRINT "MOMENT ABOUT Y-AXIS = ";SY
390  XB = SY / S
400  YB = SX / S
404  XB =  INT (1000 * XB) / 1000
406  YB =  INT (1000 * YB) / 1000
410  PRINT : PRINT "CENTROID IS AT (";XB;",";YB;")"
420  PRINT : PRINT "DO YOU WISH TO RUN THIS PROGRAM AGAIN? Y/N"
430  INPUT B$
440  HOME
450  IF B$ = "Y" THEN 90
460  GOTO 500
470  PRINT "TYPE"
480  INVERSE : PRINT "250 DEF FN F(X) = (YOUR FUNCTION) <RET>"
490  PRINT "RUN 180 <RET>": NORMAL
500  END
```

■ 7.4 VOLUMES OF REVOLUTION

The Disk Method

Consider the segment of the curve $y = x^{1/3}$ from $x = 0$ to $x = 8$. Imagine this curve being revolved about the x-axis so that the solid figure shown in Figure 7.17(a) is generated. The volume of this figure can be found by use of the definite integral if we proceed as follows: Divide the interval from $x = 0$ to $x = 8$ into n parts each dx units wide, and form rectangles of area, as we have done previously. When each of these areas is revolved about the x-axis, they generate cylinders similar to the one shown in Figure 7.17(b). The volume of the solid figure can then be found by summing the volumes of these typical cylinders. This sum is found by integration.

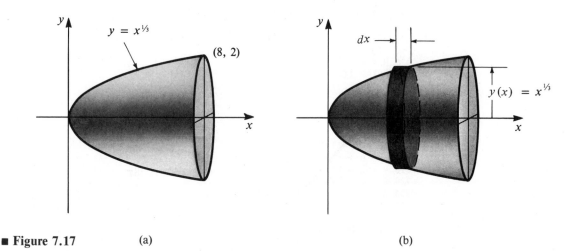

■ **Figure 7.17**   (a)   (b)

The volume of a cylinder is $V = \pi r^2 h$, where r is the radius and h is the width. Therefore, the volume of a typical cylinder in Figure 7.17(b) is

$$dV = \pi[y(x)]^2 \, dx \doteq \pi(x^{1/3})^2 \, dx = \pi x^{2/3} \, dx$$

And the total volume is

$$V = \pi \int_0^8 x^{2/3} \, dx = \pi \left[\frac{3}{5} x^{5/3} \right]_0^8$$

$$= \frac{96\pi}{5} \approx 60.3 \text{ cubic units}$$

EXAMPLE 1   ■ **Problem**

Find the volume of the reservoir generated by revolving the segment of the curve $y = x^{1/3}$ from $x = 0$ to $x = 8$ about the y-axis. See Figure 7.18.

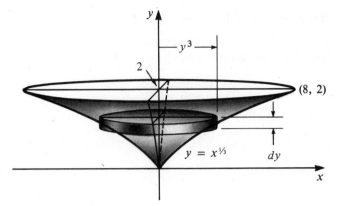

■ **Figure 7.18**

■ **Solution**

Since the volume is generated about the y-axis, we divide the interval $y = 0$ to $y = 2$ into n horizontal elements of height dy. The volume of a typical element of this reservoir is expressed as

$$dV = \pi (y^3)^2 \, dy$$

where y^3 is the radius of the typical cylinder. The desired volume is then

$$V = \pi \int_0^2 (y^3)^2 \, dy = \pi \int_0^2 y^6 \, dy = \pi \left[\frac{y^7}{7} \right]_0^2 = \frac{128\pi}{7}$$

$$\approx 57.4 \text{ cubic units} \qquad ■$$

EXAMPLE 2   ■ **Problem**

Find the volume generated by revolving the area bounded by $y = \sqrt{x}$, $x = 4$, and $y = 0$ about the line $x = 4$. See Figure 7.19.

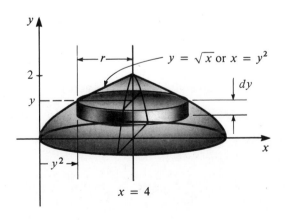

■ **Figure 7.19**

■ **Solution**

The key to the solution is the expression for the radius r of the typical element. From Figure 7.19, the desired radius is

$$r = 4 - y^2$$

The volume of this element is then

$$dV = \pi(4 - y^2)^2 \, dy = \pi(16 - 8y^2 + y^4) \, dy$$

Therefore, the volume of revolution is

$$V = \int_0^2 \pi(16 - 8y^2 + y^4) \, dy = \pi \left[16y - \frac{8y^3}{3} + \frac{y^5}{5} \right]_0^2$$

$$= \pi \left(32 - \frac{64}{3} + \frac{32}{5} \right) = \frac{256\pi}{15} \approx 53.6 \text{ cubic units} \qquad ■$$

The Cylindrical-Shell Method

Another approach that can be used to compute volumes of revolution involves the sum of cylindrical shells. If the rectangular element of area shown in Figure 7.20 is revolved about the y-axis, a cylindrical shell is formed. The volume of this element is equal to the circumference of the circular base times the area of the element, that is, $V = C \cdot A$. The volume of the shell in Figure 7.20 is

$$\overbrace{dV = (2\pi x)}^{\text{circumference}} \cdot \overbrace{(y(x) \, dx)}^{\text{area}} = 2\pi x y(x) \, dx \qquad (7.5)$$

The limit of the finite sum of these concentric shells is the desired volume of revolution.

EXAMPLE 3   ■ **Problem**

Use the method of cylindrical shells to compute the volume obtained by revolving the area bounded by $y = x^3$, $x = 2$ and $y = 0$ about the y-axis. See Figure 7.20.

■ **Solution**

The volume of the typical shell shown in the figure is

$$dV = 2\pi x(x^3 \, dx) = 2\pi x^4 \, dx$$

Summing these elements from $x = 0$ to $x = 2$, we have

$$V = 2\pi \int_0^2 x^4 \, dx = 2\pi \left[\frac{x^5}{5} \right]_0^2 = \frac{64\pi}{5} \approx 40.2 \text{ cubic units} \qquad ■$$

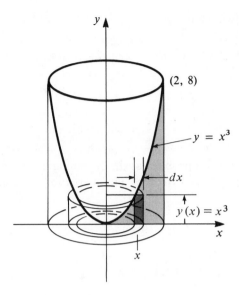

■ **Figure 7.20**

EXAMPLE 4 ■ **Problem**

Use cylindrical shells to find the volume generated by revolving, about the x-axis, the area bounded by $y^3 = x$, $x = 8$, and $y = 0$. See Figure 7.21.

■ **Solution**

The volume of the typical shell shown in the figure is

$$dV = 2\pi y(8 - y^3)\, dy$$

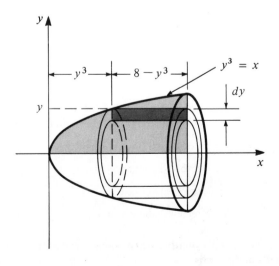

■ **Figure 7.21**

Therefore, the desired volume is

$$V = 2\pi \int_0^2 y(8 - y^3)\, dy = 2\pi \int_0^2 (8y - y^4)\, dy$$

$$= 2\pi \left[4y^2 - \frac{y^5}{5} \right]_0^2$$

$$= 2\pi \left[16 - \frac{32}{5} \right] = \frac{96\pi}{5} \approx 60.3 \text{ cubic units}$$

Notice that we solved this problem at the beginning of the section by using elements in the form of right circular cylinders. The answers are the same. ∎

EXERCISES FOR SECTION 7.4

In Exercises 1–8, use the disk method to find the volume by revolving the given area about the x-axis.

1. $3y = x, x = 6, y = 0$
2. $y^2 = x, x = 4, y = 0$
3. $y^2 = 9x, x = 4, y = 0$
4. $y^2 = x^3, x = 4, y = 0$
5. $y = x^3, x = 2, y = 0$
6. $y = \frac{1}{2}x^2, x = 8, y = 0$
7. $y = 9 - x^2, y = 0$
8. $y = 8 - 2x^2, y = 0$

In Exercises 9–16, use the disk method to find the volume generated by revolving the given area about the y-axis.

9. $y = \frac{1}{5}x, x = 0, y = 2$
10. $y^2 = \frac{1}{2}x, x = 0, y = 2$
11. $y^2 = x^3, x = 0, y = 8$
12. $y = x^2, x = 0, y = 9$
13. $y = x^3, x = 0, y = 1$
14. $y = 3, y = 0, x = 0, x = 2$
15. $y^2 = x + 4, x = 0$
16. $3y^2 = x + 27, x = 0$

In Exercises 17–20, use the disk method to find the volume generated by revolving the given area about the indicated line. (The key is the expression that you write for the radius.)

17. $y^2 = x^3, x = 0, y = 8$; revolved about $y = 8$.

18. $y = x^2, x = 0, y = 9$; revolved about $y = 9$.

19. $y = x^2, x = 2, y = 0$; revolved about $x = 2$.

20. $y = x^3, x = 2, y = 0$; revolved about $x = 2$.

In Exercises 21–28, use the cylindrical-shell method to find the volume obtained by revolving the given area about the x-axis.

21. $y = 3 - x, x = 0, y = 0$

22. $y = 2 - x, x = 0, y = 0$

23. $y = \sqrt{x}, x = 0, y = 2$

24. $y = x^2, x = 0, y = 4$

25. $y^2 = 4 - x, x = 0, y = 0$

26. $y^3 = x, x = 0, y = 1$

27. $y = \sqrt[3]{x}, x = 8, y = 0$

28. $y = 2\sqrt{x}, x = 1, y = 0$

In Exercises 29–36, use the cylindrical-shell method to find the volume obtained by revolving the given area about the *y*-axis.

29. $y = x^2, x = 3, y = 0$

30. $y = x^3, x = 1, y = 0$

31. $y = 2x - x^2, y = 0$

32. $y = 3x - x^2, y = 0$

33. $y = \frac{1}{2}x^2, x = 0, y = 8$

34. $y = x^{3/2}, x = 0, y = 8$

35. $y = x^{2/3}, x = 0, y = 4$

36. $y = 3x + x^2, x = 0, y = 10$

7.5 MOMENTS OF INERTIA

Moments of Inertia of Areas

Associated with any plane area and any axis in its plane is a physical concept called the moment of inertia of the area. This concept is also referred to as the second moment of the area. The concept of the moment of an area introduced in Section 7.3 is then referred to as the first moment of the area. The second moment or the moment of inertia is used extensively in certain engineering areas, such as strength of materials and machine design.

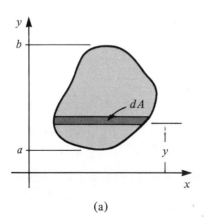

(a)

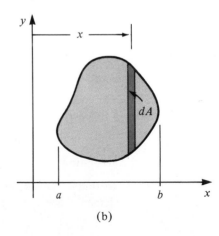

(b)

■ **Figure 7.22**

The **moment of inertia** of an area with respect to an axis is defined as the product of the area times the square of the distance from its centroid to the axis. In

Figure 7.22(a), we consider the moment of inertia of the area with respect to the x-axis. We begin by choosing an element of area parallel to the x-axis of area dA and y units from the x-axis. The moment of inertia of the element with respect to the x-axis is denoted by dI_x and defined by

$$dI_x = y^2 \, dA$$

We state, without proof, that *the moment of inertia of an area is equal to the sum of the moments of inertia of its parts.* Thus, the moment of inertia I_x of the area in Figure 7.22(a) is the integral of the elements $y^2 \, dA$ from $y = a$ to $y = b$.

To compute the moment of inertia I_y of the area with respect to the y-axis, we choose an element parallel to, and x units from, the y-axis, as shown in Figure 7.22(b). The moment of inertia of this element is

$$dI_y = x^2 \, dA$$

And the moment of inertia I_y of the area is the integral of the elements $x^2 \, dA$ from $x = a$ to $x = b$.

We have the following formulas for computing moments of inertia of an area.

FORMULA

> **Moments of Inertia of an Area**   The moment of inertia of an area A with respect to the x-axis is
>
> $$I_x = \int_a^b y^2 \, dA \tag{7.6}$$
>
> where dA is the area of an element parallel to and y units from the x-axis. See Figure 7.22(a). The moment of inertia of an area A with respect to the y-axis is
>
> $$I_y = \int_a^b x^2 \, dA \tag{7.7}$$
>
> where dA is the area of the element parallel to and x units from the y-axis. See Figure 7.22(b).

EXAMPLE 1   ■ **Problem**

Find I_x and I_y for the area bounded by $y = x^2$, $x = 2$, and $y = 0$.

■ **Solution**

To find I_x we take elements parallel to the x-axis, as shown in Figure 7.23(a). The elemental area dA is given by

$$dA = (2 - y^{1/2}) \, dy$$

So, the moment of inertia of the element is

$$dI_x = y^2(2 - y^{1/2}) \, dy$$

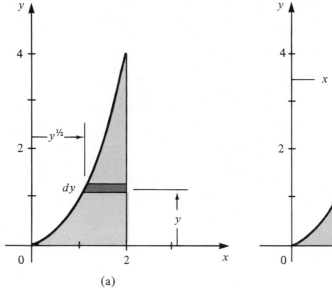

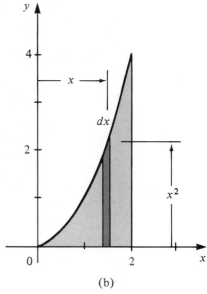

■ **Figure 7.23**   (a)   (b)

The moment of inertia I_x is then given by

$$I_x = \int_0^4 y^2(2 - y^{1/2}) \, dy = \int_0^4 (2y^2 - y^{5/2}) \, dy$$

$$= \left[\frac{2}{3}y^3 - \frac{2}{7}y^{7/2}\right]_0^4 = \frac{128}{3} - \frac{256}{7} = \frac{128}{21}$$

To find I_y, we choose elements parallel to the y-axis, as shown in Figure 7.23(b). The area of this element is given by

$$dA = y \, dx = x^2 \, dx$$

So, the moment of inertia of the element is

$$dI_y = x^2 \, dA = x^2(x^2 \, dx) = x^4 \, dx$$

The desired moment of inertia is then

$$I_y = \int_0^2 x^4 \, dx = \left[\frac{x^5}{5}\right]_0^2 = \frac{32}{5}$$ ■

Moments of Inertia of Solids of Revolution

Finding the moment of inertia of a solid of revolution with respect to its axis of rotation is a relatively easy step once the concept has been explained for areas. The *moment of inertia of a solid* of mass m is defined as the product of the mass and the square of the distance from the center of mass to the axis of rotation.

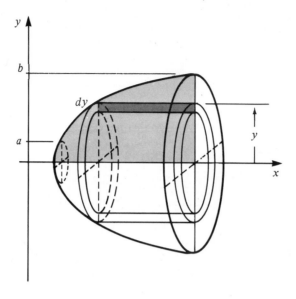

■ **Figure 7.24**

The method of cylindrical shells introduced in the previous section is usually best suited for these computations. Consider the solid of revolution shown in Figure 7.24. To find the volume of this solid by cylindrical shells, we choose an elemental shell with sides parallel to and y units from the x-axis. Let dV be the volume of the shell, and let ρ be the density of the solid. Then, the mass dm of the shell is

$$dm = \rho \, dV$$

The moment of inertia of the shell with respect to the x-axis is then

$$dI_x = y^2 \, dm$$

Expressing the moment of inertia of the solid of revolution as the sum of the moments of inertia of these shells, we have

$$I_x = \int_a^b y^2 \, dm = \int_a^b y^2 (\rho \, dV)$$

The moment of inertia of a solid about the y-axis is found in a similar way. Thus, we have the following formulas.

FORMULA

> **Moments of Inertia of a Solid of Revolution** If the density ρ is constant, which is the case for homogeneous substances, the moment of inertia I_x of a solid generated by revolving an area about the x-axis is
>
> $$I_x = \rho \int_a^b y^2 \, dV \qquad (7.8)$$
>
> The moment of inertia I_y of a solid generated by revolving an area about the y-axis is
>
> $$I_y = \rho \int_a^b x^2 \, dV \qquad (7.9)$$

EXAMPLE 2   ▪ **Problem**

Determine the moment of inertia about the y-axis of the solid of density ρ obtained by revolving the area bounded by $y = x^3$, $x = 0$, and $y = 1$ about the y-axis. See Figure 7.25.

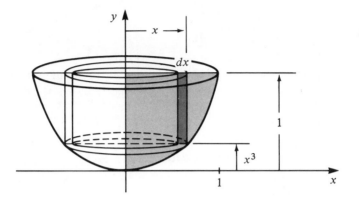

▪ **Figure 7.25**

▪ **Solution**

The volume of the elemental shell is

$$dV = 2\pi x(1 - x^3)\, dx$$

Since $dI_y = x^2\, dm = x^2(\rho\, dV)$, the moment of inertia of the shell is

$$dI_y = x^2[2\pi\rho x(1 - x^3)\, dx] = 2\pi\rho(x^3 - x^6)\, dx$$

Then, the moment of inertia of the solid of revolution is given by

$$I_y = 2\pi\rho \int_0^1 (x^3 - x^6)\, dx = 2\pi\rho \left[\frac{x^4}{4} - \frac{x^7}{7}\right]_0^1$$

$$= 2\pi\rho \left[\frac{1}{4} - \frac{1}{7}\right] = \frac{3\pi\rho}{14}$$

■

Radius of Gyration

Consider n particles having masses $m_1, m_2, \ldots, m_n$ and revolving about a common axis at respective distances of $d_1, d_2, \ldots, d_n$, as shown in Figure 7.26. The total mass of the n particles is

$$M = \sum_{i=1}^{n} m_i$$

and the total moment of inertia of the n particles is

$$I = \sum_{i=1}^{n} m_i\, d_i^2$$

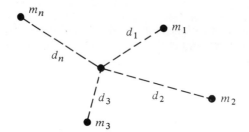

■ **Figure 7.26**

A number R such that $RM = I$ is called the **radius of gyration** of the n particles.

Radius of Gyration   If the moment of inertia and the mass of a solid are known, the radius of gyration is

$$R = \frac{I}{M} \tag{7.10}$$

If I is the moment of inertia of a plane area, the radius of gyration is given by

$$R = \frac{I}{A} \tag{7.11}$$

where A is the area of the region.

EXAMPLE 3   ■ **Problem**

Find the radius of gyration of the solid of density ρ obtained by revolving the area bounded by $y = x^2$, $x = 0$, and $y = 1$ about the x-axis, as shown in Figure 7.27.

■ **Solution**

The volume of a typical shell is $dV = 2\pi y(y^{1/2})\, dy = 2\pi y^{3/2}\, dy$. So, the volume of the solid is

$$V = 2\pi \int_0^1 y^{3/2}\, dy = 2\pi \left[\frac{2}{5} y^{5/2} \right]_0^1 = \frac{4\pi}{5}$$

The mass of the solid is equal to the product of density and volume. Thus,

$$M = \rho V = \frac{4\pi\rho}{5}$$

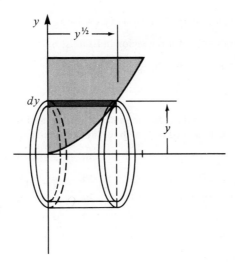

■ **Figure 7.27**

The moment of inertia of a typical shell is

$$dI_x = y^2 \, dm = y^2(\rho \, dV) = y^2(2\pi\rho y^{3/2}) \, dy = 2\pi\rho y^{7/2} \, dy$$

Thus, I_x is given by

$$I_x = 2\pi\rho \int_0^1 y^{7/2} \, dy$$

$$= 2\pi\rho \left[\frac{2}{9} y^{9/2} \right]_0^1 = \frac{4\pi\rho}{9}$$

The radius of gyration is then

$$R = \frac{I_x}{M} = \frac{4\pi\rho/9}{4\pi\rho/5} = \frac{5}{9}$$

■

EXERCISES FOR SECTION 7.5

In Exercises 1–10, find the moment of inertia of the region bounded by the given curves and with respect to the indicated axis.

1. $y = \frac{1}{2}x$, $x = 4$, $y = 0$; find I_y.

2. $y = \frac{1}{2}x$, $x = 0$, $y = 2$; find I_x.

3. $y = x$, $x = 0$, $y = 3$, $y = 6$; find I_x.

4. $y = x^2$, $x = 0$, $y = 4$; find I_x and I_y.

5. $y = x^2$, $x = 0$, $y = 1$; find I_x and I_y.

6. $y = 4 - x^2$, $x = 0$, $y = 0$; find I_y.

7. $y^2 = x$, $y = x - 2$; find I_x.

8. $y = x^2$, $y = x + 2$; find I_y.

9. $y^2 = x$, $y = \frac{1}{2}x$; find I_y.

10. $3x + y = 9$, $y + x^2 = 9$; find I_y.

11. Find the radii of gyration for the region in Exercise 5.

12. Find the radius of gyration for the region in Exercise 6.

13. Find the radius of gyration for the region in Exercise 7.

14. Find the radius of gyration for the region in Exercise 8.

In Exercises 15–22, determine the mass, moment of inertia, and radius of gyration of the solid obtained by revolving the indicated region about the indicated axis.

15. $y = \frac{1}{3}x^2, x = 0, y = 3$; x-axis

16. $y = 2 - x, x = 0, y = 0$; x-axis

17. $y = \sqrt{x}, x = 0, y = 2$; x-axis

18. $y = \sqrt[3]{x}, x = 8, y = 0$; x-axis

19. $y = x^2, x = 3, y = 0$; y-axis

20. $y = \frac{1}{2}x^2, x = 0, y = 8$; y-axis

21. $y = 2x - x^2, y = 0$; y-axis

22. $y = x^{2/3}, x = 0, y = 4$; y-axis

▬▬▬ 7.6 FORCE AND WORK

The **work** done by a constant force F in moving an object through a distance x is defined to be the product of force times distance. Symbolically, we write

$$W = F \cdot x$$

For example, if a constant force of 10 lb is required to move an object 3 ft, the work done on the object is

$$W = 10(3) = 30 \text{ ft-lb}$$

Work Done by a Variable Force

In practice, constant forces are seldom encountered. If the force applied to an object varies with the displacement of the object, the foregoing formula for work is not applicable. However, we can use integral calculus to adapt this formula to variable-force problems. Consider a force F that is a function of displacement and acting on an object, as shown in Figure 7.28. We proceed, as in previous sections of our discussion of integration, to approximate the work done by subdividing the distance that the object moves into n parts, each of width dx. If we choose n to be sufficiently large, dx will be small enough that we can assume the force over the interval to be constant. Under this assumption, the work done by the force in moving the object dx units is

$$dW = F \cdot dx$$

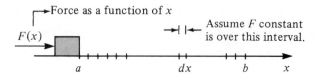

■ **Figure 7.28**

The integration of these elements of work yields the following formula.

FORMULA

> **Work Done by a Variable Force** If F is a force that is a function of displacement x, then the work W done by the force in moving an object from $x = a$ to $x = b$ is given by
>
> $$W = \int_a^b F\,dx \qquad\qquad\qquad\qquad\qquad (7.12)$$

EXAMPLE 1 ■ **Problem**

Find the work done in moving an object from $x = 0$ to $x = 5$ ft if the applied force (in pounds) varies with displacement according to $F = (x + 4)^{1/2}$.

■ **Solution**

Using the formula for the work done by a variable force, we have

$$W = \int_0^5 (x + 4)^{1/2}\,dx = \left[\frac{2(x + 4)^{3/2}}{3}\right]_0^5 = 18 - \frac{16}{3} = \frac{38}{3}\ \text{ft-lb} \qquad ■$$

Of practical importance in mechanics is the computation of the work done by a force in stretching (or compressing) a spring. We know from **Hooke's law** that the force of a stretched spring is proportional to its change in length. Specifically, if x represents the displacement of the end of a spring caused by a force F, then Hooke's law can be written as

$$F = kx$$

where k is a constant called the **spring constant.**

Comment: In using Hooke's law, we assume that the elastic limit of the spring is not exceeded.

EXAMPLE 2 ■ **Problem**

A spring has a spring constant of 2 lb/in. and a free length of 3 in. Calculate the work done in stretching the spring to a length of 7 in. Then, calculate the work done in stretching the spring from 7 to 11 in. See Figure 7.29.

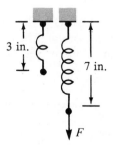

■ **Figure 7.29**

■ Solution

Before computing the work done, we note from Hooke's law that the force of the spring is given by

$$F = 2x$$

To find the work done in stretching the spring from its free length of 3 in. to a length of 7 in., we integrate from 0 to 4. Thus,

$$W = \int_0^4 2x \, dx = \left[x^2\right]_0^4 = 16 \text{ in.-lb}$$

The work done in stretching the spring from 7 to 11 in. is found by integrating from 4 to 8. Thus,

$$W = \int_4^8 2x \, dx = \left[x^2\right]_4^8 = 64 - 16 = 48 \text{ in.-lb} \qquad ■$$

Work Done in Lifting Liquids

The work done in lifting an object is

$$W = F \cdot h$$

where F is the weight of the object and h is the height through which it is lifted. As noted previously, this formula is true as long as F and h are constant. If either F or h is a variable quantity, the formula is invalid. For instance, the work done in pumping a liquid out of a container cannot be calculated by $W = F \cdot h$ because as the liquid level is lowered, the height that the next layer must be raised is increased. This problem can be solved by the use of the definite integral. To illustrate, we consider the following example.

EXAMPLE 3   **■ Problem**

Compute the work done in pumping the water out of a conical reservoir to a height of 3 ft above the reservoir. The vertical height of the reservoir is 2 ft, and the diameter across its top is 8 ft. The density of water is 62.5 lb/ft^3.

■ Solution

We first represent the cone by a geometric model in the coordinate plane. In Figure 7.30, the vertex of the cone is located at the origin. The cone is generated by revolving a straight line about the y-axis. The slope of the line is determined from the given dimensions to be

$$m = \frac{\Delta y}{\Delta x} = \frac{2}{4} = \frac{1}{2}$$

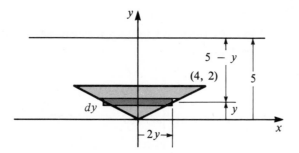

■ **Figure 7.30**

Hence, the equation of the line of revolution is $y = \frac{1}{2}x$.

To compute the work, we first divide the volume of the cone into cylindrical elements each of height dy. The radius of a typical cylindrical element is $2y$, so its volume is

$$dV = \pi(2y)^2\, dy$$

The weight of each cylindrical element is the product of the density ρ times the volume, or

$$dw = \rho\pi(2y)^2\, dy$$

To compute the work required to lift a typical cylindrical element to a height of 3 ft above the top of the reservoir, we must determine how far each element is raised. Referring to Figure 7.30, we see that the element is y units above the x-axis. Therefore each element must be raised $5 - y$ units. The work done in lifting each element will then be

$$(5 - y)\, dw = \rho\pi(2y)^2(5 - y)\, dy$$

The total work done in lifting the contents of the reservoir to a height of 3 ft above its top is, as a definite integral,

$$W = \rho\pi \int_0^2 (2y)^2(5 - y)\, dy = 62.5\pi \int_0^2 (2y)^2(5 - y)\, dy$$

$$= 62.5\pi \int_0^2 (20y^2 - 4y^3)\, dy = 62.5\pi\left[\frac{20y^3}{3} - y^4\right]_0^2$$

$$\approx 7330 \text{ ft-lb} \qquad\qquad ■$$

Comment: In general, the upper limit of the integral corresponds to the surface of the liquid in the reservoir, and the lower limit corresponds to the bottom of the reservoir. Furthermore, the solution is not dependent on the location of the reservoir in the plane. To illustrate this fact, Example 3 is restated next with the reservoir located differently.

EXAMPLE 4   ▪ **Problem**

Solve the problem described in Example 3, but locate the vertex of the cone at the point $(0, -2)$. See Figure 7.31.

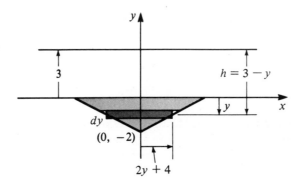

▪ **Figure 7.31**

▪ **Solution**

With the cone located in this position, the line of revolution has a y-intercept of -2. Hence, its equation is $y = \frac{1}{2}x - 2$. From Figure 7.31, we see that the surface of the water lies on the x-axis, and each of the elements is y units below the x-axis. Each element must be raised a distance from the line $y = \frac{1}{2}x - 2$ to $y = 3$. Thus, $h = 3 - y$, and the work done on each element is

$$dW = hw = (3 - y)[\rho\pi(2y + 4)^2]\,dy$$

The total work done in emptying the reservoir can therefore be written as

$$W = 62.5\pi \int_{-2}^{0} (3 - y)(2y + 4)^2\,dy$$

$$= 62.5\pi \int_{-2}^{0} (-4y^3 - 4y^2 + 32y + 48)\,dy$$

$$= 62.5\pi \left[-y^4 - \frac{4y^3}{3} + 16y^2 + 48y \right]_{-2}^{0} = 7330 \text{ ft-lb} \qquad ▪$$

Force Exerted by a Fluid

The **pressure** at a point below the surface of a fluid is equal to the product of the depth of the point and the density of the fluid. That is,

pressure = depth · density

The total force exerted by a fluid on a vertical wall can be found by the methods of integral calculus. We subdivide the area of the given wall into horizontal rectangles, as shown in Figure 7.32. The shaded area in the figure is the area against which the fluid is acting. The force exerted on each rectangular element is the prod-uct of the pressure at that depth and the area of the element. Thus, the force on the typical element shown in Figure 7.32 is

$$dF = p \cdot A = (\rho h)(x \, dy)$$

where ρ is the density of the fluid. The total force on the given area is then the sum of these elemental forces from $y = a$ to $y = b$.

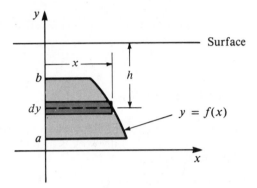

■ **Figure 7.32**

EXAMPLE 5 ■ **Problem**

Find the total force on the triangular area shown in Figure 7.33. Assume the fluid is water ($\rho = 62.5 \, \text{lb/ft}^3$).

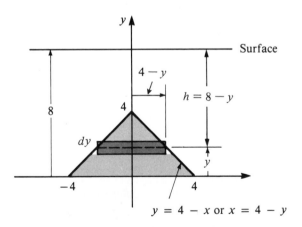

■ **Figure 7.33**

■ **Solution**

The base of the triangle is 8 ft below the surface, and the top of the triangle is 4 ft below the surface. The typical element is $h = 8 - y$ feet below the surface, and its area (in square feet) is $dA = 2(4 - y)\, dy$. The factor 2 accounts for the fact that the figure is symmetric about the y-axis. The force exerted on the typical element is then

$$dF = 62.5(8 - y)[2(4 - y)\, dy] = 125(32 - 12y + y^2)\, dy$$

Summing all such forces from $y = 0$ to $y = 4$ ft, we have

$$F = 125 \int_0^4 (32 - 12y + y^2)\, dy = 125 \left[32y - 6y^2 + \frac{y^3}{3} \right]_0^4$$

$$= 125 \left(128 - 96 + \frac{64}{3} \right) = 125 \left(\frac{160}{3} \right) \approx 6670 \text{ lb} \qquad \blacksquare$$

EXERCISES FOR SECTION 7.6

1. The force (in pounds) on a charged particle varies with the displacement (in feet) of the particle by the formula $F = 7x^3$. Find the work done on the particle by the force F in moving the particle a distance of 6 ft.

2. The force (in dynes) on a proton varies with displacement as $F = 4x^2 + 100$. How much work is done by the proton in moving 2 cm?

3. How much work is done in compressing a spring with a spring constant of 25 lb/in. a distance of 2 in.?

4. A force of 200 dynes is required to compress a spring with a free length of 10 cm to a length of 8 cm. Find the work done in compressing the spring from its free length to a length of 5 cm.

5. A spring with a free length of 12 in. stretches 3 in. under a weight of 8 lb. Find the work done in stretching the spring from its free length to a length of 20 in.

6. Work of 100 dyn-cm is needed to stretch a spring from a length of 5 to 7 cm, and another 200 dyn-cm is needed to stretch the spring from 7 to 9 cm. Find the spring constant and the free length of the spring.

7. A reservoir in the form of an inverted right circular cone is full of water. What is the work required to empty the reservoir if the altitude is 10 ft and the radius is 4 ft?

8. Solve Exercise 7 by using an altitude of 4 ft and a diameter of 10 ft.

9. A right circular cylindrical tank has a diameter of 6 ft and an altitude of 10 ft. If the tank is half-full of gasoline, how much work is required to pump the gasoline to the top of the tank? (Use 60 lb/ft^3 for the density of gasoline.)

10. A tank in the shape of a right circular cylinder has a radius of 5 ft and an altitude of 2 ft. Find the work required to empty the tank if it is full of water.

11. A reservoir is formed by revolving the segment of the curve $y = \frac{1}{3}x^2$, from $x = 0$ to $x = 3$ ft, about the y-axis. The reservoir is filled with oil (50 lb/ft^3). What is the work done in lifting the oil to a point 6 ft above the top?

12. The segment of the curve $y^2 = 4x$, from $x = 0$ to $x = 4$ ft, is revolved about the y-axis to form a reservoir. Find the work expended in emptying the reservoir, given that it is full of water.

13. A tank is formed by revolving the straight-line segment passing through $(2, 0)$ and $(4, 6)$ about the y-axis. Assume the coordinates are in feet. If the tank is full of water, how much work is done in emptying the tank to a depth of 3 ft?

14. Assume that the tank in Exercise 13 is initially empty and sitting on the ground. How much work is done in filling the tank with water, if the water is pumped over the top edge? Explain.

15. A tank in the shape of a right circular cylinder, 18 ft deep and 6 ft in diameter, is filled with water. With the base of the tank on the x-axis, calculate the work done in pumping out the water.

16. Solve Exercise 15 given that the top of the tank is located on the x-axis.

17. The inner surface of a tank has the form $y = x^3$; the diameter of the top of the tank is 4 ft. How much work is done in filling the tank with water if the water enters the tank at the bottom?

18. The curve of revolution of a tank is $y = x^2 - 4$. The surface of the water in the tank is at the x-axis. What is the work required to lift the water to a height 5 ft above the initial surface?

19. A tank is formed by revolving the curve $y^3 = x^2$ above the y-axis. The tank is filled with water to a height of 4 ft. Determine the work done in pumping all of the water to a point 3 ft above the initial surface.

20. A reservoir, which is 8 ft deep, is generated by the curve $y^2 = x$. Assume the reservoir is filled with water. Use four right circular cylinders to estimate the work done in emptying the reservoir. Take the radius of each cylinder to be the midpoint of each subinterval.

21. Find the work done for Exercise 20 by using the definite integral.

22. Find the force on the triangle shown in the accompanying figure. This problem is the same problem solved in Example 5 of this section with the triangle at a different place.

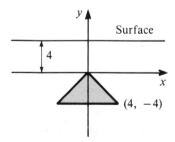

23. The vertical wall of a dam is in the form of a right triangle, as shown in the accompany-ing figure. What is the force on the dam if the reservoir behind it is full of water?

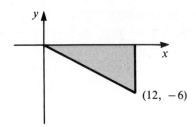

$(12, -6)$

24. A rectangular window in a sea aquarium is 10 ft wide and 6 ft high. Compute the force on the window given that the water level is 2 ft above the top of the window.

25. A gate in a dam is a rectangle 20 ft wide and 10 ft high. Compute the force that this gate must withstand given that the water level is 25 ft above the top of the gate.

26. The end of a water trough is an isoceles trapezoid, as shown in the accompanying figure. Compute the force on the end given that the trough is full of water.

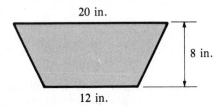

20 in.

8 in.

12 in.

REVIEW EXERCISES FOR CHAPTER 7

1. Find the total area bounded by $y = 1 - x^3$ and the x-axis from $x = 0$ to $x = 2$. Sketch the area.

2. Find the total area bounded by $y = 3x + 2$ and the x-axis over the interval $-2 \le x \le 0$. Sketch the area.

3. Find the total area bounded by $y = x^{1/3}$ and the x-axis over the interval $-1 \le x \le 1$. Sketch the area. Compute the average height of the curve on this interval.

4. Find the total area bounded by the curve $y = 3x + x^2$ and the x-axis from $x = -4$ to $x = 0$. Sketch the area. Compute the average height of the curve.

5. Compute the area bounded by $y = x^2$ and the y-axis from $y = 1$ to $y = 4$. Sketch the area.

6. Compute the area bounded by $y = x^{1/3}$ and the y-axis on the interval $0 \le y \le 2$. Sketch the area.

7. Compute the area bounded by $y = 3x + x^2$ and $y = 4$. Sketch the area.

8. Compute the area bounded by $y = x^2$, $x = 1$, and $y = 9$. Sketch the area.

9. Locate the centroid of the area bounded by $y = x^2$, $x = 1$, $x = 4$, and $y = 0$. Sketch the area.

10. Locate the centroid of the area bounded by $y = 3 - \frac{1}{2}x^2$, $x = 0$, $x = 2$, and $y = 0$. Sketch the area.

11. Locate the centroid of the area bounded by $y = x^2$, $y = 1$, $y = 9$, and $x = 1$. Sketch the figure.

12. Locate the centroid of the area bounded by $y = x^2$, $x + y = 6$, $x = 0$, and $x = 2$. Sketch the figure.

13. Use the disk method to find the volume of the solid of revolution generated by revolving the area bounded by $y - x = -2$, $x = 0$, $y = 0$, and $y = 4$ about the y-axis. Draw the figure.

14. Use the disk method to find the volume of the solid of revolution generated by revolving the area bounded by $y = x^2$, $x = 0$, $y = 1$, and $y = 4$ about the y-axis. Sketch the figure.

15. Use the disk method to find the volume of the solid of revolution generated by revolving the area bounded by $y = x^{1/2}$, $x = 1$, $x = 9$, and $y = 0$ about the x-axis. Sketch the figure.

16. Use the disk method to find the volume of the solid of revolution generated by revolving the area bounded by $y = 2x - x^2$ and $y = 0$ about the x-axis. Sketch the figure.

17. Use the cylindrical-shell method to find the volume of the solid of revolution generated by revolving the area bounded by $y = 3x - x^2$ and $y = 0$ about the y-axis. Sketch the figure.

18. Use the cylindrical-shell method to find the volume of the solid of revolution generated by revolving the area bounded by $y = x^2$, $x = 0$, $y = 1$, and $y = 4$ about the x-axis. Sketch the figure.

19. Use the cylindrical-shell method to find the volume of the solid of revolution generated by revolving the area bounded by $y = x^{1/3}$, $x = 1$, and $y = 0$ about the x-axis. Sketch the figure.

20. Use the cylindrical-shell method to find the volume of the solid of revolution generated by revolving the area bounded by $y^2 = 4x$, $x = 0$, and $y = 4$ about the y-axis. Sketch the figure.

21. Find the moment of inertia with respect to the x- and y-axes of the region bounded by $y = x^{1/2}$, $x = 0$, and $y = 1$.

22. Find the moment of inertia with respect to the x- and y-axes of the region bounded by $y = 4 - x$, $x = 0$, and $y = 0$.

23. Find the moment of inertia and the radius of gyration for the solid of revolution in Exercise 17. Make the computation with respect to the y-axis.

24. Find the moment of inertia and the radius of gyration of the solid of revolution in Exercise 18. Make the computation with respect to the x-axis.

25. A spring with a free length of 5 in. has a spring constant of 3.5 lb/in. Calculate the work done in stretching the spring to a length of 9 in.

26. Calculate the work done in stretching a spring 10 in. if it has a spring constant of 0.5 lb/in.

27. A tank is formed by revolving the area bounded by $y - 2x = -5$, $x = 0$, $y = 0$, and $y = 3$ about the y-axis. Assume units are in feet. If the tank is full of water, how much work is done to pump the contents to a point 5 ft above the top of the tank? Draw the figure.

28. A reservoir in the form of an inverted right circular cone is full of water. Find the work done in emptying the entire reservoir out the top. The reservoir is 12 ft high and 12 ft in diameter.

29. The end of a swimming pool is a rectangle 25 ft wide and 9 ft deep. What is the force on the wall if the pool is full of water?

30. A storage tank is filled with water at a rate $dV/dt = \sqrt{t}$ (in cubic feet per minute). Compute the volume of water pumped into the tank from $t = 0$ to $t = 30$ min.

31. The power input (in watts) to a system varies with time according to $p = t^{2/3}$. How much work is done from $t = 0$ to $t = 7$ sec?

32. The force (in pounds) applied to a block in a physics experiment is increased, as the block moves, according to $F = 10 + x$, where x is the distance (in feet). Compute the work done on the block in moving it from $x = 0$ to $x = 3$ ft.

CHAPTER 8

Exponential and Logarithmic Functions

Do you know what electric circuits, population growth, and atmospheric pressure have in common? They are all described mathematically by using exponential and logarithmic functions. In this chapter, we examine some of the properties of these two functions, especially those that are best understood within the framework of calculus.

8.1 REVIEW TOPICS

Exponential Functions

One of the most useful functions for describing physical phenomena is the exponential function, that is, a fixed positive number to a variable power.

DEFINITION

> **Exponential Function**   If a is any fixed constant greater than zero, $a \neq 1$, then the *exponential function with base a* is defined by
>
> $$f(x) = a^x$$
>
> The domain of the exponential function is all real numbers, and the range is the positive real numbers.

Probably the most important base is the number e, defined by

$$e = \lim_{x \to 0} (1 + x)^{1/x}$$

The following table (compiled with a calculator) suggests that the value of e is approximately 2.718. Remember that in evaluating this limit, we are not concerned with the value of $(1 + x)^{1/x}$ when $x = 0$ but its value for x arbitrarily close to zero.

x	10	1	0.1	0.01	0.001	0.0001
$(1 + x)^{1/x}$	1.271	2.000	2.594	2.705	2.717	2.718

By taking positive values of x smaller than 0.0001, we can improve the accuracy of our estimate of e. However, no matter how small the value of x, the value of the first four digits will be unaffected.

We can make the following observations about the function $f(x) = a^x$ for $a > 1$.

PROPERTIES

> **Exponential function $f(x) = a^x$**
>
> 1. $a^x > 0$ for all values of x.
> 2. When $x = 0$, $y = a^0 = 1$.
> 3. When $x = 1$, $y = a^1 = a$.
> 4. As x increases without bound, a^x increases without bound.
> 5. As x decreases without bound, a^x approaches zero.
> 6. If $x_1 < x_2$, then $a^{x_1} < a^{x_2}$.

If $0 < a < 1$, then properties 4–6 are as follows:

4.´ As x increases without bound, a^x approaches zero.
5.´ As x decreases without bound, a^x increases without bound.
6.´ If $x_1 < x_2$, then $a^{x_1} > a^{x_2}$.

Property 6 states that if $a > 1$, then a^x is an *increasing function*. Property 6´ states that if $0 < a < 1$, then a^x *is a decreasing function*. An increasing function has a graph that rises from left to right, while a decreasing function has a graph that falls from left to right.

A different exponential function is obtained for each value of a, although the shape of the graph remains basically the same, and the same kind of functional properties as 1–6 continue to hold.

Figure 8.1 shows the graph of a^x for $a = e^{-1}, \frac{1}{2}, 1, 2,$ and e. The value of $a = 1$ is excluded from exponential functions since its graph is not of exponential shape. Furthermore, $f(x) = 1^x$ does not obey the functional properties 1–6 that are associated with exponential functions. Thus, the only acceptable bases are $a > 0$, $a \neq 1$.

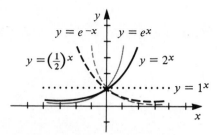

■ **Figure 8.1**

Be aware that $-e^x$ and $(-e)^x$ have different meanings. The latter is an "unacceptable" exponential function since $-e < 0$, while $-e^x$ is merely the negative of e^x. See Figure 8.2.

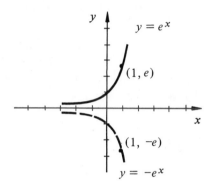

■ **Figure 8.2**

The exponential function is useful for describing physical and social phenomena such as the following:

- Electrical current,
- Atmospheric pressure,
- Decomposition of uranium,
- Population growth,
- The learning process,
- Compound interest

If a physical quantity obeys a law that is described by an exponential function and is increasing, it is said to *increase exponentially*. If it is exponential in character but is decreasing, it is said to *decrease, or decay, exponentially.*

EXAMPLE 1   ■ **Problem**

An approximate rule for atmospheric pressure at altitudes less than 50 mi is as follows: Standard atmospheric pressure, 14.7 lb/in.², is halved for each 3.25 mi of vertical ascent. Write an exponential function to express this rule. Compute the atmospheric pressure at an altitude of 19.5 mi.

■ **Solution**

Letting P denote the atmospheric pressure at altitudes less than 50 mi and h the altitude (in miles), we have

$$P = 14.7\left(\frac{1}{2}\right)^{h/3.25}$$

Using the expression for P and letting $h = 19.5$, we obtain

$$P = 14.7\left(\frac{1}{2}\right)^{19.5/3.25} = 14.7\left(\frac{1}{2}\right)^{6} = \frac{14.7}{64} = 0.23 \text{ lb/in.}^2 \qquad ■$$

EXAMPLE 2   ■ **Problem**

An unmanned satellite has a power supply whose power output (in watts) is given by the equation

$$p = 50e^{-t/260}$$

where t is the time in days that the battery has been in operation. How much power will be available at the end of one year?

■ **Solution**

Applying the given formula with $t = 365$, we have

$$p = 50e^{-365/260} = 50e^{-1.4}$$

From a calculator, we obtain

$$e^{-1.4} = 0.2466$$

Hence,

$$p = (50)(0.2466) = 12.33 \text{ W} \qquad ■$$

EXAMPLE 3   ■ **Problem**

The growth of current i (in amperes) in an inductance coil of L henrys connected in series to a resistor of R ohms is given by

$$i = \frac{V}{R}(1 - e^{-Rt/L})$$

where V is the input voltage (in volts) and t is the time (in seconds). Figure 8.3 shows the circuit. Make a sketch of the current as a function of time given that $R = 2$, $L = 4$, and $V = 6$.

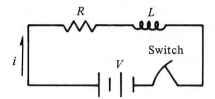

■ **Figure 8.3**

■ **Solution**

Substituting the indicated values for R, L, and V, we get

$$i = 3(1 - e^{-0.5t})$$

The graph of this function is given in Figure 8.4. The current is initially zero and becomes close to $V/R = 3$ as t increases. Hence, after sufficient time, the current is, for practical purposes, constant.

t	$e^{-0.5t}$	$3(1 - e^{-0.5t})$
0	1.000	0
1	0.606	1.18
2	0.368	1.89
3	0.223	2.33
4	0.135	2.59

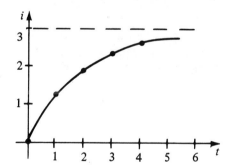

■ **Figure 8.4**

■

Logarithmic Functions

The graph of the exponential function $y = a^x$ shows that any horizontal line will intersect the graph only once. Hence, the function is one-to-one. That is, for each value of x, there is at most one value of y, and for each value of y, there is a unique x. Using this property, we can define another important function.

DEFINITION

> **Logarithmic Function**   Let $a > 0$ and $a \neq 1$. Then,
>
> $$y = \log_a x \quad \text{means} \quad x = a^y$$
>
> The function $\log_a x$ is called the *logarithmic function* of x to the base a.

Thus, the value of $y = \log_a x$ is the exponent to which the base a must be raised to give the value x. In other words, *every logarithm is an exponent of the base*.

EXAMPLE 4   From the definition we have

$$\log_2 16 = 4 \quad \text{because} \quad 2^4 = 16$$
$$\log_2 \left(\tfrac{1}{8}\right) = -3 \quad \text{because} \quad 2^{-3} = \tfrac{1}{8}$$
$$\log_{10} \left(\tfrac{1}{1000}\right) = -3 \quad \text{because} \quad 10^{-3} = \tfrac{1}{1000} \quad \blacksquare$$

EXAMPLE 5   ■ **Problem**

Find $\log_2 32$.

■ **Solution**

Let $y = \log_2 32$. Then $32 = 2^y$, or $y = 5$. Hence,

$$\log_2 32 = 5 \quad \blacksquare$$

Since, in practice, we seldom use bases other than 10 and e, we introduce special notation for these two logarithmic functions. Logarithms to the base 10, called **common logarithms**, will be denoted by **log** x; the base is understood to be 10 if no base is written. Logarithms to the base e, called **natural logarithms**, will be denoted by **ln** x.

Figure 8.5 demonstrates the following functional characteristics of the logarithmic function for $a > 1$. Any function that obeys these properties is said to *behave logarithmically*.

PROPERTIES

> **Logarithmic Function** $y = \log_a x$
>
> 1. $\text{Log}_a x$ is not defined for $x \leq 0$.
> 2. $\text{Log}_a 1 = 0$.
> 3. $\text{Log}_a a = 1$.
> 4. $\text{Log}_a x$ is negative for $0 < x < 1$ and positive for $x > 1$.
> 5. As x approaches zero, y decreases without bound.
> 6. As x increases without bound, y increases without bound.

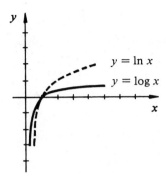

$y = \ln x$

$y = \log x$

■ Figure 8.5

The domain and the range of the logarithmic function can be inferred from the preceding discussion. We see that the domain of $y = \log_a x$ is $x > 0$, and the range is $-\infty < y < \infty$.

Comment: The inverse nature of the exponential and logarithmic functions can be obtained directly from the definition of the logarithm. Thus, since $y = \log_a x$ means $x = a^y$, we have

$$x = a^{\log_a x}$$

For example,

$$3^{\log_3 8} = 8 \qquad e^{\ln(\sin x)} = \sin x \qquad 10^{\log x^2} = x^2$$

The algebraic rearrangement of logarithmic expressions is a frequent requirement. For your reference, the rules for manipulating logarithmic expressions are listed next.

RULE 1

Logarithm of a Product

$$\log_b MN = \log_b M + \log_b N$$

RULE 2

Logarithm of a Quotient

$$\log_b \frac{M}{N} = \log_b M - \log_b N$$

RULE 3

Logarithm of a Power

$$\log_b M^n = n \log_b M$$

EXAMPLE 6   ■ **Problem**

Write the expression $\log x - 2 \log x + 3 \log(x + 1) - \log(x^2 - 1)$ as a single term.

■ **Solution**

Proceed as follows:

$$\log x - 2 \log x + 3 \log(x + 1) - \log(x^2 - 1)$$
$$= \log x - \log x^2 + \log(x + 1)^3 - \log(x^2 - 1) \qquad \text{using Rule 3}$$
$$= \log \frac{x(x + 1)^3}{x^2(x^2 - 1)} \qquad \begin{array}{c}\text{using Rules 1}\\ \text{and 2}\end{array}$$
$$= \log \frac{(x + 1)^2}{x(x - 1)} \qquad \text{cancellation law} \qquad ■$$

Exponential and Logarithmic Equations

Equations in which the variable occurs as an exponent are called **exponential equations**. To solve these equations, we use the fact that the logarithm is a one-to-one function. Thus, if $\log x = \log y$, then $x = y$. Hence, if both sides may be assumed positive, taking the logarithm of both sides of an equation is an admissible operation.

EXAMPLE 7   ■ **Problem**

Solve the exponential equation $3^x = 2^{2x+1}$.

■ **Solution**

Taking the common logarithm of both sides, we write

$$\log 3^x = \log 2^{2x+1}$$
$$x \log 3 = (2x + 1) \log 2 \qquad \text{using Rule 2}$$
$$x(\log 3 - 2 \log 2) = \log 2 \qquad \text{collecting like terms}$$
$$x = \frac{\log 2}{\log 3 - 2 \log 2} \qquad \text{solving for } x$$
$$= \frac{0.3010}{-0.1249} = -2.41 \qquad ■$$

Equations involving logarithms are called **logarithmic equations**. The use of the rules of logarithms frequently yields a solution of such an equation.

EXAMPLE 8   ■ **Problem**

Solve the logarithmic equation $\log(x^2 - 1) - \log(x - 1) = 3$.

■ Solution

Simplifying the left-hand side by combining the two logarithm terms yields

$$\log \frac{x^2 - 1}{x - 1} = 3$$

$$\log(x + 1) = 3$$

Thus, $x + 1 = 10^3$. (Can you state why?) Or

$$x = -1 + 10^3 = 999 \qquad ■$$

EXAMPLE 9 ■ **Problem**

Solve the equation $x^{\log x} = \dfrac{x^3}{100}$.

■ Solution

Taking the common logarithm of both sides, we have

$$(\log x)(\log x) = \log x^3 - \log 100$$

$$(\log x)^2 - \log x^3 + \log 100 = 0 \qquad \text{collecting terms}$$

$$(\log x)^2 - 3\log x + 2 = 0 \qquad \log x^3 = 3\log x$$

$$(\log x - 2)(\log x - 1) = 0 \qquad \text{factoring}$$

$$\log x = 2 \quad \text{or} \quad \log x = 1 \qquad \text{solving for } \log x$$

Thus, $x = 100$ or 10. ■

EXAMPLE 10 ■ **Problem**

A certain power supply has a power output (in watts) governed by the equation

$$p = 50e^{-t/250}$$

where t is the time (in days). The equipment aboard a satellite requires 10 W of power to operate properly. What is the operational life of the satellite?

■ Solution

Solving the equation $10 = 50e^{-t/250}$ for t gives

$$\frac{-t}{250} = \ln \frac{10}{50} = \ln 0.2$$

$$= -1.609$$

or

$$t = 250 \cdot 1.609 = 402 \text{ days}$$

Hence, the operational life of the satellite is 402 days. ■

EXERCISES FOR SECTION 8.1

In Exercises 1–14, solve for x.

1. $2^{x-5} = 3^x$

2. $x^{-1.2} = 18$

3. $2^{x^2-3x} = 16$

4. $3^{\log x} = 2$

5. $3^{1-x} = 5$

6. $\log 2^x = x^2 - 2$

7. $\log x = 3$

8. $\log_2 x = 5$

9. $\log_3 81 = x$

10. $\log_5 (\frac{1}{25}) = x$

11. $\log_x 64 = 6$

12. $\log_x 27 = 3$

13. $x^{4.3} = 2.1$

14. $\log_x(0.01) = -2$

In Exercises 15–24, make a sketch of the graph of the given function.

15. $y = e^{-(1-x)}$

16. $y = 2e^{-x} + 3$

17. $y = 2^2 2^t$

18. $y = (0.5)^{-x}$

19. $y = 2/3^x$

20. $y = 2 - 5e^{-(x-1)}$

21. $y = -\ln(x - 2)$

22. $y = 1 + \ln x$

23. $y = 2 - \ln x^3$

24. $y = \ln(-x) - 3$

In Exercises 25–28, tell how the graph of the given function compares with the graph of $y = e^x$.

25. e^{-x}

26. $e^{|x|}$

27. $|e^x|$

28. e^{x+3}

In Exercises 29–34, tell how the graph of the given function compares with the graph of $y = \ln x$.

29. $-\ln x$

30. $\ln(-x)$

31. $\ln(x + 3)$

32. $\ln|x|$

33. $\ln(-x)$

34. $3 + \ln x$

In Exercises 35–40, write the given expression as a single logarithmic term.

35. $\log x + \log x^2$

36. $\log 2x - \log x$

37. $\log_2 2x + 3 \log_2 x$

38. $3 \log_2 x - \log_2 x$

39. $3 \log_5 x - \log_5(2x - 3)$

40. $5 \log(x + 2) - 2 \log x$

41. Solve $2^{x+1} > 1$ for x.

42. Solve $\log x + \log(x - 3) = 1$ for x.

43. Solve $\log(x + 1) > 0$ for x.

44. A substance decays according to the law $M = M_0 e^{-kt}$, where k is the decay constant and t is the time (in hours). What would the decay constant be if $M_0 = 200$ and $M = 150$ when $t = 2$ days?

45. Find the half-life of a radioactive material that decays exponentially according to the equation $A(t) = A_0 e^{-t/4}$, where t is the time (in years).

46. The current in an RL circuit with zero initial current and DC (constant) voltage input is given by

$$i(t) = \frac{E}{R}(1 - e^{-Rt/L})$$

(a) Graph i as a function of t.
(b) Solve for t in terms of i.

47. The current in an RL circuit with zero initial current and unit square-wave input is given by

$$i(t) = \begin{cases} \dfrac{1}{R}(1 - e^{-Rt/L}), & \text{if } 0 \le t \le 1 \\[2mm] \dfrac{1}{R}(e^{-R(t-1)/L} - e^{-Rt/L}), & \text{if } 1 \le t \end{cases}$$

Graph i as a function of t. (Use $R = L = 1$.)

48. A colony of bacteria increases according to the Malthusian law

$$N(t) = N(0)e^{kt}$$

If the colony doubles in 5 hr, find the time for the colony to triple in size.

49. A bacteria colony population is given by $N(t) = 2000(2.2)^{1.3t}$. Plot a graph of N as a function of t on semilogarithmic graph paper.

50. A principal of $5000 is invested at the rate of 9.2% per year compounded continuously. What will be the amount after 20 years? The law governing continuous compounding is $P = P_0 e^{rt}$, where t is the time (in years) and r is the percentage rate.

51. In the dye dilution procedure for measuring cardiac output, the amount of dye in the heart at any time t is given by

$$D(t) = D(0)e^{-rt/V}$$

where $D(0)$ is the amount of dye injected, r is a constant representing the outflow of blood and dye (in liters per minute), and V is the volume of the heart (in liters, L). Find the amount of dye in the heart after 5 sec given that $V = 450$ mL, $r = 1.4$ L/min, and $D(0) = 2.3$ milligrams (mg). (*Hint*: Use consistent units.)

52. A large brine tank is cleaned by running pure water into the tank at a slightly slower rate than the brine is removed. At any time t ($t < 100$), the amount of brine is given approximately by $B = (100 - t)^{1.4}$. Sketch the graph of B as a function of t on logarithmic graph paper. (*Note*: The graph is *not* linear.)

53. A mass is attached to a spring and released. If its motion is characterized by the equation

$$y(t) = e^{-at}(c_1 + c_2t)$$

it is said to be *critically damped*. Compute $y(t)$ for several values of t, and graph the motion as a function of time. Use $a = 1$, $c_1 = 2$, and $c_2 = 1$.

8.2 DERIVATIVES OF LOGARITHMIC FUNCTIONS

Consider the logarithmic function

$$y = \log_b x \qquad x > 0$$

The derivative of this function with respect to x can be found by applying the definition of the derivative with $f(x) = \log_b x$. Thus, if $\Delta y = f(x + \Delta x) - f(x)$, we have, for $\Delta x > 0$,

$$\Delta y = \log_b(x + \Delta x) - \log_b x$$

or

$$\Delta y = \log_b \frac{x + \Delta x}{x} = \log_b\left(1 + \frac{\Delta x}{x}\right)$$

Using this expression, we obtain

$$\frac{\Delta y}{\Delta x} = \frac{1}{\Delta x} \log_b\left(1 + \frac{\Delta x}{x}\right)$$

To evaluate the limit, we first multiply the right member by x/x:

$$\frac{\Delta y}{\Delta x} = \frac{1}{x} \cdot \frac{x}{\Delta x} \log_b\left(1 + \frac{\Delta x}{x}\right) = \frac{1}{x} \log_b\left(1 + \frac{\Delta x}{x}\right)^{x/\Delta x}$$

Therefore,

$$\frac{dy}{dx} = \lim_{\Delta x \to 0} \frac{1}{x} \log_b\left(1 + \frac{\Delta x}{x}\right)^{x/\Delta x}$$

As $\Delta x \to 0$, $\Delta x/x \to 0$ and $x/\Delta x \to \infty$, so

$$\lim_{\Delta x \to 0} \left(1 + \frac{\Delta x}{x}\right)^{x/\Delta x} = e$$

and

$$\frac{dy}{dx} = \frac{1}{x} \log_b e$$

Since $y = \log_b x$, this expression may be rewritten as

$$\frac{d}{dx}(\log_b x) = \frac{1}{x}\log_b e \qquad (8.1)$$

By applying the chain rule, we may extend this result to include logarithmic functions of the form $\log_b u$, where u is a function of the variable x.

FORMULA

> **The Derivative of $\log_b u$**
>
> $$\frac{d}{dx}(\log_b u) = \frac{1}{u}\frac{du}{dx}\log_b e \qquad (8.2)$$

An important special case of Formula (8.2) is the case in which base e is used. The right member of (8.2) becomes $(1/u)(du/dx)(\ln e)$, and since $\ln e = 1$, we have the following formula.

FORMULA

> **The Derivative of $\ln u$**
>
> $$\frac{d}{dx}(\ln u) = \frac{1}{u}\frac{du}{dx} \qquad (8.3)$$

The simplicity of Formula (8.3) is one reason natural logarithms are used almost exclusively in calculus.

Comment: A function such as $y = \log(x - 10)$ is not defined for $x \le 10$. So, to be technically correct, we should qualify the function by indicating that it is only defined for $x > 10$. To avoid writing a qualifier for each logarithmic function, we will assume that the range of the function u is the domain of $\log_a u$.

EXAMPLE 1   ■ **Problem**

Differentiate $y = \log(3x + 2)$.

■ **Solution**

Here, we use Formula (8.2) with $u = 3x + 2$. The required derivative is then

$$\frac{dy}{dx} = \frac{1}{3x + 2}\left[\frac{d}{dx}(3x + 2)\right](\log e)$$

$$= \frac{1}{3x + 2}(3)(\log e) = \frac{3\log e}{3x + 2} \qquad ■$$

EXAMPLE 2   ■ **Problem**

Differentiate $y = \ln\sqrt{5 - x^3}$.

■ **Solution**

This equation may be rewritten as

$$y = \frac{1}{2}\ln(5 - x^3)$$

Using Formula (8.3), we have

$$\frac{dy}{dx} = \frac{1}{2}\left(\frac{1}{5 - x^3}\right)(-3x^2) = \frac{-3x^2}{2(5 - x^3)}$$ ■

EXAMPLE 3   ■ **Problem**

Differentiate $y = \ln^3 2x$.

■ **Solution**

Note that $\ln^3 2x$ can be written as $(\ln 2x)^3$, which is a function of x raised to a power. The derivative of such a function is given by

$$\frac{d}{dx}(v^n) = nv^{n-1}\frac{dv}{dx}$$

Letting $n = 3$ and $v = \ln 2x$, we have

$$\frac{dy}{dx} = 3(\ln 2x)^2\frac{d}{dx}(\ln 2x)$$

Using Formula (8.3) to evaluate d/dx $(\ln 2x)$, we have

$$\frac{dy}{dx} = 3(\ln 2x)^2\left(\frac{1}{2x} \cdot 2\right) = \frac{3}{x}\ln^2 2x$$ ■

EXAMPLE 4   ■ **Problem**

Differentiate $p = \ln\dfrac{t^3}{t^2 - 1}$.

■ **Solution**

The trick here is to simplify the expression algebraically *before* differentiating. Thus, the given expression may be written as

$$p = 3\ln t - \ln(t^2 - 1)$$

The desired derivative is then

$$\frac{dp}{dt} = 3\left(\frac{1}{t}\right) - \frac{1}{t^2 - 1}(2t)$$

$$= \frac{3}{t} - \frac{2t}{t^2 - 1} = \frac{t^2 - 3}{t(t^2 - 1)}$$

∎

EXAMPLE 5   ∎ **Problem**

The tensile strength (in pounds) of a new plastic is found to vary with temperature according to

$$S = 500 \ln(T + 50) - 4T + 2500$$

At what temperature does the plastic have its maximum tensile strength?

∎ **Solution**

The maximum value of the function is found by setting the first derivative equal to zero and solving for T. The derivative of tensile strength with respect to temperature is

$$\frac{dS}{dT} = \frac{500}{T + 50} - 4$$

Setting this derivative equal to zero, we obtain

$$\frac{500}{T + 50} - 4 = 0$$

$$500 - 4T - 200 = 0$$

$$T = 75°$$

To check that this value is truly a temperature at which a maximum occurs, we use the second derivative test, as follows:

$$\frac{d^2S}{dT^2} = \frac{-500}{(T + 50)^2}$$

Since this expression is negative for $T = 75$, we conclude that the tensile strength is a maximum at this temperature.   ∎

EXERCISES FOR SECTION 8.2

Differentiate each of the expressions in Exercises 1–20.

1. $y = \log 2x$

2. $y = \log \frac{1}{2}x$

3. $s = \ln(2t - 1)$

4. $p = \ln 2(r + 1)$

5. $z = \ln p^2$

6. $y = \log(x^2 + 1)$

7. $q = 3 \ln(x - x^2)$

8. $r = 3 + \ln(1 - s^2)$

9. $f(x) = \ln [x^3(4 - x)^2]$

10. $B = \ln[I(I^2 + 4)^3]$

11. $s = \ln \sqrt{t^3 - 1}$

12. $w = \log(t^2 + 3t + 1)^{2/3}$

13. $g(r) = \ln\dfrac{r^2 + 1}{(r^2 - 1)^3}$

14. $i = \ln\sqrt{\dfrac{2 - v}{2 + v}}$

15. $Q = \ln^4(3T + 2)$

16. $m = \sqrt{\ln(p^2 - 1)}$

17. $C = \dfrac{R^2}{\ln(R^2 - 2)}$

18. $y = (3x + 2)^3 \ln(8x^2 - 3)$

19. $E = V^2 \ln(3V + 2)$

20. $G(x) = 3x + \dfrac{x}{\ln(x - 1)}$

21. A coaxial cable consists of an inner cylindrical conductor of radius r inside a thin-walled conducting tube of radius R. If the two conductors are oppositely charged, the potential difference between the conductors is $V = 100 \ln(R/r)$. What is the rate of change of V with respect to r if R is held constant?

22. In Exercise 21, find the rate of change of V with respect to R given that r is held constant.

23. The Chezy discharge coefficient for water flowing in a wide channel is $C = 42 \ln (R/p)$, where R is the hydraulic radius and p is the roughness of the channel. Find dC/dp assuming R to be constant.

24. The displacement (in feet) of a car varies with time according to $s = \ln(t^3 + 1)$. Find the expression for the velocity of the car.

25. Find the expression for the acceleration of the car in Exercise 24.

26. The current (in amperes) in a 5-H inductor varies as $i = \ln (2t - 1)$. Find the expression for induced voltage in the coil. What is the voltage when $t = 3$ sec?

27. The tensile strength S (in pounds per square inch) of a viscoelastic material, used for damping mechanical vibrations, is a function of temperature T. What is the rate of change of tensile strength with respect to temperature if $S = \ln[3/(1 + 3T)]$?

28. The Link Trucking Company has a fleet of trucks that are serviced by company mechanics. The cost of maintenance is given by $C = \ln(3m + 2)$, where m is the interval (in miles) between repairs. Find the rate of change of maintenance cost with respect to the repair interval.

29. The energy dissipated (in joules) by a resistor during an experiment is given by $E = \ln(t + 1) - \frac{1}{2}t + 10$. At what time is the energy output a maximum?

30. What is the maximum energy output of the resistor in Exercise 29?

8.3 LOGARITHMIC DIFFERENTIATION

By combining the results of the previous section with that of implicit differentiation, we may expand the set of functions we can differentiate. To illustrate, we consider differentiating the function $y = x^x$. This function does not fit the formula for a power function, because this expression is a variable raised to a variable power. The derivative of functions of this type may be found by a technique known as *logarithmic differentiation*. The basic steps in the process are as follows:

PROCEDURE

Logarithmic Differentiation

1. Take the natural logarithm of both sides of the given relation.
2. Simplify the logarithmic expressions.
3. Differentiate the resulting equation implicitly with respect to x.
4. Solve for the required derivative.

EXAMPLE 1

■ **Problem**

Find the derivative of $y = x^x$.

■ **Solution**

1. Taking the natural logarithm gives

$$\ln y = \ln x^x.$$

2. Simplifying, we obtain

$$\ln y = x \ln x$$

Before we differentiate, we note that y is a function of x, and therefore,

$$d/dx\,(\ln y) = (1/y)(dy/dx).$$

3. Differentiating implictly yields

$$\frac{1}{y}\frac{dy}{dx} = x\frac{1}{x} + \ln x(1) = 1 + \ln x$$

4. To solve for dy/dx, we multiply both sides of this equation by y:

$$\frac{dy}{dx} = (1 + \ln x)y$$

But from the given equation, $y = x^x$. So,

$$\frac{dy}{dx} = (1 + \ln x)\, x^x$$

■

EXAMPLE 2   ■ **Problem**

Differentiate $y = \dfrac{(x^2 - 3)^4}{(3x + 1)^2}$ by using logarithmic differentiation.

■ **Solution**

1.   Take the natural logarithm:

$$\ln y = \ln \frac{(x^2 - 3)^4}{(3x + 1)^2}$$

2.   Simplify:

$$\ln y = \ln(x^2 - 3)^4 - \ln(3x + 1)^2 = 4\ln(x^2 - 3) - 2\ln(3x + 1)$$

3.   Differentiate:

$$\frac{1}{y}\frac{dy}{dx} = \frac{8x}{x^2 - 3} - \frac{6}{3x + 1} = \frac{2(9x^2 + 4x + 9)}{(x^2 - 3)(3x + 1)}$$

4.   Solve:

$$\frac{dy}{dx} = \frac{2(9x^2 + 4x + 9)}{(x^2 - 3)(3x + 1)} \cdot y = \frac{2(9x^2 + 4x + 9)}{(x^2 - 3)(3x + 1)} \cdot \frac{(x^2 - 3)^4}{(3x + 1)^2}$$

$$= \frac{2(9x^2 + 4x + 9)(x^2 - 3)^3}{(3x + 1)^3}$$

■

EXAMPLE 3   ■ **Problem**

The amplitude (in centimeters) of a damped oscillation varies with time according to $A = e^{-t}$. Find the expression for the rate of change of amplitude with respect to time.

■ **Solution**

To find the required rate, we must use logarithmic differentiation:

$$\ln A = \ln e^{-t}$$
$$\ln A = -t \ln e = -t$$

$$\frac{1}{A}\frac{dA}{dt} = -1$$

$$\frac{dA}{dt} = -A = -e^{-t} \text{ cm/sec} \quad \blacksquare$$

EXERCISES FOR SECTION 8.3

Find the derivative of the indicated function by using logarithmic differentiation.

1. $y = e^{2x}$

2. $y = x^{3x}$

3. $y = (2x - 3)^{1/3}(2x + 3)^{2/3}$

4. $s = (t^2 + 4)^3(3t + 4)^5$

5. $v = t^2 e^{2t}$

6. $y = \dfrac{x}{e^x}$

7. $i = 5e^{-3t}$

8. $s = 10e^{2/t}$

9. $y = \sqrt{\dfrac{x^2 + 1}{x^2 - 1}}$

10. $z = \sqrt[3]{\dfrac{1 - 2s}{1 + 2s}}$

11. $B = (4L^2 - 5)^{\sqrt{L^2 + 1}}$

12. $y = (1 - x^3)^{(1 - x^3)}$

13. $y = \dfrac{(2x + 1)(3x + 2)}{4x + 3}$

14. $C = T^3(T^2 + 1)^4(T^2 - 1)^4$

8.4 DERIVATIVES OF EXPONENTIAL FUNCTIONS

Functions of the form

$$y = b^u$$

where b is a positive constant ($\neq 1$) and u is a function of the variable x, are *exponential functions*. The formula for the derivative of the exponential function is derived by first taking the natural logarithm of both sides of the expression $y = b^u$ and then differentiating both sides implicitly with respect to x, as follows:

$$y = b^u$$
$$\ln y = \ln b^u \qquad \text{taking natural logarithm}$$
$$\ln y = u \ln b \qquad \log_b M^n = n \log_b M$$
$$\frac{1}{y}\frac{dy}{dx} = \frac{du}{dx} \ln b \qquad \text{Formula (8.3)}$$
$$\frac{dy}{dx} = y \frac{du}{dx} \ln b \qquad \text{solving for } \frac{dy}{dx}$$

Replacing y with b^u, we have the desired formula.

FORMULA

> **The Derivative of b^u**
>
> $$\frac{d}{dx}(b^u) = b^u \frac{du}{dx} \ln b \qquad b > 0 \text{ and } b \neq 1 \qquad\qquad \textbf{(8.4)}$$

Letting $b = e$, we have an important special case.

FORMULA

> **The Derivative of e^u**
>
> $$\frac{d}{dx}(e^u) = e^u \frac{du}{dx} \qquad\qquad \textbf{(8.5)}$$

EXAMPLE 1   ■ **Problem**

Find the derivative of $s = e^{-t^4}$.

■ **Solution**

By Formula (8.5), we have

$$\frac{ds}{dt} = e^{-t^4} \frac{d}{dt}(-t^4) = -4t^3 e^{-t^4}$$ ■

EXAMPLE 2   ■ **Problem**

Evaluate $\dfrac{d}{dx}(10^{6x+2})$.

■ **Solution**

Using Formula (8.4), we have

$$\frac{d}{dx}(10^{6x+2}) = 10^{6x+2} \frac{d}{dx}(6x + 2)(\ln 10)$$
$$= 6(10^{6x+2})(\ln 10)$$ ■

EXAMPLE 3   ■ **Problem**

The current (in amperes) in a 3-H inductance coil is governed by the expression $i = 6 - e^{-2t}$. Find the expression for the induced voltage. What is the voltage when $t = 2$?

■ **Solution**

The voltage induced in a coil is

$$v = L \frac{di}{dt}$$

Therefore,

$$v = 3 \frac{d}{dt} (6 - e^{-2t}) = 3[-e^{-2t}(-2)] = 6e^{-2t}$$

When $t = 2$, we have

$$v(2) = 6e^{-4} \approx 6(0.01832) \approx 0.11 \text{ V}$$ ■

EXAMPLE 4   ■ **Problem**

When a heated object is immersed in a cooling medium, the temperature of the object at any time after immersion is approximated by the formula $T = T_0 + Ce^{kt}$, where T_0 is the temperature of the cooling medium, C is the difference in temperature between the object and the cooling medium before immersion, and k is a constant of proportionality. For $T_0 = 10$, $C = 75$, and $k = -0.25$, find the temperature of the object 4 sec after immersion. Find the rate at which the temperature is changing at this time.

■ **Solution**

The desired temperature equation is obtained by replacing T_0, C, and k with their numerical values. Thus,

$$T = 10 + 75e^{-0.25t}$$

Substituting $t = 4$ into this equation, we get

$$T = 10 + 75e^{-1} \approx 10 + 75(0.3679) \approx 37.6°$$

The rate of change of temperature is

$$\frac{dT}{dt} = \frac{d}{dt} (10 + 75e^{-0.25t}) = -18.75e^{-0.25t}$$

Letting $t = 4$ yields

$$\frac{dT}{dt} = -18.75e^{-1} \approx -18.75(0.3679) \approx -6.9°/\text{sec}$$ ■

EXAMPLE 5   ■ **Problem**

Find the derivative of $y = (1 - e^{-2x})^3$.

■ **Solution**

Observe that the given function is a composite function of the form u^n, in which $u = 1 - e^{-2x}$ and $n = 3$. Therefore,

$$\frac{dy}{dx} = 3(1 - e^{-2x})^2 \frac{d}{dx} (1 - e^{-2x}) = 3(1 - e^{-2x})^2 (2e^{-2x})$$

$$= 6e^{-2x}(1 - e^{-2x})^2 \qquad\qquad ■$$

EXERCISES FOR SECTION 8.4

In Exercises 1–18, find the first derivative of the indicated function.

1. $s = e^{2t}$

2. $m = 10^{2r-1}$

3. $y = 4e^{x^2}$

4. $i = 5e^{3t/2}$

5. $v = \dfrac{4}{e^{3x}}$

6. $J = \dfrac{1}{10^K}$

7. $y = e^{\sqrt{x}}$

8. $z = 3e^{(1-x)}$

9. $B = T^3 e^T$

10. $G(x) = \dfrac{x}{e^{2x}}$

11. $w = \dfrac{e^v}{v^2}$

12. $y = \dfrac{1}{xe^x}$

13. $\theta = \dfrac{e^r + 1}{e^r - 1}$

14. $f(t) = \dfrac{1}{2}(e^t - e^{-t})^2$

15. $I = \sqrt{2 + e^{3x}}$

16. $q = \ln(2 + e^{2x})$

17. $P = e^t \ln t$

18. $x = e^y \sqrt{1 + \ln y}$

In Exercises 19–24, find the second derivative of the indicated function.

19. $y = e^{-x}$

20. $y = e^{3x}$

21. $p = e^{t^2}$

22. $Q = e^{\sqrt{s}}$

23. $y = te^{t^2}$

24. $p(f) = f^2 e^{-f}$

25. The growth of the current in an inductance coil connected in series to a resistor is given by $i = (V/R)(1 - e^{-Rt/L})$, where R, L, and V are constants and t is time. Find di/dt.

26. The saturation current density J at the surface of a cathode is a function of the temperature T of the cathode. The relationship between saturation current density and cathode temperature was found by Dushman to be $J = AT^2 e^{-K/T}$, where A and K are constants. Find the rate of change of J with respect to T.

27. The shear s of a simple beam is equal to the derivative of the bending moment M with respect to the distance x, measured from one end of the beam. Find the shear equation of a beam whose bending moment is $M = xe^x$.

28. When a cable hangs between two poles, it forms a curve known in mathematics as a *catenary*. The equation of a catenary is $y = (H/2)(e^{x/H} + e^{-x/H})$, where H is the height of the cable at the center point and x is the horizontal distance from the center point, as in the accompanying figure. What is the slope of the cable at $x = 20$ ft if $H = 100$ ft?

$$y = \frac{H}{2}(e^{x/H} + e^{-x/H})$$

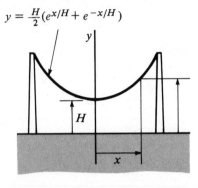

29. The atmospheric pressure varies with altitude according to $p = p_0 e^{-kz}$, where p_0 and k are constants and z is the altitude. Find the rate of change of p with respect to z.

30. The reliability function $R(t)$ of a collection of objects that are subject to failure is defined as the derivative of the failure function $F(t)$, where t is a unit of time. If $F(t) = 1 - e^{-3t}$ is the failure function of Shurfire spark plugs, what is the reliability function for the spark plugs?

31. The force F required to accelerate an object of weight W is given by Newton's second law to be $F = (W/g)a$, where g is the acceleration of gravity. If W is given in pounds and a and g in feet per second squared, then F will be in pounds. What is the expression for the force exerted on an object by a resisting medium if the velocity of the object is given by $v = 10(1 - e^{-0.1t})$? What is the magnitude of the force when $t = 20$ sec if the object weighs 96 lb?

32. In a critically damped vibrating system, the displacement of the system from its equilibrium position is given by $x = (A + Bt)e^{-pt}$, where A, B, and p are constants and t represents time. Determine the expression for the velocity of the system, dx/dt.

■ 8.5 INTEGRALS OF RECIPROCAL FUNCTIONS

In Chapter 5, when the formula

$$\int u^n \, du = \frac{u^{n+1}}{n + 1} + C$$

was introduced, we indicated that the formula was invalid for $n = -1$. In this section, we develop a formula for integrals of the form

$$\int u^{-1} \, du = \int \frac{du}{u}$$

From Formula (8.3) of Section 8.2,

$$\frac{d}{dx} (\ln u) = \frac{1}{u} \frac{du}{dx}$$

Since differentiation and antidifferentiation are inverse operations, we have

$$\int \left(\frac{1}{u} \frac{du}{dx} \right) dx = \ln u + C \tag{8.6}$$

Or letting $du = (du/dx) \, dx$, we have the following formula.

FORMULA

The Integral of $\frac{1}{u} \, du$

$$\int \frac{1}{u} \, du = \ln u + C \qquad u > 0 \tag{8.7}$$

Formula (8.7) is merely a compact form of (8.6), and hence, the symbol du must be clearly understood. Thus, if the integrand function is assumed to be of the form $1/u$, then this integration formula may only be applied if there is an accompanying du/dx. The form of the integrand may be altered only by multiplying and dividing by constants.

Comment: Formula (8.7) can be used to integrate $1/u$ for $u < 0$ by writing it in the form

$$\int \frac{1}{u} \, du = \ln |u| + C \qquad u \neq 0$$

EXAMPLE 1   ▪ **Problem**

Find $\int \dfrac{dx}{3x + 2}$.

▪ **Solution**

In this problem, if we let $u = 3x + 2$, then $du/dx = 3$. Hence, we multiply inside the integral by 3 and outside by $\frac{1}{3}$ to obtain

$$\int \frac{dx}{3x + 2} = \frac{1}{3} \int \frac{3\,dx}{3x + 2} = \frac{1}{3} \ln (3x + 2) + C$$

■

EXAMPLE 2   ■ **Problem**

Show that $\int \dfrac{dx}{x^2 + 4}$ is not of the form $\int \dfrac{du}{u}$.

■ **Solution**

Letting $u = x^2 + 4$, then $du/dx = 2x$. Hence, the numerator must have a term of this form. Since multiplication by a variable expression requires the multiplication by its reciprocal outside the integral sign, and this operation is not permitted, we conclude that the given antiderivative is not of the form $\int du/u$. In the next chapter, you will learn a formula whose standard form matches this antiderivative.   ■

EXAMPLE 3   ■ **Problem**

During the combustion of a certain flammable material, the rate at which heat is radiated (in Btu/sec) is

$$\frac{dH}{dt} = \frac{3t}{t^2 + 4}$$

Find the amount of heat generated from $t = 0$ to $t = 2$ sec.

■ **Solution**

The amount of heat radiated from $t = 0$ to $t = 2$ sec is given by the definite integral

$$H = \int_0^2 \left(\frac{dH}{dt}\right) dt = \int_0^2 \frac{3t}{t^2 + 4}\, dt$$

To evaluate this integral, we multiply the integrand function by 2 and the integral by $\frac{1}{2}$. Thus,

$$H = \frac{3}{2} \int_0^2 \frac{2t}{t^2 + 4}\, dt = \frac{3}{2}\Big[\ln (t^2 + 4)\Big]_0^2 = \frac{3}{2} (\ln 8 - \ln 4) = \frac{3}{2} \ln 2$$
$$= 1.04 \text{ Btu}$$

■

EXAMPLE 4   ■ **Problem**

Find the area bounded by the curve $y = 1/x$ and the x-axis from $x = -5$ to $x = -2$.

■ **Solution**

Since the region is below the x-axis, the indicated area is given by

$$A = -\int_{-5}^{-2} \frac{dx}{x} = -\Big[\ln |x|\Big]_{-5}^{-2}$$

$$= -[\ln 2 - \ln 5] \approx 0.916 \text{ square units}$$

See Figure 8.6. Note that ln the absolute value of x is used since the endpoints of the interval are -5 and -2.

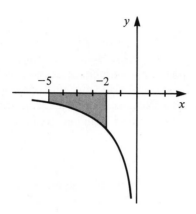

■ **Figure 8.6**   ■

EXERCISES FOR SECTION 8.5

In Exercises 1–20, find the indicated integrals.

1. $\displaystyle\int \frac{2\,dx}{x}$

2. $\displaystyle\int \frac{dx}{3x}$

3. $\displaystyle\int_{-1}^{0} \frac{dy}{1 - 2y}$

4. $\displaystyle\int_{0}^{1} \frac{ds}{3s + 4}$

5. $\displaystyle\int \frac{t\,dt}{t^2 + 4}$

6. $\displaystyle\int \frac{z\,dz}{1 - z^2}$

7. $\displaystyle\int_{0}^{1} \frac{e^x\,dx}{e^x + 1}$

8. $\displaystyle\int_{1}^{3} \frac{(2x + 1)\,dx}{x^2 + x}$

9. $\displaystyle\int \frac{x\,dx}{(x^2 - 3)^2}$

10. $\displaystyle\int \frac{B^{1/2}\,dB}{1 + B^{3/2}}$

11. $\displaystyle\int \frac{\ln^2 x\,dx}{x}$

12. $\displaystyle\int \frac{(x^2 - 1)\,dx}{x^3 - 3x + 5}$

13. $\displaystyle\int \frac{dx}{x \ln x}$

14. $\displaystyle\int \frac{x\,dx}{x^2 + 2}$

15. $\displaystyle\int \frac{e^x - e^{-x}}{e^x + e^{-x}}\,dx$

16. $\displaystyle\int \frac{dy}{\sqrt{y}(1 + \sqrt{y})}$

17. $\displaystyle\int \frac{dI}{I(2 + 3 \ln I)}$

18. $\displaystyle\int \frac{da}{a(1 + \ln a^2)}$

19. $\int \dfrac{dz}{z(1 + \ln z)^2}$

20. $\int \dfrac{(x^2 + 2x + 3)\, dx}{x + 2}$

21. When a rocket of initial mass m_0 is fired, the mass is constantly changing because the fuel is being burned. Through the principle of impulse and momentum, the velocity of the rocket at any time is given by $v = \int [qu/(m_0 - qt) - g]\, dt$, where q is the rate of fuel consumption, u is the exhaust of the burned fuel, and g is the acceleration of gravity. Determine the expression for v if the rocket starts from rest.

22. In the study of fluid resistance, the rate of change of the velocity of the fluid with respect to the distance from the boundary layer is $dv/dy = 2.5\, s/y$, where s is the shear velocity of the given fluid. Assuming s to be constant, find the velocity of the fluid as a function of the distance y. Assume that $v = 10$ when $y = 1$.

23. From $t = 0$ to $t = 0.5$ sec, the force (in pounds) acting on an object is $F = 1/(1 + 2t)$. What is the impulse of the force during this time interval if $I = \int_{t_1}^{t_2} F\, dt$?

24. The current in a capacitor is found to vary as $i = t/(t^2 + 1)$ from $t = 0$ to $t = 2$ sec. Find the charge transferred to the capacitor during this time period.

25. The rate (in cubic centimeters per second) at which oxygen is consumed by a guinea pig after being given an experimental drug is found to be $dV/dt = 2 + 1/(t + 1)$. How much oxygen does a guinea pig consume in the first 5 sec after an injection?

26. The work done in moving an object is $W = \int_{s_1}^{s_2} F\, ds$, where F is the force and s is the displacement of the object. What is the work done in moving an object 10 ft if the force varies with displacement according to $F = 1/(2s + 3)$?

27. The velocity (in centimeters per second) of a charged particle in a magnetic field is found to be $v = 10/(12t + 1)$. How far does the particle move from $t = 0$ to $t = 1$ sec?

28. Find the expression for the acceleration of the particle in Exercise 27. What are the velocity and the acceleration of the particle when $t = 1$ sec?

29. Find a function whose derivative is $f'(x) = 1/2x$.

30. Find a function whose derivative is $g'(t) = t/(t^2 + 4)$.

8.6 INTEGRALS OF EXPONENTIAL FUNCTIONS

We conclude this chapter with a discussion of integrals of the form

$$\int b^u\, du$$

The desired formula is a direct result of the differentiation formula for b^u. Since the derivative of b^u with respect to x is

$$\frac{d}{dx}(b^u) = b^u \frac{du}{dx}(\ln b)$$

we have the following integration formula.

FORMULA

> **The Integral of $b^u \, du$**
>
> $$\int b^u \, du = \frac{b^u}{\ln b} + C \qquad b > 0 \text{ and } b \neq 1 \tag{8.8}$$

When $b = e$, we have the following important formula.

FORMULA

> **The Integral of $e^u \, du$**
>
> $$\int e^u \, du = e^u + C \tag{8.9}$$

The same cautionary remarks that were made for other integration formulas apply to these two. Specifically, the du is merely shorthand for the fact that there must be a du/dx in the integrand before the formula can be applied.

EXAMPLE 1 ■ **Problem**

Evaluate $\int 10e^{2x} \, dx$.

■ **Solution**

Letting $u = 2x$, we see that $du/dx = 2$. Writing 2 inside and $\frac{1}{2}$ outside the integral, we have

$$\int 10e^{2x} \, dx = \frac{10}{2} \int e^{2x}(2 \, dx) = 5e^{2x} + C \qquad\blacksquare$$

EXAMPLE 2 ■ **Problem**

Show that $\int e^{x^2} \, dx$ is not of the form where Formula (8.9) can be applied.

■ **Solution**

Letting $u = x^2$, we have $du/dx = 2x$. Hence, there would have to be a factor of x in the integrand in order for (8.9) to be applied. Since no permissible operation will allow us to place an x there, we conclude that $\int e^{x^2} \, dx$ is not of the required form. ■

EXAMPLE 3   ■ **Problem**

Evaluate $\int e^{3x} (e^{3x} + 2)^5 \, dx$.

■ **Solution**

Although this integrand contains exponential functions, it is primarily of the form $\int u^n \, du$. To see that it is, let $u = e^{3x} + 2$. Then, $du/dx = 3e^{3x}$. We multiply and divide by 3 to obtain the proper form. Thus,

$$\int e^{3x}(e^{3x} + 2)^5 \, dx = \frac{1}{3} \int (e^{3x} + 2)^5 (3e^{3x}) \, dx$$

$$= \frac{(e^{3x} + 2)^6}{18} + C$$

■

EXAMPLE 4   ■ **Problem**

Find the distance traveled by an alpha particle during the time interval $t = 0$ to $t = 6$ sec if its velocity (in centimeters per second) is $v = e^{t/4}$.

■ **Solution**

Since displacement is the integral of the velocity equation, we evaluate the definite integral

$$s = \int_0^6 e^{t/4} \, dt$$

Letting $u = t/4$, $du/dt = \frac{1}{4}$. So,

$$s = 4 \int_0^6 e^{t/4} \left(\frac{1}{4}\right) dt = 4 \left[e^{t/4}\right]_0^6 = 4(e^{1.5} - e^0) \approx 4(4.48 - 1)$$

$$= 13.9 \text{ cm}$$

■

EXERCISES FOR SECTION 8.6

In Exercises 1–20, find the indicated integral.

1. $\int 2e^x \, dx$

2. $\int 10^{2y} \, dy$

3. $\int 2^{-x} \, dx$

4. $\int \frac{dz}{e^z}$

5. $\int (e^{2x} + e^{5x}) \, dx$

6. $\int \left(1 + \frac{1}{e^x}\right) dx$

7. $\int se^{s^2} \, ds$

8. $\int (x - 1)e^{(x^2 - 2x)} \, dx$

9. $\int x^2 (e^{x^3} + x) \, dx$

10. $\int (1 + xe^{x^2}) \, dx$

11. $\int \dfrac{e^{\sqrt{y+1}} dy}{\sqrt{y+1}}$

12. $\int e^{3x} (e^{3x} + 2)^4 \, dx$

13. $\int \dfrac{\sqrt{1 - e^{-x}} \, dx}{e^x}$

14. $\int \dfrac{e^{1/x} \, dx}{x^2}$

15. $\int (e^x + e^{-x})^2 \, dx$

16. $\int \left(2 + \dfrac{1}{e^x} \right)^2 dx$

17. $\int \sqrt[3]{e^s} \, ds$

18. $\int \sqrt{e^{2x}} \, dx$

19. $\int \dfrac{(1 + e^{3x}) \, dx}{e^x}$

20. $\int (e^{2x} - e^{-x}) e^{3x} \, dx$

21. The transient current in a capacitor is given by $i = (V/R)e^{-t/RC}$. Find the expression for the voltage across the plates of the capacitor.

22. What is the average force (in pounds) that is exerted on a piston during the time interval $1 \le t \le 2$ sec if $F = 15 - e^{-t}$ is the relation between force and time?

23. The velocity (in miles per hour) of an object is given by $v = 4e^{2t}$ from $t = 0$ to $t = 2$ hr. What is the displacement of the object during this interval?

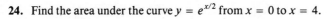

24. Find the area under the curve $y = e^{x/2}$ from $x = 0$ to $x = 4$.

25. Find the moment about the x-axis of the area given in Exercise 24.

26. The current in a 1000-Ω resistor is $i = e^{5t}$. Find the energy dissipated by the resistor from $t = 0$ to $t = 0.5$ sec.

27. The population N of virus in a test tube increases at a rate of $dN/dt = 2e^{2t}$. Find the population when $t = 1.5$ given that the initial population is 100.

28. Experiments show that when a casting is placed in an oven, the temperature of the casting increases at a rate $dT/dt = 5e^{0.05t}$. Find the temperature of the casting after it has been in the oven for 5 min given that its initial temperature is $70°$.

29. How long does it take to heat the casting in Exercise 28 to $250°$?

30. Find the volume generated by revolving the segment of the curve $y = e^{x/2}$ from $x = 0$ to $x = 2$, about the x-axis.

31. The Laplace transform of a square wave is $L\{F(t)\} = \int_0^\pi e^{-st}\, dt$, where π is the half period of the wave and s is a constant. Evaluate $L\{F(t)\}$.

REVIEW EXERCISES FOR CHAPTER 8

Find the derivative of the functions in Exercises 1–12.

1. $y = \ln (3x + 6)$

2. $y = \ln (5x^2 - 8x)$

3. $y = [\ln(x^2 - 3)]^4$

4. $y = \dfrac{1}{\ln(2x - 5x^2)}$

5. $y = e^{5t^2}$

6. $s = e^{-3x} \ln 7x$

7. $y = 5^{-x}$

8. $y = (2)^{x^2 + 1}$

9. $y = \ln (e^{x^2 + 4x})$

10. $y = e^{\ln(5x^2 - 7x + 2)}$

11. $y = e^{3x} \sqrt{x^2 + 4}$

12. $y = \dfrac{1}{(1 + e^{-x})^2}$

Find the integrals in Exercises 13–20.

13. $\displaystyle\int \dfrac{8\,dt}{2t - 5}$

14. $\displaystyle\int 4e^{7x + 10}\, dx$

15. $\displaystyle\int e^{3x - 1}\, dx$

16. $\displaystyle\int \dfrac{2t}{3t^2 + 1}\, dt$

17. $\displaystyle\int \dfrac{dx}{x \ln 2x}$

18. $\displaystyle\int \dfrac{dx}{x \ln x^2}$

19. $\displaystyle\int x e^{x^2}\, dx$

20. $\displaystyle\int (x^2 + 1)e^{(x^3 + 3x)}\, dx$

21. Find y' given that $y^{1/5} = x^2 - 1$.

22. Explain why $\int 3xe^{x^3}\, dx$ is not of the form $\int e^u\, du$.

23. The derivative of a function is $y' = 2/(3 - 2x)$. Find the function given that its graph passes through the point $(1, 0)$.

24. Find the area under the curve $y = \frac{1}{2}e^x$ from $x = -1$ to $x = 2$.

25. The rate at which a radioactive material decays is found to vary with time (in years) according to the equation $dm/dt = -1.45e^{-0.029t}$, where m is the remaining mass (in grams). The initial mass was 50 g. How many grams remained when $t = 4$ years?

26. Let $f(x) = (\ln x)/x$. Find $f'(x)$. Sketch the graph of $y = f(x)$.

27. The current in a circuit is given by $i = 5(1 - e^{-2t})$. Find di/dt at $t = 1$.

28. The velocity of an object is given by $v(t) = 2e^{-(t-1)}$ for $1 < t < 3$. What is the distance traveled during this interval?

29. Find the area bounded by the curve $y = 1/x$, the x-axis, the line $x = 1$, and the line $x = 3$.

CHAPTER 9

Trigonometric Functions

Trigonometry, which is one of the oldest branches of mathematics, was originally restricted to the study of triangle relationships. In the eighteenth century, trigonometry was developed in a completely different direction that emphasized its functional nature. It is the functional approach to trigonometry that we use to describe periodic phenomena such as simple harmonic motion and alternating current. In this chapter, we derive the calculus formulas for the trigonometric functions.

9.1 REVIEW TOPICS

Definitions of the Trigonometric Functions

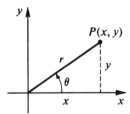

■ Figure 9.1

Consider an angle θ in standard position in the coordinate plane as shown in Figure 9.1. The terminal side of θ could lie in any one of the four quadrants. Let P be a point on the terminal side of the angle θ, having coordinates (x, y), and let the distance from the origin to the point P be r. By the Pythagorean theorem, we know that $r = \sqrt{x^2 + y^2}$.

The trigonometric functions of the angle θ in Figure 9.1 are as follows:

DEFINITIONS

Trigonometric Functions

$$\sin \theta = \frac{\text{ordinate of } P}{\text{radius of } P} = \frac{y}{r} \qquad \csc \theta = \frac{\text{radius of } P}{\text{ordinate of } P} = \frac{r}{y}$$

$$\cos \theta = \frac{\text{abscissa of } P}{\text{radius of } P} = \frac{x}{r} \qquad \sec \theta = \frac{\text{radius of } P}{\text{abscissa of } P} = \frac{r}{x}$$

$$\tan \theta = \frac{\text{ordinate of } P}{\text{abscissa of } P} = \frac{y}{x} \qquad \cot \theta = \frac{\text{abscissa of } P}{\text{ordinate of } P} = \frac{x}{y}$$

EXAMPLE 1   ■ **Problem**

An angle θ in standard position has the point $(-6, 3)$ on its terminal side. Find the values of the six trigonometric functions of the angle θ. See Figure 9.2.

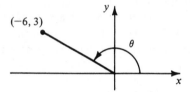

■ **Figure 9.2**

■ **Solution**

Using the definitions, with $x = -6$, $y = 3$, and $r = \sqrt{(-6)^2 + 3^2} = \sqrt{45} = 3\sqrt{5}$, we get

$$\sin \theta = \frac{y}{r} = \frac{3}{3\sqrt{5}} = \frac{1}{\sqrt{5}} \qquad \csc \theta = \frac{r}{y} = \frac{3\sqrt{5}}{3} = \sqrt{5}$$

$$\cos \theta = \frac{x}{r} = \frac{-6}{3\sqrt{5}} = -\frac{2}{\sqrt{5}} \qquad \sec \theta = \frac{r}{x} = -\frac{3\sqrt{5}}{-6} = -\frac{\sqrt{5}}{2}$$

$$\tan \theta = \frac{y}{x} = \frac{3}{-6} = -\frac{1}{2} \qquad \cot \theta = \frac{x}{y} = \frac{-6}{3} = -2$$

■

The values of the six trigonometric functions are available in tables and from a calculator or a computer. Even so, students are ordinarily expected to know, without reference to a table or a calculator, the values of the six trigonometric functions for $\theta = 0°$, $30°$, $45°$, $60°$, $90°$, and other angles symmetric to these in the other three quadrants. The first quadrant values are tabulated next.

θ	$\sin \theta$	$\cos \theta$	$\tan \theta$	$\cot \theta$	$\sec \theta$	$\csc \theta$
0°	0	1	0	Undefined	1	Undefined
30°	$\dfrac{1}{2}$	$\dfrac{\sqrt{3}}{2}$	$\dfrac{\sqrt{3}}{3}$	$\sqrt{3}$	$\dfrac{2}{\sqrt{3}}$	2
45°	$\dfrac{\sqrt{2}}{2}$	$\dfrac{\sqrt{2}}{2}$	1	1	$\sqrt{2}$	$\sqrt{2}$
60°	$\dfrac{\sqrt{3}}{2}$	$\dfrac{1}{2}$	$\sqrt{3}$	$\dfrac{\sqrt{3}}{3}$	2	$\dfrac{2}{\sqrt{3}}$
90°	1	0	Undefined	0	Undefined	1

Most trigonometric formulas in calculus require that the argument (that is, the domain variable) be stated in terms of real numbers. The domain of the trigonometric functions is extended to real numbers by simply matching real numbers with radian measure.* Thus, in the expression sin 1.2, the 1.2 could mean either 1.2 rad or the real number 1.2. Either way, the same numerical value is obtained.

Warning: When evaluating trigonometric functions of real numbers on a calculator, you must *first switch to the radian mode.*

As a matter of practice, you should know the degree measure of angles whose measure in radians is a simple multiple of π. The following table gives you an idea of which ones you should know without using a calculator.

Radians	$\dfrac{\pi}{6}$	$\dfrac{\pi}{4}$	$\dfrac{\pi}{3}$	$\dfrac{\pi}{2}$	$\dfrac{2\pi}{3}$	$\dfrac{3\pi}{4}$	$\dfrac{5\pi}{6}$	π	$\dfrac{3\pi}{2}$	2π
Degrees	30	45	60	90	120	135	150	180	270	360

Graphs of Trigonometric Functions

Functions that repeat themselves at regular intervals are called *periodic functions.* The intervals required for the function to make one complete cycle is called the **period** of the function. The trigonometric functions are familiar examples of periodic

---

*Recall that one radian (rad) is defined as the measure of a central angle that intercepts an arc equal in length to the radius. See the adjoining figure. Thus, 1 rad = 180/π°, and 1° = π/180 rad.

functions. The functions $y = \sin x$, $y = \cos x$, $y = \sec x$, and $y = \csc x$ all have periods of 2π, while $y = \tan x$ and $y = \cot x$ have periods of π.

The graphs of the six basic trigonometric functions are shown in Figure 9.3. The curve for $y = \sin x$ is often called a *sine wave*.

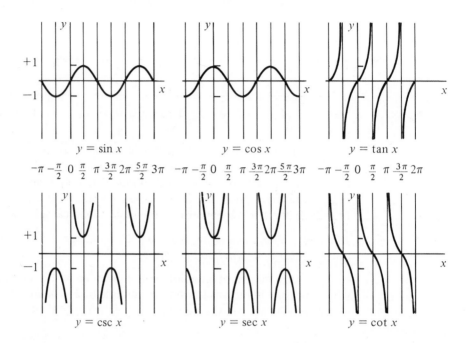

■ **Figure 9.3**

In practice, we must deal with the basic trigonometric functions altered in four possible ways:

- Multiplication of the function by a constant;
- Multiplication of the argument by a constant;
- Addition of a constant to the argument;
- Addition of a constant to the value of the function.

We use $y = \sin x$ to demonstrate each of these cases in the following paragraphs.

Multiplication of a function by a constant: If we multiply the function $y = \sin x$ by a positive constant A, we write $y = A \sin x$. Since $y = \sin x$ is bounded by $-1 \le \sin x \le 1$, $A \sin x$ is bounded by

$$-A \le A \sin x \le A$$

The value A is called the **amplitude** of the sine wave. If A is greater than one, the amplitude of the basic wave is increased. If A is less than one, the amplitude is decreased. Figure 9.4 shows $y = A \sin x$ for $A = 1$, $A = \frac{1}{2}$, and $A = 2$.

Multiplication of the argument by a constant: Assuming that $y = \sin x$ is the basic function, the function $y = \sin Bx$ represents a sine function in which the argument has been multiplied by the positive constant B. The argument of $\sin Bx$ is now Bx,

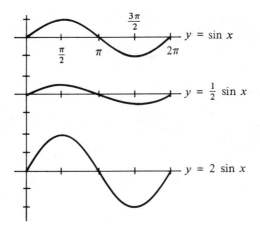

■ **Figure 9.4**

and since the sine function repeats itself for every increase in the argument of 2π, one period of sin Bx is contained in the interval $0 \le Bx \le 2\pi$, or $0 \le x \le 2\pi/B$. Therefore, multiplying the argument by a constant has the effect of altering the period of the basic function. The period of $y = \sin Bx$ is then given by $2\pi/B$. Notice that the period will be decreased if B is greater than one and increased if it is less than one. These alterations are illustrated in Figure 9.5 for $B = 1$, $B = 2$, and $B = \frac{1}{2}$.

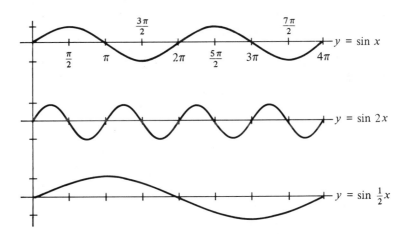

■ **Figure 9.5**

Addition of a constant to the argument: The effect of adding a constant to the argument is best handled by noticing that $y = \sin(Bx + C)$ is zero when $x = -(C/B)$. That is, the graph of $\sin(Bx + C)$ is exactly the same as sin Bx but displaced to the left by the amount $x = -(C/B)$. The amount that the graph of the function is moved to the left (or right) is called the **phase shift**. When $B = 1$, we get $y = \sin(x + C)$, which will displace the graph of sin x C units to the left if C is positive and C units to the right if C is negative. The amplitude and the period of the wave are unaffected by the addition of a constant term to the argument. The values $C = 0$, $C = \pi/4$, and $C = -\pi/4$ are illustrated in Figure 9.6.

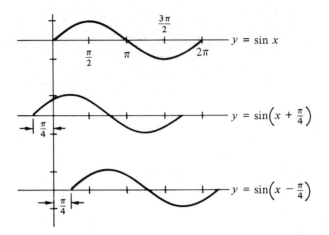

■ **Figure 9.6**

Addition of a constant to the value of the function: The graph of $y = D + \sin x$ is obtained by adding D to each value of $\sin x$. Therefore, the graph of $D + \sin x$ is the graph of $\sin x$ displaced D units up or down. The graph is translated up if D is positive and down if it is negative. The constant D is called the **mean value** of the function. Figure 9.7 shows the graph of $y = 2 + \sin x$.

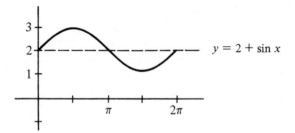

$y = 2 + \sin x$

■ **Figure 9.7**

In the more general case, the effects of changes in amplitude, period, and phase shift are all combined. If A, B, and C are positive constants, the function

$$y = A \sin(Bx + C)$$

has an amplitude of A, a period of $2\pi/B$, and a phase shift corresponding to the value of x given by $Bx + C = 0$, that is, $x = -C/B$. Figure 9.8 shows a graph of $y = 3 \sin(2x - \frac{1}{3}\pi)$.

In all cases, the distinctive shape of the sine curve remains unaltered. This basic shape is expanded or contracted vertically by the multiplication by the amplitude constant A, expanded or contracted horizontally by the constant B, and shifted to the right or left by the constant C.

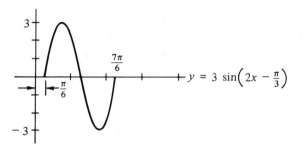

y = 3 sin(2x − π/3)

■ **Figure 9.8**

EXAMPLE 2 ■ **Problem**

Sketch the graph of $y = 3 \cos (\frac{1}{2}x + \frac{1}{4}\pi)$.

■ **Solution**

The amplitude is 3 since the basic cosine function is multiplied by 3. The period is $2\pi/(\frac{1}{2}) = 4\pi$. The phase shift is found from the equation $\frac{1}{2}x + \frac{1}{4}\pi = 0$, that is, for $x = -\pi/2$. Hence, the phase shift is $\pi/2$ units to the left. The graph of this function is shown in Figure 9.9.

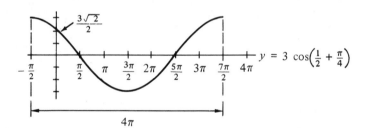

y = 3 cos(½ + π/4)

■ **Figure 9.9**   ■

Fundamental Identities

If the solution set to a trigonometric equation is the complete domain of the independent variable, the equation is called an **identity.** Identities in trigonometry are important in simplifying expressions. Here, they are given in summary form.

FORMULAS

Trigonometric Identities

1. $\sin A = \dfrac{1}{\csc A}$ **2.** $\cos A = \dfrac{1}{\sec A}$

3. $\tan A = \dfrac{1}{\cot A}$ **4.** $\tan A = \dfrac{\sin A}{\cos A}$

5. $\cot A = \dfrac{\cos A}{\sin A}$ **6.** $\sin^2 A + \cos^2 A = 1$

7. $1 + \tan^2 A = \sec^2 A$ **8.** $1 + \cot^2 A = \csc^2 A$

9. $\sin(A + B) = \sin A \cos B + \cos A \sin B$

10. $\sin(A - B) = \sin A \cos B - \cos A \sin B$

11. $\cos(A + B) = \cos A \cos B - \sin A \sin B$

12. $\cos(A - B) = \cos A \cos B + \sin A \sin B$

13. $\tan(A + B) = \dfrac{\tan A + \tan B}{1 - \tan A \tan B}$

14. $\tan(A - B) = \dfrac{\tan A - \tan B}{1 + \tan A \tan B}$

15. $\sin 2A = 2 \sin A \cos A$

16. $\cos 2A = \cos^2 A - \sin^2 A = 2 \cos^2 A - 1 = 1 - 2 \sin^2 A$

17. $\tan 2A = \dfrac{2 \tan A}{1 - \tan^2 A}$

18. $\sin \dfrac{1}{2} A = \pm \sqrt{\dfrac{1 - \cos A}{2}}$

19. $\cos \dfrac{1}{2} A = \pm \sqrt{\dfrac{1 + \cos A}{2}}$

20. $\tan \dfrac{1}{2} A = \dfrac{\sin A}{1 + \cos A}$

EXAMPLE 3   ■ Problem

Write the expression $\dfrac{\tan x \csc^2 x}{1 + \tan^2 x}$ as a single trigonometric term.

■ Solution

From Identity 7, the denominator may be written as $\sec^2 x$. Thus,

$$\frac{\tan x \csc^2 x}{1 + \tan^2 x} = \frac{\tan x \csc^2 x}{\sec^2 x}$$

We now express $\tan x$, $\csc x$, and $\sec x$ in terms of the sine and cosine functions:

$$\frac{\tan x \csc^2 x}{1 + \tan^2 x} = \frac{(\sin x)/(\cos x) \cdot 1/\sin^2 x}{1/\cos^2 x}$$

$$= \frac{\cos^2 x \sin x}{\sin^2 x \cos x} \qquad \text{inverting } \frac{1}{\cos^2 x} \text{ and multiplying}$$

$$= \frac{\cos x}{\sin x} \qquad \text{cancellation law}$$

$$= \cot x \qquad \cot x = \frac{\cos x}{\sin x} \qquad \blacksquare$$

EXAMPLE 4   ■ **Problem**

Simplify the expression $(\sec x + \tan x)(1 - \sin x)$.

■ **Solution**

Write each of the functions in terms of the sine and cosine functions, as follows:

$$(\sec x + \tan x)(1 - \sin x) = \left(\frac{1}{\cos x} + \frac{\sin x}{\cos x}\right)(1 - \sin x)$$

$$= \frac{(1 + \sin x)(1 - \sin x)}{\cos x} \qquad \text{adding fractions}$$

$$= \frac{(1 - \sin^2 x)}{\cos x} \qquad \begin{array}{l}\text{multiplying}\\ \text{in the numerator}\end{array}$$

$$= \frac{\cos^2 x}{\cos x} \qquad \cos^2 x = 1 - \sin^2 x$$

$$= \cos x \qquad \text{cancellation} \qquad \blacksquare$$

EXERCISES FOR SECTION 9.1

1. Find the six trigonometric functions of the angle whose terminal side passes through $(-2, 5)$.

2. What are the other five trigonometric functions of x if $\csc x = \frac{13}{5}$ and $\cos x < 0$?

3. What are the other five trigonometric functions of θ if $\cos \theta = -\frac{4}{5}$ and $\tan \theta > 0$?

4. Convert degrees to radians:
   a. $45°$
   b. $120°$
   c. $85°$
   d. $335°$
   e. $780°$

5. Convert radians to degrees:
   a. 1 rad
   b. 1.2 rad
   c. $\pi/6$ rad
   d. 10 rad
   e. -2 rad

Sketch the graph of the functions in Exercises 6–16, giving period and phase shift. Also give, where applicable, the amplitude or asymptotes.

6. $s = 3 \cos \frac{1}{3}t$

7. $y = 20 \cos 7x$

8. $y = 5 + 2 \sin x$

9. $v = 2 + 0.3 \sin 2x$

10. $i = 8.2 \cos(x - \frac{1}{8}\pi)$

11. $y = \tan(2x + \frac{1}{8}\pi)$

12. $y = -\sin(x + \frac{1}{6}\pi)$

13. $y = 2 + \sin(x - \frac{1}{3}\pi)$

14. $y = \tan(\pi x + \pi) - 1$

15. $a = 4 \cot(\frac{1}{2}t + \frac{1}{8}\pi)$

16. $y = 2 \cos(2 - x) + \frac{1}{2}$

Reduce the expressions in Exercises 17–30 to a single trigonometric function or a constant.

17. $\cos \theta + \tan \theta \sin \theta$

18. $\csc \theta - \cot \theta \cos \theta$

19. $(\tan x + \cot x)(\sin x)$

20. $\dfrac{1 + \cos x}{1 + \sec x}$

21. $\dfrac{(\tan x)(1 + \cot^2 x)}{1 + \tan^2 x}$

22. $\sec x - \sin x \tan x$

23. $\cos x \csc x$

24. $(\cos x)(\tan x + \cot x)$

25. $(\cos^2 x - 1)(\tan^2 x + 1)$

26. $\dfrac{\sec^2 x - 1}{\sec^2 x}$

27. $\dfrac{\sec x - \cos x}{\tan x}$

28. $\dfrac{1 + \tan^2 x}{\tan^2 x}$

29. $(\sin^2 x + \cos^2 x)^3$

30. $\dfrac{1 + \sec x}{\tan x + \sin x}$

9.2 DERIVATIVE FORMULAS FOR THE SINE AND THE COSINE FUNCTIONS

To find the derivative formula for $y = \sin x$, we first evaluate the derivatives of $\sin x$ and $\cos x$ at $x = 0$. Figure 9.10 shows the graphs of $y = \sin x$ and $y = \cos x$. Tangent lines have been drawn to these graphs at $x = 0$. A careful examination of these tangents suggests that the slope of $\sin x$ at $x = 0$ is 1 and that of $\cos x$ at $x = 0$ is 0. Since the derivative of a function at a given point is equal to the slope of the curve at that point, we conclude that $d/dx(\sin x) = 1$ at $x = 0$ and $d/dx(\cos x) = 0$ at $x = 0$.

Using the information from Figure 9.10 and expressing the derivative of $\sin x$ at $x = 0$ in terms of the difference quotient, we have

$$\frac{d}{dx}(\sin x)\bigg|_{x=0} = \lim_{\Delta x \to 0} \frac{\sin(0 + \Delta x) - \sin 0}{\Delta x} = \lim_{\Delta x \to 0} \frac{\sin \Delta x}{\Delta x} = 1$$

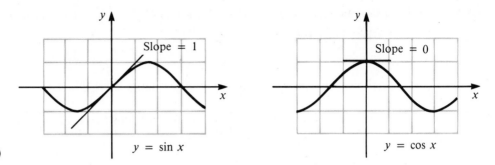

■ **Figure 9.10**

Similarly, since the derivative of cos x is 0 at $x = 0$, we have

$$\frac{d}{dx}(\cos x)\bigg|_{x=0} = \lim_{\Delta x \to 0} \frac{\cos(0 + \Delta x) - \cos 0}{\Delta x}$$

$$= \lim_{\Delta x \to 0} \frac{\cos \Delta x - 1}{\Delta x} = 0$$

The Derivative of sin u

The derivative of sin x is

$$\frac{d}{dx}(\sin x) = \lim_{\Delta x \to 0} \frac{\sin(x + \Delta x) - \sin x}{\Delta x}$$

Using the identity $\sin(A + B) = \sin A \cos B + \cos A \sin B$, we write

$$\frac{d}{dx}(\sin x) = \lim_{\Delta x \to 0} \frac{\sin x \cos \Delta x + \cos x \sin \Delta x - \sin x}{\Delta x}$$

$$= \lim_{\Delta x \to 0} \frac{\cos x \sin \Delta x + (\sin x)(\cos \Delta x - 1)}{\Delta x}$$

$$= \cos x \lim_{\Delta x \to 0} \frac{\sin \Delta x}{\Delta x} + \sin x \lim_{\Delta x \to 0} \frac{\cos \Delta x - 1}{\Delta x}$$

$$= \cos x \left[\frac{d}{dx}(\sin x)\bigg|_{x=0} \right] + \sin x \left[\frac{d}{dx}(\cos x)\bigg|_{x=0} \right]$$

$$= \cos x$$

Thus, the derivative of sin x is cos x. If u represents any differentiable function of x, the chain rule yields the following more general formula.

FORMULA

The Derivative of sin u

$$\frac{d}{dx}(\sin u) = \cos u \,\frac{du}{dx} \qquad\qquad (9.1)$$

EXAMPLE 1   ▪ **Problem**

Given $y = \sin (x^2 + 3)$, find y'.

▪ **Solution**

In this example, $u = x^2 + 3$. Therefore, using (9.1) we obtain

$$y' = \cos (x^2 + 3) \frac{d}{dx} (x^2 + 3) = 2x \cos (x^2 + 3)$$

▪

The Derivative of cos u

The formula for the derivative of $y = \cos u$ can be found by making use of Formula (9.1) and the trigonometric identities

$$\cos u = \sin\left(\frac{\pi}{2} - u\right) \quad \text{and} \quad \sin u = \cos\left(\frac{\pi}{2} - u\right)$$

Using the first of these identities with (9.1), we obtain

$$\frac{d}{dx} (\cos u) = \frac{d}{dx} \sin\left(\frac{\pi}{2} - u\right) = \cos \left(\frac{\pi}{2} - u\right) \frac{d}{dx} \left(\frac{\pi}{2} - u\right)$$

$$= - \cos\left(\frac{\pi}{2} - u\right) \frac{du}{dx}$$

Finally, using the second of the identities, we get the following formula.

FORMULA

The Derivative of cos u

$$\frac{d}{dx} (\cos u) = - \sin u \frac{du}{dx} \qquad\qquad (9.2)$$

EXAMPLE 2   ▪ **Problem**

If $y = \cos 3x^4$, find y'.

▪ **Solution**

Using Formula (9.2) with $u = 3x^4$, we have

$$y' = - \sin 3x^4 \frac{d}{dx} (3x^4) = - 12x^3 \sin 3x^4$$

▪

EXAMPLE 3   ▪ **Problem**

Differentiate $y = \cos^3 2x$ with respect to x.

■ **Solution**

The given expression may be regarded as a function of x raised to the third power; that is, $y = v^3$, where $v = \cos 2x$. Using the power rule, we get

$$\frac{dy}{dx} = 3v^2 \frac{dv}{dx}$$

Replacing v with $\cos 2x$, we obtain

$$\frac{dy}{dx} = 3\cos^2 2x \frac{d}{dx}(\cos 2x)$$

By Formula (9.2),

$$\frac{d}{dx}(\cos 2x) = -\sin 2x \frac{d}{dx}(2x) = -2\sin 2x$$

Therefore,

$$\frac{dy}{dx} = 3\cos^2 2x(-2\sin 2x) = -6\cos^2 2x \sin 2x$$

■

EXAMPLE 4 ■ **Problem**

Find the expression for the velocity of an object given that its displacement (in feet) is $s = e^{-t} \sin \frac{1}{2}t$ when t is time (in seconds). Find the displacement and the velocity at $t = 0.3$ sec.

■ **Solution**

The velocity is given by ds/dt. To differentiate the expression for s, we use the product rule, as follows:

$$v = \frac{ds}{dt} = e^{-t} \frac{d}{dt}\left(\sin \frac{1}{2}t\right) + \sin \frac{1}{2}t \frac{d}{dt}(e^{-t})$$

$$= e^{-t}\left(\frac{1}{2}\cos \frac{1}{2}t\right) + \sin \frac{1}{2}t(-e^{-t}) = e^{-t}\left(\frac{1}{2}\cos \frac{1}{2}t - \sin \frac{1}{2}t\right)$$

The displacement of the object at $t = 0.3$ sec is

$$s = e^{-0.3} \sin 0.15 \approx (0.7408)(0.1494) \approx 0.11 \text{ ft}$$

The velocity at $t = 0.3$ sec is

$$v = e^{-0.3}\left(\frac{1}{2}\cos 0.15 - \sin 0.15\right)$$

$$\approx (0.7408)(0.4944 - 0.1494)$$

$$\approx 0.26 \text{ ft/sec}$$

■

EXERCISES FOR SECTION 9.2

In Exercises 1–24, find the first derivative of the given function.

1. $v = \sin \frac{1}{2}\pi t$

2. $y = \pi \sin \pi t$

3. $z = 2 \cos 3t$

4. $i = \cos \pi t$

5. $p = \frac{1}{2} \cos (t + 1)$

6. $s = 2 \sin(3r + 4)$

7. $P = \sin T^3$

8. $f(y) = 3 \cos\sqrt{2y + 1}$

9. $s = \cos e^t$

10. $y = x^2 + \cos x^2$

11. $w = \cos^3 t$

12. $C = \cos^2 x$

13. $Z = \sqrt{\sin 2X}$

14. $h(x) = \dfrac{1}{\sqrt{\cos \pi x}}$

15. $i = e^t \sin t$

16. $q = \cos 2t \sin 2t$

17. $y = (\ln \sqrt{x})(\cos 2x)$

18. $y = x^3 \sin^3 x$

19. $p = \ln \sin 5s$

20. $L = \ln \cos z^2$

21. $G(t) = e^{\cos 2t}$

22. $I = V^2 e^{\sin V}$

23. $y = \sin(\cos x)$

24. $R = \dfrac{e^{\sin \theta}}{\sin \theta}$

25. Find the second derivative of $y = \sin x^2$.

26. Find the second derivative of $y = \cos x^3$.

27. If a force F is parallel to the plane of motion, the magnitude of the force necessary to cause impending motion up an incline is $F = W \cos \theta$, where W is the weight of the object and θ is the angle of inclination of the plane. Find the rate of change of F with respect to θ.

28. The horizontal displacement of a simple oscillator is given by $x = A \cos 2\pi ft$, where A and f are constants and t is time. Find the expression for the hoirzontal velocity of the oscillation.

29. The current in an RL series curcuit is $i = I_m \sin (\omega t - \phi)$, where I_m is the peak current, ϕ is the phase angle, ω is the angular frequency, and t is the elasped time. Find the expression for the induced voltage in the inductance coil.

30. The torque T acting on a loop of wire, carrying a current i in a magentic field of flux density B, is given by $T = iBA \cos \alpha$, where A is the area of the loop and α is the angle between the plane of the loop and the magnetic field. Find $dT/d\alpha$.

31. The expression for the shearing stress on a plane located by the angle ϕ and caused by a system of stresses is $S = -\frac{1}{2}(\sigma_x - \sigma_y) (\sin 2\phi) + T \cos 2\phi$. The quantities σ_x, σ_y, and T are initially known constants. Find $dS/d\phi$.

32. The equation of motion of a damped vibration is $s = e^{-at} \cos bt$. Show that the velocity is given by $v = -e^{-at} (b \sin bt + a \cos bt)$.

9.3 DERIVATIVES OF THE OTHER TRIGONOMETRIC FUNCTIONS

The derivative formulas for the remaining four trigonometric functions are obtained by using Formulas (9.1) and (9.2) and some basic trigonometric identities.

The Derivative of tan u

Writing $\tan u = \sin u / \cos u$, we have

$$\frac{d}{dx} (\tan u) = \frac{d}{dx} \left(\frac{\sin u}{\cos u} \right)$$

Differentiating the right side by using the formula for the derivative of a quotient, we get

$$\frac{d}{dx} (\tan u) = \frac{\cos u[\cos u \ (du/dx)] - \sin u[-\sin u \ (du/dx)]}{\cos^2 u}$$

$$= \frac{(\cos^2 u + \sin^2 u)(du/dx)}{\cos^2 u}$$

Using the fact that $\cos^2 u + \sin^2 u = 1$ and $(1/\cos u) = \sec u$, we obtain the following formula.

FORMULA

> **The Derivative of tan u**
>
> $$\frac{d}{dx} (\tan u) = \sec^2 u \frac{du}{dx}$$
>
> (9.3)

The Derivative of cot u

The formula for the derivative of cot u is found by a procedure analogous to that used to find the derivative of tan u.

FORMULA

> **The Derivative of cot u**
>
> $$\frac{d}{dx} (\cot u) = -\csc^2 u \frac{du}{dx}$$
>
> (9.4)

The Derivative of sec u

Letting $\sec u = 1/\cos u$, and applying the quotient rule, we get

$$\frac{d}{dx}(\sec u) = \frac{d}{dx}\left(\frac{1}{\cos u}\right) = \frac{0 - (-\sin u)(du/dx)}{\cos^2 u}$$

$$= \left(\frac{1}{\cos u}\right)\left(\frac{\sin u}{\cos u}\right)\frac{du}{dx}$$

Hence, since $1/\cos u = \sec u$ and $\sin u/\cos u = \tan u$, we have the following formula.

FORMULA

> **The Derivative of sec u**
>
> $$\frac{d}{dx}(\sec u) = \sec u \tan u \frac{du}{dx} \qquad\qquad (9.5)$$

The Derivative of csc u

By using a procedure similar to the procedure for obtaining the derivative of secant u, we get the following formula for the derivative of csc u.

FORMULA

> **The Derivative of csc u**
>
> $$\frac{d}{dx}(\csc u) = -\csc u \cot u \frac{du}{dx} \qquad\qquad (9.6)$$

EXAMPLE 1   ▪ **Problem**

Differentiate $s = \sec(3t + 2)$ with respect to t.

▪ **Solution**

By Formula (9.5),

$$\frac{ds}{dt} = \sec(3t + 2)\tan(3t + 2)\frac{d}{dt}(3t + 2)$$

$$= 3\sec(3t + 2)\tan(3t + 2) \qquad\qquad ▪$$

EXAMPLE 2   ▪ **Problem**

Differentiate $y = 2\tan 5x^2$.

■ **Solution**

We obtain

$$\frac{dy}{dx} = 2\frac{d}{dx}(\tan 5x^2) = 2\sec^2 5x^2 (10x) = 20x \sec^2 5x^2$$ ■

EXAMPLE 3   ■ **Problem**

Evaluate $\dfrac{d}{d\theta}(\theta^3 \csc 2\theta)$.

■ **Solution**

We have

$$\frac{d}{d\theta}(\theta^3 \csc 2\theta) = \theta^3 \frac{d}{d\theta}(\csc 2\theta) + \csc 2\theta \frac{d}{d\theta}(\theta^3)$$
$$= \theta^3(-2 \csc 2\theta \cot 2\theta) + (\csc 2\theta)(3\theta^2)$$
$$= \theta^2 \csc 2\theta(3 - 2\theta \cot 2\theta)$$ ■

EXAMPLE 4   ■ **Problem**

Differentiate $y = \sec^4 3x$.

■ **Solution**

Note that the function is of the form u^n. Hence,
$$y' = 4 \sec^3 3x \frac{d}{dx}(\sec 3x) = 4 \sec^3 3x (3 \sec 3x \tan 3x)$$
$$= 12 \sec^4 3x \tan 3x$$ ■

EXAMPLE 5   ■ **Problem**

The magnetic induction B of a charged particle moving through a uniform magnetic field at an angle θ is given by $B = (F/qv)(\csc \theta)$, where F, q, and v are constants. Find the equation for the rate of change of B with respect to θ.

■ **Solution**

To obtain the derivative of the given equation with respect to θ, we use Formula (9.6):

$$\frac{dB}{d\theta} = \frac{F}{qv}\frac{d}{d\theta}(\csc \theta)$$
$$= -\frac{F}{qv} \csc \theta \cot \theta$$ ■

EXERCISES FOR SECTION 9.3

In Exercises 1–20, find the first derivative of the given function.

1. $y = \tan 3x$

2. $y = 2 \cot 3x$

3. $s = 2 \sec \pi t$

4. $i = \tan \pi t$

5. $y = \frac{1}{2} \cot(2t + 1)$

6. $f(x) = 2 \csc(3x + 4)$

7. $C = \sec x^2$

8. $m = 5 \tan y^3$

9. $g(x) = 5 - 2 \csc 3\sqrt{x}$

10. $q = t^2 + \csc \sqrt{t}$

11. $y = \tan^3 2x$

12. $y = \sec^2(3x - 5)$

13. $s = \sqrt{1 + \cot 2t}$

14. $R = (x - \sec \frac{1}{2}x)^5$

15. $P = e^{2t} \sec \pi t$

16. $y = \sin 2x \cot 3x$

17. $F(x) = \ln \sec x^3$

18. $I = \cos(\pi + \sec \pi t)$

19. $V = \dfrac{r^3}{\cot 2r}$

20. $y = \dfrac{\tan 2x}{\sin 5x}$

21. Find the second derivative of $y = \tan 2x$.

22. Find the second derivative of $y = \sec 3x$.

23. Derive Formula (9.4).

24. Derive Formula (9.6).

25. Evaluate $d/dx (\tan 2x)$ at $x = \pi/8$.

26. Evaluate $d/dx (\sec \frac{1}{2}x)$ at $x = \pi/3$.

27. Find the velocity of an object at $t = 0.5$ sec given that its displacement (in centimeters) is $s = 2 \sec 3t$.

28. What is the current in a circuit at $t = 0.1$ sec if the electric charge is given by $q = 0.05 \tan 0.5t$?

29. The total bending B of electromagnetic waves received from outside the atmosphere is described by $B = N \cot E$ when the elevation angle E is greater than $5°$. N is a constant called the *surface refractivity*. What is the rate of change of total bending with respect to elevation angle?

30. The azimuth error Δ_a of the angle tracker of a radar will increase as the target rises in elevation E according to $\Delta_a = \Delta_t \sec E$, where Δ_t is a constant bias error. Find $d\Delta_a/dE$ for the angle tracker.

9.4 INVERSE TRIGONOMETRIC FUNCTIONS

Suppose you were asked to design a calculator to solve the equation $\sin x = \frac{1}{2}$. What value would you choose to display as the solution? This question is an important one, because as Figure 9.11 shows, there are infinitely many possible solutions. To avoid this ambiguity, mathematicians restrict the domain of $\sin x$ to the interval $[-\pi/2, \pi/2]$. Thus, $x = \pi/6$ is the only value on $[-\pi/2, \pi,/2]$ for which $\sin x = \frac{1}{2}$.

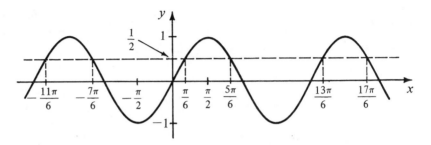

■ **Figure 9.11**

The situation described for the sine function is typical of the trigonometric functions. Hence, we restrict the domain of each of the trigonometric functions so that there is one and only one number in the domain of the function for a given number in its range. The restricted domains for the sine, the cosine, and the tangent are shown in the following table. There are restricted domains defined for the other trigonometric functions but they are of little practical value and are not presented here.

Function	$\sin x$	$\cos x$	$\tan x$
Restricted domain	$-\frac{1}{2}\pi \le x \le \frac{1}{2}\pi$	$0 \le x \le \pi$	$-\frac{1}{2}\pi < x < \frac{1}{2}\pi$

Definitions of the Inverse Trigonometric Functions

A function obtained from another function by interchanging the domain and range in the original function is called an **inverse function.** Thus, if we start with the sine function,

$$y = \sin x \qquad -\tfrac{1}{2}\pi \le x \le \tfrac{1}{2}\pi$$

and interchange x and y, we obtain the inverse sine function,

$$x = \sin y \qquad -\tfrac{1}{2}\pi \le y \le \tfrac{1}{2}\pi$$

We cannot algebraically solve $x = \sin y$ for y, but we know that the solution should state that "y is a number or angle whose sine is x." To convey this idea, we also denote the inverse sine function $x = \sin y$ by

$$y = \text{Arcsin } x \quad \text{or} \quad y = \text{Sin}^{-1} x$$

Both Arcsin x and Sin^{-1} x are used to represent the inverse sine function. Note that Sin^{-1} $x \neq 1/\sin x$. The -1 is not an exponent but part of the function name. Since the notation Sin^{-1} x is sometimes confusing, we will use Arcsin x. Furthermore, this notation reflects our understanding that y is the arc length on the unit circle whose sine is x. Thus, Arcsin $\frac{1}{2}$ means the arc or angle whose sine is $\frac{1}{2}$, that is, $\pi/6$.

The definitions of the other inverse trigonometric functions parallel that of Arcsin x. The three inverse functions of interest are defined next.

DEFINITION

Inverse Trigonometric Functions

$$y = \text{Arcsin } x \quad \text{means} \quad x = \sin y \quad -\tfrac{1}{2}\pi \leq y \leq \tfrac{1}{2}\pi$$

$$y = \text{Arccos } x \quad \text{means} \quad x = \cos y \quad 0 \leq y \leq \pi$$

$$y = \text{Arctan } x \quad \text{means} \quad x = \tan y \quad -\tfrac{1}{2}\pi \leq y \leq \tfrac{1}{2}\pi$$

Again, we note that the inverse function is obtained from the given function by interchanging the domains and ranges. Thus, the domain of the sine function becomes the range of the inverse sine function, and vice versa. The following table lists the domains and the ranges for Arcsin x, Arccos x, and Arctan x. The range of each inverse trigonometric function comprises the **principal values** of the function.

Function	Domain	Range (principal values)
$y = \text{Arcsin } x$	$-1 \leq x \leq 1$	$-\tfrac{1}{2}\pi \leq y \leq \tfrac{1}{2}\pi$
$y = \text{Arccos } x$	$-1 \leq x \leq 1$	$0 \leq y \leq \pi$
$y = \text{Arctan } x$	$-\infty < x < \infty$	$-\tfrac{1}{2}\pi < y < \tfrac{1}{2}\pi$

EXAMPLE 1   ■ **Problem**

Evaluate $y = \text{Arcsin } \dfrac{\sqrt{3}}{2}$.

■ **Solution**

We want the principal value of the number y for which $\sin y = \sqrt{3}/2$. Recalling the special-angle values, we see that $y = \text{Arcsin } \sqrt{3}/2 = \pi/3$. If y is restricted to its principal values the equation $y = \text{Arcsin } (\sqrt{3}/2)$ is equivalent to $\sin y = \sqrt{3}/2$.   ■

EXAMPLE 2   ■ **Problem**

Find Arccos $\dfrac{1}{3}$.

■ **Solution**

Let $y = $ Arccos $\frac{1}{3}$. Then,

$$\frac{1}{3} = \cos y \qquad 0 \le y \le \pi$$

Using a calculator, we find that $y = 1.23$.   ■

From the properties of functions and their inverses, we have

$$\sin(\text{Arcsin } x) = x \qquad \cos(\text{Arccos } x) = x \qquad \tan(\text{Arctan } x) = x$$

Also, if x is limited to the principal values of the function, then

$$\text{Arcsin}(\sin x) = x \qquad \text{Arccos}(\cos x) = x \qquad \text{Arctan}(\tan x) = x$$

EXAMPLE 3 From the preceding discussion, we have

$$\sin (\text{Arcsin } 0.3) = 0.3$$

and

$$\text{Arctan}(\tan 1.2) = 1.2$$ ■

The following examples are concerned with taking a trigonometric function of some inverse trigonometric function. In these circumstances, showing the functional value by drawing a right triangle is helpful, always keeping track of the principal values.

EXAMPLE 4 ■ **Problem**

Find $\sin\left(\text{Arccos } \dfrac{1}{2}\right)$.

■ **Solution**

We let $\theta = $ Arccos $\frac{1}{2}$. Then, θ is the angle as shown in Figure 9.12, from which it is easy to see that

$$\sin\left(\text{Arccos } \frac{1}{2}\right) = \sin \theta = \frac{\sqrt{3}}{2}$$

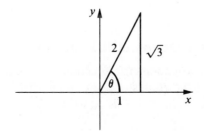

■ **Figure 9.12**   ■

EXAMPLE 5 ■ **Problem**

Find cos (Arcsin x).

■ **Solution**

Letting θ = Arcsin x, we want to find cos θ. Since the range values of Arcsin x are $[-\frac{1}{2}\pi, \frac{1}{2}\pi]$, the angle θ will be one of the angles shown in Figure 9.13. In either case, cos $\theta = \sqrt{1 - x^2}$.

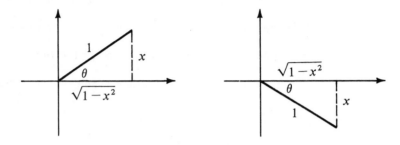

■ **Figure 9.13**   ■

EXAMPLE 6 ■ **Problem**

Find x:

a. If Arccos $x = 0.578$
b. If Arccos $x = 2\pi$

■ **Solution**

a. If Arccos $x = 0.578$, then $x = \cos 0.578$ if 0.578 is one of the principal values, that is, if 0.578 is between 0 and π. Since it is, we have

$$x = \cos 0.578 = 0.8376$$

b. Observe that 2π is not in the set of principal values for the cosine function, so the equation

$$\text{Arccos } x = 2\pi$$

has no solution. If you were not on the lookout for the principal values, you might conclude that since Arccos $x = 2\pi$, then $x = \cos 2\pi = 1$, which is incorrect.   ■

Graphs of the Inverse Trigonometric Functions

The graphs of the inverse trigonometric functions are found by using their definitions and a knowledge of the graphs of the trigonometric functions. For example, y = Arcsin x if, and only if, $x = \sin y$, where $-\frac{1}{2}\pi \le y \le \frac{1}{2}\pi$. Thus, $y =$

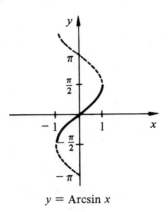

$y = \text{Arcsin } x$

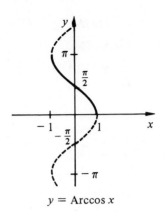

$y = \text{Arccos } x$

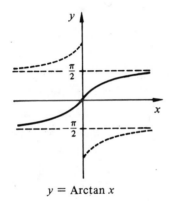

■ **Figure 9.14** $y = \text{Arctan } x$

Arcsin x looks like a portion of the relation $x = \sin y$. See Figure 9.14.

The other parts of Figure 9.14 show the graph of $y = \text{Arccos } x$ and the graph of $y = \text{Arctan } x$. In each case, think of the graph of the original function wrapped around the y-axis, and then consider the portion that corresponds to the principal values.

EXERCISES FOR SECTION 9.4

Find the exact values of Exercises 1–19 without the use of tables or a calculator.

1. Arcsin $\frac{1}{2}$

2. Arcsin 1

3. Arctan 1

4. Arccos $\sqrt{3}/2$

5. Arccos $(-\sqrt{3}/2)$

6. Arcsin $(-\sqrt{2}/2)$

7. Arccos $(-\frac{1}{2})$

8. Arctan $(-1/\sqrt{3})$

9. Arccos 1

10. Arcsin $(1/\sqrt{2})$

11. Arctan $(-\sqrt{3})$

12. sin [Arccos $(-\frac{3}{5})$]

13. $\cos\left[\text{Arcsin}\left(-\frac{5}{13}\right)\right]$

14. $\sin\left(\text{Arcsin } 1\right)$

15. $\sin\left(\text{Arctan } 2\right)$

16. $\sec\left(\text{Arccos }\frac{1}{3}\right)$

17. $\cos\left(\text{Arcsin }\frac{1}{4}\right)$

18. $\sin\left(2 \text{ Arcsin }\frac{1}{3}\right)$

19. $\cos\left(2 \text{ Arcsin }\frac{1}{4}\right)$

In Exercises 20–29, use your calculator to determine the values. Give your answer in radians or real numbers.

20. Arcsin 0.7863

21. Arctan 2.659

22. Arccos 0.3547

23. Arctan 5.78

24. Arctan 2.76

25. Arcsin 0.9866

26. Arccos 0.9034

27. Arccos 0.8966

28. Arcsin -0.5548

29. Arctan -1.593

30. See how your calculator reacts to the problem of finding Arcsin 2. What is wrong?

In Exercises 31–40, solve for x or show that there is no solution.

31. Arccos $x = 0.241$

32. Arcsin $x = -0.314$

33. Arccos $x = -0.5$

34. Arcsin $x = -\pi$

35. Arctan $x = 1.2$

36. Arctan $x = 2.43$

37. Arctan $x = -1.34$

38. Arccos $x = \frac{3}{4}\pi$

39. Arcsin $x = 0.8947$

40. Arccos $x = 2.815$

9.5 DERIVATIVES OF INVERSE TRIGONOMETRIC FUNCTIONS

The graphs of the inverse trigonometric functions are a ready, intuitive source of information about the value of the derivatives of these functions. See Figure 9.15. For example, we do not expect the derivative of $y = \text{Arcsin } x$ to exist at $x = -1$ or $x = 1$ because the tangent line is vertical at each of these points. Furthermore, we expect the sign of d/dx (Arcsin x) to be positive, since the slope of the tangent line is positive throughout the interval. By similar observations, we can see that d/dx (Arccos x) is undefined at $x = \pm 1$ and is negative over the interval. The values of d/dx (Arctan x) are positive and bounded, reaching a maximum when $x = 0$ and becoming arbitrarily close to zero when x is large. All this information is obtained without a formula.

The Derivative of Arcsin u

The formula for the derivative of the inverse sine function follows immediately from the definition of Arcsin u and the use of implicit differentiation. Thus, by the function

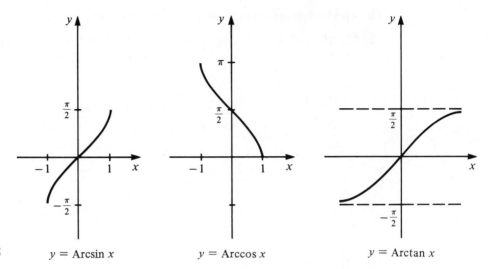

■ **Figure 9.15** $y = \text{Arcsin } x$ $y = \text{Arccos } x$ $y = \text{Arctan } x$

$$y = \text{Arcsin } u \tag{9.7}$$

we mean

$$u = \sin y \tag{9.8}$$

Differentiating both sides of this equality implicitly with respect to x, we get

$$\frac{du}{dx} = \cos y \, \frac{dy}{dx}$$

Solving for dy/dx, we obtain

$$\frac{dy}{dx} = \frac{1}{\cos y} \frac{du}{dx}$$

Using the identity $\cos y = \sqrt{1 - \sin^2 y}$, we write

$$\frac{dy}{dx} = \frac{1}{\sqrt{1 - \sin^2 y}} \frac{du}{dx}$$

But from Equation (9.8), $u = \sin y$. So, we may express the required derivative as shown in the following formula.

FORMULA

> **The Derivative of Arcsin u**
>
> $$\frac{d}{dx} (\text{Arcsin } u) = \frac{1}{\sqrt{1 - u^2}} \frac{du}{dx} \qquad |u| < 1 \tag{9.9}$$

The Derivative of Arccos u

The derivation of the following formula is similar to that used for Arcsin u.

FORMULA

> **The Derivative of Arccos u**
>
> $$\frac{d}{dx}(\text{Arccos } u) = -\frac{1}{\sqrt{1 - u^2}}\frac{du}{dx} \qquad |u| < 1 \qquad\qquad (9.10)$$

Can you explain why (9.9) and (9.10) are so much alike? (See Exercise 26 at the end of this section.)

The Derivative of Arctan u

By the definition of the inverse tangent function, we know that if

$$y = \text{Arctan } u$$

then

$$u = \tan y$$

and

$$\frac{du}{dx} = \sec^2 y \frac{dy}{dx}$$

Solving for dy/dx, we get

$$\frac{dy}{dx} = \frac{1}{\sec^2 y}\frac{du}{dx}$$

Using the facts that $\sec^2 y = 1 + \tan^2 y$, and $u = \tan y$, we have the following formula.

FORMULA

> **The Derivative of Arctan u**
>
> $$\frac{d}{dx}(\text{Arctan } u) = \frac{1}{1 + u^2}\frac{du}{dx} \qquad\qquad (9.11)$$

EXAMPLE 1   ■ **Problem**

Find y' given that $y = \text{Arccos } 2x^3$.

■ **Solution**

Letting $u = 2x^3$ and applying Formula (9.10), we have

$$y' = -\frac{1}{\sqrt{1 - (2x^3)^2}} \frac{d}{dx} (2x^3) = -\frac{6x^2}{\sqrt{1 - 4x^6}}$$ ■

EXAMPLE 2   ■ **Problem**

Find the slope of the curve defined by $y = 4x$ Arctan $2x$ at $x = \frac{1}{2}$.

■ **Solution**

Since the given function is the product of two functions of x, the product rule is used to obtain

$$\frac{dy}{dx} = 4x \frac{1}{1 + (2x)^2} \frac{d}{dx} (2x) + \text{Arctan } 2x \frac{d}{dx} (4x)$$

$$= \frac{8x}{1 + 4x^2} + 4 \text{ Arctan } 2x$$

The slope at $x = \frac{1}{2}$ is then

$$\frac{dy}{dx}\bigg|_{x=1/2} = \frac{4}{2} + 4 \text{ Arctan } 1$$

$$= 2 + 4\left(\frac{\pi}{4}\right) \approx 5.14$$ ■

EXAMPLE 3   ■ **Problem**

Find y' given that $y = \text{Arctan } \dfrac{x}{1 + x}$.

■ **Solution**

The derivative is

$$y' = \frac{1}{1 + [x/(1 + x)]^2} \frac{d}{dx} \left(\frac{x}{1 + x}\right) = \frac{(1 + x)^2}{(1 + x)^2 + x^2} \cdot \frac{(1 + x) - x}{(1 + x)^2}$$

$$= \frac{1 + x - x}{1 + 2x + x^2 + x^2} = \frac{1}{1 + 2x + 2x^2}$$ ■

EXAMPLE 4   ■ **Problem**

A winch on a loading dock is used to drag a large container along the ground. The winch winds in the cable at the rate of 2 ft/sec, and it is 5 ft above the ground. At what rate is the angle θ changing when there is 10 ft of cable out?

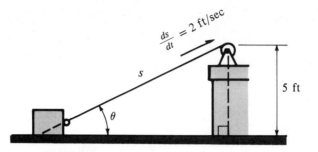

■ **Figure 9.16**

■ **Solution**

Referring to Figure 9.16, we see that the angle θ may be written as

$$\theta = \text{Arcsin } \frac{5}{s}$$

The rate at which θ is changing is then found by differentiating this function with respect to time. Note that s is a function of time. Thus,

$$\frac{d\theta}{dt} = \frac{1}{\sqrt{1 - (25/s^2)}} \frac{d}{dt}\left(\frac{5}{s}\right) = \frac{1}{\sqrt{(s^2 - 25)/s^2}}\left(-\frac{5}{s^2}\right)\frac{ds}{dt}$$

$$= \frac{-5 \, ds/dt}{s \sqrt{s^2 - 25}}$$

Substituting $ds/dt = -2$ and $s = 10$ into this expression gives

$$\frac{d\theta}{dt} = \frac{-5(-2)}{10 \sqrt{75}} = \frac{1}{\sqrt{75}} \approx 0.115 \text{ rad/sec}$$

Note that ds/dt is negative, indicating that the cable is getting shorter.   ■

EXERCISES FOR SECTION 9.5

Differentiate the functions in Exercises 1–17.

1. $\theta = \text{Arcsin } 3t$

2. $y = \text{Arcsin } x^3$

3. $z = \text{Arctan } \frac{1}{2}y$

4. $B = \text{Arccos } q^2$

5. $s = \text{Arcsin } \sqrt{t}$

6. $T = \text{Arctan } \frac{1}{z}$

7. $y = \text{Arctan } \frac{x}{3}$

8. $h(x) = \text{Arccos } \sqrt{1 - x^2}$

9. $p = \text{Arctan } e^t$

10. $f(x) = x^2 \text{ Arcsin } 3x$

11. $m = \sqrt{x} \text{ Arcsin } \sqrt{x}$

12. $i = \dfrac{\text{Arccos } 3t}{t}$

13. $q = \dfrac{\text{Arctan } 2t}{e^t}$

14. $y = \sqrt{\text{Arctan } 2x}$

15. $H = \dfrac{1}{\sqrt{\text{Arccos } \sqrt{\phi}}}$

16. $y = x \text{ Arcsin } x - \sqrt{1 - x^2}$

17. $y = x \text{ Arctan } x - \ln \sqrt{1 + x^2}$

18. Derive Formula (9.10).

19. Find the slope of the graph of $y = x \text{ Arcsin } x$ at $x = \frac{1}{2}$.

20. Find the slope of the graph of $y = e^x \text{ Arccos } x$ at $x = 0$.

 21. When an AC generator is applied to a series RLC circuit, the voltage and the current are out of phase by some angle ϕ. The magnitude of ϕ is described by $\phi = \text{Arctan } X/R$, where X is the reactance of the circuit and R is the resistance. Find $d\phi/dR$.

 22. In Exercise 21, find $d\phi/dX$.

 23. When a thermocouple is placed in a gas stream, the angle by which the peak thermocouple temperature lags the peak temperature of the gas stream is $\theta = \text{Arctan } T/\omega$, where T is a constant and ω is the angular frequency of the temperature variation. Find $d\theta/d\omega$.

 24. A balloon rises vertically from a point that is 1000 ft from an observer. At what rate is the angle of elevation changing when the balloon is 200 ft high if the balloon is rising at a rate of 15 ft/sec?

 25. An airplane flying at an altitude of 10,000 ft and a velocity of 500 ft/sec passes directly over a radar station. At what rate is the angle of elevation of the radar changing 20 sec later?

26. Note that from Formulas (9.9) and (9.10), we may conclude that

$$\frac{d}{dx} (\text{Arcsin } u) = -\frac{d}{dx} (\text{Arccos } u)$$

What can you conclude from this result?

■■■■ 9.6 INTEGRALS OF TRIGONOMETRIC FUNCTIONS

Since integration is the inverse of differentiation, every differentiation formula has a corresponding integration formula. The formulas listed on page 332 are a direct consequence of the differentiation formulas.

In the general situation, u is a function of some other variable, say x, and then the du is just a shorthand way of writing $(du/dx)dx$. The given formulas may be applied only when a given antiderivative precisely fits the form as given. For example, $\int \cos x^3 \, dx$ may not be evaluated by (9.13) because if $u = x^3$, then $du/dx = 3x^2$, and $3x^2$ is not a part of the integrand.

FORMULAS

Integrals of the Sine, Cosine, and Related Functions

$$\int \sin u \, du = -\cos u + C \qquad\qquad\qquad (9.12)$$

$$\int \cos u \, du = \sin u + C \qquad\qquad\qquad (9.13)$$

$$\int \sec^2 u \, du = \tan u + C \qquad\qquad\qquad (9.14)$$

$$\int \csc^2 u \, du = -\cot u + C \qquad\qquad\qquad (9.15)$$

$$\int \sec u \tan u \, du = \sec u + C \qquad\qquad\qquad (9.16)$$

$$\int \csc u \cot u \, du = -\csc u + C \qquad\qquad\qquad (9.17)$$

EXAMPLE 1   ■ **Problem**

Find $\int 12\, x \sin 3x^2 \, dx$.

■ **Solution**

Letting $u = 3x^2$, we get $du/dx = 6x$. Writing 12 as $6 \cdot 2$, bringing the 2 outside the integral sign, and applying (9.12), we have

$$\int 12\, x \sin 3x^2 \, dx = 2 \int 6x \sin 3x^2 \, dx$$
$$= 2 \int \sin u \, du = -2 \cos 3x^2 + C \qquad\qquad ■$$

EXAMPLE 2   ■ **Problem**

Find $\int \sec \frac{1}{4}\phi \tan \frac{1}{4}\phi \, d\phi$.

■ **Solution**

Letting $u = \phi/4$, we get $du/d\phi = \frac{1}{4}$. Therefore (9.16) may be used if we first multiply the integrand by $\frac{1}{4}$ and multiply outside the integral sign by 4. Hence,

$$\int \sec \frac{1}{4}\phi \tan \frac{1}{4}\phi \, d\phi = 4 \int \sec \frac{1}{4}\phi \tan \frac{1}{4}\phi(\frac{1}{4}\, d\phi)$$

$$= 4 \sec \frac{1}{4}\phi + C \qquad\qquad ■$$

EXAMPLE 3   ■ **Problem**

Find $\int \sin^3 \theta \cos \theta \, d\theta$.

■ **Solution**

This antiderivative is not one of the standard forms given in Formulas (9.12) through (9.17). However, by letting $u = \sin \theta$ and $n = 3$, we have $du/d\theta = \cos \theta$. Thus, the given antiderivative is in the standard form $\int u^n \, du$. So,

$$\int \sin^3 \theta \cos \theta \, d\theta = \frac{\sin^4 \theta}{4} + C$$

∎

To conclude this section, we derive the antiderivative formulas for tan u, cot u, sec u, and csc u.

The formula for the antiderivative of tan u can be found by using the elementary trigonometric identity tan $u = \sin u/\cos u$. Thus,

$$\int \tan u \, du = \int \frac{\sin u}{\cos u} \, du$$

$$= -\int \frac{(-\sin u \, du)}{\cos u}$$

By letting $v = \cos u$, we see that since $dv/du = -\sin u$, we may use the integration formula for dv/v to obtain

$$\int \tan u \, du = -\ln (\cos u) + C$$
$$= \ln (\cos u)^{-1} + C$$

Therefore, we have the following formula.

FORMULA

> **The Integral of tan u du**
>
> $$\int \tan u \, du = \ln \sec u + C \qquad\qquad (9.18)$$

By a procedure similar to that used in deriving (9.18), we have the next formula.

FORMULA

> **The Integral of cot u du**
>
> $$\int \cot u \, du = \ln \sin u + C \qquad\qquad (9.19)$$

Rather than derive the secant formula, we will state and verify it.

FORMULA

> **The Integral of sec u du**
>
> $$\int \sec u \, du = \ln (\sec u + \tan u) + C \qquad\qquad (9.20)$$

To verify Formula (9.20), we must show that the derivative of the function $\ln (\sec u + \tan u)$ is $\sec u$.

$$\frac{d}{du} [\ln (\sec u + \tan u)] = \frac{1}{\sec u + \tan u} (\sec u \tan u + \sec^2 u)$$

$$= \frac{\sec u (\tan u + \sec u)}{\sec u + \tan u} = \sec u$$

The Cosecant formula may be verified just as (9.20) was.

FORMULA

> **The Integral of csc u du**
>
> $$\int \csc u \, du = \ln (\csc u - \cot u) + C \qquad \text{(9.21)}$$

EXAMPLE 4   ■ **Problem**

Find $\int x \sec(x^2 + 3) \, dx$.

■ **Solution**

If we let $u = x^2 + 3$, then $du/dx = 2x$. Multiplying inside the integral sign by 2 and outside by $\frac{1}{2}$, we have, from Formula (9.20),

$$\int x \sec(x^2 + 3) \, dx = \frac{1}{2} \int \sec (x^2 + 3)(2x \, dx)$$

$$= \frac{1}{2} \ln[\sec(x^2 + 3) + \tan(x^2 + 3)] + C \qquad ■$$

EXAMPLE 5   ■ **Problem**

The current in an AC circuit is given by $i = I_m \sin \omega t$, where I_m is the maximum amplitude of the current and ωt is the angular displacement of the generator. Show that the average current over half a cycle is $I_{avg} = 0.637 I_m$. See Figure 9.17.

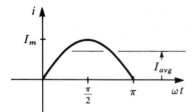

■ **Figure 9.17**

■ **Solution**

If the period of $\sin \omega t$ is 2π, then half a cycle is the interval $0 \le \omega t \le \pi$. Recalling from Section 6.4, the formula for the average value of a function, we have

$$I_{avg} = \frac{1}{\pi - 0} \int_0^\pi I_m \sin \omega t \, d(\omega t) = \frac{I_m}{\pi} \left[- \cos \omega t \right]_0^\pi = \frac{I_m}{\pi} (2) = 0.637 I_m \qquad ■$$

EXAMPLE 6   ▪ **Problem**

The angle of twist θ of the free end of the cylindrical bar shown in Figure 9.18 is given by

$$\theta = \frac{1}{GJ} \int_0^L T \, dx$$

where L is the length, T is the torque, and G and J are constants. Find the equation for the angle of twist given that the torque applied to the bar varies with distance from the fixed end according to $T = \tan(\pi x/3L)$.

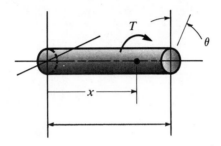

▪ **Figure 9.18**

▪ **Solution**

Substituting $T = \tan(\pi x/3L)$ into the given integral, we get

$$\theta = \frac{1}{GJ} \int_0^L \tan\left(\frac{\pi x}{3L}\right) dx$$

Letting $u = (\pi x/3L)$, we get $du/dx = \pi/3L$. Therefore,

$$\theta = \frac{3L}{\pi GJ} \int_0^L \tan\left(\frac{\pi x}{3L}\right)\left(\frac{\pi}{3L}\right) dx = \left[\frac{3L}{\pi GJ}\left(\ln \sec \frac{\pi x}{3L}\right)\right]_0^L$$

$$= \frac{3L}{\pi GJ}\left(\ln \sec \frac{\pi}{3} - \ln \sec 0\right) = \frac{3L \ln 2}{\pi GJ}$$

▪

EXERCISES FOR SECTION 9.6

Find the integrals in Exercises 1–24.

1. $\displaystyle\int \sin 3x \, dx$

2. $\displaystyle\int \sec^2 \tfrac{1}{2} x \, dx$

3. $\displaystyle\int \tan 2t \, dt$

4. $\displaystyle\int \sec 5x \, dx$

5. $\displaystyle\int \cot \omega t \, d(\omega t)$

6. $\displaystyle\int x \cos x^2 \, dx$

7. $\displaystyle\int 6x \sec x^2 \tan x^2 \, dx$

8. $\displaystyle\int (x - 1) \cos(x^2 - 2x + 1) \, dx$

9. $\displaystyle\int \frac{\sin e^{-x}}{e^x} \, dx$

10. $\displaystyle\int \frac{\tan\sqrt{\theta} \, d\theta}{\sqrt{\theta}}$

11. $\displaystyle\int \csc \frac{x}{2} \, dx$

12. $\displaystyle\int \csc^2 3x \, dx$

13. $\displaystyle\int \csc \omega t \cot \omega t \, dt$

14. $\displaystyle\int \frac{25}{x^2} \cos \frac{1}{x} \, dx$

15. $\displaystyle\int \sin^5 \theta \cos \theta \, d\theta$

16. $\displaystyle\int \frac{\sin x \, dx}{\sqrt{\cos x}}$

17. $\displaystyle\int \sec^3 \phi \tan \phi \, d\phi$

18. $\displaystyle\int \csc^4 x \cot x \, dx$

19. $\displaystyle\int e^{\sin 2\phi} \cos 2\phi \, d\phi$

20. $\displaystyle\int \frac{\sec^2 y \, dy}{1 + \tan y}$

21. $\displaystyle\int \frac{\sec^2 x \, dx}{(1 + \tan x)^2}$

22. $\displaystyle\int \frac{dz}{\sin^2 z}$

23. $\displaystyle\int (1 + \sec \omega t)^2 \, d(\omega t)$

24. $\displaystyle\int \frac{\cos 2\omega t + \sin 2\omega t}{\cos 2\omega t} \, d(\omega t)$

25. The x-coordinate of the centroid of a circular arc of radius r that subtends an angle $\phi = \beta$ at the center is given by $\bar{x} = 2/\beta \int_0^{\beta/2} r \cos \phi \, d\phi$. Find $\bar{x}$ for a circular arc with $r = 10$ in. and $\beta = \pi/2$ rad.

26. Find the average height of the graph of $y = \tan x$ from $x = 0$ to $x = \pi/3$.

27. The work done dW in moving a block through a distance ds is $dW = F \cos \theta \, ds$, where F is the applied force (in pounds), and θ is the angle the force makes with the plane. Find the work done in moving the block 10 ft if $\theta = 30°$ and $F = s^2$.

28. The work done dW in rotating a magnet in a magnetic field through an angle $d\phi$ is $dW = MH \cos \phi \, d\phi$, where M and H are constants. Find the work done in rotating the magnet from $\phi = 0$ to $\phi = 90°$.

29. The total radiation through a hemispherical surface is given by

$$R = 2\pi h \int_0^{\pi/2} \cos \theta \sin \theta \, d\theta$$

where h is a constant and θ is the angle of radiation. Show that the total radiation is $R = \pi h$.

30. The amount of energy E that is scattered by electrons in a cubic centimeter of matter is

$$E = \frac{\pi I n e^4}{m^2 c^4} \int_0^\pi (1 + \cos^2 \theta) \sin \theta \, d\theta$$

where θ is the angle of scatter. Show that

$$E = \frac{8\pi n e^4 I}{3m^2 c^4}$$

31. When the bar shown in the accompanying figure is rotated about point A, work is done on the spring. The work done in rotating an 8-ft bar to point C is given by

$$W = 960 \int_0^{\pi/4} (3 \cos \alpha - 2) \sin \alpha \, d\alpha$$

Evaluate W.

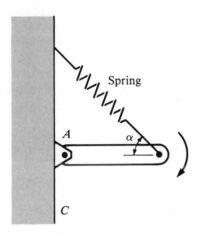

■ 9.7 USING IDENTITIES TO SIMPLIFY TRIGONOMETRIC INTEGRANDS

Integrals with trigonometric terms in the integrand cannot always be evaluated by using the formulas of the previous section. Sometimes, the substitution of an appropriate identity simplifies the integrand to a form whose antiderivative we recognize. For example, terms of the form $\sin^2 u$ and $\cos^2 u$ may be replaced by the identities $\sin^2 u = \frac{1}{2}(1 - \cos 2u)$ and $\cos^2 u = \frac{1}{2}(1 + \cos 2u)$, respectively.

EXAMPLE 1   ■ **Problem**

Find $\int \cos^2 3x \, dx$.

■ **Solution**

We have

$$\int \cos^2 3x \, dx = \frac{1}{2} \int (1 + \cos 6x) \, dx = \frac{1}{2}x + \frac{1}{12} \sin 6x + C$$

■

Comment: Some beginning students make the serious error of believing that $\int \cos^2 u \, du$ is $\frac{1}{3} \cos^3 u$, which is not true. Take the time to convince yourself that the power rule is not applicable to integrals of this type by showing that $d/dx \, (\frac{1}{3} \cos^3 u) \neq \cos^2 u$.

EXAMPLE 2   ■ **Problem**

The instantaneous value of an alternating current varies continuously from a maximum in one direction to a maximum in the opposite. In practice, it is more convenient to describe alternating currents by their *effective values*. The effective value of an alternating current is defined as the steady current that develops the same quantity of heat. Show that for a sinusoidal current of the form $i = I_m \sin \omega t$, the effective current I_{eff} is equal to $1/\sqrt{2}$ times the maximum current I_m.

■ **Solution**

The power in a resistance R is given by $p = i^2 R$, and therefore, the average power over one cycle of the alternating current $i = I_m \sin \omega t$ is

$$P = \frac{1}{2\pi - 0} \int_0^{2\pi} (I_m \sin \omega t)^2 \, R \, d(\omega t) = \frac{I_m^2 R}{2\pi} \int_0^{2\pi} \sin^2 \omega t \, d(\omega t)$$

$$= \frac{I_m^2 R}{2\pi} \int_0^{2\pi} \frac{1}{2}(1 - \cos 2\omega t) \, d(\omega t) = \frac{I_m^2 R}{4\pi} \left[\omega t - \frac{1}{2} \sin 2\omega t \right]_0^{2\pi}$$

$$= \frac{I_m^2 R}{4\pi} (2\pi) = \frac{I_m^2 R}{2}$$

Now, if we let I_{eff} represent the steady-state current that produces the same power, we can write

$$I_{eff}^2 R = \frac{1}{2} I_m^2 R$$

or

$$I_{eff} = \frac{I_m}{\sqrt{2}}$$

Hence, for sinusoidal currents, the effective current is equal to $1/\sqrt{2}$ times the maximum current.   ■

EXAMPLE 3   ■ **Problem**

Find $\int \sin^2 u \cos^2 u \, du$.

■ **Solution**

Using the facts that $\cos^2 u = \frac{1}{2}(1 + \cos 2u)$ and $\sin^2 u = \frac{1}{2}(1 - \cos 2u)$, we have

$$\int \sin^2 u \cos^2 u \, du = \int \frac{1}{4}(1 - \cos 2u)(1 + \cos 2u) \, du$$

$$= \int \left(\frac{1}{4} - \frac{1}{4} \cos^2 2u \right) du$$

We now use the fact that $\cos^2 2u = \frac{1}{2}(1 + \cos 4u)$. Thus,

$$\int \sin^2 u \cos^2 u \, du = \int \left[\frac{1}{4} - \frac{1}{8}(1 + \cos 4u) \right] du = \int \left[\frac{1}{8} - \frac{1}{8} \cos 4u \right] du$$

$$= \frac{1}{8} u - \frac{1}{32} \sin 4u + C \qquad \blacksquare$$

Integrals of the form $\int \sin^m x \cos^n x \, dx$ may be expressed in the form $\int v^n \, dv$ if either m or n is a positive odd integer. In this case, factor either $\sin x$ or $\cos x$, and then use the identity $\sin^2 x + \cos^2 x = 1$ to obtain $v^n \, dv$.

EXAMPLE 4 ■ **Problem**

Find $\sqrt{\sin x} \cos^5 x \, dx$.

■ **Solution**

The exponent on the cosine term is odd and positive, so we write

$$\int \sqrt{\sin x} \cos^5 x \, dx = \int \sqrt{\sin x} \cos^4 x \, (\cos x \, dx)$$

Writing $\cos^4 x$ as $(\cos^2 x)^2$, we have

$$\int \sqrt{\sin x} \cos^5 x \, dx = \int \sin^{1/2} x \, (\cos^2 x)^2 (\cos x \, dx)$$

$$= \int \sin^{1/2} x (1 - \sin^2 x)^2 (\cos x \, dx)$$

$$= \int (\sin^{1/2} x - 2 \sin^{5/2} x + \sin^{9/2} x)(\cos x \, dx)$$

Since $\cos x$ is the derivative of $\sin x$, it follows that each of these terms is of the form $\int \sin^k x \, d (\sin x)$. So,

$$\int \sqrt{\sin x} \cos^5 x \, dx = \frac{2}{3} \sin^{3/2} x - \frac{4}{7} \sin^{7/2} x + \frac{2}{11} \sin^{11/2} x + C \qquad \blacksquare$$

Integrals of the form $\int \tan^m u \sec^n u \, du$ (where n is an even, positive integer) can be evaluated by factoring $\sec^n u$ into $\sec^{n-2} u \sec^2 u$ and changing $\sec^{n-2} u$ into tangents, using $\sec^2 u = 1 + \tan^2 u$. The factor $\sec^2 u$ is preserved as the derivative of $\tan u$. Similarly, $\int \cot^m u \csc^n u \, du$ (where n is an even, positive integer) can be evaluated by factoring $\csc^n u$ into $\csc^{n-2} u \csc^2 u$ and using $\csc^2 u = 1 + \cot^2 u$ to change $\csc^{n-2} u$ into cotangent terms.

EXAMPLE 5 ■ **Problem**

Find $\int \frac{\sec^4 x}{\tan^2 x} \, dx$.

■ **Solution**

Rewriting the integrand by associating $\sec^2 x$ with dx, we have

$$\int \frac{\sec^4 x}{\tan^2 x} \, dx = \int \tan^{-2} x \sec^2 x \, (\sec^2 x \, dx)$$

$$= \int \tan^{-2} x (1 + \tan^2 x)(\sec^2 x \, dx)$$

$$= \int (\tan^{-2} x + 1)(\sec^2 x \, dx) = -\cot x + \tan x + C \qquad ■$$

EXERCISES FOR SECTION 9.7

By using appropriate trigonometric identities, find the integrals in Exercises 1–17.

1. $\displaystyle\int \cos^2 \theta \, d\theta$

2. $\displaystyle\int \tan^2 x \, dx$

3. $\displaystyle\int \sin^2 3\phi \, d\phi.$

4. $\displaystyle\int \sin^3 x \cos^2 x \, dx$

5. $\displaystyle\int \sin^4 2t \cos^3 2t \, dt$

6. $\displaystyle\int \sin^5 x \sqrt{\cos x} \, dx$

7. $\displaystyle\int \tan \theta \sec^4 \theta \, d\theta$

8. $\displaystyle\int \frac{\csc^4 y \, dy}{\cot^2 y}$

9. $\displaystyle\int \cos^3 \omega t \, d(\omega t)$

10. $\displaystyle\int \frac{\sin^3 x \, dx}{\cos x}$

11. $\displaystyle\int \csc t (\sin^3 t + \cos t) \, dt$

12. $\displaystyle\int \left(\frac{\sec \phi}{\tan \phi} \right)^4 d\phi$

13. $\displaystyle\int \frac{\cos^2}{\tan x} \, dx$

14. $\displaystyle\int \sec^4 t \, dt$

15. $\displaystyle\int (1 + \csc^2 \theta)^2 \, d\theta$

16. $\displaystyle\int \sin^2 3t \cos^2 3t \, dt$

17. $\displaystyle\int x \sin^3 x^2 \, dx$

18. The moment of inertia I_x of a circular area of radius R is

$$I_x = \frac{R^4}{4} \int_0^{2\pi} \sin^2 \theta \, d\theta$$

Determine the expression for I_x.

19. The elemental moment of inertia of a sine curve is $dI = \frac{1}{12}\sin^3 x\, dx$. Determine the value of I for a half cycle of the sine curve.

20. The velocity (in feet per second) of a particle varies as $v = \sin^2 t \cos^3 t$. What is the expression for the displacement of the particle if $s = 0$ when $t = 0$?

21. The power (in watts) in a resistor is $p = \tan^2 3t$. Find the energy dissipated by the resistor from $t = 0$ to $t = (\pi/12)$ sec.

9.8 INTEGRALS THAT YIELD INVERSE TRIGONOMETRIC FUNCTIONS

Since the derivative of Arcsin (u/a) is

$$\frac{d}{du}\left(\text{Arcsin }\frac{u}{a}\right) = \frac{1}{\sqrt{1 - (u/a)^2}}\left(\frac{1}{a}\right) = \frac{1}{\sqrt{a^2 - u^2}}$$

We have the following formula.

FORMULA

> **An Integral that Gives an Arcsin Function**
>
> $$\int \frac{du}{\sqrt{a^2 - u^2}} = \text{Arcsin }\frac{u}{a} + C \qquad |u| < |a| \qquad\qquad \textbf{(9.22)}$$

Similarly, we have the next formula.

FORMULA

> **An Integral that Gives an Arctan Function**
>
> $$\int \frac{du}{a^2 + u^2} = \frac{1}{a}\text{Arctan }\frac{u}{a} + C \qquad\qquad \textbf{(9.23)}$$

EXAMPLE 1 ■ **Problem**

Find $\displaystyle\int \frac{ds}{\sqrt{4 - 9s^2}}$.

▪ Solution

If we let $a = 2$ and $u = 3s$ in Formula (9.22), the given integrand may be written as

$$\int \frac{ds}{\sqrt{4 - 9s^2}} = \int \frac{ds}{\sqrt{2^2 - (3s)^2}}$$

Since $du/ds = 3$, we multiply inside the integral sign by 3 and outside by $\frac{1}{3}$. Thus,

$$\int \frac{ds}{\sqrt{4 - 9s^2}} = \frac{1}{3} \int \frac{3\,ds}{\sqrt{2^2 - (3s)^2}} = \frac{1}{3} \text{Arcsin} \frac{3s}{2} + C$$

▪

EXAMPLE 2   ▪ **Problem**

Find $\displaystyle\int \frac{e^x\,dx}{1 + e^{2x}}$.

▪ Solution

We use Formula (9.23) with $a = 1$ and $u = e^x$. Then, $du/dx = e^x$ and we have

$$\int \frac{e^x\,dx}{1 + e^{2x}} = \int \frac{e^x\,dx}{1 + (e^x)^2} = \text{Arctan } e^x + C$$

▪

EXAMPLE 3   ▪ **Problem**

Find $\displaystyle\int \frac{dx}{\sqrt{1 + 4x - x^2}}$.

▪ Solution

The expression under the radical may be written in the form $a^2 - u^2$ by completing the square. Thus,

$$1 + 4x - x^2 = 1 - (x^2 - 4x) = 1 + 4 - (x^2 - 4x + 4)$$
$$= (\sqrt{5})^2 - (x - 2)^2$$

Using (9.22) with $a = \sqrt{5}$ and $u = x - 2$, we have

$$\int \frac{dx}{\sqrt{1 + 4x - x^2}} = \int \frac{dx}{\sqrt{(\sqrt{5})^2 - (x - 2)^2}} = \text{Arcsin} \frac{x - 2}{\sqrt{5}} + C$$

▪

EXAMPLE 4   ▪ **Problem**

Find the volume obtained by rotating the area that is bounded by $y = 1/\sqrt{1 + x^2}$, $x = 0$, $x = 1$, and $y = 0$ about the x-axis.

■ **Solution**

The area to be rotated is shown in Figure 9.19. The element of volume is $\pi (1/\sqrt{1 + x^2})^2 dx = \pi\, dx/(1 + x^2)$. So, the desired integral is

$$V = \pi \int_0^1 \frac{dx}{1 + x^2} = \pi \left[\text{Arctan } x \right]_0^1 = \pi \text{ Arctan } 1 = \frac{\pi^2}{4} \text{ cubic units}$$

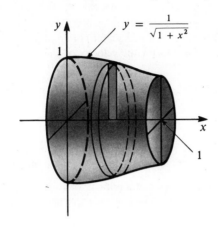

$$y = \frac{1}{\sqrt{1 + x^2}}$$

■ **Figure 9.19**

EXERCISES FOR SECTION 9.8

Find the integrals in Exercises 1–20.

1. $\displaystyle\int \frac{dy}{4 + y^2}$

2. $\displaystyle\int \frac{dx}{\sqrt{1 - 4x^2}}$

3. $\displaystyle\int \frac{dv}{\sqrt{3 - 9v^2}}$

4. $\displaystyle\int \frac{dz}{3z^2 + 5}$

5. $\displaystyle\int \frac{dx}{2 + 3x^2}$

6. $\displaystyle\int \frac{y\,dy}{4 + y^4}$

7. $\displaystyle\int \frac{dx}{x^2 + 4x + 8}$

8. $\displaystyle\int \frac{3\,dx}{x^2 + 2x + 5}$

9. $\displaystyle\int \frac{z\,dz}{3z^2 + 5}$

10. $\displaystyle\int \frac{z\,dz}{(3z^2 + 5)^2}$

11. $\displaystyle\int \frac{dx}{\sqrt{4x - x^2}}$

12. $\displaystyle\int \frac{dx}{\sqrt{5 + 4x - x^2}}$

13. $\displaystyle\int \frac{dy}{4y^2 + 8y + 13}$

14. $\displaystyle\int \frac{\cos \theta\, d\theta}{\sqrt{4 - \sin^2 \theta}}$

15. $\int \dfrac{(x + 2)\, dx}{x^2 + 1}$

16. $\int \dfrac{e^{2x}\, dx}{1 + e^{4x}}$

17. $\int \dfrac{dy}{5(y^2 + 3)}$

18. $\int \dfrac{2\, dp}{\sqrt{1 - 4p}}$

19. $\int \dfrac{(x + 2)\, dx}{x^2 + 2x + 3}$

20. $\int \dfrac{(3 - x)\, dx}{9 + 4x^2}$

21. The acceleration (in meters per second squared) of an object in a resisting medium is $a = 1/(1 + 9t^2)$. Find an expression for the velocity of the object. Assume $v = 100$ m/sec when $t = 0$.

22. Find the average height of the curve $y = 1/\sqrt{1 - x^2}$ over the interval from $x = 0$ to $x = \frac{1}{2}$.

23. The current (in amperes) in a circuit is $i = 1/[4(t^2 + 1)]$. What is the average current in the circuit from $t = 0$ to $t = 1$ sec?

24. For $dy/dx = 4y^2 + 3$, determine an expression for x as a function of y.

25. A vertical column begins to buckle at a critical, or Euler, load. The slope of the column at this load is given by $(dy/dx)^2 = -k^2y^2 + c^2$. What is the expression for x if k and c are constants?

26. Find the volume obtained by rotating the area bounded by $y = 1/\sqrt{x^2 + 4x + 13}$, $x = -2$, $x = 1$, and $y = 0$ about the x-axis.

REVIEW EXERCISES FOR CHAPTER 9

In Exercises 1–14, find the derivative of the given function.

1. $y = \sin(x^2 + 1)$

2. $y = 3 \cos x^2$

3. $m = 2 \cot(3x + 2)$

4. $y = \sec \sqrt{2 - x}$

5. $y = x \tan^3(2x - 1)$

6. $y = \sqrt{\tan (x^2 - 3)}$

7. $v = \ln (\sin e^t)$

8. $y = e^{\cos 3t}$

9. $s = 3 \operatorname{Arcsin} 4t$

10. $w = \operatorname{Arccos} 2\sqrt{x}$

11. $y = \operatorname{Arctan} \sqrt{1 - x^2}$

12. $y = \dfrac{1}{\operatorname{Arcsin} x}$

13. $y = \sec^5 3x^2$

14. $u = e^{2v} \operatorname{Arctan} 3v$

In Exercises 15–30, find the indicated integrals.

15. $\int \cos 6t\, dt$

16. $\int \sin 0.2t\, dt$

17. $\int x \tan 3x^2\, dx$

18. $\int \sqrt[3]{\cos x} \, \sin x\, dx$

19. $\displaystyle\int \sin^2 5x \, dx$

20. $\displaystyle\int x \tan^2(3x^2) \, dx$

21. $\displaystyle\int \cot x(\ln \sin x) \, dx$

22. $\displaystyle\int \sin^{-3} t \cos t \, dt$

23. $\displaystyle\int \sqrt{\sin x} \cos x \, dx$

24. $\displaystyle\int \frac{dt}{\sqrt{15 - 16t^2}}$

25. $\displaystyle\int \sin^3 x \sqrt{\cos x} \, dx$

26. $\displaystyle\int \cot^2 3\theta \, d\theta$

27. $\displaystyle\int \frac{2 \, dx}{4 + 9x^2}$

28. $\displaystyle\int \frac{dy}{y^2 + 2y + 10}$

29. $\displaystyle\int \frac{ds}{s^2 + 16s + 100}$

30. $\displaystyle\int \frac{\sec^2 x}{\tan^4 x} \, dx$

31. What is the electric charge transferred in a circuit from $t = 0$ to $t = 0.5$ sec if the current is given by $i = 15 \cos 2t$?

32. Find the equation of the line tangent to the curve $y = x^2 \sin \pi x$ at the point $(\frac{1}{2}, \frac{1}{4})$.

33. What is the velocity of a particle at $t = \frac{1}{2}$ sec if the equation for the displacement of the particle is $s = 5 \sin \pi t^2$ (in feet)?

34. The current (in amperes) in a circuit is $i = 1/(9 + 4t^2)$. What is the average current in the circuit for $0 < t < \frac{3}{2}$?

35. Find the area bounded by $y = \sin^2 x$ and the x-axis from $x = 0$ to $x = \pi$.

36. Find the volume obtained by rotating the area bounded by $y = \tan x$ and the x-axis about the x-axis from $x = 0$ to $x = \pi/4$.

37. Find the moment about the y-axis of the area bounded by the curve $y = \sin x^2$ and the x-axis from $x = 0$ to $x = \sqrt{\pi}$.

CHAPTER 10

Integration Techniques and Improper Integrals

The integral formulas developed in the previous chapters are fundamental to the study of integral calculus, but they are by no means exhaustive. Large lists of additional formulas are available in various mathematics or engineering handbooks under the heading of "Table of Integrals" or "Table of Antiderivatives." An example of such a table is given at the back of this book. This table includes a number of integral formulas not covered in the previous chapters. The formulas in such a table are called **standard forms.**

No matter which table of integrals you use, you will, at times, encounter integrands that do not precisely fit one of the standard forms. Hence, you must know how to manipulate integrands so that the formulas of the table may be applied. In this chapter, we introduce some of the techniques used for this purpose.

■ 10.1 CHANGING THE VARIABLE OF INTEGRATION

Integrals that cannot be found by the direct application of a table of integrals can sometimes be changed to a standard form by changing the variable of integration. There are many substitutions that can be made on any given integral, but only those that lead to one of the standard integral forms are of interest. We will show the technique for two specific types of integrands. Once you understand the mechanics of these substitutions, you should be able to do others.

Integrands that Contain $(ax + b)^{p/q}$

For integrands that contain a factor of the form $(ax + b)^{p/q}$, the substitution to try is $u = (ax + b)^{1/q}$. The first two examples illustrate the approach.

EXAMPLE 1　■ **Problem**

Find $\int \dfrac{x^2}{\sqrt{x - 3}} \, dx$.

■ **Solution**

The radical in the integrand suggests that we make the substitution $u = (x - 3)^{1/2}$. Therefore, $x = u^2 + 3$, and

$$dx = \frac{dx}{du} \, du = 2u \, du$$

Substituting these values into the given integrand, we obtain

$$\int \frac{x^2}{\sqrt{x - 3}} \, dx = \int \frac{(u^2 + 3)^2}{u} (2u \, du) = \int (2u^4 + 12u^2 + 18) \, du$$

$$= \frac{2}{5} u^5 + 4u^3 + 18u + C$$

Reversing the substitution, we have

$$\int \frac{x^2}{\sqrt{x - 3}} \, dx = \frac{2}{5} (x - 3)^{5/2} + 4(x - 3)^{3/2} + 18(x - 3)^{1/2} + C \qquad ■$$

When evaluating definite integrals by the method of substitution, we often find it convenient to change the limits of integration to agree with the substituted variable. By rewriting the limits of integration in terms of the new variable, we can find the definite integral without reversing the original substitution.

EXAMPLE 2 ■ **Problem**

Evaluate $\int_1^6 x\sqrt{x + 3} \, dx$.

■ **Solution**

If we let $z = (x + 3)^{1/2}$, then $x = z^2 - 3$ and $dx = 2z \, dz$. Also, we note that when $x = 1, z = 2$; and when $x = 6, z = 3$. The given integral may then be written as

$$\int_1^6 x\sqrt{x + 3} \, dx = \int_2^3 (z^2 - 3)(z)(2z \, dz) = 2\int_2^3 (z^4 - 3z^2) \, dz$$

$$= 2\left[\frac{z^5}{5} - z^3 \right]_2^3 = 2\left[\left(\frac{243}{5} - 27 \right) - \left(\frac{32}{5} - 8 \right) \right] = \frac{232}{5} \qquad ■$$

EXAMPLE 3 ■ **Problem**

Suppose the power (in watts) in a circuit is given by the formula

$$p(t) = \frac{1}{t\sqrt{1 - (\ln t)^2}}$$

Find the average power between $t = 1$ and $t = e$ seconds.

■ **Solution**

By definition, the average power is given by the definite integral

$$p_{avg} = \frac{1}{e-1} \int_1^e \frac{1}{t\sqrt{1-(\ln t)^2}} \, dt$$

We make the substitution $u = \ln t$. When $t = 1$, $u = 0$; and when $t = e$, $u = 1$. Also, $t = e^u$ and $dt = e^u \, du$. Therefore,

$$p_{avg} = \frac{1}{e-1} \int_0^1 \frac{du}{\sqrt{1-u^2}} = \frac{1}{e-1} \, (\text{Arcsin } 1 - \text{Arcsin } 0)$$

$$= \frac{\pi/2}{e-1} \text{ watts}$$

■

Integrands that Contain $\sqrt{a^2 \pm x^2}$

When terms of the type $\sqrt{a^2 \pm x^2}$ appear in the integrand, change the variable as follows:

1. When $\sqrt{a^2 - x^2}$ occurs, let $x = a \sin \theta$.
2. When $\sqrt{a^2 + x^2}$ occurs, let $x = a \tan \theta$.

In each case, the substitution will eliminate the radical. Then, the integral is transformed in the usual manner, and the resulting trigonometric integrations are performed. The reversal of the substitution is best described by using a right triangle to relate the variables x and θ, as shown in the next example.

EXAMPLE 4   ■ **Problem**

Find $\int \dfrac{x^2}{\sqrt{9-x^2}} \, dx$.

■ **Solution**

Let $x = 3 \sin \theta$. Then, since $dx/d\theta = 3 \cos \theta$, we have

$$\int \frac{x^2}{\sqrt{9-x^2}} \, dx = \int \frac{9 \sin^2 \theta}{\sqrt{9 - 9 \sin^2 \theta}} \, (3 \cos \theta) \, d\theta = \int \frac{9 \sin^2 \theta}{3 \cos \theta} \, (3 \cos \theta) \, d\theta$$

$$= 9 \int \sin^2 \theta \, d\theta$$

This last integral may be found by using Formula (24) from the Table of Common Integrals in the appendix. Thus,

$$\int \frac{x^2}{\sqrt{9-x^2}} \, dx = \frac{9}{2} \theta - \frac{9}{4} \sin 2\theta + C$$

To express this result in terms of x, we use the fact that $\sin \theta = x/3$, or $\theta = \arcsin(x/3)$, to generate the right triangle in Figure 10.1. Recalling the identity $\sin 2\theta = 2 \sin \theta \cos \theta$, and from Figure 10.1, noting that $\cos \theta = \sqrt{9 - x^2}/3$, we have

$$\int \frac{x^2}{\sqrt{9 - x^2}} \, dx = \frac{9}{2} \operatorname{Arcsin} \frac{x}{3} - \frac{1}{2} x \sqrt{9 - x^2} + C$$

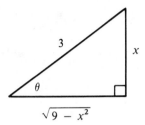

■ **Figure 10.1**

EXAMPLE 5 ■ **Problem**

Find $\displaystyle\int \frac{(1 + x^2)^{3/2} \, dx}{x^6}$.

■ **Solution**

Since $(1 + x^2)^{3/2} = (\sqrt{1 + x^2})^3$, we use the substitutions

$$x = \tan \theta \qquad \text{and} \qquad dx = \sec^2 \theta \, d\theta$$

to obtain

$$\int \frac{(1 + x^2)^{3/2} \, dx}{x^6} = \int \frac{(1 + \tan^2 \theta)^{3/2} \sec^2 \theta \, d\theta}{\tan^6 \theta} = \int \frac{\sec^3 \theta \sec^2 \theta \, d\theta}{\tan^6 \theta}$$

$$= \int \frac{\sec^5 \theta \, d\theta}{\tan^6 \theta} = \int \frac{1}{\cos^5 \theta} \cdot \frac{\cos^6 \theta}{\sin^6 \theta} \, d\theta$$

$$= \int \sin^{-6} \theta \cos \theta \, d\theta = \frac{\sin^{-5} \theta}{-5} + C$$

To express the answer in terms of the original variable, we refer to Figure 10.2. Noting that $\sin \theta = x/\sqrt{1 + x^2}$, we get

$$\int \frac{(1 + x^2)^{3/2} \, dx}{x^6} = -\frac{(1 + x^2)^{5/2}}{5x^5} + C$$

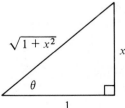

■ **Figure 10.2**

EXERCISES FOR SECTION 10.1

Find the integrals in Exercises 1–6 by using an appropriate substitution.

1. $\int x \sqrt{x + 2} \, dx$

2. $\int x \sqrt{2x + 3} \, dx$

3. $\int (t + 2)\sqrt{t - 3} \, dt$

4. $\int (x - 1)\sqrt{2x - 1} \, dx$

5. $\int \frac{3s^2 \, ds}{\sqrt{1 - s}}$

6. $\int \frac{x \, dx}{(x - 1)^{1/3}}$

7. $\int \frac{dy}{(4y + 13)\sqrt{y - 1}}$

8. $\int \frac{(3x - 1) \, dx}{\sqrt{2x + 1}}$

9. $\int \frac{dx}{x + x^{1/2}}$

10. $\int \frac{dy}{(1 + y)^{1/2}}$

11. $\int x \sqrt[3]{x - 1} \, dx$

12. $\int \frac{dt}{t + t^{2/3}}$

13. $\int \frac{3t \, dt}{(2t + 1)^{3/4}}$

14. $\int \frac{x \, dx}{(3x + 4)^{2/3}}$

15. $\int x(2x + 1)^{2/3} \, dx$

16. $\int \frac{dw}{\sqrt[3]{w} + w}$

17. $\int \sqrt{4 - x^2} \, dx$

18. $\int \frac{x^2 \, dx}{\sqrt{16 - x^2}}$

19. $\int \frac{dx}{\sqrt{9 + x^2}}$

20. $\int \frac{dx}{x^2 \sqrt{25 + x^2}}$

21. $\int x^3 \sqrt{1 - x^2} \, dx$

22. $\int \frac{\sqrt{5 + x^2} \, dx}{x^4}$

23. $\int \frac{dz}{(9 + z^2)^{3/2}}$

24. $\int \frac{dr}{(1 - r^2)^{3/2}}$

25. $\int \frac{x^2 \, dx}{(9 - x^2)^{3/2}}$

26. $\int \frac{u^2 \, du}{(7 - u^2)^{3/2}}$

27. The power (in watts) being generated by a system is given by $p = (2t - 1)\sqrt{t + 3}$. Find the work done by the system from $t = 1$ to $t = 6$ sec.

28. The current (in amperes) in a resistor varies as $i = t/(t + 1)^{2/3}$. Find the amount of charge transferred through the resistor from $t = 0$ to $t = 7$ sec.

29. The acceleration (in feet per second squared) of a shaper head during its return stroke varies according to $a = t/(t^{1/2} + t^{3/2})$. Find the equation for the velocity of the head given that $v = 0$ when $t = 0$.

30. The force (in pounds) required to motivate a particular mechanism varies with displacement according to $F = 3s/\sqrt{s - 1}$. Find the work done in moving from $s = 2$ to $s = 5$ ft.

31. Find the area under the curve $y = \sqrt{1 - x^2}$ from $x = 0$ to $x = 1$. Draw the figure.

■ 10.2   INTEGRATION BY PARTS

A very powerful tool for evaluating antiderivatives comes from the formula for the derivative of the product of two functions. Let u and v represent any two differentiable functions. Then,

$$\frac{d}{dx}(uv) = u\frac{dv}{dx} + v\frac{du}{dx}$$

which, in differential form, may be written as

$$d(uv) = u\,dv + v\,du$$

Rearranging, we get

$$u\,dv = d(uv) - v\,du \qquad\qquad \textbf{(10.1)}$$

Finally, integrating each term, we obtain the following formula.

FORMULA

> **Integration by Parts**
>
> $$\int u\,dv = uv - \int v\,du \qquad\qquad \textbf{(10.2)}$$

In the formula, the product $u\,dv$ indicates that the integrand is to be separated into two factors, or *parts*: a function u and a differential of another function, dv. Once the u part and the dv part have been identified (usually on a trial basis), Formula (10.2) is applied. The success of the method depends on our ability to find the antiderivative of the part labeled dv and the resulting term $v\,du$. Surprisingly, there are many cases in which $v\,du$ can be found in the table of integrals when $u\,dv$ cannot.

The major problem in using Formula (10.2) is in choosing the parts labeled u and dv. There are no general rules that govern the choice of parts, but the following hints may be useful.

> **Using Integration by Parts**
>
> 1. The term dv must be a function that can be integrated.
> 2. If an extra x is present, let $u = x$.
> 3. If $\ln x$ is present, let $u = \ln x$.

EXAMPLE 1   ■ **Problem**

Find $\int 2x \cos x \, dx$.

■ **Solution**

This integral may be found by letting

$$u = 2x \quad \text{and} \quad dv = \cos x \, dx$$

Then, by differentiating u and integrating dv, we get

$$du = 2 \, dx \quad \text{and} \quad v = \sin x$$

We need not add an arbitrary constant to the function v. All that is required is that a constant be added after the final integration.

Substituting the appropriate values into the parts formula, we get

$$\int u \quad dv \;\; = \;\; u \quad v \;\; - \int v \quad du$$
$$\downarrow \quad \downarrow \qquad \downarrow \quad \downarrow \qquad \downarrow \quad \downarrow$$
$$\int (2x)(\cos x \, dx) = (2x)(\sin x) - \int (\sin x)(2 \, dx)$$
$$\int 2x \cos x \, dx = 2x \sin x + 2 \cos x + C \qquad\qquad ■$$

Sometimes, the integration-by-parts formula must be applied repeatedly before $v \, du$ is a standard integral form. The following example illustrates a case where two applications are needed.

EXAMPLE 2   ■ **Problem**

Find $\int 3x^2 e^{6x} \, dx$.

■ **Solution**

Choose the parts

$$u = 3x^2 \quad \text{and} \quad dv = e^{6x} \, dx$$

Then

$$du = 6x \, dx \quad \text{and} \quad v = \tfrac{1}{6} e^{6x}$$

Substituting into Formula (10.2), we get

$$\int 3x^2\, e^{6x}\, dx = \tfrac{1}{2}x^2\, e^{6x} - \int xe^{6x}\, dx$$

Since $\int xe^{6x}\, dx$ is not a standard form, we must apply Formula (10.2) again—this time choosing the parts

$$u = x \qquad \text{and} \qquad dv = e^{6x}\, dx$$

so that

$$du = dx \qquad \text{and} \qquad v = \tfrac{1}{6}e^{6x}$$

Thus,

$$\int 3x^2\, e^{6x}\, dx = \frac{1}{2}\, x^2 e^{6x} - \left[\frac{1}{6}\, xe^{6x} - \frac{1}{6}\int e^{6x}\, dx\right]$$

$$= \frac{1}{2}\, x^2 e^{6x} - \frac{1}{6}\, xe^{6x} + \frac{1}{36}\, e^{6x} + C$$

$$= \frac{1}{2}\, e^{6x}\!\left(x^2 - \frac{1}{3}\, x + \frac{1}{18}\right) + C \qquad \blacksquare$$

EXAMPLE 3   ■ **Problem**

Evaluate $\displaystyle\int_2^4 t \ln t\, dt$.

■ **Solution**

Integration by parts is directly applicable to definite integrals. Letting

$$u = \ln t \qquad \text{and} \qquad dv = t\, dt$$

we have

$$du = \frac{dt}{t} \qquad \text{and} \qquad v = \frac{t^2}{2}$$

Substituting into Formula (10.2), we get

$$\int_2^4 t \ln t\, dt = \left[\frac{t^2}{2}\ln t\right]_2^4 - \int_2^4 \left(\frac{t^2}{2}\right)\frac{dt}{t} = \left[\frac{t^2}{2}\ln t\right]_2^4 - \frac{1}{2}\int_2^4 t\, dt$$

$$= \left[\frac{t^2}{2}\ln t - \frac{t^2}{4}\right]_2^4 = (8 \ln 4 - 4) - (2 \ln 2 - 1)$$

$$= 8 \ln 4 - 2 \ln 2 - 3 \approx 8(1.3863) - 2(0.6931) - 3$$

$$\approx 6.7042 \qquad \blacksquare$$

EXAMPLE 4   ■ **Problem**

Find $\int e^{4x} \sin 2x\, dx$. Integrals like this one arise in the study of Fourier series.

■ **Solution**

This integral can be found by applying the parts formula twice. We proceed by letting

$$u = e^{4x} \quad \text{and} \quad dv = \sin 2x \, dx$$

then,

$$du = 4e^{4x} \, dx \quad \text{and} \quad v = -\tfrac{1}{2} \cos 2x$$

By the parts formula, we get

$$\int e^{4x} \sin 2x \, dx = -\tfrac{1}{2}e^{4x} \cos 2x + 2 \int e^{4x} \cos 2x \, dx$$

To evaluate the integral on the right side, we repeat the above process with

$$u = e^{4x} \quad \text{and} \quad dv = \cos 2x \, dx$$

We have

$$du = 4e^{4x} \, dx \quad \text{and} \quad v = \tfrac{1}{2} \sin 2x$$

so that

$$\int e^{4x} \sin 2x \, dx = -\tfrac{1}{2}e^{4x} \cos 2x + 2[\tfrac{1}{2}e^{4x} \sin 2x - 2 \int e^{4x} \sin 2x \, dx]$$
$$= -\tfrac{1}{2}e^{4x} \cos 2x + e^{4x} \sin 2x - 4 \int e^{4x} \sin 2x \, dx$$

We see that the integral on the right is the same as the one given. Hence, solving for $\int e^{4x} \sin 2x \, dx$, we get

$$5 \int e^{4x} \sin 2x \, dx = -\tfrac{1}{2}e^{4x} \cos 2x + e^{4x} \sin 2x$$

or

$$\int e^{4x} \sin 2x \, dx = \tfrac{1}{10}e^{4x} (2 \sin 2x - \cos 2x) + C \qquad ■$$

EXERCISES FOR SECTION 10.2

In Exercises 1–20 find the indicated integrals.

1. $\int x \sin x \, dx$

2. $\int x \cos 3x \, dx$

3. $\int te^{-t} \, dt$

4. $\int ye^{y} \, dy$

5. $\int \theta^2 \sin \theta \, d\theta$

6. $\int 3x^2 \cos x \, dx$

7. $\int z^3 \ln z \, dz$

8. $\int x^2 \sin x^3 \, dx$

9. $\int \text{Arctan } y \, dy$

10. $\int \text{Arcsin } t \, dt$

11. $\int \text{Arccos } 3x \, dx$

12. $\int x\sqrt{x+1} \, dx$

13. $\int t(3t^2 - 2)^{-4} \, dt$

14. $\int x \sec^2 x^2 \, dx$

15. $\int \dfrac{x \, dx}{(2x-5)^{1/4}}$

16. $\int y \tan y \sec^2 y \, dy$

17. $\int \phi^3 e^{\phi^2} \, d\phi$

18. $\int \cos^2 \theta \sin \theta \, d\theta$

19. $\int e^{-x} \cos 3x \, dx$

20. $\int e^{x/2} \sin \dfrac{x}{2} \, dx$

In Exercises 21–24, evaluate the given definite integral.

21. $\int_1^2 \ln t \, dt$

22. $\int_0^{\pi/2} \theta \sin \theta \, d\theta$

23. $\int_0^{1/2} \text{Arccos } r \, dr$

24. $\int_0^{1/2} y \, e^{2y} \, dy$

25. Find the function whose derivative is $dy/dx = x^{1/2} \ln x$, given that $y = 0$ when $x = 1$.

26. Find the area bounded by the curve $y = (x+2)e^x$ from $x = 0$ to $x = 1$.

 27. The velocity (in centimeters per second) of a particle is given by $v = \arctan t$. Find the displacement of the particle from $t = 0$ to $t = 1$ sec.

 28. Find the centroid of the area bounded by the curve $y = e^x$ and the ordinates $x = 0$ and $x = 2$.

 29. The resistance (in ohms) of a resistor varies as $r = t + 10$ over the interval $t = 0$ to $t = 1$. Find the energy dissipated by the resistor during this time interval if the current (in amperes) varies according to $i = e^{-2t}$.

▦ 10.3 PARTIAL FRACTIONS

A **rational function** is one that can be expressed as the ratio of two polynomial functions. If the degree of the polynomial in the numerator is less than the degree of the polynomial in the denominator, it is called a **proper rational function**; otherwise, the function is said to be **improper**. Any improper rational function can be reduced to a form consisting of the sum of a polynomial and a proper rational function. For example,

$$\frac{6x^4 + 7x^3 + 6x^2 + 32x - 7}{3x^2 + 5x - 2} = 2x^2 - x + 5 + \frac{5x + 3}{3x^2 + 5x - 2}$$

A proper rational function can frequently be written or decomposed into a sum of functions that turn out to be one of the standard forms. This sum is often called the **partial-fractions decomposition** of the rational function. It is literally the reverse of what we learn in elementary algebra. That is, there we learn that

$$\frac{2}{x-3} - \frac{1}{x+1} = \frac{x+5}{x^2-2x-3}$$

Now, we will show how to "recover" the elements of the sum making up

$$\frac{x+5}{x^2-2x-3}$$

Parenthetically, we note that if we had to evaluate

$$\int \frac{x+5}{x^2-2x-3}\,dx$$

we could write

$$\int \frac{x+5}{x^2-2x-3}\,dx = \int \frac{2}{x-3}\,dx - \int \frac{1}{x+1}\,dx$$

$$= 2\ln(x-3) - \ln(x+1) + C = \ln\frac{(x-3)^2}{x+1} + C$$

Our problem, then, is to describe the process by which a proper fraction can be decomposed into its partial fractions. In higher-algebra courses, we prove that a proper fraction can always be expressed as the sum of partial fractions if the denominator of the proper fraction can be expressed as the product of factors of the form $ax + b$ and $ax^2 + bx + c$. Furthermore, the form of the partial-fractions decomposition is completely determined by the form of the factors making up the denominator. The following rules govern the form of the partial-fractions decomposition of a given rational function. In each case, the factors of the denominator of the rational function are what are being considered.

RULE 1

Denominator with ($ax + b$) Unrepeated   If the denominator has an unrepeated linear factor of the form ($ax + b$), a fraction of the form

$$\frac{A}{ax+b}$$

must be included in the partial-fraction decomposition.

RULE 2

Denominator with ($ax + b$) Repeated   If the denomiantor has a linear factor ($ax + b$) that is repeated n times, a sum of fractions of the form

$$\frac{A_1}{ax+b} + \frac{A_2}{(ax+b)^2} + \cdots + \frac{A_n}{(ax+b)^n}$$

must be included in the partial-fraction decomposition.

RULE 3

Denominator with $(ax^2 + bx + c)$ Unrepeated   If the denominator has an unrepeated quadratic factor $(ax^2 + bx + c)$, a fraction of the form

$$\frac{Ax + B}{ax^2 + bx + c}$$

must be included in the partial-fraction decomposition.

The following examples show how to utilize the knowledge of the form of the partial-fractions decomposition to find the actual one. The examples are stated within the context of finding antiderivatives of rational functions since that is our only application at this time.

EXAMPLE 1   ■ **Problem**

Find $\int \dfrac{(x - 11)\, dx}{2x^2 + 5x - 3}$.

■ **Solution**

The factors of the denominator are $x + 3$ and $2x - 1$. Hence, the given fraction can be expressed as the sum of the partial fractions

$$\frac{A}{x + 3} \quad \text{and} \quad \frac{B}{2x - 1}$$

where A and B are constants to be determined. Thus, we write

$$\frac{x - 11}{(x + 3)(2x - 1)} = \frac{A}{x + 3} + \frac{B}{2x - 1}$$

To solve for A and B, we proceed by clearing the fractions from the equation to get

$$x - 11 = A(2x - 1) + B(x + 3)$$

Expanding gives

$$x - 11 = 2Ax - A + Bx + 3B$$

Collecting like terms in x, we have

$$x - 11 = (2A + B)x + (3B - A)$$

For this equality to hold, the coefficients of corresponding powers of x must be equal, so

$$2A + B = 1 \quad \text{and} \quad 3B - A = -11$$

Solving these equations simultaneously, we get $A = 2$ and $B = -3$. Hence,

$$\frac{x - 11}{(x + 3)(2x - 1)} = \frac{2}{x + 3} - \frac{3}{2x - 1}$$

Substituting into the given integral, we have

$$\int \frac{(x - 11)\, dx}{(x + 3)(2x - 1)} = \int \frac{2\, dx}{x + 3} - \int \frac{3\, dx}{2x - 1}$$

$$= 2 \ln(x + 3) - \frac{3}{2} \ln(2x - 1) + C$$

$$= \ln \frac{(x + 3)^2}{(2x - 1)^{3/2}} + C$$

■

EXAMPLE 2   ■ **Problem**

Find $\displaystyle \int \frac{x^2\, dx}{(x + 1)^3}$.

■ **Solution**

We assume that the partial fractions may be written as

$$\frac{x^2}{(x + 1)^3} = \frac{A}{x + 1} + \frac{B}{(x + 1)^2} + \frac{C}{(x + 1)^3}$$

Clearing the fractions gives

$$x^2 = A(x + 1)^2 + B(x + 1) + C$$

Expanding yields

$$x^2 = Ax^2 + 2Ax + A + Bx + B + C$$

Collecting like terms in x, we have

$$x^2 = Ax^2 + (2A + B)x + (A + B + C)$$

Equating the like powers of x, we get

$$A = 1 \qquad 2A + B = 0 \qquad A + B + C = 0$$

These equations yield the values $A = 1$, $B = -2$, and $C = 1$. Making these substitutions, we have

$$\frac{x^2}{(x + 1)^3} = \frac{1}{x + 1} - \frac{2}{(x + 1)^2} + \frac{1}{(x + 1)^3}$$

The given integral may then be written as

$$\int \frac{x^2\, dx}{(x + 1)^3} = \int \frac{dx}{x + 1} - \int \frac{2\, dx}{(x + 1)^2} + \int \frac{dx}{(x + 1)^3}$$

$$= \ln(x + 1) + \frac{2}{x + 1} - \frac{1}{2(x + 1)^2} + C \qquad \blacksquare$$

EXAMPLE 3   ■ **Problem**

Find $\displaystyle\int \frac{(2x + 8)\, dx}{x^3 + 4x^2 + 4x}$.

■ **Solution**

The factors of the denominator are x and $(x + 2)^2$. Therefore, the partial fractions are

$$\frac{2x + 8}{x(x + 2)^2} = \frac{A}{x} + \frac{B}{x + 2} + \frac{C}{(x + 2)^2}$$

Clearing fractions yields

$$2x + 8 = A(x + 2)^2 + Bx(x + 2) + Cx$$

By the method used in the preceding examples, we obtain the simultaneous equations

$$A + B = 0 \qquad 4A + 2B + C = 2 \qquad 4A = 8$$

The solution of these equations yields $A = 2$, $B = -2$, and $C = -2$. The integral is then

$$\int \frac{(2x + 8)\, dx}{x(x + 2)^2} = \int \frac{2\, dx}{x} - \int \frac{2\, dx}{x + 2} - \int \frac{2\, dx}{(x + 2)^2}$$

$$= 2 \ln x - 2 \ln(x + 2) + \frac{2}{x + 2} + C$$

$$= \ln\left(\frac{x}{x + 2}\right)^2 + \frac{2}{x + 2} + C \qquad \blacksquare$$

EXAMPLE 4   ■ **Problem**

Find $\displaystyle\int \frac{(9x + 14)\, dx}{(x - 2)(x^2 + 4)}$.

■ **Solution**

The denominator is composed of a linear factor $(x - 2)$ and a quadratic factor $(x^2 + 4)$. The required partial fractions are therefore

$$\frac{9x + 14}{(x - 2)(x^2 + 4)} = \frac{A}{x - 2} + \frac{Bx + C}{x^2 + 4}$$

Clearing fractions gives

$$9x + 14 = A(x^2 + 4) + (Bx + C)(x - 2)$$

Expanding the right side and collecting like terms, we get

$$9x + 14 = (A + B)x^2 + (C - 2B)x + (4A - 2C)$$

Equating the coefficients of corresponding powers of x, we find

$$A + B = 0 \qquad C - 2B = 9 \qquad 4A - 2C = 14$$

from which $A = 4$, $B = -4$, and $C = 1$. Substituting these values in the partial fractions, we obtain

$$\int \frac{(9x + 14)\,dx}{(x - 2)(x^2 + 4)} = \int \frac{4\,dx}{x - 2} + \int \frac{(-4x + 1)\,dx}{x^2 + 4}$$

The first integral on the right-hand side is easy to find, but the second is a little harder. To find its value, we proceed in the following manner:

$$\int \frac{(-4x + 1)\,dx}{x^2 + 4} = -\int \frac{4x\,dx}{x^2 + 4} + \int \frac{dx}{x^2 + 4}$$

$$= -2 \ln(x^2 + 4) + \frac{1}{2} \operatorname{Arctan} \frac{x}{2}$$

The desired integral is then

$$\int \frac{(9x + 14)\,dx}{(x - 2)(x^2 + 4)} = 4 \ln(x - 2) - 2 \ln(x^2 + 4) + \frac{1}{2} \operatorname{Arctan} \frac{x}{2} + C$$

$$= \ln \frac{(x - 2)^4}{(x^2 + 4)^2} + \frac{1}{2} \operatorname{Arctan} \frac{x}{2} + C \qquad \blacksquare$$

Rational functions with denominators that contain repeated quadratic factors of the form $(ax^2 + bx + c)^m$ may also be resolved into partial fractions. However, we choose to exclude this type, since repeated quadratic factors occur much less frequently than the other three types.

EXERCISES FOR SECTION 10.3

In Exercises 1–21, evaluate the indicated integral.

1. $\displaystyle \int \frac{(x + 1)\,dx}{x^2 + 4x - 5}$

2. $\displaystyle \int \frac{(x + 2)\,dx}{x^2 - x - 6}$

3. $\displaystyle \int \frac{dx}{x^2 + x}$

4. $\displaystyle \int \frac{2\,dx}{x^3 - 4x}$

5. $\displaystyle\int \frac{(2z^2 + 5z + 5)\,dz}{z(z + 5)(z + 2)}$

6. $\displaystyle\int \frac{(x + 12)\,dx}{x^3 + x^2 - 6x}$

7. $\displaystyle\int \frac{(x + 3)\,dx}{x^2 + 6x + 5}$

8. $\displaystyle\int \frac{dz}{(2z + 3)^3}$

9. $\displaystyle\int \frac{(y^2 + 3y - 6)\,dy}{y(y - 1)^2}$

10. $\displaystyle\int \frac{(x + 1)\,dx}{x^3 - x^2}$

11. $\displaystyle\int \frac{ds}{s^3 + 2s^2 + s}$

12. $\displaystyle\int \frac{(t^2 + 11)\,dt}{(t - 5)(t + 1)^2}$

13. $\displaystyle\int \frac{(w^2 - 3w + 6)\,dw}{(w^2 - 1)(3 - 2w)}$

14. $\displaystyle\int \frac{dr}{(r + 4)^3}$

15. $\displaystyle\int \frac{(2x - 1)\,dx}{(x - 1)^2}$

16. $\displaystyle\int \frac{2(x^2 + 1)\,dx}{x^3 + 2x}$

17. $\displaystyle\int \frac{t^2 + t + 2)\,dt}{t^3 + 2t}$

18. $\displaystyle\int \frac{(2x^2 - x + 5)\,dx}{(x + 1)(x^2 + 4)}$

19. $\displaystyle\int \frac{(x^2 + 1)\,dx}{(2x + 3)(x^2 + 2x + 4)}$

20. $\displaystyle\int \frac{(5 + 3s - s^2)\,ds}{(s^2 + 2s - 3)(s^2 + 5)}$

21. $\displaystyle\int \frac{(x^3 + x^2 - x + 1)\,dx}{(x - 1)^2(x^2 + 1)}$

In Exercises 22–25, evaluate the indicated definite integrals.

22. $\displaystyle\int_1^2 \frac{(2x + 3)\,dx}{x(x + 3)}$

23. $\displaystyle\int_0^1 \frac{(3y + 8)\,dy}{y^2 + 5y + 6}$

24. $\displaystyle\int_0^2 \frac{dx}{x^2 + 4x + 20}$

25. $\displaystyle\int_0^1 \frac{(x + 3)\,dx}{(x + 1)^2}$

■■■■ 10.4 IMPROPER INTEGRALS

All of the definite integrals that you have encountered so far have had finite limits of integration and continuous integrand functions. Both of these properties are required for the integral to be defined as **proper.** If either is missing, the integral is, strictly speaking, not defined and is referred to as being **improper.** We now wish to show how to give meaning to improper integrals.

Improper Integrals of the First Kind

In the first kind of improper integral, one or both of the limits of integration become infinite. We write

$$\int_a^\infty f(x)\,dx$$

to describe the situation. By this notation, we mean the limit of the proper integral $\int_a^k f(x)\, dx$ as k increases without bound. That is, we have the following definition.

DEFINITION

Integral with ∞ as an Upper Limit

$$\int_a^\infty f(x)\, dx = \lim_{k \to \infty} \int_a^k f(x)\, dx \qquad\qquad \textbf{(10.3)}$$

Similarly, we have the following definition for an infinite lower limit.

DEFINITION

Integral with $-\infty$ as a Lower Limit

$$\int_{-\infty}^a f(x)\, dx = \lim_{k \to -\infty} \int_k^a f(x)\, dx \qquad\qquad \textbf{(10.4)}$$

If the limit on the right side exists, the improper integral is said to **converge**. Otherwise, it is said to **diverge**.

EXAMPLE 1   ■ **Problem**

The graph of the function $y = x^{-2}$ is presented in Figure 10.3. Show that a value can be assigned to the area under the curve as x increases without bound from the point $x = 1$.

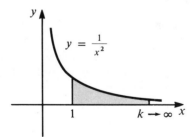

■ **Figure 10.3**

■ **Solution**

The desired area is given by

$$\int_1^\infty x^{-2}\, dx$$

This integral is evaluated as

$$\int_1^\infty x^{-2}\, dx = \lim_{k\to\infty} \int_1^k x^{-2}\, dx = \lim_{k\to\infty} \left[-x^{-1} \right]_1^k$$

$$= \lim_{k\to\infty} \left(-\frac{1}{k} + 1 \right) = 1$$

Therefore, the improper integral converges and has the value 1.   ■

EXAMPLE 2   ■ **Problem**

Show that $\displaystyle\int_1^\infty \frac{dx}{x}$ does not exist.

■ **Solution**

By definition, we have

$$\int_1^\infty \frac{dx}{x} = \lim_{k\to\infty} \int_1^k \frac{dx}{x} = \lim_{k\to\infty} \left[\ln x \right]_1^k = \lim_{k\to\infty} (\ln k - \ln 1) = \infty$$

Since $\ln k \to \infty$ as $k \to \infty$, the integral diverges.   ■

EXAMPLE 3   ■ **Problem**

The force between two charged particles is given by Coulomb's law to be $F = Qq/4\pi ks^2$, where Q and q are the respective charges, k is a constant, and s is the distance separating the charges. Find the formula for the work done in bringing a charge q of 1 C from a great distance to a point d meters from a charge of Q coulombs.

■ **Solution**

Since $q = 1$, the force between Q and q can be written as $F = Q/4\pi ks^2$. The work done by a force is $W = \int_{s_1}^{s_2} F\, ds$. Thus,

$$W = \int_d^\infty \frac{Q}{4\pi ks^2}\, ds = \frac{Q}{4\pi k} \int_d^\infty \frac{ds}{s^2} = \lim_{m\to\infty} \frac{Q}{4\pi k} \int_d^m \frac{ds}{s^2} = \lim_{m\to\infty} \left[-\frac{Q}{4\pi ks} \right]_d^m$$

$$= \lim_{m\to\infty} \left(-\frac{Q}{4\pi km} + \frac{Q}{4\pi kd} \right) = \frac{Q}{4\pi kd}$$

In physics, the work done in bringing a charge of 1 C into an electric field is called **electric potential.** The formula derived in Example 3 is then the formula for the electric potential d meters from a charge of Q coulombs.   ■

Improper Integrals of the Second Kind

In some instances, definite integrals arise that have finite limits of integration but are improper because the integrand function becomes infinite (called an **infinite dis-**

continuity) for some value of x within the interval of integration. Improper integrals of this kind are also defined as the limit of a proper integral. Thus, if the function $f(x)$ has an infinite discontinuity at the upper limit of integration $x = b$, we define the value of the integral as follows:

DEFINITION

Integral of a Function with an Infinite Discontinuity at the Upper Limit

$$\int_a^b f(x)\, dx = \lim_{q \to b^-} \int_a^q f(x)\, dx \qquad (10.5)$$

where q approaches b from within the interval of integration.

Likewise, if the function $f(x)$ has an infinite discontinuity at the lower limit of integration $x = a$, we define the value of the integral from a to b as follows:

DEFINITION

Integral of a Function with an Infinite Discontinuity at the Lower Limit

$$\int_a^b f(x)\, dx = \lim_{q \to a^+} \int_q^b f(x)\, dx \qquad (10.6)$$

where q approaches a from within the interval of integration.

As in the case of improper integrals of the first kind, these integrals converge if the limit on the right exists and diverge if it does not.

EXAMPLE 4   ■ **Problem**

Evaluate $\int_0^2 \dfrac{dx}{\sqrt{x}}$.

■ **Solution**

Since the integrand has an infinite discontinuity at $x = 0$, we choose a value $x = p$ within the interval of integration for which the integrand is defined. And then, we take the limit of the integral as $p \to 0$. See Figure 10.4. Hence, the integral can be written as

$$\int_0^2 \frac{dx}{\sqrt{x}} = \lim_{p \to 0^+} \int_p^2 x^{-1/2}\, dx = \lim_{p \to 0^+} \left[2x^{1/2} \right]_p^2$$

$$= \lim_{p \to 0^+} (2\sqrt{2} - 2\sqrt{p}) = 2\sqrt{2}$$

Geometrically, this result is interpreted to mean that the area under the curve $y = 1/\sqrt{x}$ approaches $2\sqrt{2}$ as a limit when p approaches arbitrarily close to zero.

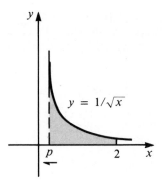

$$y = 1/\sqrt{x}$$

■ **Figure 10.4**

The function $f(x)$ may have an infinite discontinuity at some value $x = c$ that lies between $x = a$ and $x = b$. Then, the integral of $f(x)$ from $x = a$ to $x = b$ can be expressed as the sum of two improper integrals. That is,

$$\int_a^b f(x)\, dx = \int_a^c f(x)\, dx + \int_c^b f(x)\, dx$$

where c is the point of infinite discontinuity. Each integral on the right can then be handled as in the previous example, since the first integral on the right is discontinuous at its upper limit and the second is discontinuous at its lower limit. Note that if either of the two integrals on the right diverges, then the integral on the left diverges.

EXAMPLE 5   ■ **Problem**

Evaluate $\displaystyle\int_{-1}^{3} \frac{dx}{x^2}$.

■ **Solution**

The integrand is discontinuous at $x = 0$, which is an interior point of the interval of integration. The given integral may then be written as

$$\int_{-1}^{3} \frac{dx}{x^2} = \int_{-1}^{0} \frac{dx}{x^2} + \int_{0}^{3} \frac{dx}{x^2}$$

Writing the improper integrals as the limit of proper integrals, we have

$$\int_{-1}^{3} \frac{dx}{x^2} = \lim_{p \to 0^-} \int_{-1}^{p} x^{-2}\, dx + \lim_{q \to 0^+} \int_{q}^{3} x^{-2}\, dx$$

$$= \lim_{p \to 0^-} \left[-\frac{1}{x} \right]_{-1}^{p} + \lim_{q \to 0^+} \left[-\frac{1}{x} \right]_{q}^{3}$$

$$= \lim_{p \to 0^-} \left(-\frac{1}{p} - 1 \right) + \lim_{q \to 0^+} \left(-\frac{1}{3} + \frac{1}{q} \right) = \infty$$

■

You should now be on the alert for improper integrals. If you are not, you may erroneously treat an improper integral as proper. For example, if we evaluate the improper integral of the previous example as if it were proper, we would have

$$\int_{-1}^{3} \frac{dx}{x^2} = \left[-\frac{1}{x} \right]_{-1}^{3} = \left[-\frac{1}{3} - 1 \right] = -\frac{4}{3}$$

Intuitively, this answer is meaningless because the integrand is always positive and such integrals *must* yield positive values.

EXERCISES FOR SECTION 10.4

Evaluate the integrals in Exercises 1–20, if they exist.

1. $\int_{1}^{\infty} \frac{dx}{x^3}$

2. $\int_{-\infty}^{-1} \frac{dz}{z^2}$

3. $\int_{0}^{4} \frac{dx}{\sqrt{x}}$

4. $\int_{-1}^{0} \frac{dx}{x^{1/3}}$

5. $\int_{0}^{\infty} \frac{dy}{y + 1}$

6. $\int_{0}^{1} \frac{t \, dt}{t^2 - 1}$

7. $\int_{0}^{2} \frac{dx}{x^{3/2}}$

8. $\int_{0}^{\infty} e^{-t} \, dt$

9. $\int_{0}^{\pi/2} \frac{\sin \theta \, d\theta}{1 - \cos \theta}$

10. $\int_{2}^{\infty} \frac{dx}{(x + 1)^3}$

11. $\int_{0}^{\infty} \frac{dy}{\sqrt{2y + 1}}$

12. $\int_{0}^{1} \frac{dx}{\sqrt{1 - x^2}}$

13. $\int_{0}^{\infty} \frac{dt}{1 + t^2}$

14. $\int_{0}^{2} \frac{e^x \, dx}{1 - e^x}$

15. $\int_{-1}^{2} \frac{dx}{\sqrt[3]{x}}$

16. $\int_{1}^{3} \frac{dz}{(z - 2)^3}$

17. $\int_{0}^{2} \frac{dx}{\sqrt[3]{x - 1}}$

18. $\int_{-1}^{0} \frac{dt}{\sqrt{t + 2}}$

19. $\int_{0}^{2} \frac{x \, dx}{\sqrt{2 - x}}$

20. $\int_{-1}^{0} \frac{x \, dx}{\sqrt{x + 1}}$

21. Find the area under the curve $y = e^{-x}$ for $x \geq 0$.

22. The Laplace transform of $f(t)$ is given by the improper integral $\int_0^\infty f(t)e^{-st}\,dt$. Find the Laplace transform of the function $f(t) = \sin 3t$. Assume $s > 0$.

23. Find the area under the curve $y = 1/\sqrt{2x}$ from $x = 0$ to $x = 2$.

REVIEW EXERCISES FOR CHAPTER 10

Find the integrals in Exercises 1–20.

1. $\int x(2x - 5)^{2/3}\,dx$

2. $\int 5(4 + x^2)^{-1/2}\,dx$

3. $\int \dfrac{dx}{\sqrt{4 + x^2}}$

4. $\int \dfrac{x\,dx}{\sqrt{x + 3}}$

5. $\int x\,\text{Arctan}\,x^2\,dx$

6. $\int x\sec^2 x\,dx$

7. $\int x(x - 3)^{1/2}\,dx$

8. $\int \sqrt{9 - x^2}\,dx$

9. $\int \pi x \sin \pi x\,dx$

10. $\int \ln 4x\,dx$

11. $\int \dfrac{\ln 3x}{x}\,dx$

12. $\int x^2 \sin \tfrac{1}{2}x\,dx$

13. $\int \sin x \sec^2 x\,dx$

14. $\int (x^2 + 4)^{3/2}\,dx$

15. $\int \dfrac{(x - 1)\,dx}{x^2 + 5x + 4}$

16. $\int \dfrac{(2x + 1)\,dx}{x^2 + x - 6}$

17. $\int \dfrac{x^2\,dx}{(x - 1)(x^2 + 1)}$

18. $\int \dfrac{(3x^2 + 8x - 20)\,dx}{(x^2 + 4)(x - 4)}$

19. $\int \dfrac{16\,dx}{(x + 1)(x - 3)^2}$

20. $\int \dfrac{(x - 1)\,dx}{x^3 + x^2}$

In Exercises 21–25, evaluate each improper integral or show that it does not exist.

21. $\int_2^\infty \dfrac{x\,dx}{x^2 - 1}$

22. $\int_{-2}^1 \dfrac{dx}{x^2}$

23. $\int_0^\infty \dfrac{dt}{(t + 9)^2}$

24. $\int_1^\infty \dfrac{dt}{t^2 + 1}$

25. $\displaystyle\int_{-1}^{2} \frac{dt}{\sqrt{t+1}}$

26. Compute the volume of a solid of revolution obtained by revolving the area bounded by $y = e^x$, $x = 2$, $x = 4$, and $y = 0$ about the x-axis.

27. The velocity (in centimeters per second) of a particle is given by $v = $ Arcsin $2t$. Find the displacement of the particle from $t = 0$ to $t = \frac{1}{2}$ sec.

28. The current (in amperes) in a circuit varies as $i = t/\sqrt{t+1}$. Find the amount of charge transferred from $t = 0$ to $t = 3$ sec.

CHAPTER 11

Conic Sections

Curves that can be formed by cutting a right circular cone with a plane are called **conic sections.** Four distinct curves are obtained by cutting a right circular cone with a plane: the circle, the ellipse, the hyperbola, and the parabola. The properties of the conics discovered by the early Greek mathematicians include those that we use as definitions in this chapter.

If two right circular cones are placed vertex to vertex with a common axis, the resulting figure is referred to as a *cone* with two *nappes*. To see how the four conics can be generated from a cone, refer to Figure 11.1. The four conics are obtained as follows:

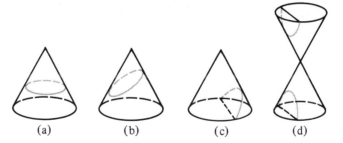

■ **Figure 11.1**

     (a)   (b)   (c)   (d)

1. A *circle* (Figure 11.1a) is obtained when the cutting plane is perpendicular to the axis, provided that it does not pass through the vertex.
2. An *ellipse* (Figure 11.1b) is obtained when the cutting plane is inclined so as to cut entirely through one nappe of the cone without cutting the other nappe.
3. A *parabola* (Figure 11.1c) is obtained when the cutting plane is parallel to one element in the side of the cone.
4. A *hyperbola* (Figure 11.1d) is obtained when the cutting plane is inclined so that it cuts through both nappes.

Analytic geometry is the study of the relation between algebra and geometry. So, in this chapter, we want to establish the algebraic representation of the four conic sections.

■ 11.1 THE PARABOLA

In this section, we define and discuss the parabola.

DEFINITION	**Parabola** A *parabola* is the set of all points in a plane that are equidistant from a fixed point and a fixed line.

The fixed point is called the **focus,** and the fixed line is called the **directrix** of the parabola. The line through the focus and perpendicular to the directrix is the **axis** of the parabola. By definition, the midpoint between the focus and the directrix is a point on the parabola known as the **vertex.** (Recall from Chapter 1 that the shape of the graph of a quadratic function is a parabola whose axis is *vertical.*)

Figure 11.2 shows a parabola with vertex located at the origin, focus at $F(a, 0)$, and directrix perpendicular to the x-axis at $D(-a, 0)$. Observe that a is the distance from the vertex to the focus, sometimes referred to as the **focal distance.**

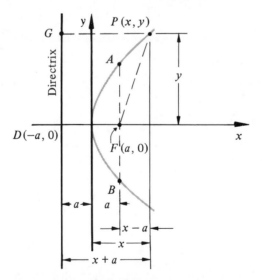

■ **Figure 11.2**

To find the algebraic equation of this parabola, consider a point $P(x, y)$ on the parabola. Then, by definition,

$$\overline{GP} = \overline{FP} \tag{11.1}$$

But from Figure 11.2, we see that

$$\overline{GP} = x + a \quad\text{and}\quad \overline{FP} = \sqrt{(x - a)^2 + y^2}$$

Substituting these expressions in Equation (11.1) yields

$$x + a = \sqrt{(x - a)^2 + y^2}$$

Or when both sides are squared, we have

$$(x + a)^2 = (x - a)^2 + y^2$$

With like terms expanded and collected, the equation of the parabola is as given next.

FORMULA

Parabola with Vertex at the Origin and Focus on the x-axis

$$y^2 = 4ax \tag{11.2}$$

As noted in the title, Equation (11.2) is referred to as the standard form of the equation of a parabola with vertex at the origin and focus on the x-axis. It is clearly symmetric about its axis. The focus is at $(a, 0)$ if the coefficient of x is positive. In this case, the parabola opens to the right. If the coefficient of x is negative, the focus is at $(-a, 0)$, and the parabola opens to the left.

The chord AB through the focus and perpendicular to the axis is called the **right chord,** or frequently, the **latus rectum.** The length of the right chord is found by letting $x = a$ in Equation (11.2). By this substitution.

$$y^2 = 4a^2$$
$$y = \pm 2a$$

The length of the right chord is numerically equal to $4|a|$. This fact helps to define the shape of the parabola, giving us an idea of the "opening" of the parabola.

The standard form of a parabola with vertex at the origin and focus on the y-axis is as follows.

FORMULA

Parabola with Vertex at the Origin and Focus on the y-axis

$$x^2 = 4ay \tag{11.3}$$

Deriving this formula, which parallels Equation (11.2), is left to you. The parabola represented by Equation (11.3) is symmetric about the y-axis. It opens upward if the coefficient of y is positive and downward if it is negative.

The equation of a parabola is characterized by having one linear variable and one squared variable. *The linear variable indicates the direction of the axis of the parabola.*

A rough sketch of the parabola can be drawn if the location of the vertex and the extremities of the right chord are known. All this information can be obtained from the standard form of the equation.

EXAMPLE 1   ■ **Problem**

Discuss and sketch the graph of the equation $x^2 = 8y$.

■ **Solution**

This equation has the form of Equation (11.3) with $4a = 8$, or $a = 2$. The y-axis is the axis of the parabola, the focus is at (0, 2), and the directrix is the line $y = -2$. The endpoints of the right chord are $(-4, 2)$ and $(4, 2)$. Figure 11.3 shows the parabola.

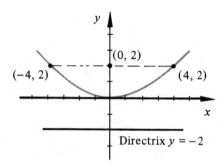

■ **Figure 11.3**

EXAMPLE 2   ■ **Problem**

Find the equation of the parabola with focus at $(-1, 0)$ and directrix $x = 1$. Sketch the curve.

■ **Solution**

The focus lies on the x-axis to the left of the directrix, so the parabola opens to the left. The desired equation is then the form of Equation (11.2) with $a = -1$. That is,

$$y^2 = -4x$$

To sketch the parabola, note that the length of the right chord is $\overline{AB} = 4$. Its extremeties are $(-1, 2)$ and $(-1, -2)$. The curve appears in Figure 11.4.

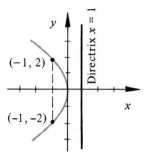

■ Figure 11.4   ■

Comment: A unique physical property of the parabola is that it will reflect any rays emitted from the focus so that they travel parallel to the axis of the parabola. This feature makes the parabola a particularly desirable shape for reflectors in spotlights, reflecting telescopes, and radar antennas.

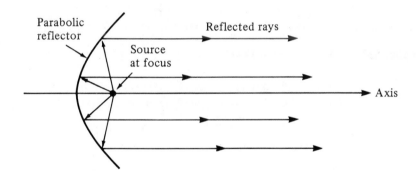

EXAMPLE 3   ■ **Problem**

A parabolic reflector is to be built with a focal distance of 2.25 ft. What is the diameter of the reflector if it is to be 1 ft deep at its axis?

■ **Solution**

To solve this problem, we need the equation of the parabola used to generate the reflector. Refer to Figure 11.5; the vertex is located at the origin, and the focus is on the x-axis. (We could have also placed the focus on the y-axis.) Then, Equation (11.2) with $a = 2.25$ yields the equation

$$y^2 = 4(2.25)x = 9x$$

The diameter of the reflector can now be found by substituting $x = 1$ into this equation. Thus,

$$y^2 = 9 \quad \text{and} \quad y = \pm 3$$

which means that the diameter of the reflector is 6 ft.

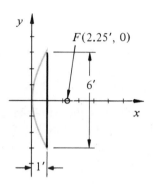

$F(2.25', 0)$

$6'$

$1'$

■ Figure 11.5

EXERCISES FOR SECTION 11.1

In Exercises 1–10, find the coordinates of the focus, the endpoints of the right chord, and the equation of the directrix of each of the parabolas. Sketch the graph of each parabola.

1. $y^2 = -8x$
2. $x^2 = 12y$
3. $2x^2 = 12y$
4. $x^2 = -24y$
5. $y^2 + 16x = 0$
6. $y + 2x^2 = 0$
7. $y^2 = 3x$
8. $3x^2 = 4y$
9. $y^2 = -2x$
10. $y^2 = 10x$

In Exercises 11–18, find the equation of the parabolas having the given properties. Sketch each curve.

11. Focus at $(0, 2)$; directrix $y = -2$.

12. Focus at $(0, -\frac{1}{2})$; directrix $y = \frac{1}{2}$.

13. Focus at $(\frac{3}{2}, 0)$; directrix $x = -\frac{3}{2}$.

14. Focus at $(-10, 0)$; directrix $x = 10$.

15. Endpoints of right chord $(2, -1)$ and $(-2, -1)$; vertex at $(0, 0)$.

16. Endpoints of right chord $(3, 6)$ and $(3, -6)$; vertex at $(0, 0)$.

17. Vertex at $(0, 0)$; vertical axis; one point of the curve $(2, 4)$.

18. Vertex at $(0, 0)$; horizontal axis; one point of the curve $(2, 4)$.

In Exercises 19–22, solve the given system of equations graphically.

19. $2x + 4y = 0$
    $x^2 - 4y = 0$

20. $y = e^x$
    $y^2 = -3x$

21. $y^2 = 12x$
    $y = \log x$

22. $y = x^3$
    $x^2 = 8y$

23. The supporting cable of a suspension bridge hangs in the shape of a parabola. See the accompanying figure. What is the equation of a cable hanging from two 400-ft-high supports that are 1000 ft apart if the lowest point of the cable is 250 ft below the top of the supports? Choose the origin in the most convenient location.

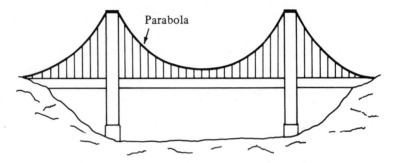

Parabola

24. A parabolic antenna is to be built by revolving the parabola $y^2 = 24x$. Sketch, in cross section, the antenna that has a circular front measuring 12 ft in diameter. Locate the focus.

25. The power loss in a resistor with resistance R is given by $P = I^2R$. Sketch the graph of P as a function of I if $R = 100\,\Omega$.

11.2 THE ELLIPSE

An ellipse can be constructed by placing two pins at F and F', as shown in Figure 11.6, and placing a loop of string over them. Pull the string taut with the point of a pencil, and then move the pencil, keeping the string taut. The figure generated is an ellipse. From this construction, observe that the sum of the distances from the two fixed points to the point P is always the same, since the loop of string is kept taut. This property characterizes the ellipse.

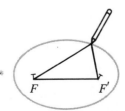

F F'

■ **Figure 11.6**

DEFINITION

> **Ellipse**   An *ellipse* is the set of all points in a plane the sum of whose distances from two fixed points in the plane is constant.

The fixed points are called the **foci** of the ellipse. The midpoint of a line through the foci is called the **center** of the ellipse. We use the center of the ellipse to locate the ellipse in the plane, as we used the vertex to locate the parabola.

Equation of the Ellipse

To get the equation of an ellipse, we consider an ellipse with foci located on the x-axis so that the origin is midway between them, as in Figure 11.7. Let the foci be the points $F(c, 0)$ and $F'(-c, 0)$ and let the sum of the distances from a point $P(x, y)$ of the ellipse to the foci be $2a$, where $a > c$. We have

$$\overline{PF} + \overline{PF'} = 2a$$

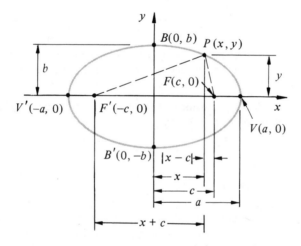

■ **Figure 11.7**

From Figure 11.7,

$$\overline{PF} = \sqrt{(x - c)^2 + y^2} \quad \text{and} \quad \overline{PF'} = \sqrt{(x + c)^2 + y^2}$$

so that

$$\sqrt{(x - c)^2 + y^2} + \sqrt{(x + c)^2 + y^2} = 2a$$

Transposing the first radical and squaring gives

$$(x + c)^2 + y^2 = 4a^2 - 4a\sqrt{(x - c)^2 + y^2} + (x - c)^2 + y^2$$

Then, if like terms are expanded and collected, we get

$$a \sqrt{(x - c)^2 + y^2} = a^2 - cx$$

If the equation is squared again and simplified, we obtain

$$(a^2 - c^2)x^2 + a^2y^2 = a^2(a^2 - c^2)$$

By substituting $b^2 = a^2 - c^2$, we get

$$b^2x^2 + a^2y^2 = a^2b^2$$

Finally, dividing through by the nonzero quantity a^2b^2, we can write the equation of an ellipse with its center at the origin and its foci on the x-axis as follows.

FORMULA

> **Ellipse with Center at the Origin and Foci on the x-axis**
>
> $$\frac{x^2}{a^2} + \frac{y^2}{b^2} = 1 \qquad\qquad\qquad (11.4)$$

Equation (11.4) is symmetric about both axes and the origin. By letting $y = 0$ in Equation (11.4), we see that the x-intercepts of the ellipse are $(a, 0)$ and $(-a, 0)$. The segment of the line through the foci from $(a, 0)$ to $(-a, 0)$ is called the **major axis** of the ellipse. The length of the major axis is $2a$, which is also the value chosen for the sum of the distances PF and PF'. The length a is then known as the **semimajor axis**, and the endpoints of the major axis are called the **vertices** of the ellipse.

The y-intercepts of the ellipse are found to be $(0, b)$ and $(0, -b)$ by letting $x = 0$ in Equation (11.4). The segment of the line perpendicular to the major axis from $(0, b)$ to $(0, -b)$ is called the **minor axis**. Since the length of the minor axis is $2b$, the length of the **semiminor axis** is b. The graph of an ellipse can readily be sketched once the semimajor and semiminor axes are known.

The foci of an ellipse are located by solving the equation $b^2 = a^2 - c^2$ for the focal distance c. Thus, the expression for c is

$$c = \sqrt{a^2 - b^2}$$

A similar derivation gives the next formula.

FORMULA

> **Ellipse with Center at the Origin and Foci on the y-axis**
>
> $$\frac{x^2}{b^2} + \frac{y^2}{a^2} = 1 \qquad\qquad\qquad (11.5)$$

Equation (11.5) is the equation of an ellipse with its center at the origin and its foci on the y-axis. (The letter a is again being used to represent the semimajor axis.)

EXAMPLE 1   ■ **Problem**

Find the equation of an ellipse centered at the origin with foci on the x-axis if the major axis is 10 and the minor axis is 4.

■ **Solution**

The major axis is on the x-axis. The semimajor axis is $a = 5$, and the semiminor axis is $b = 2$. Substituting these values into Equation (11.4) gives the required equation:

$$\frac{x^2}{5^2} + \frac{y^2}{2^2} = 1 \qquad \text{or} \qquad \frac{x^2}{25} + \frac{y^2}{4} = 1$$

■

EXAMPLE 2   ■ **Problem**

Find the equation of the ellipse with vertices at $(0, 5)$ and $(0, -5)$ and foci at $(0, 4)$ and $(0, -4)$.

■ **Solution**

The foci are on the y-axis, the center of the ellipse is at the origin, and the semimajor axis is $a = 5$. To find the semiminor axis, we use the relation $b^2 = a^2 - c^2$. Thus, $b = \sqrt{25 - 16} = \sqrt{9} = 3$. Substituting $a = 5$ and $b = 3$ into Equation (11.5) yields

$$\frac{x^2}{9} + \frac{y^2}{25} = 1$$

The ellipse is sketched in Figure 11.8.

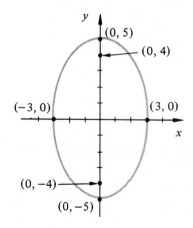

■ **Figure 11.8**

■

EXAMPLE 3   ■ **Problem**

Sketch the graph of the ellipse $4x^2 + 16y^2 = 64$.

■ **Solution**

Dividing by 64, we get

$$\frac{x^2}{16} + \frac{y^2}{4} = 1$$

The major axis lies along the x-axis, since the denominator of the x term in the equation is larger than the denominator of the y term. Consequently, the semimajor axis is $a = 4$, and the semiminor axis is $b = 2$. The vertices are $(4, 0)$ and $(-4, 0)$, and the endpoints of the minor axis are $(0, 2)$ and $(0, -2)$. The ellipse is sketched in Figure 11.9.

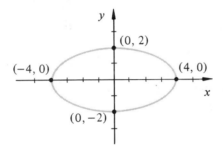

■ **Figure 11.9**   ■

One of the first scientific applications of the ellipse was in astronomy. The astronomer Kepler (about 1600) discovered that the planets moved in elliptical orbits about the sun with the sun at one focus. Artificial satellites also move in elliptical orbits about the earth. Another application is in the design of machines, in which elliptic gears are used to get a slow, powerful movement with a quick return. A third application of the ellipse is found in electricity, in which the magnetic field of a single-phase induction motor is elliptical under normal operating conditions.

Comment: An ellipse has the property that rays emitted from one focus are reflected off its surface through the other focus. This property is used in the construction of whispering galleries—buildings in which a person standing at one focus can converse with a person at the other in a normal voice, even though the rest of the building is noisy.

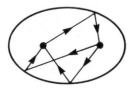

The Circle

We can obtain the equation of a **circle** centered at the origin by noting that the ellipse approaches a circle as the foci move closer together. If the foci F and F' coincide, the ellipse becomes a circle, and $b = a = r$ in Formula (11.4). Thus, the equation of a circle is

$$\frac{x^2}{r^2} + \frac{y^2}{r^2} = 1$$

Multiplying both sides of this equation by r^2, we obtain the following formula.

FORMULA

> **Circle with Center at the Origin and Radius r**
>
> $$x^2 + y^2 = r^2 \tag{11.6}$$

Equation (11.6) is the standard form of the equation of a circle with center at the origin and radius equal to r.

EXAMPLE 4   ▪ **Problem**

Find the equation of the circle centered at the origin and with a radius of 4.

▪ **Solution**

Substituting $r = 4$ into Formula (11.6), we get

$$x^2 + y^2 = 16$$

as the desired equation.   ▪

EXERCISES FOR SECTION 11.2

In Exercises 1–10, discuss each of the given equations, and sketch their graphs.

1. $5x^2 + y^2 = 25$

2. $4x^2 + 9y^2 = 36$

3. $16x^2 + 4y^2 = 16$

4. $3x^2 + 9y^2 = 27$

5. $x^2 + y^2 = 9$

6. $25x^2 + 4y^2 = 100$

7. $2x^2 + 3y^2 - 24 = 0$

8. $5x^2 + 20y^2 = 20$

9. $9x^2 + 4y^2 = 4$

10. $4x^2 + y^2 = 25$

In Exercises 11–20, find the equation of the ellipses having the given properties. Sketch each curve.

11. Vertices at $(\pm 4, 0)$; minor axis 6.

12. Vertices at $(0, \pm 1)$; minor axis 1.

13. Vertices at $(0, \pm 5)$; semiminor axis $\frac{3}{2}$.

14. Vertices at $(\pm 6, 0)$; semiminor axis 2.

15. Major axis 10; foci at $(\pm 4, 0)$.

16. Major axis 10; foci at $(0, \pm 3)$.

17. Foci at $(\pm 1, 0)$; semimajor axis 4.

18. Vertices at $(0, \pm 7)$; foci at $(0, \pm\sqrt{28})$.

19. Vertices at $(\pm\frac{5}{2}, 0)$; one point of the curve at $(1, 1)$.

20. Vertices at $(\pm 3, 0)$; one point of the curve at $(\sqrt{3}, 2)$.

In Exercises 21–24, solve the given system of equations graphically.

21. $\dfrac{x^2}{4} + y^2 = 4$

$2y + 3x = 0$

22. $\dfrac{x^2}{9} + \dfrac{y^2}{9} = 1$

$y = e^{x+2}$

23. $y = x^2$

$x^2 + 4y^2 = 4$

24. $y^2 - 12x = 0$

$y^2 + 9x^2 = 9$

 25. An elliptical cam with a horizontal major axis of 10 in. and a minor axis of 3 in. is to be machined by a numerically controlled vertical mill. Find the equation of the ellipse to be used in programming the control device.

 26. An elliptical cam having the equation $9x^2 + y^2 = 81$ revolves against a pushrod. See the accompanying figure. What is the maximum travel of the pushrod?

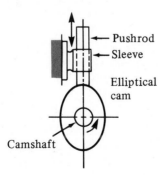

Pushrod

Sleeve

Elliptical cam

Camshaft

 27. The magnetic field curves of a single-phase induction motor are given by the set of ellipses $x^2 + 4y^2 = c^2$. Sketch some of the curves.

28. A *ripple tank* is a tank of water in the form of an ellipse. When water is disturbed at one focus, ripples radiate outward, and eventually, a drop of water spurts up at the other focus. Suppose such a tank is of the shape given by the equation $3.1x^2 + 4.5y^2 = 15.6$. Determine where to poke your finger into the water and where the water will spurt up.

29. The elliptical orbit of the earth is very nearly circular. In fact, it is much like the ellipse $x^2 + (y/1.1)^2 = 1$. Make a sketch of this ellipse.

■■■■ 11.3 THE HYPERBOLA

The final conic that we consider is the **hyperbola**.

DEFINITION

> **Hyperbola**   A *hyperbola* is the set of all points in a plane the difference of whose distances from two fixed points in the plane is constant.

Figure 11.10 is a hyperbola with foci at $F(c, 0)$ and $F'(-c, 0)$. The origin is at the midpoint between the foci, which corresponds to the **center** of the hyperbola. The points $V(a, 0)$ and $V'(-a, 0)$ are called the **vertices**, and the segment $\overline{VV'}$ is called the **transverse axis** of the hyperbola. The length of the transverse axis is $2a$. The segment $\overline{BB'}$, which is perpendicular to the transverse axis at the center of the hyperbola, is called the **conjugate axis** and has length $2b$.

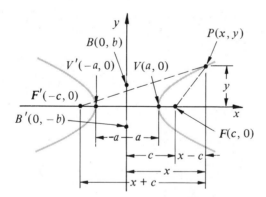

■ **Figure 11.10**

To find the algebraic equation of a hyperbola, we consider a point $P(x, y)$ on the hyperbola. Then, by definition,

$$\overline{F'P} - \overline{FP} = 2a$$

which, in turn, can be written as

$$\sqrt{(x + c)^2 + y^2} - \sqrt{(x - c)^2 + y^2} = 2a$$

Using the same procedure here that we used for the ellipse, we can simplify this equation to

$$\frac{x^2}{a^2} - \frac{y^2}{c^2 - a^2} = 1$$

Or if we let $b^2 = c^2 - a^2$, the equation of the hyperbola is as follows:

FORMULA

Hyperbola with Center at the Origin and Foci on the x-axis

$$\frac{x^2}{a^2} - \frac{y^2}{b^2} = 1 \qquad\qquad\qquad \textbf{(11.7)}$$

Like the ellipse, this hyperbola is symmetric about both axes and the origin. Letting $y = 0$, the x-intercepts of Equation (11.7) are $x = \pm a$. We can get additional information about the shape of the hyperbola by solving Equation (11.7) for y:

$$y = \pm \frac{b}{a} \sqrt{x^2 - a^2}$$

This equation shows that the curve does not exist for $x^2 < a^2$. Consequently, the hyperbola consists of two separate curves, or *branches*—one to the right of $x = a$ and a similar one to the left of $x = -a$.

The shape of the hyperbola is constrained by two straight lines called the **asymptotes** of the hyperbola. The asymptotes of a hyperbola are the extended diagonals of the rectangle formed by drawing lines parallel to the coordinate axes through the endpoints of both the transverse axis and the conjugate axis. Referring to Figure 11.11, note that the slopes of the diagonals of this rectangle are

$$m = \pm \frac{b}{a}$$

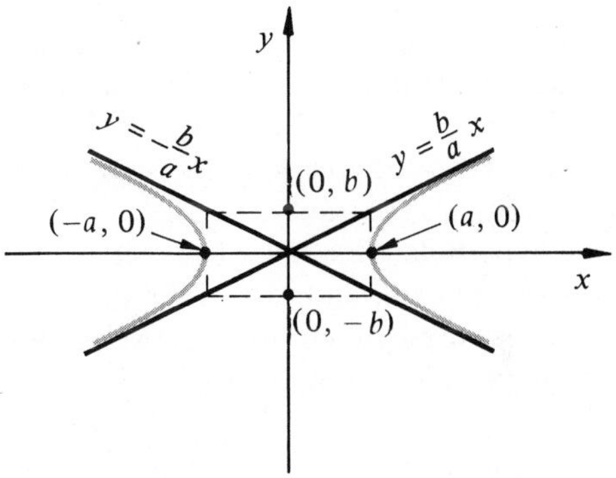

■ **Figure 11.11**

Therefore, the asymptotes are given by the lines

$$y = \pm \frac{b}{a}x$$

Now, we want to show that these lines are asymptotes of the hyperbola. That is, we want to show that the hyperbola approaches the lines as x increases without bound. Solving Equation (11.7) for y^2, we can write it in the form

$$y^2 = \frac{b^2 x^2}{a^2}\left(1 - \frac{a^2}{x^2}\right)$$

If we take the square root of both sides, then

$$y = \pm \frac{b}{a}x\sqrt{1 - \frac{a^2}{x^2}} \tag{11.8}$$

Now, consider the value of the right member as x becomes large. The quantity a^2/x^2 becomes small. Therefore,

$$\pm \frac{b}{a}x\sqrt{1 - \frac{a^2}{x^2}}$$

approaches $\pm(b/a)x$, which means that the hyperbola $(x^2/a^2) - (y^2/b^2) = 1$ is asymptotic to the lines $y = \pm(b/a)x$.

If we begin with the foci of the hyperbola on the y-axis and the center at the origin, the standard form of the equation of a hyperbola is as follows:

FORMULA

Hyperbola with Center at the Origin and Foci on the y-axis

$$\frac{y^2}{a^2} - \frac{x^2}{b^2} = 1 \tag{11.9}$$

The vertices of this hyperbola are on the y-axis. *The positive term always indicates the direction of the transverse axis.* For the hyperbola, like the ellipse, standard form demands that the coefficients of x^2 and y^2 be in the denominator and that the number on the right side be 1. For the hyperbola, the sign of the term, *not* the magnitude of the denominator, determines the transverse axis.

To sketch the hyperbola, first draw the rectangle through the extremities of the transverse and conjugate axes, and extend the diagonals of the rectangle. Then, draw the hyperbola so that it passes through the vertex and comes closer to the extended diagonals as x moves away from the origin.

EXAMPLE 1 ▪ **Problem**

Discuss and sketch the graph of $4x^2 - y^2 = 16$.

▪ **Solution**

Dividing by 16, we have

$$\frac{x^2}{4} - \frac{y^2}{16} = 1$$

which is the equation of a hyperbola with center at the origin and foci on the x-axis. It has vertices at $(\pm 2, 0)$ and its conjugate axis extends from $(0, 4)$ to $(0, -4)$. The foci are found from the equation $b^2 = c^2 - a^2$. Thus,

$$c = \sqrt{a^2 + b^2} = \sqrt{4 + 16} = \sqrt{20} = 2\sqrt{5}$$

and the foci are located at $(\pm 2\sqrt{5}, 0)$. We plot these points and draw the rectangle and its extended diagonals to get the hyperbola shown in Figure 11.12.

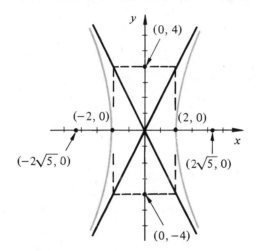

▪ **Figure 11.12**

EXAMPLE 2 ▪ **Problem**

Discuss and sketch the graph of $\dfrac{y^2}{13} - \dfrac{x^2}{9} = 1$.

▪ **Solution**

This hyperbola has a vertical transverse axis with vertices located at $(0, \sqrt{13})$ and $(0, -\sqrt{13})$. The extremes of the conjugate axis are $(3, 0)$ and $(-3, 0)$. The foci are at $(0, \sqrt{22})$ and $(0, -\sqrt{22})$. The hyperbola is shown in Figure 11.13.

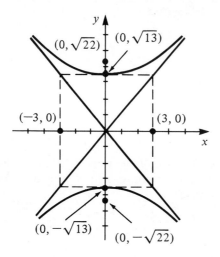

■ **Figure 11.13**   ■

EXAMPLE 3   ■ **Problem**

Determine the equation of the hyperbola centered at the origin with foci at $(\pm 6, 0)$ and a transverse axis 8 units long.

■ **Solution**

Here, $a = 4$ and $c = 6$. Since $c^2 = a^2 + b^2$, we have $b^2 = c^2 - a^2 = 36 - 16 = 20$. Substituting $a^2 = 16$ and $b^2 = 20$ into Equation (11.7) gives

$$\frac{x^2}{16} - \frac{y^2}{20} = 1$$

■

EXERCISES FOR SECTION 11.3

In Exercises 1–10, discuss the properties of the graph of each of the given equations, and then sketch the graph.

1. $x^2 - y^2 = 16$
2. $y^2 - x^2 = 9$
3. $4x^2 - 9y^2 = 36$
4. $9x^2 - y^2 = 9$
5. $4y^2 - 25x^2 = 100$
6. $3x^2 - 3y^2 = 9$
7. $4x^2 - 16y^2 = 25$
8. $4y^2 - x^2 = 9$
9. $y^2 + 1 = x^2$
10. $x^2 - 25 = 5y^2$

In Exercises 11–20, find the equation of the hyperbolas having the given properties. Sketch each curve.

11. Vertices at $(\pm 4, 0)$; foci at $(\pm 5, 0)$.

12. Vertices at $(0, \pm 3)$; foci at $(0, \pm 5)$.

13. Conjugate axis 4; vertices at $(0, \pm 1)$.

14. Conjugate axis 1; vertices at $(\pm 4, 0)$.

15. Transverse axis 6; foci at $(\pm \frac{7}{2}, 0)$.

16. Transverse axis 3; foci at $(\pm 2, 0)$.

17. Vertices at $(0, \pm 4)$; asymptotes $y = \pm (\frac{1}{2})x$.

18. Vertices at $(\pm 3, 0)$; asymptotes $y = \pm 2x$.

19. Vertices at $(0, \pm 3)$; one point of the curve at $(2, 7)$.

20. Vertices at $(\pm 3, 0)$; one point of the curve at $(7, 2)$.

21. In the study of electrostatic potential with particular boundary conditions, the equipotential curves are found to be

$$\frac{x^2}{\sin^2 c} - \frac{y^2}{\cos^2 c} = 1$$

Show that every member of this family has foci at $(-1, 0)$ and $(1, 0)$.

22. Curves of the form $xy = c$ are hyperbolas, but their foci lie along the lines $y = \pm x$ instead of along the coordinate axes. Make a sketch of the hyperbola $xy = 2$.

23. Any two variables x and y that are related by the equation $xy = c$ are said to *vary inversely* with one another. Sketch the inverse variation $xy = -1$, and note the hyperbolic shape.

▬▬ 11.4 TRANSLATION OF AXES

The equation of a circle centered at the origin has the form $x^2 + y^2 = r^2$. Sometimes, a circle is centered at some other point in the plane besides the origin. We can write the equation of this circle with respect to a new pair of axes that are parallel to the original axes. Changing from one pair of axes to another in this way is called a **translation of axes**. We will investigate how translating axes affects the form of the standard equations of the conic sections.

Consider a point $P(x, y)$ in the xy-coordinate plane. Under a translation of axes, the coordinates of P are changed to agree with a new pair of axes. Construct an x'-axis parallel to the x-axis and a y'-axis parallel to the y-axis, as shown in Figure 11.14. Let the intersection of the new axes be at the point (h, k) in the original coordinate system. Points referring to the x'-axis and y'-axis will be designated by a prime $(')$ notation. Thus, the coordinates of the point P, with respect to the translated axes, are (x', y'). To transform (x, y) into (x', y'), we must know the relation between the new and old axes. From Figure 11.14, we see that the two coordinate systems are related by the equations

$$x = x' + h \quad \text{and} \quad y = y' + k \tag{11.10}$$

or

$$x' = x - h \quad \text{and} \quad y' = y - k \tag{11.11}$$

These equations are called the **equations for the translation of axes**.

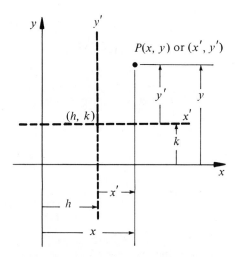

■ **Figure 11.14**

Now, let us consider a circle with its center at (h, k) in the xy-coordinate plane, as shown in Figure 11.15. If (h, k) is taken to be the intersection of the $x'y'$-axes, the equation of the circle that refers to these axes can be written as

$$(x')^2 + (y')^2 = r^2$$

To express this equation in xy-coordinates, we apply Equation (11.11). Thus,

$$(x - h)^2 + (y - k)^2 = r^2 \tag{11.12}$$

is the standard equation of a circle of radius r centered at a point (h, k).

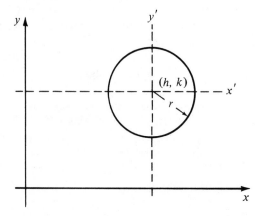

■ **Figure 11.15**

The procedure for getting the standard form of the equation of a circle centered at a point (h, k) can be extended to the other conic sections when they have been displaced from the origin.

The equation of the parabola that has its vertex at the origin and a vertical axis is $x^2 = 4ay$. Hence, the equation of the same parabola with its vertex at (h, k) can be written in terms of the $x'y'$-coordinate plane as

$$(x')^2 = 4ay'$$

or, by the translating Equation (11.11), as

$$(x - h)^2 = 4a(y - k) \tag{11.13}$$

Similarly, the parabola with a horizontal axis and vertex at (h, k) is given by

$$(y - k)^2 = 4a(x - h) \tag{11.14}$$

By a similar procedure, the standard equation of an ellipse centered at a point (h, k) with a horizontal major axis is

$$\frac{(x - h)^2}{a^2} + \frac{(y - k)^2}{b^2} = 1 \tag{11.15}$$

If the major axis is vertical, the equation is

$$\frac{(x - h)^2}{b^2} + \frac{(y - k)^2}{a^2} = 1 \tag{11.16}$$

Finally, the standard equation of the hyperbola centered at a point (h, k) with a horizontal transverse axis is

$$\frac{(x - h)^2}{a^2} - \frac{(y - k)^2}{b^2} = 1 \tag{11.17}$$

If the transverse axis is vertical, the equation is

$$\frac{(y - k)^2}{a^2} - \frac{(x - h)^2}{b^2} = 1 \tag{11.18}$$

EXAMPLE 1   ■ **Problem**

Write the equation of the ellipse centered at $(2, -3)$ with a horizontal axis of 10 units and a vertical axis of 4 units. See Figure 11.16.

■ **Solution**

Since the longer of the two axes is the horizontal axis, use Equation (11.15) with $a = 5$ and $b = 2$. Also, $h = 2$ and $k = -3$. Making these substitutions, we have, as the equation of the ellipse,

$$\frac{(x - 2)^2}{25} + \frac{(y + 3)^2}{4} = 1$$

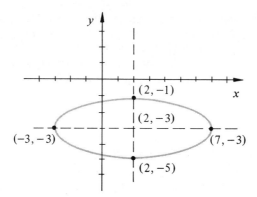

■ **Figure 11.16**   ■

EXAMPLE 2   ■ **Problem**

Write the equation of the parabola whose directrix is the line $y = -2$ and whose vertex is at (3, 1). See Figure 11.17.

■ **Solution**

The parabola has a vertical axis and opens upward, since the directrix is horizontal and lies below the vertex. The vertex lies midway between the focus and the directrix, so $a = 3$. Let $h = 3$, $k = 1$, and $a = 3$ in Equation (11.13). Then, the equation of the parabola is

$$(x - 3)^2 = 12(y - 1)$$

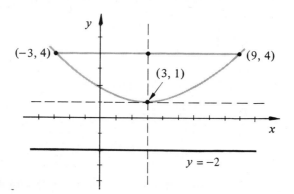

■ **Figure 11.17**   ■

EXAMPLE 3   ■ **Problem**

Discuss and sketch the graph of the hyperbola

$$\frac{(x + 5)^2}{16} - \frac{(y - 2)^2}{9} = 1$$

■ Solution

The hyperbola is centered at $(-5, 2)$. By Equation (11.17), it has a horizontal transverse axis with vertices at $(-9, 2)$ and $(-1, 2)$. The endpoints of the conjugate axis are at $(-5, 5)$ and $(-5, -1)$. Also, the foci are $(-10, 2)$ and $(0, 2)$ since $c = \sqrt{9 + 16} = 5$. The graph appears in Figure 11.18.

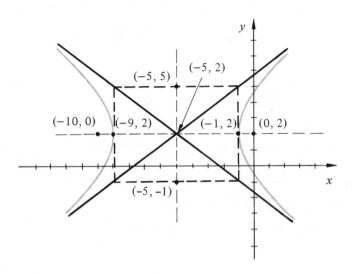

■ **Figure 11.18**   ■

         EXERCISES FOR SECTION 11.4

In Exercises 1–6, write the equations of the *parabolas* having the given properties. Sketch each graph.

1. Vertex at $(3, 1)$; focus at $(5, 1)$.

2. Vertex at $(-2, 3)$; focus at $(-2, 0)$.

3. Directrix $y = 2$; vertex at $(1, -1)$.

4. Directrix $x = -1$; vertex at $(0, 4)$.

5. Endpoints of right chord at $(2, 4)$ and $(2, 0)$; opening to the right.

6. Endpoints of right chord at $(-1, -1)$ and $(5, -1)$; opening upward.

In Exercises 7–12, write the equations of the *ellipses* having the given properties. Sketch each graph.

7. Major axis 8; foci at $(5, 1)$ and $(-1, 1)$.

8. Minor axis 6; vertices at $(2, -1)$ and $(10, -1)$.

9. Minor axis 2; vertices at $(\frac{1}{2}, 0)$ and $(\frac{1}{2}, -8)$.

10. Semimajor axis $\frac{3}{2}$; foci at $(1, 1)$ and $(1, -1)$.

11. Vertices at $(-6, 3)$ and $(-2, 3)$; foci at $(-5, 3)$ and $(-3, 3)$.

12. Center at $(1, -3)$; major axis 10; minor axis 6; vertical axis.

In Exercises 13–18, write the equations of the *hyperbolas* having the given properties. Sketch each graph.

13. Center at $(-1, 2)$; transverse axis 7; conjugate axis 8; vertical axis.

14. Center at $(3, 0)$; transverse axis 6; conjugate axis 2; horizontal axis.

15. Vertices at $(5, 1)$ and $(-1, 1)$; foci at $(6, 1)$ and $(-2, 1)$.

16. Vertices at $(2, \pm 4)$; conjugate axis 2.

17. ·Vertices at $(-4, -2)$ and $(0, -2)$; asymptotes $m = \pm\frac{1}{2}$.

18. Vertices at $(3, 3)$ and $(5, 3)$; asymptotes $m = \pm 3$.

Write the equation of the family of curves indicated in Exercises 19–24.

19. Circles with center on the x-axis.

20. Parabolas with vertical axis and vertex on the x-axis.

21. Parabolas with vertex and focus on the x-axis.

22. Ellipses with center on the y-axis and horizontal major axis.

23. Circles passing through the origin with center on the x-axis.

24. Circles tangent to the x-axis.

25. The path of a projectile is given by $y = 20x - \frac{1}{10}x^2$, where y is the vertical distance and x is the horizontal distance away from the initial point. Locate the vertex, and sketch the path.

26. Locate the center and foci of an elliptical satellite orbit whose equation is given by $(x - 0.1)^2 + 1.1y^2 = 100$.

27. Compute the (hyperbolic) lines of magnetic force for a magnetic field whose center is at $(2, 3)$ with one vertex at $(0, 3)$ and one focus at $(-1, 3)$.

11.5 THE GENERAL SECOND-DEGREE EQUATION

The **general second-degree equation** is of the form

$$Ax^2 + Bxy + Cy^2 + Dx + Ey + F = 0 \qquad (11.19)$$

where A, B, C, D, E, and F are constants and either A, B, or C is nonzero. Each conic described in Sections 11.1 through 11.4 can be expressed in the form of Equation (11.19) with $B = 0$, as can be seen by expanding the standard form of each conic. If we assume that $B = 0$, the following statements are true.

PROPERTIES

> **General Second-Degree Equation**
>
> 1. If $A = C = 1$, Equation (11.19) represents a circle.
> 2. If $A \neq C$ and A and C have the same numerical sign, Equation (11.19) represents an ellipse.
> 3. If A and C have different numerical signs, Equation (11.19) represents a hyperbola.
> 4. If A or $C = 0$ (but not both), Equation (11.19) represents a parabola.
> 5. Special cases such as a single point or no graph may result.

If $B = 0$, the general form of a conic can be reduced to one of the standard forms by completing the square on x and y. Two examples of this technique follow.

Comment: Before proceeding to the examples, we will review **completing the square** on a quadratic function of the form $ax^2 + bx + c$. Note the expression $x^2 + 6x + 9$ is said to be a *perfect square* since it factors to $(x + 3)^2$. The process of finding the constant c that makes $ax^2 + bx + c$ a perfect square is called completing the square. The procedure is based on the rule

$$(x + k)^2 = x^2 + 2kx + k^2$$

in which the constant term k^2 is exactly the square of one-half the coefficient of x. Thus, $x^2 - 8x + c$ will be a perfect square if $c = [\tfrac{1}{2}(8)]^2 = 16$.

EXAMPLE 1 ■ **Problem**

Discuss and sketch the graph of $x^2 - 4y^2 + 6x + 24y - 43 = 0$.

■ **Solution**

This equation is the equation of a hyperbola since the coefficients of the x^2 and y^2 terms have unlike signs. To sketch the hyperbola, we rewrite the given equation in standard form by rearranging the terms and completing the square on the x terms and the y terms. Thus, $x^2 - 4y^2 + 6x + 24y - 43 = 0$ can be written as

$$(x^2 + 6x \quad) - 4(y^2 - 6y \quad) = 43$$

Completing the square on each variable, we get

$$(x^2 + 6x + 9) - 4(y^2 - 6y + 9) = 43 + 9 - 36$$

$$(x + 3)^2 - 4(y - 3)^2 = 16$$

$$\frac{(x + 3)^2}{16} - \frac{(y - 3)^2}{4} = 1$$

The center of the hyperbola is the point $(-3, 3)$. The transverse axis is horizontal, with vertices at $(1, 3)$ and $(-7, 3)$. The endpoints of the conjugate axis are at $(-3, 5)$ and $(-3, 1)$. Finally, the foci are at $(-3 + 2\sqrt{5}, 3)$ and $(-3 - 2\sqrt{5}, 3)$ since $c = \sqrt{16 + 4} = 2\sqrt{5}$. The graph appears in Figure 11.19.

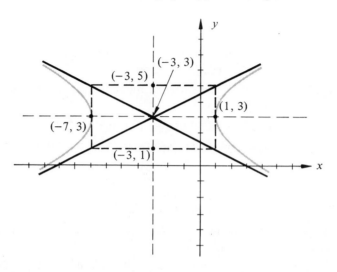

■ **Figure 11.19**   ■

EXAMPLE 2   ■ **Problem**

Discuss and sketch the graph of $2y^2 + 3x - 8y + 9 = 0$.

■ **Solution**

If the square on the y-variable is completed, this equation can be rewritten in the form of Equation (11.14). Thus,

$$2y^2 + 3x - 8y + 9 = 0$$
$$2(y^2 - 4y) = -3x - 9$$
$$2(y^2 - 4y + 4) = -3x - 9 + 8$$
$$2(y - 2)^2 = -3x - 1$$
$$2(y - 2)^2 = -3(x + \tfrac{1}{3})$$
$$(y - 2)^2 = -\tfrac{3}{2}(x + \tfrac{1}{3})$$

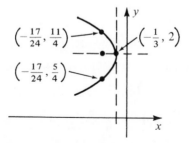

■ **Figure 11.20**   ■

This equation is the standard form of the equation of a parabola with horizontal axis and vertex at $(-\frac{1}{3}, 2)$. Since $4a = -\frac{3}{2}$, $a = -\frac{3}{8}$. Therefore, the focus is at $(-\frac{17}{24}, 2)$, and the endpoints of the right chord are at $(-\frac{17}{24}, \frac{11}{4})$ and $(-\frac{17}{24}, \frac{5}{4})$. The parabola, which opens to the left, is shown in Figure 11.20.

EXERCISES FOR SECTION 11.5

Transform each of the equations in Exercises 1–20 into standard form, and sketch their graphs.

1. $x^2 + y^2 + 4x + 6y + 4 = 0$ 2. $x^2 - 2x - 8y + 25 = 0$

3. $9x^2 + 4y^2 + 18x + 8y - 23 = 0$ 4. $2x^2 + 2y^2 - 4x - 16 = 0$

5. $x^2 - y^2 - 4x - 21 = 0$ 6. $y^2 + 4y + 6x - 8 = 0$

7. $x^2 - 6x - 3y = 0$ 8. $x^2 + 4y^2 + 8x = 0$

9. $2y^2 - 2y + x - 1 = 0$ 10. $3x^2 + 4y^2 - 18x + 8y + 19 = 0$

11. $4x^2 - y^2 + 8x + 2y - 1 = 0$ 12. $y^2 - 2x - 4y + 10 = 0$

13. $9x^2 + 4y^2 - 18x + 16y - 11 = 0$ 14. $y^2 - 25x^2 + 50x - 50 = 0$

15. $x^2 - 2y^2 + 2y = 0$ 16. $y^2 - y - \frac{1}{2}x + \frac{1}{4} = 0$

17. $x(x + 4) = y^2 + 3$ 18. $x^2 + y(4 + 2y) = 0$

19. $y = x^2 + 5x + 7$ 20. $y = 2x^2 + 10x$

21. In the analysis of the heat distribution on a hot plate, the isotherms are found to be included in the family of curves $c^2x^2 - 8c^2x + 4y^2 + 12c^2 = 0$. What type of curve is each of the isotherms? Sketch a few of them for $y > 0$.

22. A space vehicle is scheduled to take off from earth and follow the curved path given by the conic $y^2 - 4y - x^2 + 2x + 7 = 0$. Assuming that earth is represented by the vertex in the part of the plane to the right of center of the conic, find the coordinates of earth. Will the vehicle ever return to the neighborhood of earth?

23. A garden is dug in the shape of a right triangle so that the hypotenuse is always 5 ft larger than one of the legs. Find the equation relating the legs. What kind of curve does it represent?

24. The outline of a lens of a camera has the approximate equation $2x^2 - 3y - 4x + 2 = 0$. Where is the focus? What kind of curve is the lens?

REVIEW EXERCISES FOR CHAPTER 11

Sketch the graph of each of Exercises 1–10. Identify all important parts.

1. $x^2 = -3y$ 2. $x = -y^2 + 2y$

3. $x^2 = -y^2 - 4y + 5$ 4. $x^2 - 3y^2 - 4 = 0$

5. $(x + 1)^2 - 2y = y^2$ 6. $2x^2 - 3y^2 - x = 4$

7. $x^2 = y^2 - 100$

8. $y = 2x^2 + 3$

9. $y^2 + 2x^2 + 6x = 7$

10. $y^2 - y = x^2$

11. Write the equation of the parabola with vertex at (2, 0) and focus at the origin.

12. Write the equation of the parabola with vertex at (7, −2) and focus at (10, −2).

13. Write the equation of the circle centered at (−1, 2) with radius equal to 8.

14. Write the equation of the circle with center at (1, 3) that passes through (6, 4).

15. Write the equation of the ellipse with center at the origin, focus at (2, 0), and major axis equal to 8.

16. Write the equation of the hyperbola with center at (2, −2), the ends of the transverse axis at (2, 0) and (2, −4), and one focus at (2, 1).

17. Write the equation of the hyperbola with vertices at (7, 4) and (−1, 4) and whose asymptotes have slope ±2.

18. Show that each ellipse in the family of ellipses

$$\frac{x^2}{\sin^2 c} + \frac{y^2}{\cos^2 c} = 1$$

has its foci at the same two points. What are the foci? Sketch a few ellipses from the family. These ellipses are said to be *confocal*.

19. Whispering galleries have cross sections that are ellipses. In such galleries, a whisper at one focus will be heard distinctly at the other focus. Find the two foci in a gallery whose cross section is a semiellipse with a height of 12 ft and a length of 30 ft.

20. Certain navigational systems use the set of hyperbolas

$$\frac{x^2}{e^c - e^{-c}} - \frac{y^2}{e^c + e^{-c}} = 2$$

as references in a coordinate system. Show that each hyperbola pair has the same foci. Sketch a few hyperbolas from this family.

21. Solve the equations $2x^2 + x + y^2 + 2y = 1$ and $y + 2x = 1$ graphically.

22. Show graphically that $x^2 - 4y^2 = 4$ and $y = 3x + 2$ do not have any common real solutions.

23. Explain how each of the conic sections may be obtained from a right circular cone.

CHAPTER 12

Calculus of Functions of Two Variables

To this point our discussion of differential and integral calculus has been restricted to functions of one variable, that is, functions of the form $y = f(x)$. In this chapter, we extend these two basic concepts to functions of two variables. We find applications that involve the calculus of two variables in areas of study such as fluid mechanics and the transfer of heat.

12.1 FUNCTIONS OF TWO VARIABLES

The concept of a function can be extended to two or more independent variables. For example, the area of a triangle is $A = \frac{1}{2}bh$, where b is the length of the base and h is the altitude. We say that the area is a function of both the base and the altitude of the triangle and write $A = f(b, h)$ to indicate the general nature of this dependency.

EXAMPLE 1 The volume of an ideal gas is a function of both temperature T and applied pressure P. To indicate this general functional dependence, we write $V = f(P, T)$. The specific formula is known to be $V = kT/P$, where k is a constant depending on the units being used. ∎

In general, we let the letter z represent the variable dependent on the two variables x and y and write $z = f(x, y)$. As in the case of a function of one real variable, a unique value of z must be given for each particular pair of (x, y) values.

EXAMPLE 2 The function given by the formula $f(x, y) = x^2 - y + 5$ is tabulated here for various values of x and y. In each case, the value of $f(x, y)$ is found by merely substituting the given values of x and y into the formula. For example, when $x = 2$ and $y = 3$ $f(2, 3) = 2^2 - 3 + 5 = 6$. Additional values of f are shown in the table.

x	0	1	2	-1	-2	0	1	2	2
y	0	0	0	1	2	1	2	-1	3
$f(x, y)$	5	6	9	5	7	4	4	10	6

■

EXAMPLE 3   ■ **Problem**

Let $f(x, y) = 2x^2 - y^2 + xy + 4y$. Find $f(t^2, y), f(y, x)$, and $f(2 + h, 3)$.

■ **Solution**

By substitution, we have

$$f(t^2, y) = 2(t^2)^2 - y^2 + t^2 y + 4y = 2t^4 - y^2 + t^2 y + 4y$$

$$f(y, x) = 2y^2 - x^2 + xy + 4x$$

$$f(2 + h, 3) = 2(2 + h)^2 - 3^2 + (2 + h)(3) + 4(3)$$
$$= 2(4 + 4h + h^2) - 9 + 6 + 3h + 12 = 2h^2 + 11h + 17 \quad ■$$

To graph functions of two real variables, we let $z = f(x, y)$. Thus, a third axis is needed to represent the z variable. This axis is taken perpendicular to the xy-plane, as shown in Figure 12.1. The particular orientation of x, y, and z, as shown in the figure, is a **right-handed, three-dimensional, rectangular coordinate system.**

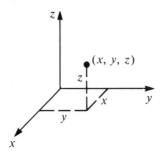

■ **Figure 12.1**

The y- and z-coordinate axes are regarded as being in the plane of the paper, and the x-axis is regarded as perpendicular to this plane with an orientation as shown in the figure. The planes formed by pairing two of the three coordinate axes are designated, respectively, as the xy-plane, the xz-plane, and the yz-plane. Over the entire xy-plane, $z = 0$; over the xz-plane, $y = 0$; and over the yz-plane, $x = 0$. Planes parallel to these coordinate planes are given by x, y, or z equal to a constant. A point is located in space by giving values to x, y, and z.

The three-dimensional rectangular coordinate system divides space into eight octants that are not ordinarily numbered except for the first octant in which x, y, and z are all considered positive. Because sketches of sets of points in three dimensions are often difficult to visualize, we are usually content with a first-octant sketch.

In making sketches of functions of the form $z = f(x, y)$, you will quickly realize that plotting the points (x, y, z) does not give even a remote idea of the shape of the graph because graphs of functions of two real variables ordinarily turn out to be surfaces. We describe a method of making a first-octant sketch of the graph of $z = f(x, y)$.

PROCEDURE

Sketching $z = f(x, y)$ by Parallel-Plane Sections

1. Let $x = 0$ to obtain the points in the yz-plane lying on the surface. These points define the **trace** of the surface on the yz-plane. Likewise, let $z = 0$ to locate the trace in the xy-plane, and let $y = 0$ to locate the trace in the xz-plane.

2. By letting $z = c$, $y = c$, or $x = c$, obtain a family of curves formed by the intersection of the given surface and planes parallel to one of the coordinate planes. These curves will give a rough idea of the shape of the surface.

EXAMPLE 4 ■ **Problem**

Make a first-octant sketch of the function $f(x, y) = x^2 + 2y^2$.

■ **Solution**

The traces in the coordinate planes are as follows: $z = x^2$ in the xz-plane (a parabola); $z = 2y^2$ in the yz-plane (a parabola); and $x^2 + 2y^2 = 0$ in the xy-plane (the point $x = y = 0$). By holding z fixed, say $z = c$, the family of curves $x^2 + 2y^2 = c$ is obtained. Each is an ellipse in a plane parallel to the xy-plane. By sketching a few of these ellipses, we obtain the outline of the surface shown in Figure 12.2.

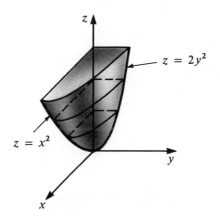

■ **Figure 12.2**

EXAMPLE 5 ▪ **Problem**

Make a first-octant sketch of the function $f(x, y)$ defined implicitly by the expression $2x + y + 3z = 6$.

▪ **Solution**

The traces in the coordinate planes are the lines $2x + y = 6$, $y + 3z = 6$, and $2x + 3z = 6$. By letting x be a constant, we obtain the family of lines $y + 3z = 6 - 2c$. This surface, which is a plane, is shown in Figure 12.3.

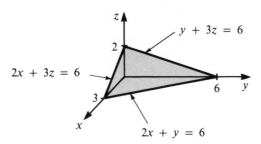

▪ **Figure 12.3**

EXAMPLE 6 ▪ **Problem**

Make a first-octant sketch of $f(x, y) = x^2$.

▪ **Solution**

By letting $x = 0$, we obtain the trace in the yz-plane, $z = 0$. This trace is also the trace in the xy-plane. Since y is not a part of the defining rule, the curve $z = x^2$ is obtained for *each* value of y, including the trace in the xz-plane. See Figure 12.4.

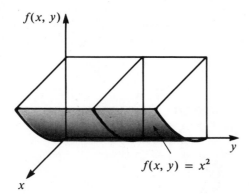

▪ **Figure 12.4**

EXAMPLE 7 ▪ **Problem**

The volume of a perfect gas varies with temperature and pressure according to

$$V = \frac{kT}{P}$$

Sketch a graph showing this variation.

■ **Solution**

To sketch the graph of this function, we can hold T fixed and allow P to vary. This technique will yield a family of curves, one curve for each value of T that we choose. If we now consider all possible values of T and all possible values of P, the resulting figure will be a surface in space, as shown in Figure 12.5. The volume of the gas for any given pair of values of T and P will be equal to the height of the surface above the TP-plane.

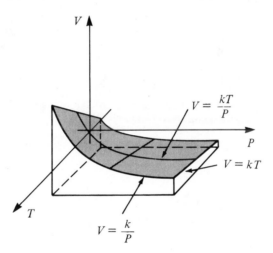

■ **Figure 12.5**

EXERCISES FOR SECTION 12.1

1. For $f(x, y) = 4xy^2 + 2x + y^2$, what is $f(3, 1)$?

2. For $g(x, y) = xy + x^2$, what is $g(0, 2)$?

3. For $h(x, y) = x^2 + 2 \sin xy + y^2$, what is $h(2, \pi)$?

4. For $z(x, y) = x + y + 3x^2 + 2$, what is $z(0, t)$?

5. For $f(x, y) = x^2$, what is $f(t, t)$?

6. For $g(x, y) = x + 3yx + 2$, what is $g(y, y)$?

7. For $h(x, y) = x^2 + y^2$, what is $h(2 + t, 1) - h(2, 1)$?

8. For $z(x, y) = y^3 + x$, what is $z(x^3, y)$?

9. For $f(x, y) = \log xy + 2^{x+y}$, what is $f(1, 1)$?

10. For $g(u, v) = u + v$, what is $g(x^2, y^2)$?

Make a first-octant sketch of the functions $z = f(x, y)$ **in Exercises 11–21.**

11. $z + x = 2$ **12.** $x + 2y + 3z = 1$

13. $x^2 + z^2 = 4$ **14.** $z = \sin x$

15. $z = x^3$ **16.** $z = x^2 + y^2$

17. $z = 4 - y^2$ **18.** $z = 5 - x - y$

19. $z = 4 - 2x - y$ **20.** $y = 4$

21. $z = 4 - x^2 - y^2$

▬▬▬ 12.2 PARTIAL DIFFERENTIATION

The concept of a limit extends in a natural way to functions of two real variables. In this case, each of the two variables may independently approach different limits.

EXAMPLE 1   ▪ **Problem**

Let $f(x, y) = 2x^2 + y + 3$. Find the limit of $f(x, y)$ as $x \to 3$ and $y \to 2$, and find the limit of $f(x, y)$ as $x \to 0$ and $y \to 2$.

▪ **Solution**

$$\lim_{\substack{x \to 3 \\ y \to 2}} f(x, y) = \lim_{\substack{x \to 3 \\ y \to 2}} (2x^2 + y + 3) = 18 + 2 + 3 = 23$$

$$\lim_{\substack{x \to 0 \\ y \to 2}} f(x, y) = \lim_{\substack{x \to 0 \\ y \to 2}} (2x^2 + y + 3) = 0 + 2 + 3 = 5$$ ■

The derivative of a function of two real variables is defined for each variable independently. We define *two* rates of change called *partial derivatives*.

DEFINITION

> **The Partial Derivative**   The *partial derivative* of $f(x, y)$ with respect to x is denoted by $\partial f/\partial x$ and is defined by
>
> $$\frac{\partial f}{\partial x} = \lim_{\Delta x \to 0} \frac{f(x + \Delta x, y) - f(x, y)}{\Delta x}$$
>
> Note that the variable y is held constant in this limit process.

The partial derivative of $f(x, y)$ with respect to y is denoted by $\partial f/\partial y$ and is defined analogously. The two partial derivatives of $f(x, y)$ are also denoted by f_x and f_y.

The process of finding the partial derivative of a function is exactly like that of differentiating a function of one real variable. To find $\partial f/\partial x$, simply treat y as a constant and apply the rules learned previously. All the rules of differentiation (for example, product, quotient, and chain rules) apply. No new formulas need be learned!

EXAMPLE 2   ■ **Problem**

Let $f(x, y) = x^2y + x \sin xy$. Find $\dfrac{\partial f}{\partial x}$ and $\dfrac{\partial f}{\partial y}$.

■ **Solution**

Treating y as a constant and differentiating with respect to x, we obtain

$$\frac{\partial f}{\partial x} = 2xy + \sin xy + xy \cos xy$$

Note that $(\partial/\partial x)(x \sin xy)$ was obtained by using the product rule. To find $\partial f/\partial y$, hold x constant and differentiate with respect to y. Thus,

$$\frac{\partial f}{\partial y} = x^2 + x^2 \cos xy$$ ■

EXAMPLE 3   ■ **Problem**

Let $z = x^2 \ln y$. Find z_x and z_y.

■ **Solution**

Treating y as a constant we have

$$z_x = 2x \ln y$$

Holding x constant, we have

$$z_y = \frac{x^2}{y}$$ ■

EXAMPLE 4   ■ **Problem**

The volume of an ideal gas is $V = kT/P$. Find the rate of change of volume with respect to pressure if the temperature remains constant during the process.

■ **Solution**

Treating T as a constant and differentiating, we obtain

$$\frac{\partial V}{\partial P} = -\frac{kT}{P^2}$$ ■

The partial derivative has a graphical interpretation. See Figure 12.6. Each figure depicts the graph of the function $z = f(x, y)$ but Figure 12.6(a) shows the surface cut by a plane $y = c$, and Figure 12.6(b) shows the surface cut by the plane $x = c$. The partial derivative $\partial z/\partial x$ considers y as a constant, which means that a point P on the surface is constrained to move along the surface where it is cut by the plane $y = c$. Hence, $\partial z/\partial x$ itself is the value of the slope of the tangent line to this curve of intersection. Similarly, $\partial z/\partial y$ treats x as a constant and, thus, is the slope of the tangent line shown in Figure 12.6(b).

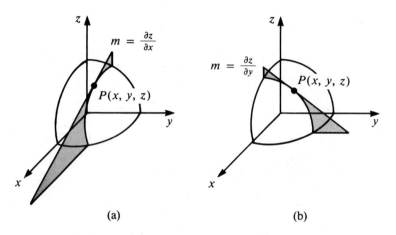

■ **Figure 12.6**   (a)   (b)

We can take partial derivatives of first partial derivatives, which results in new functions called **partial derivatives of higher order.** From a function of two variables, there are four possible second partial derivatives. First,

$$\frac{\partial}{\partial x}\left(\frac{\partial z}{\partial x}\right) = \frac{\partial^2 z}{\partial x^2} = f_{xx}(x, y)$$

which is called the second partial derivative of z with respect to x. Second,

$$\frac{\partial}{\partial y}\left(\frac{\partial z}{\partial x}\right) = \frac{\partial^2 z}{\partial y\,\partial x} = f_{xy}(x, y)$$

which is called a second mixed partial derivative. Third,

$$\frac{\partial}{\partial x}\left(\frac{\partial z}{\partial y}\right) = \frac{\partial^2 z}{\partial x\,\partial y} = f_{yx}(x, y)$$

which is called a second mixed partial derivative. Fourth,

$$\frac{\partial}{\partial y}\left(\frac{\partial z}{\partial y}\right) = \frac{\partial^2 z}{\partial y^2} = f_{yy}(x, y)$$

which is called the second partial derivative of z with respect to y.

EXAMPLE 5   ▪ **Problem**

Find all of the second partial derivatives of $z = f(x, y) = x^3y^2 + y^3 + xy$.

▪ **Solution**

To find the second partial derivatives, we obtain the first partial derivatives:

$$\frac{\partial z}{\partial x} = 3x^2y^2 + y \quad \text{and} \quad \frac{\partial z}{\partial y} = 2x^3y + 3y^2 + x$$

Therefore,

$$\frac{\partial^2 z}{\partial x^2} = 6xy^2 \qquad \frac{\partial^2 z}{\partial y^2} = 2x^3 + 6y$$

$$\frac{\partial^2 z}{\partial y\, \partial x} = 6x^2y + 1 \qquad \frac{\partial^2 z}{\partial x\, \partial y} = 6x^2y + 1$$

▪

EXAMPLE 6   ▪ **Problem**

Find all of the second partial derivatives of $f(x, y) = x^2y^2 + x \sin y$.

▪ **Solution**

The first partial derivatives are

$$f_x(x, y) = 2xy^2 + \sin y \quad \text{and} \quad f_y(x, y) = 2x^2y + x \cos y$$

Hence,

$$f_{xx}(x, y) = 2y^2 \qquad\qquad f_{yy}(x, y) = 2x^2 - x \sin y$$

$$f_{xy}(x, y) = 4xy + \cos y \qquad f_{yx}(x, y) = 4xy + \cos y$$

▪

Notice that in both of these examples, the mixed partial derivatives have identical values; that is, $f_{xy}(x, y) = f_{yx}(x, y)$. These results suggest that the order in which mixed partial derivatives are formed is immaterial to the final answer. This conjecture is true for functions whose successive derivatives are continuous functions of the variables involved. Fortunately, most functions encountered in physical problems satisfy this condition. We are therefore justified in using mixed partial derivatives interchangeably. This result can be extended to higher partial derivatives as long as the number of differentiations with respect to each independent variable is the same. For instance, the following third-order mixed partial derivatives are equal:

$$f_{xxy} = f_{xyx} = f_{yxx} \quad \text{and} \quad f_{xyy} = f_{yxy} = f_{yyx}$$

In fluid mechanics, the volume flow rate associated with streamlines is called the **stream function** and is designated by u. The stream function of an ideal fluid must satisfy the equation

$$\frac{\partial^2 u}{\partial x^2} + \frac{\partial^2 u}{\partial y^2} = 0$$

which is known as **Laplace's equation.**

EXAMPLE 7   ■ **Problem**

Show that $u = e^{2x} \sin 2y$ satisfies Laplace's equation.

■ **Solution**

The required partial derivatives of the given stream function are

$$u_x = 2e^{2x} \sin 2y \qquad u_y = 2e^{2x} \cos 2y$$
$$u_{xx} = 4e^{2x} \sin 2y \qquad u_{yy} = -4e^{2x} \sin 2y$$

Substituting into Laplace's equation, we get

$$\frac{\partial^2 u}{\partial x^2} + \frac{\partial^2 u}{\partial y^2} = 4e^{2x} \sin 2y - 4e^{2x} \sin 2y = 0$$

■

EXERCISES FOR SECTION 12.2

Evaluate the limits in Exercises 1–8. Let $f(x, y) = x^2 + xy - y^3 + 4$.

1. $\lim\limits_{\substack{x \to 2 \\ y \to 1}} f(x, y)$

2. $\lim\limits_{\substack{x \to -1 \\ y \to 2}} f(x, y)$

3. $\lim\limits_{\substack{x \to 4 \\ y \to 0}} f(x, y)$

4. $\lim\limits_{\substack{x \to 0 \\ y \to 0}} f(x, y)$

5. $\lim\limits_{x \to 1} f(x, y)$

6. $\lim\limits_{y \to 0} f(x, y)$

7. $\lim\limits_{x \to 2} f(x, y)$

8. $\lim\limits_{x \to 4} f(x, y)$

Find both of the partial derivatives of the functions in Exercises 9–28.

9. $z(x, y) = x^3 y^3 + x^2$

10. $z(x, y) = y^{1/2} - xy^4$

11. $f(x, t) = x^2 \cos t$

12. $h(s, t) = s^2 t^2 - \sin st$

13. $G(x, y) = e^{xy}$

14. $R(V, T) = V \ln T + \cos T$

15. $F(u, v) = 1 + 6uv^2$

16. $V(r, h) = \frac{1}{2} \pi r^2 h$

17. $z(x, y) = \sin x \cos y$

18. $Z(X, R) = \sqrt{X^2 + R^2}$

19. $f(x, y) = \dfrac{e^x}{\sin y}$

20. $i(R, t) = R \sin 2t - 2 \cos 2t$

21. $A(r, x) = r \operatorname{Arctan} x$

22. $p(\omega, t) = \sin^3 \omega t$

23. $v(x, y) = \sqrt{xy} + 3x$

24. $\theta(\alpha, \beta) = \alpha^2 \tan \alpha\beta$

25. $H(r, s) = \sec^2 rs$

26. $v(\omega, t) = e^t \sin \omega t$

27. $z(x, y) = \sqrt{1 + xy}$

28. $z(x, y) = \dfrac{\ln xy}{\sqrt{x^2 + y^2}}$

In Exercises 29–38, find all of the second partial derivatives of the given functions.

29. $z(x, y) = x^5 y^2 + \sqrt{x}$

30. $w(p, t) = pe^{pt}$

31. $h(x, y) = x^2 \tan y$

32. $Z(R, X) = (R^2 + X^2)^{1/2}$

33. $h(s, t) = \sin st + s^2$

34. $p(x, y) = \sin x \cos y$

35. $F(z, y) = z \tan zy$

36. $R(s, t) = \text{Arcsin } st$

37. $z(x, y) = y^2 + \ln x$

38. $h(x, y) = y^2 \sin(x^2 + y^2)$

39. Find all the third partial derivatives of the function $z(x, y) = x^5 y^3 + y^2 - x^4$.

40. The voltage v across the plates of a discharging capacitor is a function of t and r, given by the expression $v = 100e^{-t/rC}$, where C is a constant. Find $\partial^2 v/\partial t\, \partial r$.

In Exercises 41 and 42, show that the given functions satisfy the *heat equation*,

$$\frac{\partial u}{\partial t} = \frac{\partial^2 u}{\partial x^2}$$

41. $u(x, t) = 10 + 50x$

42. $u(x, t) = e^{-4t} \sin 2x$

Show that the functions of Exercises 43–48 are valid stream functions of an ideal fluid.

43. $u(x, y) = 3 - x^2 + y^2$

44. $u(x, y) = x^2 - y^2$

45. $u(x, y) = e^x \cos y$

46. $u(x, y) = e^x \sin y$

47. $u(x, y) = 3x + 2y - 1$

48. $u(x, y) = 7 + 3y - x$

49. Show that $u(x, y) = (x - y)^n$ satisfies the equation $u_x + u_y = 0$ for any positive integer n.

50. Show that $u(x, t) = \sin \pi x \cos \pi t$ satisfies the *wave equation* $u_{xx} = u_{tt}$.

51. The hydraulic diameter of a noncircular pipe is defined by the equation $D = 4A/P$, where A is the flow cross section and P is the wetted perimeter. Find the rate at which D varies (a) as a function of A and (b) as a function of P.

52. The torque acting on a single loop of wire in a magnetic field is given by $T = KI \cos \theta$, where K is a constant, I is the current in the coil, and θ is the angle that the coil makes with the magnetic field. Find the rate at which the torque changes as a function of the angle θ.

53. The work done by a moving body is given by $W = wv^2/2g$, where w is the weight of the body, v is its velocity, and g is the acceleration of gravity. Find the rate at which W changes as a function of the velocity.

54. The mutual conductance g_m of a triode electronic vacuum tube is defined as $g_m = \partial i_b / \partial e_c$, where i_b is the plate current and e_c is the grid voltage. For a certain triode, the plate current is given by $i_b = (e_b + 10e_c)^{5/2} \times 10^{-6}$, where e_b is the plate voltage. Find the expression for the mutual conductance of this triode.

55. What is the mutual conductance of the triode in Exercise 54 when $e_b = 350$ V and $e_c = -25$ V? Mutual conductance is measured in mhos, which is the name given to reciprocal ohms.

56. The stress x inches from the fixed end of a uniformly loaded cantilever beam is given by $s = (W/2ZL)(L - x)^2$, where W is the load, L is the length, and Z is the section modulus of the beam. Find $\partial s / \partial x$. If $W = 1000$ lb, $L = 50$ in., and $Z = 1.7$ in.^{3}, at what rate is the stress changing 30 in. from the fixed end?

57. Given that $f(x, y) = x \sin(2x + y)$, show that $f_x(0, \pi/2) = 1$.

58. Given that $f(s, t) = s^2 e^{2t}$, show that $f_t(2, 0) = 8$.

■ 12.3 TOTAL DIFFERENTIALS AND DERIVATIVES

Recall that the differential of a function $y = f(x)$ was defined by $dy = f'(x)\, dx$. The idea of a differential may be extended to functions of two variables.

Total Differentials

Consider a function of two variables $z = f(x, y)$. The **total differential** dz of this function is given by the following formula.

FORMULA

> **The Total Differential of a Function**   The total differential of $z = f(x, y)$ is
>
> $$dz = \frac{\partial z}{\partial x}\, dx + \frac{\partial z}{\partial y}\, dy \qquad\qquad (12.1)$$
>
> where dx and dy may represent any real number.

In applications, we let $dx = \Delta x$ and $dy = \Delta y$. In this case (intuitively), the product $(\partial z / \partial x)\, dx$ represents the variation in z due to a change dx in the variable x, and $(\partial z / \partial y)\, dy$ represents the variation in z due to a change dy in the variable. The quantity dz is then equal to the variation due to a change in x alone plus the variation due to y alone. This use explains why we call dz the *total* differential of the function.

EXAMPLE 1   ■ **Problem**

Find the total differential of $z = x^3 y^2 - e^{2x}$.

■ **Solution**

We have

$$dz = \frac{\partial}{\partial x}(x^3y^2 - e^{2x})\,dx + \frac{\partial}{\partial y}(x^3y^2 - e^{2x})\,dy$$

$$= (3x^2y^2 - 2e^{2x})\,dx + 2x^3y\,dy \qquad\qquad ■$$

To show a physical example of the concept of the total differential of a function, we consider what happens to the area of a rectangle when its width w and its length l increase simultaneously. The area of the rectangle in Figure 12.7 can be written as

$$A = wl$$

The total differential of the area is then given by

$$dA = \frac{\partial A}{\partial w}\,dw + \frac{\partial A}{\partial l}\,dl \qquad\qquad (12.2)$$

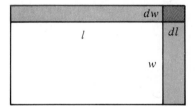

■ **Figure 12.7**

Using the fact that $A = wl$, we have

$$\frac{\partial A}{\partial w} = l \qquad \text{and} \qquad \frac{\partial A}{\partial l} = w$$

Substituting these values into Equation (12.2), we get

$$dA = l\,dw + w\,dl \qquad\qquad (12.3)$$

This result has a geometric interpretation when related to the rectangle shown in Figure 12.7. Referring to this figure, we see that the first term in Equation 12.3 is equal to the increase in area due to an increase dw in width, under the assumption that the length remains constant. Similarly, the second term in Equation 12.3 is equal to the increase in area due to an increase dl in length, assuming that the width remains constant. Thus, the value of dA is equal numerically to the sum of the lightly shaded areas shown in the figure.

The *exact* increase ΔA is given by

$$\Delta A = (l + \Delta l)(w + \Delta w) - lw = l(\Delta w) + w(\Delta l) + (\Delta l)(\Delta w) \qquad (12.4)$$

where Δl and Δw are the same changes represented, respectively, by dl and dw in the preceding discussion. Clearly, by comparing Equation 12.3 and 12.4, ΔA and dA differ by an amount equal to the area of the small rectangle in the upper right-hand corner of Figure 12.7. However, dA will often be a good approximation to ΔA as long as dl and dw are both small. We conclude from this discussion that the differential can be used to estimate small changes in the dependent variable in much the same way as differentials of functions of one independent variable.

EXAMPLE 2   ■ **Problem**

A cylindrical piece of steel is initially 8 in. long and has a diameter of 8 in. See Figure 12.8. During heat treating, the length and the diameter increase by 0.01 in. Find the approximate increase in the volume of the piece during heat treatment.

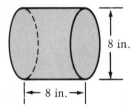

■ **Figure 12.8**

■ **Solution**

The volume of a cylinder is given by

$$V = \pi r^2 h$$

where r is the radius and h is the height of the cylinder. Writing the total differential of this function, we have

$$dV = \frac{\partial V}{\partial r} dr + \frac{\partial V}{\partial h} dh$$

But

$$\frac{\partial V}{\partial r} = 2\pi rh \quad \text{and} \quad \frac{\partial V}{\partial h} = \pi r^2$$

Therefore,

$$dV = 2\pi rh \, dr + \pi r^2 \, dh$$

Using the initial values of r and h and the values $dr = 0.005$ and $dh = 0.01$ for the respective differentials, we get

$$dV = 2\pi(4)(8)(0.005) + \pi(4)^2(0.01) = 1.005 + 0.503 = 1.508 \text{ in.}^3$$

The exact change in volume can be found by taking the difference in the volume when $r = 4.005$ and $h = 8.01$, and the volume when $r = 4$ and $h = 8$. The exact value of ΔV is

$$\Delta V = 1.510 \text{ in.}^3$$ ■

Total Derivatives

Suppose the variables x and y in the function $z = f(x, y)$ are, in turn, functions of some other variable (say t). Then, we can assume, from the form of the total differential, that the derivative of z with respect to t is given by the following formula.

FORMULA

> **Total Derivative of a Function**   The *total derivative* of $z = f(x, y)$ is
>
> $$\frac{dz}{dt} = \frac{\partial z}{\partial x}\frac{dx}{dt} + \frac{\partial z}{\partial y}\frac{dy}{dt} \qquad\qquad (12.5)$$

This equation is referred to as the total derivative of the function because dz/dt includes the effects of both component rates dx/dt and dy/dt.

EXAMPLE 3   ■ **Problem**

Find the total derivative of $f(x, y) = x^2 + y^3$, where $x = \sin t$ and $y = \cos t$.

■ **Solution**

We have

$$\frac{df}{dt} = \frac{\partial}{\partial x}(x^2 + y^3)\frac{dx}{dt} + \frac{\partial}{\partial y}(x^2 + y^3)\frac{dy}{dt} = 2x\frac{dx}{dt} + 3y^2\frac{dy}{dt}$$

Substituting $x = \sin t$ and $y = \cos t$, we have

$$\frac{df}{dt} = 2\sin t\,\frac{d}{dt}(\sin t) + 3\cos^2 t\,\frac{d}{dt}(\cos t)$$

$$= 2\sin t\cos t - 3\cos^2 t\sin t = (\sin t\cos t)(2 - 3\cos t)$$ ■

EXAMPLE 4   ■ **Problem**

If a coil of wire is placed in a uniform magnetic field of density B, and a current i is sent through the coil, the resulting torque is given by

$$T = NBAi\cos\theta$$

where A is the area of the coil, N is the number of turns of wire, and θ is the angle the coil makes with the magnetic field. Find the expression for dT/dt if i and θ are changed simultaneously. Find dT/dt if $i = \sin 2t$ and $\theta = t^3$.

■ **Solution**

By Equation (12.5), dT/dt is

$$\frac{dT}{dt} = \frac{\partial T}{\partial i}\frac{di}{dt} + \frac{\partial T}{\partial \theta}\frac{d\theta}{dt} = NBA \cos\theta \frac{di}{dt} - NBAi \sin\theta \frac{d\theta}{dt}$$

$$= NBA \left[\cos\theta \frac{di}{dt} - i \sin\theta \frac{d\theta}{dt}\right]$$

Using $i = \sin 2t$ and $\theta = t^3$, we have

$$\frac{dT}{dt} = NBA [\cos t^3 (2 \cos 2t) - \sin 2t \sin t^3 (3t^2)]$$

$$= NBA(2 \cos t^3 \cos 2t - 3t^2 \sin t^3 \sin 2t)$$ ■

EXERCISES FOR SECTION 12.3

Find the total differential of each of the functions in Exercises 1–10.

1. $f(x, y) = x^2 y^3$

2. $f(x, y) = y^3 + \cos x^2$

3. $f(x, y) = xe^x + \log y$

4. $f(x, y) = x \tan y + y^4 e^x$

5. $f(x, y) = e^y + x \sin y$

6. $f(x, y) = \sqrt{xy} + x^2 y^2$

7. $f(x, y) = \sec y^2 + \tan e^{xy}$

8. $f(x, y) = ye^x - xy$

9. $f(x, y) = (x + y)^2$

10. $f(x, y) = e^{xy} \sin(x + y)$

Find the total derivative of each of the functions in Exercises 11–16.

11. $f(x, y) = x^2 y^3; x = \sin t; y = \cos t$

12. $f(x, y) = y^3 + \cos x^2; x = t, y = t^2$

13. $f(x, y) = xe^x + \ln y; x = t^2; y = t$

14. $f(x, y) = x \tan y + y^4 e^x; x = \ln t; y = t$

15. $f(x, y) = e^y + x \sin y; x = t^2; y = \ln t$

16. $f(x, y) = \sqrt{xy} + x^2 y^2; x = \sin^2 t; y = \cos^2 t$

17. A triangle has an initial base $b = 10$ in. and an initial height $h = 4$ in. Find the approximate change in area that occurs when b is increased to 10.1 in. and h is increased to 4.01 in.

18. Find the exact change in the area of the triangle in Exercise 17.

19. The volume of a right circular cone is given by $V = \frac{1}{3}\pi r^2 h$. Find the expression for the total differential dV.

20. Assuming $r = 10$ and $h = 10$, find the approximate change in the volume of the right circular cone in Exercise 19 when $dr = 0.2$ and $dh = -0.1$.

21. The dynamic pressure of an object moving with a velocity v through a medium of density ρ is given by $P = \frac{1}{2}\rho v^2$. Find the formula that expresses the approximate error in P due to small errors in the measurement of ρ and v.

22. In designing a cone clutch, the normal pressure between the cone surfaces is found to be $P = S/\sin \alpha$, where S is the spring pressure and α is the angle that the clutch surface makes with the axis of the shaft. Find the formula for the approximate change in P caused by simultaneous small changes in S and α.

23. The velocity of an object sliding down an inclined plane can be expressed in the form $v = 32t \sin \alpha$, where t is the time (in seconds) and α is the angle of inclination of the plane. Find the equation for the approximate error in the velocity due to small errors in measuring t and α.

24. When an axial load is applied to a slender square column, there is a tendency for the column to buckle. The axial load that will cause failure is given by Euler's formula. $F = \pi^2 d^4 E/12L^2$, where L is the length of the column, d is the length of its sides, and E is the modulus of elasticity.

    a. What is the failure load of a column having $L = 300$ in., $d = 10$ in., and $E = 10^6$ lb/in.?
    b. By how much is this failure load in error if L may be in error by 0.2 in. and E may be in error by 600 lb/in.?

25. When friction is ignored, the velocity of a freely falling object is $v = \sqrt{2gh}$, where g is the acceleration of gravity and h is the distance the object has fallen from rest.

    a. Find the velocity of an object if $g = 32$ ft/sec^2 and $h = 100$ ft.
    b. Suppose the true value of g is 32.2 ft/sec^2, and h can be measured to an accuracy of 0.1 ft. Find the approximate error in the computed velocity.

26. The power in a resistor is given by $p = i^2 r$, where i is the current and r is the resistance. For a certain circuit $i = 2.25$ A and $r = 10^4\, \Omega$.

    a. Compute the power p for this circuit.
    b. By how much may this answer be in error if i and r are subject to errors of 0.02 A and 200 Ω, respectively?

27. Find the formula for the rate of change of the volume of a cylinder that results when the radius and the height of the cylinder change.

28. Suppose that the height of the cylinder in Exercise 27 is 100 in. and increases at a rate of 3 in./sec., and that the radius is 10 in. and increases at 2 in./sec. What is the rate of change of the volume?

29. Repeat Exercise 28, but assume that the radius decreases at 2 in./sec.

30. As a ballistic missile rises through the atmosphere, the velocity of the missile and the atmosphere density are changing simultaneously. Find a formula for the rate of change of dynamic pressure with respect to time due to changes in ρ and v. (See Exercise 21.)

31. During the compression stroke of a certain internal combustion engine, the increase in the temperature of the gasoline–air mixture can be found by using $T = 0.01PV$. When the piston is 6° before top dead center, the volume of the mixture is 20 in.^{3} and the pressure is 110 psi.

  a.   Find the increase in the temperature of the mixture at this point.
  b.   Find the rate at which the temperature is increasing if the volume is decreasing at a rate of 500 in.^{3}/sec and the pressure is increasing at a rate of 1000 psi/sec. (Assume dv/dt to be positive.)

▬▬▬ 12.4 ITERATED INTEGRALS

Consider the function $z = xy^3 + x^2 + y^2$. The partial derivative of z with respect to x is given by

$$\frac{\partial z}{\partial x} = y^3 + 2x$$

Now, suppose we were given this partial derivative and told to reconstruct the original function. We would proceed by integrating $y^3 + 2x$ with respect to x while holding y constant. Using the same notation for partial integration that we have used previously for functions of a single variable, we obtain

$$z = \int (y^3 + 2x)\, dx = y^3x + x^2 + G(y)$$

where $G(y)$ is some function of y only. Since the y^2 term in the given function vanished when we found the partial derivative of z with respect to x, we can only reconstruct a function to within an "arbitrary function" of one of the variables.

EXAMPLE 1   ■ **Problem**

Find z, given that $z_y = 3x^2 + y^3$.

■ **Solution**

We have

$$z = \int (3x^2 + y^3)\, dy = 3x^2y + \frac{y^4}{4} + F(x)$$

The arbitrary function in this case is a function of x only.   ■

Sometimes, more than one partial integration is needed to reconstruct the function.

EXAMPLE 2   ■ **Problem**

Find z, given that $\dfrac{\partial^2 z}{\partial x^2} = x^2$.

■ **Solution**

The first integration yields

$$\frac{\partial z}{\partial x} = \int x^2 \, dx = \frac{x^3}{3} + G(y)$$

and the second yields

$$z = \int \left[\int x^2 \, dx \right] dx = \int \left[\frac{x^3}{3} + G(y) \right] dx = \frac{x^4}{12} + xG(y) + H(y) \qquad ■$$

EXAMPLE 3 ■ **Problem**

Find z, given that $z_{xy} = x \sin xy$.

■ **Solution**

Here, $\partial z / \partial x$ is given by

$$\frac{\partial z}{\partial x} = \int (x \sin xy) \, dy = -\cos xy + F(x)$$

Then, z is given by

$$z = \int \left[\int x \sin xy \, dy \right] dx = \int [-\cos xy + F(x)] \, dx$$

$$= -\frac{1}{y} \sin xy + \int F(x) \, dx + Q(y) \qquad ■$$

Integrals of the type arising in Example 3 are usually written without brackets. Thus, we write

$$\int \int x \sin xy \, dy \, dx$$

with the understanding that this notation means to integrate with respect to y first and then with respect to x.

The same type of convention is used for **repeated**, or **iterated, definite integration.** Iterated integrals are usually written in the form

$$\int_a^b \int_{f(x)}^{g(x)} S(x, y) \, dy \, dx \qquad \text{or} \qquad \int_c^d \int_{F(y)}^{G(y)} H(x, y) \, dx \, dy$$

Notice that the limits on the inner integral may be a function of x when dy is the differential and a function of y when dx is the differential.

EXAMPLE 4 ■ **Problem**

Evaluate the iterated integral $\int_{-1}^{3} \int_{1}^{2} 3x^2 y \, dx \, dy$.

■ **Solution**

By definition, the notation means

$$\int_{-1}^{3} \left[\int_{1}^{2} 3x^2y \, dx \right] dy$$

Since

$$\int_{1}^{2} 3x^2y \, dx = \left[x^3y \right]_{1}^{2} = 8y - y = 7y$$

and

$$\int_{-1}^{3} 7y \, dy = \left[\frac{7y^2}{2} \right]_{-1}^{3} = \frac{63}{2} - \frac{7}{2} = 28$$

we have

$$\int_{-1}^{3} \int_{1}^{2} 3x^2y \, dx \, dy = 28$$

■

EXAMPLE 5   ■ **Problem**

Evaluate $\int_{0}^{1}\int_{0}^{2} (x + y) \, dy \, dx$.

■ **Solution**

We have

$$\int_{0}^{1} \int_{0}^{2} (x + y) \, dy \, dx = \int_{0}^{1} \left[\int_{0}^{2} (x + y) \, dy \right] dx = \int_{0}^{1} \left[xy + \frac{y^2}{2} \right]_{0}^{2} dx$$

$$= \int_{0}^{1} (2x + 2) \, dx = \left[x^2 + 2x \right]_{0}^{1} = 3$$

■

EXAMPLE 6   ■ **Problem**

Evaluate $\int_{\pi/2}^{\pi} \int_{0}^{1} x \sin y \, dx \, dy$.

■ **Solution**

We obtain

$$\int_{\pi/2}^{\pi} \int_{0}^{1} x \sin y \, dx \, dy = \int_{\pi/2}^{\pi} \left[\int_{0}^{1} x \sin y \, dx \right] dy = \int_{\pi/2}^{\pi} \left[\frac{x^2}{2} \sin y \right]_{0}^{1} dy$$

$$= \int_{\pi/2}^{\pi} \frac{1}{2} \sin y \, dy = \left[-\frac{1}{2} \cos y \right]_{\pi/2}^{\pi}$$

$$= -\left[-\frac{1}{2} - 0 \right] = \frac{1}{2}$$

■

EXAMPLE 7   ■ **Problem**

Evaluate $\int_0^1 \int_0^{x^2} dy\ dx$.

■ **Solution**

We have

$$\int_0^1 \int_0^{x^2} dy\ dx = \int_0^1 \left[\int_0^{x^2} dy \right] dx = \int_0^1 \left[y \right]_0^{x^2} dx = \int_0^1 x^2\ dx$$

$$= \left[\frac{x^3}{3} \right]_0^1 = \frac{1}{3}$$

■

EXERCISES FOR SECTION 12.4

Find the function $z(x, y)$ for the conditions in Exercises 1–10.

1. $z_x = x^2 + y$

2. $z_y = y$

3. $z_y = x$

4. $z_{xx} = \sin y$

5. $z_{xy} = xy$

6. $z_{yy} = e^x + x^2$

7. $z_x = x^2 + 2e^{xy}$

8. $z_{yx} = x^2 + x \sin y$

9. $z_{yy} = \cos xy$

10. $z_{xx} = \cos xy$

Evaluate each of the following iterated integrals.

11. $\int_0^2 \int_0^3 xy\ dx\ dy$

12. $\int_0^2 \int_0^3 xy\ dy\ dx$

13. $\int_0^\pi \int_0^2 \cos y\ dx\ dy$

14. $\int_{-1}^2 \int_0^1 se^t\ dt\ ds$

15. $\int_0^3 \int_1^2 3x^2y\ dx\ dy$

16. $\int_1^4 \int_0^5 (x^2 + y^2)\ dx\ dy$

17. $\int_{-1}^0 \int_0^4 (z^{1/2}y + 2)\ dz\ dy$

18. $\int_0^2 \int_0^2 Te^{RT}\ dR\ dT$

19. $\int_1^3 \int_0^\pi x \sin xy\ dy\ dx$

20. $\int_{-1}^1 \int_{-1}^1 e^{x+y}\ dx\ dy$

21. $\int_{-1}^2 \int_0^{y^2} x\ dx\ dy$

22. $\int_0^2 \int_0^{\sqrt{y}} xe^{y^2}\ dx\ dy$

23. $\int_1^2 \int_v^{v^2} (v + 2u)\ du\ dv$

24. $\int_0^\pi \int_0^{\cos\theta} r \sin\theta\ dr\ d\theta$

25. $\int_0^{\pi/2} \int_0^x \cos y \sin x\ dy\ dx$

26. $\int_0^9 \int_0^{\sqrt{25-x}} dy\ dx$

■■■■ **12.5 AREA AS AN ITERATED INTEGRAL**

In Chapter 6, we showed that the area of a plane figure could be approximated by a summation process that led to the definite integral. We would now like to show how a plane area can be represented by an iterated integral.

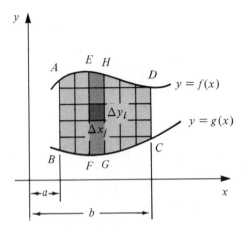

■ **Figure 12.9**

Consider the plane area $ABCD$ shown in Figure 12.9. Divide the interval on the x-axis into n equal parts, and draw a line from each point of division parallel to the y-axis. Let the width of each of these elements be Δx. Now, do the same thing for the interval on the y-axis, representing the width of these elements by Δy. In this way, the area $ABCD$ is divided into a network of elementary rectangles. Clearly, the area of each of these elementary rectangles is given by

$$\Delta A_{ij} = \Delta y_i \Delta x_j$$

If we now choose a typical Δx and sum all of the areas formed by the product of this Δx and the corresponding Δy's, the result will be the area of a rectangle like $EFGH$ shown in the figure. This area may be written symbolically as

$$\text{area } EFGH \approx \sum_{i=1}^{n} \Delta y_i \, \Delta x_j = \left(\sum_{i=1}^{n} \Delta y_i \right) \Delta x_j \qquad \textbf{(12.6)}$$

We may regard the exact area as the limit of (12.6) as $\Delta y \to 0$; that is,

$$\text{area } EFGH = \left(\lim_{n \to \infty} \sum_{i=1}^{n} \Delta y_i \right) \Delta x_j$$

The Δy's in this equation are being summed from $y = g(x)$ to $y = f(x)$. Therefore, by the definition of a definite integral, we can write

$$\text{area } EFGH = \left(\int_{g(x)}^{f(x)} dy \right) \Delta x_j$$

Observing that area *EFGH* is a typical element, we conclude that the desired area is the limit of the sum of these areas from $x = a$ to $x = b$ as $\Delta x \to 0$. Hence,

$$\text{area } ABCD = \lim_{m \to \infty} \sum_{j=1}^{m} \left(\int_{g(x)}^{f(x)} dy \right) \Delta x_j$$

Or we have the following formula for the area shown in Figure 12.9.

FORMULA

Area as an Iterated Integral

$$\text{area } ABCD = \int_{a}^{b} \left(\int_{g(x)}^{f(x)} dy \right) dx = \int_{a}^{b} \int_{g(x)}^{f(x)} dy \, dx$$

For areas like the one shown in Figure 12.10, it is more convenient to choose a typical Δy and sum horizontally first. Under these circumstances, the area is given by the following formula.

$$\text{area } NMOP = \int_{c}^{d} \int_{G(y)}^{F(y)} dx \, dy$$

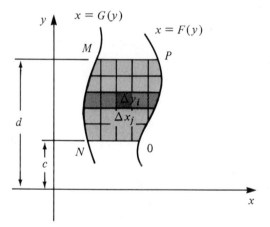

■ **Figure 12.10**

PROCEDURE

Evaluating Area Integrals

1.   To evaluate

$$\int_{a}^{b} \int_{g(x)}^{f(x)} dy \, dx$$

we integrate first with respect to y. That is, we sum the elements $dy \, dx$ to get the area of a vertical strip. The vertical strips are then summed to get the desired area.

2. To evaluate

$$\int_a^b \int_{G(y)}^{F(y)} dx\, dy$$

we integrate first with respect to x. That is, we sum the elements $dx\, dy$ to get the area of a horizontal strip. The horizontal strips are then summed to get the desired area.

Comment: The order of the iteration is ordinarily determined by the nature of the boundaries of the area. If both iterations can be formed, choose the one easiest to evaluate.

EXAMPLE 1   ▪ **Problem**

Find the area bounded by $y = \sin x$, $y = 0$, and $x = \pi/2$, using iterated integrals.

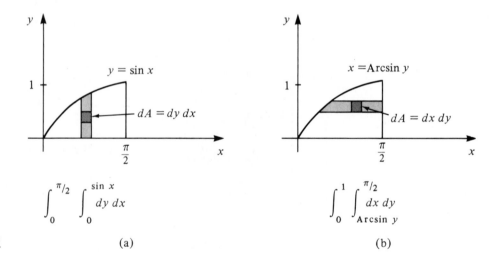

▪ **Figure 12.11**   (a)   (b)

▪ **Solution**

See Figure 12.11. The area may be expressed as

$$A = \int_0^{\pi/2} \int_0^{\sin x} dy\, dx \quad \text{or} \quad A = \int_0^1 \int_{\text{Arcsin } y}^{\pi/2} dx\, dy$$

The first of these integrals is the easiest to evaluate. Thus,

$$A = \int_0^{\pi/2} \int_0^{\sin x} dy\, dx = \int_0^{\pi/2} \Big[y \Big]_0^{\sin x} dx = \int_0^{\pi/2} \sin x\, dx$$

$$= \Big[-\cos x \Big]_0^{\pi/2} = 1$$

▪

EXAMPLE 2   ▪ **Problem**

Use iterated integration to find the area bounded by the curves $y = x$ and $y = x^2$. See Figure 12.12.

▪ **Solution**

The area is given by

$$A = \int_a^b \int_{g(x)}^{f(x)} dy \, dx$$

The limits on the inner integral sign are $f(x) = x$ and $g(x) = x^2$. The limits on the outer integral can be found by solving the two equations simultaneously. Thus,

$$x^2 = x$$
$$x^2 - x = 0$$
$$x = 0, 1$$

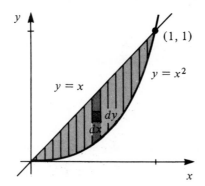

▪ **Figure 12.12**

The numerical values of a and b are then $a = 0$ and $b = 1$. The required area is then

$$A = \int_0^1 \int_{x^2}^x dy \, dx = \int_0^1 \left[y \right]_{x^2}^x dx = \int_0^1 (x - x^2) \, dx$$

$$= \left[\frac{x^2}{2} - \frac{x^3}{3} \right]_0^1 = \left[\frac{1}{2} - \frac{1}{3} \right] = \frac{1}{6}$$ ▪

EXAMPLE 3   ▪ **Problem**

Use iterated integration to find the area bounded by the two curves $y = 5x$ and $y = x^2$, and the straight lines $y = 1$ and $y = 4$. See Figure 12.13.

▪ **Solution**

Here, we use the iterated integral

$$A = \int_c^d \int_{G(y)}^{F(y)} dx \, dy$$

with $c = 1$, $d = 4$, $G(y) = y/5$, and $F(y) = y^{1/2}$. The required area is therefore

$$A = \int_1^4 \int_{y/5}^{y^{1/2}} dx\, dy = \int_1^4 \left[x\right]_{y/5}^{y^{1/2}} dy = \int_1^4 \left(y^{1/2} - \frac{y}{5}\right) dy$$

$$= \left[\frac{2y^{3/2}}{3} - \frac{y^2}{10}\right]_1^4 = \left[\frac{16}{3} - \frac{16}{10}\right] - \left[\frac{2}{3} - \frac{1}{10}\right] = \frac{19}{6}$$

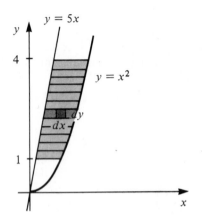

■ **Figure 12.13**

■ **EXERCISES FOR SECTION 12.5**

Use iterated integration to find the area bounded by the given curves.

1. $y = 2 - x^2$; $y = x$

2. $y = x^2$; $y = 8x$

3. $y = x^2$; $y = 4x$; $y = 1$; $y = 2$

4. $y = x$; $y = -\frac{1}{2}x$; $x = 1$; $x = 3$

5. $y = x^2 + 4x$; $y = x$

6. $y = x - x^2$; $y = -x$

7. $y^2 = 4x$; $x^2 = 4y$

8. $y^2 = 8x$; $y = \frac{1}{2}x$

9. $y = x$; $y = x^{3/2}$

10. $y = 3$; $y = x$; $x = -1$; $x = 2$

11. $y = 9 - x^2$; $y = x + 7$

12. $y^2 = 2x$; $y = x - x^2$

■ **12.6 DOUBLE INTEGRATION**

If f is a function of one real variable, then $\int_a^b f(x)\, dx$ is a *number* defined as the limit of approximating sums. We now wish to show the analogous concept for functions of two real variables.

Let $H(x, y)$ represent a function defined on some region of the xy-plane, for example, the region shown in Figure 12.14. We subdivide the region R into elementary areas, where the subdivision process may be carried out in *any* manner, not necessarily into rectangular regions parallel to the coordinate axes. Call the area of each of the subregions ΔA_i. Evaluate $H(x, y)$ at some point (x_i, y_i) in each of the sub-regions and form the products $H(x_i, y_i) \, \Delta A_i$. Now, if we sum these products, we will have a sum analogous to that obtained for functions of one variable; that is,

$$\sum_{i=1}^{n} H(x_i, y_i) \, \Delta A_i \qquad\qquad (12.7)$$

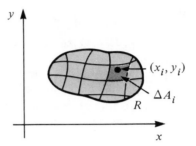

■ **Figure 12.14**

EXAMPLE 1   ■ **Problem**

Compute the sum in (12.7) if $H(x, y) = x^2 + 7y$ and R is the region bounded by the coordinate axes and the lines $x = 4$ and $y = 8$. Use four equal rectangular sub-regions, and evaluate $H(x, y)$ at the midpoint of each elementary area. See Figure 12.15.

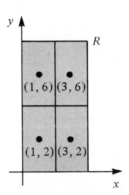

■ **Figure 12.15**

■ **Solution**

Each of the subregions shown in the figure have an area of $\Delta A_i = 2(4) = 8$. The functional values of H at the midpoints are

$$H(1, 2) = 1^2 + 7(2) = 15 \qquad H(1, 6) = 1^2 + 7(6) = 43$$
$$H(3, 2) = 3^2 + 7(2) = 23 \qquad H(3, 6) = 3^2 + 7(6) = 51$$

The desired sum is then

$$\sum_{i=1}^{4} H(x_i, y_i)\, \Delta A_i = (15 + 23 + 43 + 51)(8) = 1056 \qquad \blacksquare$$

As in the case of sums for functions of one variable, we are interested in the limit of (12.7) as n increases without bound; that is,

$$\lim_{n \to \infty} \sum_{i=1}^{n} H(x_i, y_i)\, \Delta A_i$$

If this limit exists, we denote it by $\int_R \int H(x, y)\, dA$, and we call it the **double integral** of the function $H(x, y)$ over the region R.

There are several elementary techniques for evaluating double integrals, all based on different methods of subdividing the region R. We will be content to show what happens when the process used is that of subdividing the region into rectangles whose sides are parallel to the coordinate axes. In this case, the area of each sub-region is given by

$$\Delta A_i = \Delta y_j\, \Delta x_i$$

The sum $\Sigma\, H(x_i, y_j)\, \Delta A_i$ may be written as the double summation

$$\sum_{i=1}^{n} \sum_{j=1}^{m} H(x_i, y_j)\, \Delta y_j\, \Delta x_i$$

Suppose the region R is the one shown in Figure 12.9. Then, if we let m approach infinity and n approach infinity independently, we arrive at the following formula.

FORMULA

> **Evaluating a Double Integral**   For the region R shown in Figure 12.9, we have
>
> $$\int_R \int H(x, y)\, dA = \int_a^b \int_{g(x)}^{f(x)} H(x, y)\, dy\, dx \qquad (12.8)$$
>
> Similarly, if the region R is as shown in Figure 12.10, then
>
> $$\int_R \int H(x, y)\, dA = \int_c^d \int_{G(y)}^{F(y)} H(x, y)\, dx\, dy \qquad (12.9)$$

Comment: Because double integrals can be evaluated in terms of iterated integrals, the two concepts are closely identified. Indeed, it is quite natural to refer to iterated integrals as double integrals.

EXAMPLE 2 ■ **Problem**

Evaluate the double integral $\int_R \int (x^2 + 7y) \, dA$, where R is the region shown in Figure 12.16.

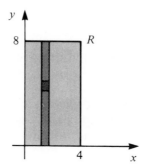

■ **Figure 12.16**

■ **Solution**

Using (12.8) to evaluate the given double integral, we have

$$\int_R \int (x^2 + 7y) \, dA = \int_0^4 \int_0^8 (x^2 + 7y) \, dy \, dx = \int_0^4 \left[x^2 y + \frac{7y^2}{2} \right]_0^8 dx$$

$$= \int_0^4 (8x^2 + 224) \, dx = \left[\frac{8x^3}{3} + 224x \right]_0^4$$

$$= \frac{512}{3} + 896 = \frac{3200}{3}$$

■

EXAMPLE 3 ■ **Problem**

Express the area of the region R bounded by the curves $y = x$ and $y = x^2$ as a double integral. This region is the same region that we considered in Example 2 of Section 12.5. See Figure 12.17.

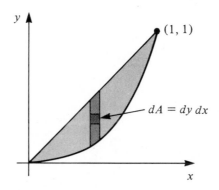

(1, 1)

$dA = dy \, dx$

■ **Figure 12.17**

■ Solution

Since the area of R can be approximated by $\sum_{i=1}^{n} \Delta A_i$, we can express this area by the double integral $\int_R \int dA$. Using (12.8) to evaluate this double integral, we have

$$A = \int_R \int dA = \int_0^1 \int_{x^2}^x dy \, dx$$

which is the same iterated integral we used in Example 2 of Section 12.5.   ■

The next example shows how double integrals are derived. Usually, we consider the effect of some physical phenomenon on an element of area A. Then, the double integral gives the cumulative effect, and the evaluation is done by iteration.

EXAMPLE 4   ■ Problem

The pressure (in pounds per square inch) exerted on the triangular plate shown in Figure 12.18 increases with distance from the y-axis according to $P = 3x$. Find the total force being exerted on the plate.

■ Solution

The force ΔF being exerted on an element of area ΔA is

$$\Delta F = P(x, y) \, \Delta A$$

So, the total force is given by the double integral $\int_R \int P(x, y) \, dA$. Thus,

$$F = \int_R \int P(x, y) \, dA = \int_0^8 \int_0^y 3x \, dx \, dy = \int_0^8 \left[\frac{3x^2}{2} \right]_0^y dy$$

$$= \int_0^8 \frac{3y^2}{2} \, dy = \left[\frac{y^3}{2} \right]_0^8 = 256 \text{ lb}$$

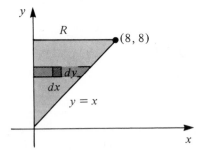

■ Figure 12.18   ■

■■■■■■■ **EXERCISES FOR SECTION 12.6**

In Exercises 1–6, evaluate the given double integral over the indicated region. Draw each region.

1. $\int_R \int x \, dA$; R is bounded by $x = 0$, $y = 0$, $x = 2$, and $y = 3$.

2. $\int_R \int y \, dA$; R is bounded by $x = 0$, $y = 0$, $x = 5$, and $y = 4$.

3. $\int_R \int (x^2 + y^2) \, dA$; R is bounded by $x = 0$, $y = 0$, $x = 4$, and $y = x$.

4. $\int_R \int xy \, dA$; R is bounded by $x = 0$, $y = 0$, and $y = 4 - 2x$.

5. $\int_R \int (2x + 3y) \, dA$; R is bounded by $y = 2x$ and $y = x^2$.

6. $\int_R \int (3 + 4y) \, dA$; R is bounded by $y = 0$ and $y = 3x - x^2$.

7. Express the area bounded by $y = 4x - x^2$ and $y = x$ as a double integral, and evaluate by using Equation (12.8).

8. The region R is bounded by $y = x^3$, $x = 2$, and $y = 0$. Find the area of R.

9. The moment of a region R about the x-axis is given by $\int_R \int y \, dA$. Find the moment about the x-axis of the region bounded by $y = x$ and $y = x^2$.

10. Express the moment of a region R about the y-axis in terms of a double integral.

11. Using the expression in Exercise 10, find the moment of the region bounded by $y = x$ and $y = x^2$ about the y-axis.

12. The pressure on the triangular region formed by the lines $x = 0$, $y = 0$, $x = 4$, and $y = \frac{1}{2}x$ increases with distance from the y-axis according to $P = x^2$. Find the total force being applied to the region.

13. Find the total force being applied to the region in Exercise 12 if the pressure (in psi) varies with distance from the x-axis according to $P = 3y$.

■■■■■■ **12.7 VOLUME AS A DOUBLE INTEGRAL**

We conclude this chapter with a discussion of how an iterated integral can be used to represent the volume under a surface.

Let $z = f(x, y)$ be a continuous function of x and y having as its graph the surface shown in Figure 12.19. Consider the volume V of the solid bounded by the given surface S, its projection on the xy-plane R, and the vertical lines through the boundaries of the surfaces S and R. In the xy-plane, draw n lines parallel to the y-axis. Let the width of each interval intercepted by these lines on the x-axis be Δx. Now, do the same thing with lines parallel to the x-axis, representing the width of each of these elements by Δy. In this way, the projection R in the xy-plane is divided into a network of elementary rectangles, each having an area of $\Delta y \, \Delta x$.

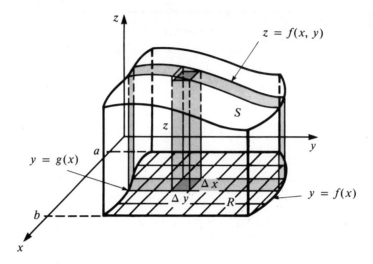

Through the lines in the xy-plane, pass planes parallel to the xz-plane and the yz-plane, respectively. This procedure divides the volume V into vertical rectangular columns. The volume of each of these rectangular columns can be written as

$$\Delta V = z_{ij}\, \Delta y_i\, \Delta x_j$$

where z_{ij} is the height of each column. Now, choose a typical Δx, and sum the volumes of all of the rectangular columns corresponding to this Δx. The limit of this sum as $\Delta y \to 0$ is equal to the volume of a slice of the figure Δx units wide. That is,

$$V_y = \left[\lim_{\Delta y \to 0} \sum_{i=1}^{m} z_{ij}\, \Delta y_i\right] \Delta x_j \qquad (12.10)$$

Observing that the Δy's are the summed between the lateral boundaries of the given surface, we may write (12.10) as

$$V_y = \left(\int_{g(x)}^{f(x)} z_j\, dy\right) \Delta x_j$$

where $y = f(x)$ and $y = g(x)$ are the lateral boundaries of the given surface in the y direction. The notation $f(x)$ and $g(x)$ is used to indicate that in general, these limits are functions of x. The desired volume is then the limit of the sum of the volume of these slices as $\Delta x \to 0$, or

$$V = \lim_{\Delta x \to 0} \sum_{j=1}^{n} \left(\int_{g(x)}^{f(x)} z_j\, dy\right) \Delta x_j$$

Thus, as a double integral, the volume is given by the following formula.

FORMULA

> **Volume as a Double Integral**
>
> $$V = \int_a^b \int_{g(x)}^{f(x)} z \, dy \, dx \qquad \qquad (12.11)$$
>
> The limits on the outer integral sign are the lateral boundaries of the given surface in the x direction. The outer limits are always constants.
>
> By a similar development, we may also write the volume as follows:
>
> $$V = \int_c^d \int_{G(y)}^{F(y)} z \, dx \, dy \qquad \qquad (12.12)$$

EXAMPLE 1   ■ **Problem**

Find the volume bounded by $z = 4 - x - y$ and the first octant of the coordinate system.

■ **Solution**

This surface can be sketched by noting the following features:

1. When $x = 0$, $z = 4 - y$, which is a straight line in the yz-plane.
2. When $y = 0$, $z = 4 - x$, which is a straight line in the xz-plane.
3. When $z = 0$, $y = 4 - x$, which is a straight line in the xy-plane.

Drawing these lines and a plane surface through them, we get the graph shown in Figure 12.20.

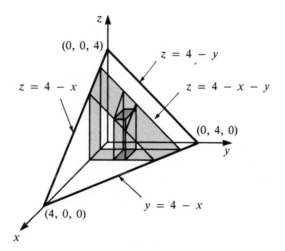

■ **Figure 12.20**

The required volume is then given by

$$V = \int_0^4 \int_0^{4-x} (4 - x - y)\, dy\, dx = \int_0^4 \left[4y - xy - \frac{1}{2}y^2 \right]_0^{4-x} dx$$

$$= \int_0^4 \left[4(4 - x) - x(4 - x) - \frac{1}{2}(4 - x)^2 \right] dx$$

$$= \int_0^4 \left[8 - 4x + \frac{1}{2}x^2 \right] dx = \left[8x - 2x^2 + \frac{x^3}{6} \right]_0^4 = \frac{32}{3}$$ ■

EXAMPLE 2   ■ **Problem**

Find the volume under the surface $z = x^2 + y^2$ over the region R bounded by $y = 1$, $x = 0$, and $y = 3 - x$. See Figure 12.21.

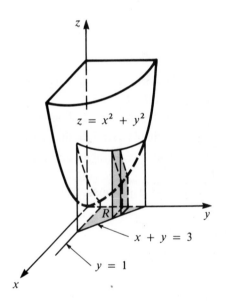

■ Figure 12.21

■ Solution

The volume is given by

$$V = \int_1^3 \int_0^{3-y} (x^2 + y^2)\, dx\, dy = \int_1^3 \left[\frac{x^3}{3} + xy^2 \right]_0^{3-y} dx$$

$$= \int_1^3 \left[\frac{(3 - y)^3}{3} + (3 - y)\, y^2 \right] dy = \left[-\frac{(3 - y)^4}{12} + y^3 - \frac{y^4}{4} \right]_1^3$$

$$= \left(0 + 27 - \frac{81}{4} \right) - \left(-\frac{4}{3} + 1 - \frac{1}{4} \right) = \frac{22}{3}$$ ■

EXERCISES FOR SECTION 12.7

1. Find the volume bounded by the plane surface $z = 2 - x - y$ and the first octant of the coordinate system.

2. Find the volume in the first octant bounded by the plane surface $z = 1 - 2x - y$.

3. Find the volume under the plane surface $z = 3 + x + 4y$ over a region R, where R is bounded by $x = 0$, $y = 0$, $y = 2$, and $x = 3$.

4. Find the volume under the surface $z = x^2 + y^2$ over the region R bounded by $x = 0$, $y = 0$, and $y = 2 - x$.

5. Find the volume under the surface $z = x + y^2$ over the region R bounded by $y = x$ and $y = x^2$.

6. Find the volume under the surface $z = y$ over the region R bounded by $y = 0$ and $y = 4x - x^2$.

7. Find the volume bounded by $z = 4/(y^2 + 1)$ and the planes $y = x$, $y = 3$, $x = 0$, and $z = 0$.

8. Find the volume in the first octant bounded by $z = 1 - y$ and $z = 2 - y$ between $x = 0$, $x = 2$, and $y = 1$.

9. Find the volume in the first octant under the surface $x - z = 0$ over the region R bounded by $x^2 + y^2 = 4$.

10. Find the volume in the first octant bounded by the cylinder $y^2 + z^2 = 4$ and the planes $y = x$, $x = 0$, and $z = 0$.

REVIEW EXERCISES FOR CHAPTER 12

1. Given that $f(x, y) = x^2y + 3xy^2 - 2$, find $f(-1, 3)$.

2. Given that $G(t, s) = t + t^3s^2 + 5s$, find $G(2, -2)$.

3. Make a first-octant sketch of the graph of $z = 2 - 2x - y$.

4. Make a first-octant sketch of the graph of $z = y^2$.

In Exercises 5–10, find both first partial derivatives of the given function.

5. $f(x, y) = 5 - x^3y^4 + x^2$

6. $f(x, y) = 3y^2 - 4x + xy$

7. $z(x, y) = e^{2x} \cos 3y$

8. $g(t, s) = 3se^{5t}$

9. $f(t, s) = t \cos 2st$

10. $z(x, y) = \sqrt{x(1 + xy)}$

In Exercises 11–14, find all of the second partial derivatives of the given functions.

11. $f(x, y) = xy^2 + x^5 + y$

12. $h(x, y) = 5x^3 - y \sin 2x$

13. $z(x, y) = 2y \sec x$

14. $g(x, y) = e^{-x} \sin 5y$

In Exercises 15–18, find the total differential of the given function.

15. $z = x^2 \sin 2y$

16. $z = \cos xy$

17. $f(x, y) = x + e^{3y}$

18. $F(x, y) = x^3 y + \sqrt{y}$

19. Find the total derivative of $z = x^2 + y^2$ if $x = e^{2t}$ and $y = \sin 3t$.

20. Find the total derivative of $z = \sin x \cos 2y$ if $x = t^2$ and $y = t$.

In Exercises 21–26, evaluate each of the given integrals.

21. $\displaystyle \int_0^3 \int_{-1}^2 x \, dy \, dx$

22. $\displaystyle \int_0^2 \int_0^1 e^y \, dx \, dy$

23. $\displaystyle \int_0^\pi \int_0^1 \sin y \, dx \, dy$

24. $\displaystyle \int_0^2 \int_0^{\pi/2} y \cos x \, dx \, dy$

25. $\displaystyle \int_{-2}^1 \int_0^x (x + y) \, dy \, dx$

26. $\displaystyle \int_1^2 \int_1^{2y} xy \, dx \, dy$

27. Evaluate $\int_R \int (x^2 + y) \, dA$; R is the region bounded by $y = 0$, $y = x$, and $x = 3$.

28. Evaluate $\int_R \int (x + y) \, dA$; R is the region bounded by $x = 0$, $y = 0$, and $x + y = 2$.

29. Find the first-octant volume bounded by $z = 0$, $z = 1$, and $y = 4 - x^2$.

30. Find the first-octant volume bounded by $z = 0$, $z = 1$, and the cylinder $y = 2x - x^2$.

31. Find the first-octant volume bounded by $z = 10 - x - y$.

32. Find the first-octant volume under the surface $z = x^2$ bounded by $x = 2$ and $y = 4$.

33. Find the volume in the first octant bounded by the surfaces $y^2 = x$, $z = 1 - x$, $y = 0$, and $z = 0$.

CHAPTER 13

Differential Equations

What does the passenger in a car feel when the tire of the car hits a bump in the road? How long will it take a polluted lake to return to its natural state once pollution is stopped? Does the quantity of fuel burned by a rocket affect its velocity? The answers to these questions are obtained by using techniques like those described in this chapter.

13.1 SOME ELEMENTARY TERMINOLOGY

An equation that includes at least one derivative of some unknown function of one variable is called an **ordinary differential equation**. Some examples of ordinary differential equations follow.

$$\frac{dy}{dx} = e^{2x} \qquad\qquad \frac{d^2y}{dx^2} + \sin y = 0$$

$$y' = y^2 + 4 \qquad\qquad xy' + y = 0$$

$$y'' + 4y = 0 \qquad\qquad yy' = x$$

$$y'' + (y')^3 - 2y = 0 \qquad x^3y'' + 3y = \sin x$$

In each of these equations, the letter y represents the unknown function, and x stands for the independent variable.

The **order** of a differential equation is the highest-order derivative occurring in the equation. For instance, the third and fourth equations in the first column and the first and fourth equations in the second column are of the second order, and the others are of the first order.

A **solution** to a differential equation is a function free of derivatives that *satisfies* the differential equation. That is, if $y = f(x)$ is a solution to a certain differential equation, then the differential equation will be reduced to an *identity* if y and the derivatives of y are replaced by $f(x)$ and its respective derivatives.

EXAMPLE 1   ■ **Problem**

Show that $y = \sin x$ satisifes the differential equation $y'' + y = 0$.

■ **Solution**

Since $y = \sin x$, we have

$$y' = \cos x \quad \text{and} \quad y'' = -\sin x$$

Substituting these values into the given differential equation yields the identity

$$y'' + y = -\sin x + \sin x = 0$$

Hence, $y = \sin x$ is a solution.   ■

EXAMPLE 2   ■ **Problem**

Show that $y = 8 + 5e^{-3x}$ is a solution of $\dfrac{d^2y}{dx^2} + 3\dfrac{dy}{dx} = 0$.

■ **Solution**

We need the first and second derivative of $y = 8 + 5e^{-3x}$. Thus,

$$y' = -15e^{-3x} \quad \text{and} \quad y'' = 45e^{-3x}$$

Substituting these values into $(d^2y/dx^2) + 3(dy/dx) = 0$, we get

$$45e^{-3x} + 3(-15e^{-3x}) = 0$$

which verifies that $y = 8 + 5e^{-3x}$ is a correct solution.   ■

In the first two examples we showed how to verify a solution of a differential equation. Naturally, we also want to be able to find such solutions. In Chapter 5, you studied differential equations of the form $dy/dx = f(x)$ and found the solutions to be

$$y = \int f(x)\, dx + C$$

where C is an arbitrary constant. We call $y = \int f(x)\, dx + C$ the **general solution** of $dy/dx = f(x)$.

EXAMPLE 3   ■ **Problem**

Solve the differential equation $\dfrac{dy}{dx} = 3x^2$.

■ **Solution**

By integration, we obtain

$$y = x^3 + C$$

as the general solution of the given differential equation.   ■

The value of the arbitrary constant is determined by imposing conditions on the solution. When sufficient information is given to determine the value of the arbitrary constant, the resulting solution is called a **particular solution**. The usual method of specifying an initial condition is by giving the value of y for a value of x. A differential equation with an initial condition is called an **initial-value problem**.

EXAMPLE 4   ■ **Problem**

Solve the differential equation $y' = \sqrt{4 - 3x}$ given $x = 1$ and $y = 5$.

■ **Solution**

The general solution is

$$y = \int (4 - 3x)^{1/2}\, dx = -\tfrac{1}{3} \int (4 - 3x)^{1/2}(-3)\, dx = -\tfrac{2}{9}(4 - 3x)^{3/2} + C$$

To find the value of C, we let $x = 1$ and $y = 5$. Thus,

$$5 = -\tfrac{2}{9}(4 - 3\cdot 1)^{3/2} + C$$
$$C = \tfrac{47}{9}$$

The particular solution is then

$$y = -\tfrac{2}{9}(4 - 3x)^{3/2} + \tfrac{47}{9}$$ ■

EXAMPLE 5   ■ **Problem**

The acceleration (in centimeters per second squared) of a particle is given by the formula $a = 6t + 10$. The initial velocity of the particle is 3 cm/sec. What is its velocity at $t = 2$ sec?

■ **Solution**

Here, you should recall that $a = dv/dt$, where $v = $ velocity and $a = $ acceleration. Hence, $dv/dt = 6t + 10$, and the equation for the velocity is

$$v = \int (6t + 10)\, dt = 3t^2 + 10t + C$$

The constant C is evaluated by using the conditions $v = 3$ when $t = 0$ in this equation. Thus,

$$3 = 3(0)^2 + 10(0) + C$$

or

$$C = 3$$

and

$$v = 3t^2 + 10t + 3$$

The velocity of the particle at $t = 2$ sec is

$$v = 3(2)^2 + 10(2) + 3 = 35 \text{ cm/sec}$$ ■

EXERCISES FOR SECTION 13.1

In Exercises 1–12, show that the equation on the right is a solution of the corresponding differential equation.

Differential equation   *Solution*

1. $\dfrac{dy}{dx} = 2x$ $y = x^2 + C$

2. $\dfrac{dy}{dx} = 3x^2 + 2$ $y = x^3 + 2x + C$

3. $t\dfrac{ds}{dt} - 2s = 0$ $s = Kt^2$

4. $x\dfrac{dy}{dx} - 2y = -x$ $y = x + cx^2$

5. $\dfrac{dV}{dr} + V = \cos r - \sin r$ $V = \cos r + ce^{-r}$

6. $\dfrac{dy}{dx} - 2y = 3e^{2x}$ $y = (3x + C)e^{2x}$

7. $\dfrac{d^2y}{dx^2} - \dfrac{dy}{dx} - 2y = 0$ $y = Ae^{-x} + Be^{2x}$

8. $\dfrac{d^2s}{dt^2} + 16s = 0$ $s = C_1 \sin 4t + C_2 \cos 4t$

9. $\dfrac{d^2y}{dx^2} + 3\dfrac{dy}{dx} = 0$ $y = A + Be^{-3x}$

10. $\dfrac{d^2s}{dt^2} - s = 1$ $s = C_1e^t + C_2e^{-t} - 1$

11. $\dfrac{d^2y}{dx^2} + \dfrac{dy}{dx} = -\cos x$ $y = c_1 + c_2e^{-x} + \frac{1}{2}\cos x - \frac{1}{2}\sin x$

12. $\dfrac{d^2i}{dt^2} - \dfrac{di}{dt} - 2i = 4t$ $i = c_1e^{-t} + c_2e^{2t} + 1 - 2t$

Solve the differential equations in Exercises 13–24.

13. $\dfrac{dy}{dx} = \dfrac{1}{x + 2}$ 14. $\dfrac{dS}{dt} = e^{2t}$

15. $\dfrac{dy}{dx} = \sin 2x$ 16. $\dfrac{dy}{dx} = 3 \tan \frac{1}{2}x$

17. $\dfrac{dy}{dx} = 2x(x^2 + 3)^{3/2}$ 18. $y' = \dfrac{x}{\sqrt{3x^2 + 2}}$

19. $y' = \sqrt{x + 3}$ 20. $\dfrac{dy}{dx} = \sec^2 x$

21. $\dfrac{dy}{dx} = \dfrac{1}{x^2 + 4}$

22. $\dfrac{ds}{dt} = \dfrac{t^2 + 2}{t^3 + 6t}$

23. $\dfrac{dy}{dx} = x \cos 3x$

24. $\dfrac{ds}{dt} = te^{-t}$

25. The graph of a function passes through the point (2, 9). What is the function if its slope is given by $dy/dx = 3x^2$?

26. The velocity (in centimeters per second) of an object is given by $30/\sqrt{6t + 4}$. How far has it traveled after 2 sec if its initial displacement is zero?

27. The shear in a beam is given by $S = dM/dx$, where M is the bending moment and x is the distance from one end of the beam. The shear equation is $S = (3 + 2x)^{-2}$. What is the equation for the bending moment? Assume $M = 10$ when $x = 0$.

28. Electric current is defined as $i = dq/dt$, where q is the electric charge and t is time. The current (in amperes) in a capacitor is $i = 4e^{-2t}$. What is the expression for the electric charge? Assume $q = 0$ when $t = 0$.

13.2 SEPARABLE EQUATIONS

A first-order differential equation that can be put into the form

$$g(y)\, dy + f(x)\, dx = 0$$

is said to be a **separable equation.** The process of associating $g(y)$ with dy and $f(x)$ with dx is called **separating the variables.** In most cases, the differential equation is not given in separated form, so we must perform some algebraic manipulation. Once the variables are separated, the general solution is given by

$$\int g(y)\, dy + \int f(x)\, dx = C$$

Warning: The coefficient of dx must be a function of x only, and the coefficient of dy must be a function of y only before the integration is performed.

EXAMPLE 1 ■ **Problem**

Solve the equation $3(y^2 + 1)\, dx + 2xy\, dy = 0$.

■ **Solution**

As written, this equation is not separated. However, the variables can be separated by dividing each term by $x(y^2 + 1)$. Performing this division, we get

$$\frac{3\, dx}{x} + \frac{2y\, dy}{y^2 + 1} = 0$$

Integrating each term, we obtain the solution

$$3 \ln x + \ln(y^2 + 1) = C$$

The solution may be simplified by using properties of logarithms. Hence,

$$\ln x^3 + \ln(y^2 + 1) = C$$
$$\ln[x^3 (y^2 + 1)] = C$$

is another form of the solution. An additional simplification can be made by writing the arbitrary constant in the form $C = \ln C_1$, so that

$$\ln x^3(y^2 + 1) = \ln C_1$$

Therefore,

$$x^3(y^2 + 1) = C_1 \qquad \blacksquare$$

EXAMPLE 2 ▪ **Problem**

Solve the equation $\dfrac{dy}{dx} = \dfrac{xy}{\sqrt{x^2 - 4}}$.

▪ **Solution**

Multiplying by dx and dividing by y yields

$$\frac{dy}{y} = x(x^2 - 4)^{-1/2} \, dx$$

Integrating both sides, we get

$$\ln y = (x^2 - 4)^{1/2} + C_1$$
$$\ln y - \ln C = (x^2 - 4)^{1/2} \qquad \text{letting } C_1 = \ln C \text{ and transposing}$$

$$\ln \frac{y}{C} = (x^2 - 4)^{1/2} \qquad\qquad \ln y - \ln C = \ln \frac{y}{C}$$

$$\frac{y}{C} = e^{(x^2-4)^{1/2}} \qquad\qquad \ln a = b \text{ means } a = e^b$$

$$y = Ce^{(x^2-4)^{1/2}} \qquad\qquad \text{multiplying both sides by } C \qquad \blacksquare$$

EXAMPLE 3 ▪ **Problem**

Find the particular solution for $\dfrac{dy}{dx} = xy^2 e^x$ given that $y = 2$ when $x = 0$.

▪ **Solution**

To separate the variables, we multiply both sides by $y^{-2} \, dx$. Performing this operation, we get

$$y^{-2}\, dy = xe^x\, dx$$

The left side is of the form $\int y^n\, dy$. To integrate the expression on the right side, we use integration by parts with $u = x$ and $dv = e^x\, dx$. From this choice of parts, we get $du = dx$ and $v = e^x$. So,

$$\int xe^x\, dx = xe^x - \int e^x\, dx = xe^x - e^x$$

Thus, integrating both sides of the separated equation yields the general solution

$$-\frac{1}{y} = xe^x - e^x + C$$

Now, letting $x = 0$ and $y = 2$, we have

$$-\tfrac{1}{2} = 0 - 1 + C$$

Solving for C gives

$$C = \tfrac{1}{2}$$

The particular solution satisfying the given conditions is then

$$-\frac{1}{y} = xe^x - e^x + \frac{1}{2} \qquad \text{or} \qquad y = \frac{2}{2e^x - 2xe^x - 1} \qquad \blacksquare$$

Separable equations arise in some very important physical problems. For example, population growth, electric current in a capacitor, and radioactive decay are all described by separable differential equations. Each of these examples has one thing in common—the time rate of change of the quantity is proportional to the quantity itself. Quantities that vary in this manner are described by the differential equation

$$\frac{dy}{dt} = ky$$

which may be solved by the method of separation of variables to give

$$y = ce^{kt}$$

The proportionality constant k is fixed by the particular differential equation, while the constant c is the arbitrary constant of integration.

EXAMPLE 4 ■ **Problem**

The electric circuit shown in Figure 13.1 containing a resistor and capacitor is called an *RC* circuit. When the switch is closed, there is an initial charging current (transient current) in the capacitor. The transient current i in a capacitor increases at a rate equal to $(-1/RC)i$, where $1/RC$ is the *time constant* of the circuit. What is the expression for the current in the capacitor at any time t if $i = V/R$ when $t = 0$?

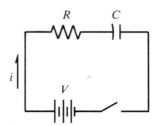

■ **Figure 13.1**

■ **Solution**

The rate of increase of the transient current is given by

$$\frac{di}{dt} = -\frac{1}{RC}i$$

Separating variables, we obtain

$$\frac{di}{i} = -\frac{1}{RC}dt$$

$$\ln i - \ln K = -\frac{t}{RC}$$ integrating both sides; adding a constant in the form $(-\ln K)$

$$\ln \frac{i}{K} = -\frac{t}{RC}$$ $\ln i - \ln K = \ln \dfrac{i}{K}$

$$i = Ke^{-t/RC}$$ $\ln a = b$ means $a = e^b$; multiplying both sides by K

Letting $t = 0$ and $i = V/R$ in this equation, we find $K = V/R$. Consequently,

$$i = \frac{V}{R}e^{-t/RC}$$

is the equation for the transient current. This equation is graphed in Figure 13.2.

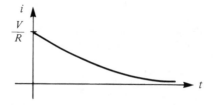

■ **Figure 13.2**   ■

We know from empirical evidence that a radioactive substance decomposes at a rate proportional to its mass. This rate is called the **decay rate.** Thus, if $M(t)$

represents the mass of a substance at any time, then dM/dt is equal to some constant k times $M(t)$. The **half-life** of a substance is the amount of time required for it to decay to one-half of its initial mass.

EXAMPLE 5 ■ **Problem**

A radioactive isotope has an initial mass of 100 mg, which two years later is 75 mg.

a. Find the expression for the amount of the isotope remaining at any time.

b. Determine the half-life of the isotope.

■ **Solution**

a. Let M be the mass of the isotope remaining after t years. Then, the rate of decomposition is given by

$$\frac{dM}{dt} = -kM \qquad\qquad \text{the sign indicates the mass is decreasing}$$

$$\frac{dM}{M} = -k\, dt \qquad\qquad \text{separating variables}$$

$$\ln M - \ln C = -kt \qquad \begin{array}{l}\text{integrating and adding a constant} \\ \text{in the form of } (-\ln C)\end{array}$$

$$\ln \frac{M}{C} = -kt \qquad\qquad \text{combining logarithms}$$

$$\frac{M}{C} = e^{-kt} \qquad\qquad \text{definition of a natural logarithm}$$

$$M = Ce^{-kt} \qquad\qquad \text{multiplying by } C \qquad\qquad\qquad\qquad \textbf{(13.1)}$$

To find the value of C, we recall that $M = 100$ when $t = 0$. Substituting these values into Equation (13.1), we get $C = 100$ and

$$M = 100e^{-kt} \qquad\qquad\qquad\qquad\qquad\qquad\qquad\qquad\qquad\qquad \textbf{(13.2)}$$

The value of k may now be determined by substituting the condition $t = 2$ and $M = 75$ into Equation (13.2). Thus,

$$75 = 100e^{-2k}$$

$$e^{-2k} = \frac{3}{4} \qquad\qquad \begin{array}{l}\text{dividing by 100; canceling} \\ \text{common factors}\end{array}$$

$$-2k = \ln \frac{3}{4} \qquad\qquad \text{taking logarithms of both sides}$$

$$k = -\frac{1}{2} \ln \frac{3}{4} \qquad\qquad \text{dividing by } -2$$

$$= -\frac{1}{2}(-0.2877) \approx 0.14 \qquad \text{evaluating } \ln \frac{3}{4}$$

The mass of the isotope remaining after t years is then given by

$$M = 100e^{-0.14t}$$

b.   The half-life t_h is the time corresponding to $M = 50\,\text{mg}$. Thus,

$$50 = 100e^{-0.14t_h}$$

$$\tfrac{1}{2} = e^{-0.14t_h}$$

$$t_h = -\frac{1}{0.14}\ln 0.5 = \frac{-0.693}{-0.14} = 4.95 \text{ years}$$

∎

The next example is concerned with the process of heating and cooling. Experiments show that if an object is immersed in a medium, and if the medium is kept at a constant temperature, a good approximation to the temperature of the object can be found using *Newton's law of cooling.*

LAW

> **Newton's Law of Cooling**   The temperature Q of an object will change at a rate proportional to the difference in temperature of the object and the temperature Q_a of the surrounding medium.

EXAMPLE 6   ▪ **Problem**

A thermometer reading $100°F$ is placed in a pan of oil maintained at $10°F$. What is the temperature of the thermometer when $t = 10\,\text{sec}$ if its temperature is $60°F$ when $t = 4\,\text{sec}$?

▪ **Solution**

The physical situation may be described mathematically by using Newton's law of cooling. If Q is the temperature of the thermometer at any time t, then dQ/dt is the time rate of change of temperature and $Q - 10$ is the difference between the temperature of the thermometer and that of the oil. The conditions are $Q = 100$ when $t = 0$ and $Q = 60$ when $t = 4$. Thus, the problem to be solved is

$$\frac{dQ}{dt} = -k(Q - 10) \qquad Q(0) = 100 \qquad Q(4) = 60$$

where k is the constant of proportionality (the minus sign is used to indicate that the temperature of the thermometer will decrease). So,

$$\frac{dQ}{Q - 10} = -k\,dt$$

Integrating, we get

$$\ln(Q - 10) = -kt + \ln C$$

which leads to

$$Q = Ce^{-kt} + 10$$

The constant C is evaluated by using $T(0) = 100$, to get

$$100 = C + 10 \quad \text{or} \quad C = 90$$

So, we have

$$Q = 90e^{-kt} + 10$$

To solve for k, we use the second condition $Q(4) = 60$. Making the substitution gives

$$60 = 90e^{-4k} + 10$$

or

$$k = \tfrac{1}{4} \ln \tfrac{9}{5} = 0.147$$

Therefore, the temperature-time equation is

$$Q = 90e^{-0.147t} + 10$$

The temperature variation is shown graphically in Figure 13.3. Notice that the limiting temperature is $10°F$.

Finally, the temperature of the thermometer at $t = 10 \sec$ is

$$Q(10) = 90e^{-1.47} + 10 = 90(0.2299) + 10 = 30.7°F$$

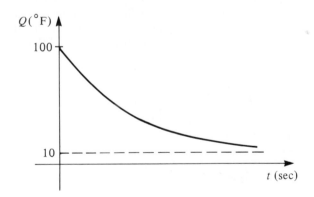

■ **Figure 13.3**

<hr>

EXERCISES FOR SECTION 13.2

In Exercises 1–20, obtain the family of solutions of each of the differential equations.

1. $y\,dx + x\,dy = 0$

2. $x^2\,dy + y^2\,dx = 0$

3. $xy\,dx + y^2\,dy = 0$

4. $xy\,dx + (1 + x^2)\,dy = 0$

5. $ds/dt = ts^3$

6. $y\, dx + (3x + 2)\, dy = 0$

7. $x(1 - y)\, dx + (x^2 - 2)\, dy = 0$

8. $\cot y\, dx = x\, dy$

9. $dx + (1 + x^2)y\, dy = 0$

10. $dp/dt = 2 - p$

11. $xy\, dx + e^x\, dy = 0$

12. $\dfrac{dq}{dt} + \dfrac{t \sin t}{q} = 0$

13. $dx + e^{x+y}\, dy = 0$

14. $(v - 1)(t + 3)\, dv$
$\qquad - (v^2 - 2v - 3)\, dt = 0$

15. $(x^2 - 1)\, dy + x(y^2 + 5)\, dx = 0$

16. $e^x \cos^2 y\, dx - (1 + e^x)\, dy = 0$

17. $\tan x\, dx + y \cos^2 x\, dy = 0$

18. $(x + 1)\, dy = 2xy\, dx$

19. $x^2\, (dy/dx) = y(1 - x)$

20. $4\, dt = \sqrt{4 - t^2}\, ds$

In Exercises 21–26, obtain the particular solution of the differential equation for the given conditions.

21. $(y + 2)\, dx + (x - 4)\, dy = 0;\ x = 2,\ y = 0$

22. $ds/dt = -t/s;\ t = -1,\ s = 3$

23. $xy\, dx + \sqrt{9 + x^2}\, dy = 0;\ x = 4,\ y = 1$

24. $\cos y\, dx + x \sin y\, dy = 0;\ x = 3,\ y = \pi/3$

25. $(1 + x^2)\, dy + (1 + y^2)\, dx = 0;\ x = 1,\ y = \sqrt{3}$

26. $dy + y\, dx = x^2\, dy;\ x = 2,\ y = e$

27. The circuit shown in the accompanying figure, consisting of a resistor and an inductor, is called an *RL* circuit. When the switch is *opened,* the current in the circuit does not instantly drop to zero but decays at a rate equal to $-Ri/L$. If the initial current is 50 A, what is the value of the current in a circuit with $R = 1\ \Omega$, $L = 0.1$ H, and $E = 50$ V, at 1 sec after the opening of the switch?

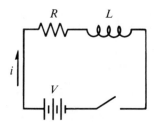

$R \qquad\qquad L$

i

V

28. A certain radioactive substance decays at a rate that is always equal to 25% of the mass of the substance. What is the formula for the mass of the substance remaining after t hours if $M = 200$ g when $t = 0$?

29. What is the half-life of a radioactive substance of which 20% disappears in 100 years?

30. After 20 years, a quantity of radioactive material has decayed to 60 g; at the end of 50 years, it has decayed to 40 g. How many grams were there in the first place?

31. The half-life of carbon 14 (C^{14}) is 5568 years.

 (a) Show that the formula for the mass at time t is

$$m = m_0 e^{-0.0001245t}$$

 One of the tools used by archeologists to estimate the date of origin of organic artifacts is *carbon dating,* a technique developed by W. F. Libby in the late 1950s. Carbon dating is based on the principles of radioactive decay of C^{14} in plants and animals that have died. Cosmic rays produce neutrons in the atomosphere that combine with nitrogen to form C^{14}. In living plants and animals, the rate of absorption of C^{14} is balanced by its rate of decay, so the amount of C^{14} is in a state of equilibrium. When the living organism dies, C^{14} atoms decay, but no new C^{14} is absorbed. The decay process at death is described by the equation given here. The date when the organism died can then be estimated by measuring the C^{14} level in the specimen and using this equation to compute the corresponding time. Carbon dating assumes that the level of C^{14} in living organisms is the same today as it was in the past.

 (b) A piece of charcoal from an archeological excavation is found to contain 15% as much C^{14} as living wood. Use the equation in part (a) to estimate the age of the specimen.

32. A steel ball is heated to a temperature of 80° and at time $t = 0$ is placed in water, maintained at 30°. At $t = 3$ min, the temperature of the ball is 55°. When is the temperature of the ball reduced to 40°?

33. A steel casting at a temperature of 20° is put into an oven that has a temperature of 200°. One minute later, the temperature of the casting is 30°.

 (a) Find the temperature of the casting 5 min after it is put into the oven.

 (b) How long will it take the temperature of the casting to reach 190°?

 34. The intensity of light I emitted by the phosphor of a television tube changes at a rate proportional to the intensity I. Find the expression for the intensity of light being emitted at any time t given that the initial intensity is $I = 25$ and the intensity at $t = 1$ is $I = 20$.

35. The graph of a function passes through the point $(-2, 3)$, and its slope is given by $y' = -x/2y$. Find the function, and sketch its graph.

36. The graph of a function passes through the point $(0, 3)$. What is the function if its slope is given by $y' = 2(2 - y)$?

■ 13.3 LINEAR EQUATIONS OF THE FIRST ORDER

A first-order differential equation that can be put into the form

$$a_1(x)y' + a_0(x)y = r(x)$$

is called **linear.** By dividing each member of this equation by the coefficient $a_1(x)$, we can write it in the following form.

EQUATION

> **Standard Form of a Linear Differential Equation of the First Order**
>
> $$y' + p(x)y = q(x) \tag{13.3}$$

　　Essential to understanding the solution of linear first-order differential equations is the concept of the *total differential* of a function. Recall from Section 12.3 that the total differential of $f(x, y)$ is given by

$$df = \frac{\partial f}{\partial x}\, dx + \frac{\partial f}{\partial y}\, dy$$

For example, the total differential of $x^2 e^{3y}$ is $2xe^{3y}\, dx + 3x^2 e^{3y}\, dy$.

EXAMPLE 1　■ **Problem**

Find a function whose total differential is $3x^2 y^2\, dy + 2xy^3\, dx$.

■ **Solution**

Since $\partial f/\partial y = 3x^2 y^2$ and $\partial f/\partial x = 2xy^3$, then $f(x, y) = x^2 y^3$. Therefore,

$$3x^2 y^2\, dy + 2xy^3\, dx = d(x^2 y^3) \qquad\qquad ■$$

EXAMPLE 2　■ **Problem**

Find a function whose total differential is $2x \sin y\, dx + x^2 \cos y\, dy$.

■ **Solution**

Here, $\partial f/\partial x = 2x \sin y$, and $\partial f/\partial y = x^2 \cos y$; so $f(x, y) = x^2 \sin y$. Consequently, we can write the given expression in the form $d(x^2 \sin y)$. ■

　　The differential equation

$$3x^2 y^2\, dy + 2xy^3\, dx = 0 \tag{13.4}$$

can be written, using the result in Example 1, as

$$d(x^2 y^3) = 0$$

Noticing that $\int d(x^2 y^3) = x^2 y^3$, we conclude that the general solution of Equation (13.4) is

$$x^2 y^3 = c$$

　　Differential equations that consist of terms that are the total differential of a function are called **exact differential equations.** The reason we point this feature out is that every first-order linear differential equation can be made exact by multiplying Equation (13.3) by an appropriate function of x, called an **integrating factor**. An integrating factor for a first-order linear differential equation in standard form, (13.3), is given by the following formula.

FORMULA

> **Integrating Factor for $y' + p(x)y = q(x)$**
>
> $$v(x) = e^{\int p(x)\,dx}$$

Multiplying Equation (13.3) by this integrating factor makes the left side the total differential of the dependent variable times the integrating factor, that is, $d(y \cdot e^{\int p(x)\,dx})$. The solution to (13.3) is then as follows:

FORMULA

> **Solution of $y' + p(x)y = q(x)$**
>
> $$ye^{\int p(x)\,dx} = \int q(x)e^{\int p(x)\,dx}\,dx$$

The technique is summarized in four steps.

PROCEDURE

> **Solving a First-Order Linear Differential Equation**
> 1. Put the equation into standard form. Note that the coefficient of y' must be 1.
> 2. Identify $p(x)$, and compute $v(x) = e^{\int p(x)\,dx}$.
> 3. Multiply the standard form of the equation by $v(x)$.
> 4. Integrate both sides, and solve for y.

EXAMPLE 3   ■ **Problem**

Solve the equation $2y' - 4y = 16e^x$.

■ **Solution**

Rearranging in standard form, we get

$$y' - 2y = 8e^x.$$

We see that $p(x) = -2$, and therefore, the integrating factor is

$$v(x) = e^{-\int 2\,dx} = e^{-2x}$$

Multiplying the given differential equation (in standard form) by e^{-2x}, we get

$$e^{-2x}y' - 2ye^{-2x} = 8e^{-x}$$

$$e^{-2x}\,dy - 2ye^{-2x}\,dx = 8e^{-x}\,dx \qquad \text{multiplying by } dx$$

$$d(ye^{-2x}) = 8e^{-x}\,dx \qquad\qquad d(ye^{-2x}) = e^{-2x}\,dy - 2ye^{-2x}\,dx$$

$$ye^{-2x} = -8e^{-x} + C \qquad\qquad \text{integrating both sides}$$

$$y = -8e^{x} + Ce^{2x} \qquad\qquad \text{multiplying by } e^{2x} \qquad\qquad \blacksquare$$

EXAMPLE 4   ▪ **Problem**

Solve $y' = \dfrac{\cos x}{x^2} - \dfrac{2y}{x}$.

▪ **Solution**

Writing the equation in standard differential form, we have

$$dy + \frac{2}{x}y\,dx = \frac{\cos x}{x^2}\,dx$$

In this form, we see that $p(x) = 2/x$ and

$$v(x) = e^{\int 2\,dx/x} = e^{2\ln x} = e^{\ln x^2} = x^2$$

You may wonder how $e^{2\ln x} = x^2$. Since integrating factors of this type occur frequently, we will demonstrate the algebra involved. Let $v = e^{\ln x^2}$. Then,

$$\ln v = \ln e^{\ln x^2} \qquad\qquad \text{taking ln of both sides}$$

$$\ln v = \ln x^2 \ln e \qquad\qquad \ln a^y = y\ln a$$

$$\ln v = \ln x^2 \qquad\qquad\quad \ln e = 1$$

$$v = x^2 \qquad\qquad\qquad \text{equating arguments of equal logarithms}$$

Applying the integrating factor x^2 to the standard form of the given equation, we get

$$x^2\,dy + 2xy\,dx = \cos x\,dx$$

The desired solution is

$$x^2 y = \sin x + C \qquad\qquad\qquad\qquad\qquad\qquad\qquad \blacksquare$$

Warning: Always change integrating factors of the form $e^{\ln f(x)}$ into $f(x)$ before applying them.

EXERCISES FOR SECTION 13.3

In Exercises 1–10, find the function whose total differential is given.

1. $x\,dy + y\,dx$

2. $2xy\,dy + y^2\,dx$

3. $xy^2\,dx + x^2y\,dy$

4. $3ye^{3x}\,dx + e^{3x}\,dy$

5. $\dfrac{y\,dx}{x} + \ln x\,dy$

6. $x^{-1}\,dy - x^{-2}y\,dx$

7. $y\cos x\,dx + \sin x\,dy$

8. $\sec x\,dy + y\sec x\tan x\,dx$

9. $(x^2 + 1)^{-1}\,dy - 2xy(x^2 + 1)^{-2}\,dx$

10. $(6x + 5)^{1/2}\,dy + 3y(6x + 5)^{-1/2}\,dx$

Find the family of solutions of the differential equations in Exercises 11–28 by using the standard integrating factor for linear first-order equations.

11. $dy + 2y\,dx = 4\,dx$

12. $2\,dy + 8xy\,dx = x\,dx$

13. $dy - 4y\,dx = 3\,dx$

14. $\dfrac{di}{dt} + i - 2 = 0$

15. $\dfrac{dy}{dx} + \dfrac{y}{x} = x + 1$

16. $x^2\,dy - 2xy\,dx = dx$

17. $(t + 2q)\,dt - dq = 0$

18. $x(3 + y)\,dx + dy = 0$

19. $(t^2 + 1)\,ds + 2ts\,dt = 3\,dt$

20. $(x^2 - 4)\,dy + xy\,dx = x\,dx$

21. $(x^2 + 1)\,dy - xy\,dx = 2x\,dx$

22. $dy + 2y\,dx = e^{-x}\,dx$

23. $\dfrac{dv}{dr} + v\tan r = \cos r$

24. $dy = (\csc x - y\cot x)\,dx$

25. $dy - y\,dx = 2e^{-x}\,dx$

26. $\dfrac{di}{dt} + i = e^{-t}$

27. $t\dfrac{ds}{dt} + s = t$

28. $(y\cot x - x)\,dx + dy = 0$

In Exercises 29–32, find the particular solution corresponding to the given conditions.

29. $x\,dy - y\,dx = x^2 e^x\,dx;\ x = 1,\ y = 0$

30. $\dfrac{dy}{dx} + 2y = 4;\ x = 0,\ y = 1$

31. $\dfrac{dy}{dx} + y\cot x = 10;\ x = \dfrac{\pi}{3},\ y = 0$

32. $\dfrac{dr}{d\theta} + r\cot\theta = \sec\theta;\ \theta = \dfrac{\pi}{3},\ r = 2$

13.4 APPLICATIONS OF LINEAR EQUATIONS

Dynamics

Dynamics is the study of the motion of objects and the forces that cause motion. In this section, we study the motion of an object through a resisting medium. We begin with a basic definition and a fundamental principle.

DEFINITION

> **Momentum**   The momentum P of a moving object is the product of its mass m and its velocity v. As an equation, we write
>
> $$P = mv$$

PRINCIPLE

> **Newton's Second Principle**   The time rate of change of the momentum of an object is equal to the applied force F. Thus,
>
> $$F = \frac{d}{dt}(mv) \tag{13.5}$$

Comment: If the mass of the object is constant, Newton's second principle is usually stated in the form

$$F = m\frac{dv}{dt} = ma$$

where a is the acceleration of the object and m is its mass. This statement of the second principle is valid only if the mass is constant.

Newton's second principle, together with a knowledge of the technique of solving first-order linear equations, enables us to analyze and solve a rather contemporary problem from dynamics involving a rocket moving through the atmosphere. Consider a rocket moving in a straight, horizontal path as shown in Figure 13.4. Assume that the propelling force (called the *thrust*) can be described by a function F. If the rocket is moving through a resisting medium, we may represent the resulting *drag force* by the function D. Hence, the effective external force on the rocket is $F - D$. From Newton's second principle, we can write

$$\frac{d}{dt}(mv) = F - D \tag{13.6}$$

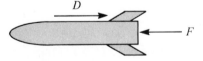

■ **Figure 13.4**

Since the mass of the rocket changes with time as the propellant is burned, and since the velocity is also changing due to the thrust, both m and v are functions of time. So, (13.6) becomes

$$m\frac{dv}{dt} + v\frac{dm}{dt} = F - D$$

This equation is linear in v and can be solved if we are given the expressions for the functions F, D, and m.

The drag force experienced by an object moving through a resisting medium typically increases as the velocity increases. Consequently, we can express drag force as a function of velocity. A common model for drag force assumes that it is directly proportional to its velocity, that is,

$$D = kv$$

This model agrees with experience as long as the velocity is not too great.

Comment: The use of Equation (13.6) requires consistent units. In all examples and exercises, length is in feet, force in pounds, time in seconds, and mass in slugs.

EXAMPLE 1   ▪ **Problem**

Consider a rocket that starts from rest and is propelled by a constant thrust of 2000 lb in a straight, horizontal path for 20 sec. Suppose as the propellant is burned, the mass (in slugs) of the rocket varies with time in accordance with $m = 50 - t$ and that the drag force is numerically equal to three times the velocity. Find the velocity-time equation of the rocket. What is its velocity when $t = 10$ sec?

▪ **Solution**

Note that $m = 50 - t$, $F = 2000$, and $D = 3v$. Using these results in the equation that follows Equation (13.6) gives

$$(50 - t)\frac{dv}{dt} + v\frac{d}{dt}(50 - t) = 2000 - 3v$$

or

$$(50 - t)\frac{dv}{dt} + 2v = 2000$$

When put into standard form, this equation becomes

$$\frac{dv}{dt} + \frac{2}{50 - t}v = \frac{2000}{50 - t}$$

Here, $p = 2/(50 - t)$, and so the integrating factor is $(50 - t)^{-2}$. Thus,

$$d[v(50 - t)^{-2}] = 2000(50 - t)^{-3}\,dt$$

Therefore,

$$v(50 - t)^{-2} = 1000(50 - t)^{-2} + C$$

so that

$$v = 1000 + C(50 - t)^2$$

Using the fact that $v = 0$ when $t = 0$, we get $C = -0.4$. Hence, the velocity equation is

$$v = 1000 - 0.4(50 - t)^2$$

This function is graphed in Figure 13.5.

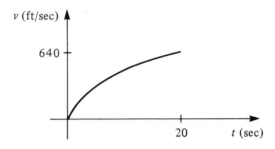

■ **Figure 13.5**

When $t = 10$ sec,

$$v(10) = 1000 - 0.4(1600) = 360 \text{ ft/sec}$$ ■

EXAMPLE 2   ■ **Problem**

A 96-lb test sled that rides on a horizontal track is propelled by a horizontal force that varies with time according to

$$F(t) = \begin{cases} t, & \text{if } 0 \le t < 3 \text{ sec} \\ 0, & \text{if } t \ge 3 \text{ sec} \end{cases}$$

See Figure 13.6. If the drag force is equal to the velocity of the sled, what is the velocity equation? Assume $v(0) = 0$.

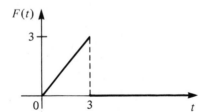

■ **Figure 13.6**

■ **Solution**

The mass of an object is obtained from its weight by

$$m = \frac{W}{g}$$

where W is the weight and g is the acceleration due to the force of gravity. This formula gives mass in slugs if weight is in pounds and g is in feet per second squared. (Throughout this book, we use $g = 32 \text{ ft/sec}^2$.)

In this problem, the mass of the sled is $m = \frac{96}{32} = 3$ slugs. The drag force experienced by the sled is equal to the velocity, so the drag force (in pounds) is given by $D = v$. Finally, since the thrust is defined by two rules, the velocity must be determined separately for $0 \le t < 3$ sec and $t \ge 3$ sec.

For $0 \le t < 3$, we must solve the initial-value problem

$$3 \frac{dv}{dt} = t - v \qquad v(0) = 0$$

Rearranging this equation, we get

$$\frac{dv}{dt} + \frac{v}{3} = \frac{t}{3}$$

The integrating factor is $e^{\int 1/3 \, dt} = e^{t/3}$. For this integrating factor, the equation becomes

$$d(v e^{t/3}) = \frac{t}{3} e^{t/3} \, dt$$

Integration by parts on the right-hand side yields the solution

$$v e^{t/3} = t e^{t/3} - 3 e^{t/3} + c$$

or

$$v = t - 3 + c e^{-t/3}$$

Since $v = 0$ when $t = 0$, the value of the arbitrary constant is $c = 3$. Thus,

$$v = t - 3(1 - e^{-t/3})$$

For $t \ge 3$ sec, $F(t) = 0$. So, we must solve the initial-value problem

$$3 \frac{dv}{dt} = -v \qquad v(3) = 3e^{-1}$$

where $v(3)$ is obtained from the solution for the preceding interval. By separating variables, we get

$$\frac{dv}{v} = -\frac{1}{3} \, dt$$

or

$$v = c_1 e^{-t/3}$$

Substituting $t = 3$, $v = 3e^{-1}$, we get $c_1 = 3$. Thus,

$$v = 3 e^{-t/3}$$

Combining these results, we write the velocity function as

$$v = \begin{cases} t - 3(1 - e^{-t/3}), & \text{if } 0 \le t < 3 \text{ sec} \\ 3e^{-t/3}, & \text{if } t \ge 3 \text{ sec} \end{cases}$$

This function is graphed in Figure 13.7.   ■

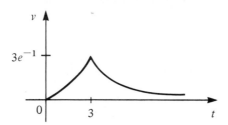

■ Figure 13.7

Mixture Problems

Another problem that is modeled with a linear differential equation involves the process of mixing liquids that contain dissolved substances such as salt or dye. A typical situation involves a tank filled with a liquid containing a concentration (the amount of the dissolved substance per unit volume of the liquid) of some substance, a liquid flowing into the tank with a different concentration of the same substance, and the simultaneous release of the thoroughly mixed solution from a valve at the bottom of the tank. The problem is to determine the amount Q of the dissolved substance in the tank at any time t. Assuming the two liquids are thoroughly mixed, we can express the rate at which the substance is accumulating in the tank by the following formula.

FORMULA

> **Mixture Formula**
>
> $$\frac{dQ}{dt} = (\text{rate of substance in}) - (\text{rate of substance out}) \qquad \textbf{(13.7)}$$

This differential equation must satisfy the condition $Q(t_0) = Q_0$.

EXAMPLE 3   ■ Problem

A tank with a capacity of 300 gal initially contains 100 gal of pure water. A salt solution containing 3 lb of salt per gallon is allowed to run into the tank at a rate of 8 gal/min, and the mixture is then removed at a rate of 6 gal/min. See Figure 13.8. Find the expression for the number of pounds of salt in the tank at any time t.

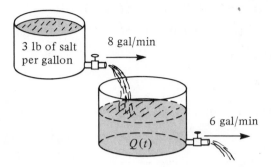

■ **Figure 13.8**

■ **Solution**

Let Q be the amount of salt in the tank at any time t. Since the water is initially pure, $Q(0) = 0$. The rate at which salt is being added to the tank is

$$(3 \text{ lb/gal})(8 \text{ gal/min}) = 24 \text{ lb/min}$$

The rate at which brine is entering the tank is greater than the rate at which the mixture is being removed, so at any time t, the amount of solution in the tank is increasing at the rate of

$$8 \text{ gal/min} - 6 \text{ gal/min} = 2 \text{ gal/min}$$

Therefore, the number of gallons of solution in the tank at any time t is given by $n = 100 + 2t$, and the concentration of salt in solution in the tank is given by

$$\frac{Q}{100 + 2t}$$

The rate at which salt is being removed from the tank is then

$$\left(\frac{Q}{100 + 2t} \text{ lb/gal}\right)(6 \text{ gal/min}) = \frac{6Q}{100 + 2t} \text{ lb/min}$$

Combining the previous results in (13.7), we write the initial-value problem as

$$\frac{dQ}{dt} = 24 - \frac{6Q}{100 + 2t}, \qquad Q(0) = 0$$

The linear equation

$$\frac{dQ}{dt} + \frac{3Q}{50 + t} = 24$$

has an integrating factor

$$v = e^{\int 3/(50+t)\,dt} = e^{3\ln(50+t)} = (50 + t)^3$$

This integrating factor yields

$$Q(50 + t)^3 = \int 24(50 + t)^3 \, dt = 6(50 + t)^4 + c$$

Solving for Q, we get

$$Q = 6(50 + t) + c(50 + t)^{-3}$$

The initial condition, $Q(0) = 0$, yields $c = -300(50^3)$. Hence, the desired equation is

$$Q = 6(50 + t) - 300\left(\frac{50}{50 + t}\right)^3$$ ∎

Electric Circuits

In elementary physics, you learned that a constant voltage V applied across a conductor produces a current i that is directly proportional to the constant applied voltage. This relationship, known as **Ohm's law**, is written as follows.

FORMULA

> **Ohm's Law**
>
> $$V = iR$$
>
> where R is a constant called the *resistance* of the circuit.

Circuits that involve constant voltage and resistance can be analyzed by using Ohm's law and do not require the solution of differential equations. However, if the circuit contains an inductance, the current changes with time, and Ohm's law does not completely describe the circuit. Because of the inductance, there is a voltage induced in the circuit, defined as follows.

FORMULA

> **Induced Voltage**
>
> $$v_L = L \frac{di}{dt}$$
>
> where L is a constant of proportionality called the *inductance* of the circuit.

A circuit including a resistance and an inductance is shown in Figure 13.9. The components are said to be in a series arrangement. A circuit of this type obeys Kirchhoff's second law.

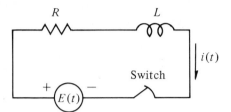

■ **Figure 13.9**

LAW

> **Kirchhoff's Second Law** At any instant, the algebraic sum of the voltage drops around a circuit is zero.

Thus, in the *RL* series circuit shown in Figure 13.9, we have

$$v_L + v_R = E(t)$$

or

$$L\frac{di}{dt} + Ri = E(t) \tag{13.8}$$

where $E(t)$ is the applied, or driving, voltage. The current $i(t)$ is then the solution of a first-order linear differential equation.

Comment: Consistent electrical units are voltage in volts, resistance in ohms, current in amperes, and inductance in henrys.

EXAMPLE 4 ■ **Problem**

Find the current in a series *RL* circuit in which resistance, inductance, and voltage are constant. Assume the initial current is zero.

■ **Solution**

Replacing $E(t)$ with V in Equation (13.8), we have

$$L\frac{di}{dt} + Ri = V$$

which can be written in standard form as

$$\frac{di}{dt} + \frac{R}{L}i = \frac{V}{L}$$

An integrating factor for this equation is $e^{Rt/L}$, which, when applied to the standard form, yields

$$d(ie^{Rt/L}) = \frac{V}{L} e^{Rt/L} \, dt$$

Integrating, we get

$$ie^{Rt/L} = \frac{V}{R} e^{Rt/L} + C$$

Or solving for i, we have

$$i = \frac{V}{R} + Ce^{-Rt/L}$$

The condition $i(0) = 0$ yields $C = -V/R$. Therefore, the current is

$$i = \frac{V}{R} (1 - e^{-Rt/L}) \qquad\qquad\qquad (13.9)$$

The graph of this function is shown in Figure 13.10.

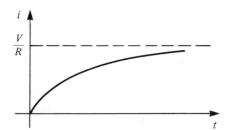

■ **Figure 13.10**

Comment: Equation (13.9) is a very famous formula for the response of a series RL circuit to a constant voltage. Notice that the equation is valid only for circuits in which R, L, and V are constants. Can you explain where, in the derivation of this formula, we must use this fact?

EXAMPLE 5 ■ **Problem**

A 6-V battery is applied in series with a resistance of 1 Ω and an inductance (in henrys) that varies with time according to

$$L = \begin{cases} 2 - t, & \text{if } 0 \le t < 2 \text{ sec} \\ 0, & \text{if } t \ge 2 \text{ sec} \end{cases}$$

Find the expression for the current in the circuit. Assume $i(0) = 0$. [Note that we may not use Equation (13.9) to solve this problem.]

■ **Solution**

On the interval $0 \leq t < 2$, the inductance is $2 - t$. So,

$$(2 - t) \frac{di}{dt} + i = 6$$

or, the standard form is

$$\frac{di}{dt} + \frac{1}{2 - t} i = \frac{6}{2 - t}$$

Using $p = 1/(2 - t)$, an integrating factor is

$$v = e^{-\ln(2-t)} = (2 - t)^{-1}$$

Multiplying the standard form by the integrating factor, we get

$$d[i(2 - t)^{-1}] = 6(2 - t)^{-2} \, dt$$

Integrating yields

$$i(2 - t)^{-1} = 6(2 - t)^{-1} + C$$

or

$$i = 6 + C(2 - t)$$

The value of C is found by using $i(0) = 0$ in this equation. Thus,

$$0 = 6 + 2C$$

or

$$C = -3$$

So, the equation for the current for $0 \leq t < 2$ is

$$i = 6 - 3(2 - t) = 3t$$

For $t \geq 2$, the inductance is equal to zero, so

$$0 \frac{di}{dt} + i = 6$$

which means $i = 6$ A for $t \geq 2$.

   Summarizing what we know about the current in this circuit, we see that

$$i = \begin{cases} 3t, & \text{if } 0 \leq t < 2 \text{ sec} \\ 6, & \text{if } t \geq 2 \text{ sec} \end{cases}$$

The graph of the current is shown in Figure 13.11.

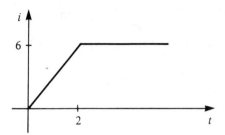

■ **Figure 13.11**   ■

EXAMPLE 6   ■ **Problem**

Find the response of a series RL circuit to a unit square wave. See Figure 13.12(a). Assume that the initial current is zero.

■ **Solution**

For the interval $0 \le t \le 1$, Equation (13.9) may be used. Substituting $V = 1$ in (13.9) yields

$$i = \frac{1}{R}(1 - e^{-Rt/L}) \qquad 0 \le t \le 1$$

For $t \ge 1$, Equation (13.9) does not apply because the initial current for this interval is not zero. In this case, we write the initial-value problem as

$$L\frac{di}{dt} + Ri = 0 \qquad i(1) = \frac{1}{R}(1 - e^{-R/L})$$

The differential equation is solved by separation of variables to yield

$$i = c_1 e^{-Rt/L}$$

Applying the initial condition for the interval $t \ge 1$,

$$\frac{1}{R}(1 - e^{-R/L}) = c_1 e^{-R/L}$$

from which

$$c_1 = \frac{1}{R}(e^{R/L} - 1)$$

Therefore, the current for $t \ge 1$ is

$$i = \frac{1}{R}(e^{R/L} - 1)e^{-Rt/L} = \frac{1}{R}(e^{-R(t-1)/L} - e^{-Rt/L})$$

See Figure 13.12(b).

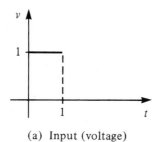

(a) Input (voltage)

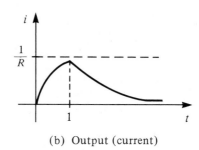

(b) Output (current)

■ **Figure 13.12**

EXERCISES FOR SECTION 13.4

Exercises 1–10.

1. Find the equation for the velocity of an object having a weight of 64 lb if the resisting force (in pounds) is $D = 2v$ and the thrust (in pounds) is $F = 10t$. Assume $v(0) = 0$ ft/sec.

2. A block weighing 32 lb is moved along a horizontal plane by a force (in pounds) of $e^{-0.01t}$. Friction produces a retarding force (in pounds) equal to $0.01v$. Describe the motion of the block. Assume that the block starts from rest.

3. A rocket starting from rest is moved in a straight-line path by a thrust of 1000 lb. During the first 3 sec of propulsion, the mass (in slugs) of the rocket decreases according to $m = 25 - t$, and the drag force (in pounds) is equal to $1.5v$. Determine the velocity equation of the rocket.

4. Draw the graph of the velocity equation found in Exercise 3.

5. A constant thrust of 2000 lb is generated by the motor of a rocket during the first 5 sec of flight. During this interval, the mass (in slugs) of the rocket varies with time according to $m = 25 - 4t$, and the drag (in pounds) is equal to the $8v$. Assuming the initial velocity of the rocket is 100 ft/sec, find its velocity at the end of the burn period.

6. A 1600-lb rocket sled is accelerated from rest by a thrust (in pounds) that varies with time according to

$$F = \begin{cases} 1000t, & \text{if } 0 \le t \le 2 \text{ sec} \\ 2000, & \text{if } t > 2 \text{ sec} \end{cases}$$

Determine the velocity equation for the sled if the mass remains constant and the drag is equal to ten times its velocity.

7. A 160-lb parachutist, including equipment, free-falling toward the earth, experiences a drag force (in pounds) equal to one-half her velocity (in feet per second). When the parachute opens, the drag is equal to five-eighths of the square of her velocity.

   (a) Find the equation for her velocity before the parachute opens.
   (b) Assuming she falls for 30 sec before her chute opens, find her velocity equation after it opens.

8. Assume the inclined plan shown in the accompanying figure is frictionless and that the only force acting on the 25-lb block is gravity. Assume $x = 5$ ft and $v = 2$ ft/sec when $t = 0$.

   (a) Write the differential equation describing the motion of the block.
   (b) Determine the equation for the velocity of the block.
   (c) Determine the equation for the displacement of the block.
   (d) Determine how long it will take the block to move 20 ft down the incline.

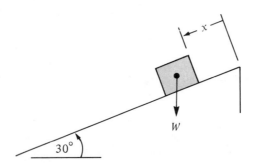

9. An object that moves over a rough surface encounters a resistance (called *friction*) due to the roughness of the surface. The friction experienced by an object is a force given by $f = \mu N$, where μ is a constant called the *coefficient of friction* and N is the normal force exerted by the plane on the object. Suppose a 64-lb block is released from the top of a plane inclined at 30° to the horizontal. The coefficient of friction of the block is 0.25.

   (a) Determine the equation for the velocity of the block.
   (b) Determine the equation for the displacement of the block.
   (c) Calculate the displacement and the velocity of the block 5 sec after it is released.

10. Refer to the discussion in Exercise 9. A 64-lb block is released from the top of a plane inclined at 30° to the horizontal. As the block slides down the plane its coefficient of friction is 0.25, and it experiences a drag force due to air resistance equal to one-half its velocity (in feet per second).

    (a) Determine the equation for the velocity of the block.
    (b) Determine the equation for the displacement of the block.
    (c) Calculate the displacement and the velocity of the block 5 sec after it is released.

Exercises 11–16

11. A tank contains a brine solution in which 5 lb of salt is dissolved in 10 gal of water. Brine containing 3 lb of salt per gallon flows into the tank at 2 gal/min, and the well-stirred mixture flows out at the same rate.

    (a) Determine the equation for the amount of salt in the tank at any time $t > 0$.
    (b) How much salt is present after a long time?

12. A tank contains 100 gal of a 25% acid solution. A 40% acid solution is allowed to enter the tank at a rate of 10 gal/hr, and the well-stirred solution is removed from the tank at the same rate. Determine the equation for the amount of acid in the tank at any time $t > 0$.

13. A tank contains a brine solution in which 10 lb of salt is dissolved in 50 gal of water. Brine containing 2 lb of salt per gallon flows into the tank at the rate of 5 gal/min, and the well-stirred solution flows out at a rate of 3 gal/min.

   (a) Determine the equation for the amount of salt in the tank at any time $t > 0$.
   (b) How much salt is in the tank after 5 min?

14. A 250-gal tank contains 100 gal of pure water. Brine containing 4 lb of salt per gallon flows into the tank at 5 gal/hr. The well-stirred mixture flows out at 3 gal/hr. Find the concentration of salt in the tank at the instant it is filled to the top.

15. Two tanks each contain 100 gal of pure water. A solution containing 3 lb/gal of a dye flows into tank 1 at 5 gal/min. The well-stirred solution flows out of tank 1 into tank 2 at the same rate. Assuming that the solution in tank 2 is well stirred and that this solution flows out of tank 2 at 5 gal/min, determine the amount of dye in tank 2 for any $t > 0$.

16. A 600-gal brine tank is to be cleaned by piping in pure water at 1 gal/min and allowing the well-stirred solution to flow out at the rate of 2 gal/min. The tank initially contains 1500 lb of salt. How much salt is left in the tank after 1 hr? After 9 hr and 59 min?

17. Water pollution is a major problem in today's world. Rivers and lakes become polluted with various waste products. Left unchecked, these waste products can kill the marine life. If the source of pollution is stopped, the rivers and lakes will clean themselves by the natural process of replacing the polluted water with clean water. The rate of change of pollution in a lake is a mixture problem and can be modeled by using Equation (13.7). In this model, we assume that water entering the lake is perfectly mixed so that the pollutants are uniformly distributed, that the volume of the lake is constant, and that the pollutants are only removed from the lake by outflow.

   If P is the pollution concentration in a lake and V is its volume, then the total amount of pollutant is $Q = VP$. Let r be the rate of water both in and out of the lake, and let P_i be the pollutant concentration of the water coming into the lake. Then, by Equation (13.7), we get

$$\frac{d}{dt}(VP) = (P_i - P)r$$

or

$$\frac{dP}{dt} + \frac{r}{V}P = \frac{P_i r}{V}$$

This equation is linear with integrating factor $e^{\int r/V\,dt} = e^{rt/V}$. Thus,

$$\frac{d}{dt}(e^{rt/V}P) = \frac{r}{V}e^{rt/V}P_i$$

Integrating and evaluating at $t = 0$ yields

$$e^{rt/V}P = \frac{r}{V}\int_0^t e^{rt/V}P_i\,dt + P(0)$$

   The fastest possible cleanup will occur if the value of the pollutant concentration P_i of the incoming water is zero. In this case, the equation becomes

$$e^{rt/V}P = P(0)$$

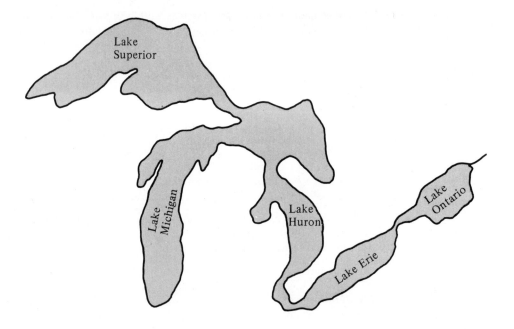

Solving this equation for t, we get

$$t = \frac{V}{r} \ln \frac{P(0)}{P}$$

as the time (in years) for the pollution level of a given lake to drop to a given percentage of its present value.

(a) Verify that $e^{rt/V}P = P(0)$ yields

$$t = \frac{V}{r} \ln \frac{P(0)}{P}$$

(b) Given that $V/r = 2.9$ for Lake Erie, determine how long it will take to reduce the pollution level to 5% of its present value.

18. If P_i is a nonzero constant in the equation for $e^{rt/V}P$ given in Exercise 17, determine an expression for $P(t)$.

Exercises 19–28

19. A series RL circuit has a resistance of 20 Ω, an inductance of 0.2 H, and an impressed voltage of 12 V. Find the current equation if the initial current is zero.

20. Find the current in a series RL circuit with $R = 2$ Ω, $L = 1$ H, and $E = e^{-t}$ (in volts). Assume the initial current is zero.

21. Find the current in a series RL circuit with $R = 10\,\Omega$ and $L = 5\,H$ if the applied voltage (in volts) varies with time according to

$$E(t) = \begin{cases} 2t, & \text{if } 0 \le t \le 2 \text{ sec} \\ 4, & \text{if } t > 2 \text{ sec} \end{cases}$$

Assume $i\,(0) = 0$.

22. Sketch the graphs of voltage and current for the circuit in Exercise 21.

23. The inductance (in henrys) in a series RL circuit is given by

$$L(t) = \begin{cases} 1 + t, & \text{if } 0 \le t \le 1 \text{ sec} \\ 2, & \text{if } t > 1 \text{ sec} \end{cases}$$

For $R = 2\,\Omega$, $E = 6\,V$, and $i(0) = 0$, find the equation for the current in the circuit. Sketch the graph of the current equation.

24. The initial current in a series RL circuit is zero. Find the current in the circuit when $t = 2$ sec if $R = 2\,\Omega$, $E = 10\,V$, and $L = 4 - t$ (in henrys).

25. A capacitor is a device for storing electric charge. The amount of charge that a capacitor will store is called its *capacitance*. When a constant voltage of V volts is applied to a resistance of R ohms and a capacitance of C farads in series, the rate at which the capacitor is charged is given by

$$R\frac{dq}{dt} + \frac{q}{C} = V$$

where q is the electric charge (in coulombs) on the left plate in the accompanying figure. Find the equation for the charge transferred if the capacitor is initially discharged.

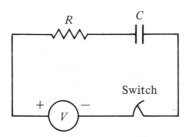

26. Using the results of Exercise 25 and the fact that $i = dq/dt$, find the equation for the current in the capacitor.

27. Referring to Exercise 25, show that for large values of t, the charge on the plates of the capacitor is given by $q = CV$.

13.5 DIFFERENTIAL EQUATIONS OF THE FORM
$a_2 y'' + a_1 y' + a_0 y = 0$

Second-order differential equations of the form

$$a_2 y'' + a_1 y' + a_0 y = 0$$

where a_0, a_1, and a_2 are constants, arise in the solution of certain types of mechanical and electrical systems such as those shown in Figure 13.13 and 13.14.

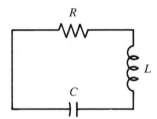

■ Figure 13.13

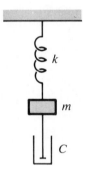

■ Figure 13.14

The general solution of a second-order differential equation must have two arbitrary constants. Furthermore, the form of the solution of $a_2 y'' + a_1 y' + a_0 y = 0$ depends on the roots of a quadratic equation and involves an exponential function of the form e^{mx}. We will demonstrate this fact by solving the equation

$$y'' - 4y' + 3y = 0$$

under the assumption that $y = e^{mx}$ is a solution of this equation. If $y = e^{mx}$ is a solution of this equation, then the left side will be zero when y and its derivatives are replaced with e^{mx} and its corresponding derivatives. Since

$$\frac{dy}{dx} = m e^{mx} \quad \text{and} \quad \frac{d^2 y}{dx^2} = m^2 e^{mx}$$

the given equation becomes

$$m^2 e^{mx} - 4me^{mx} + 3e^{mx} = 0$$

or

$$(m^2 - 4m + 3)e^{mx} = 0$$

Since e^{mx} is never zero, we have

$$m^2 - 4m + 3 = 0$$

which is called the **auxiliary equation.** The roots of this equation are $m = 1$ and $m = 3$. Therefore, both $y_1 = e^x$ and $y_2 = e^{3x}$ are solutions.

We conclude from this example that the auxiliary equation can be obtained from the given differential equation by replacing y'' with m^2 and y' with m. Thus, $a_2 m^2 + a_1 m + a_0 = 0$ is the auxiliary equation for $a_2 y'' + a_1 y' + a_0 y = 0$. The solution to this differential equation is dependent on the roots of the auxiliary equation. There are three possibilities, as presented next.

RULE

> **Roots of the Auxiliary Equation and Solutions of $a_2 y'' + a_1 y' + a_0 y = 0$**
>
> 1. *distinct real roots* $(m_1 \neq m_2)$ of the auxiliary equation, the solution of the differential equation is
>
> $$y = c_1 e^{m_1 x} + c_2 e^{m_2 x}$$
>
> 2. *repeated real roots* $(m_1 = m_2)$ of the auxiliary equation, the solution of the differential equation is
>
> $$y = (c_1 + c_2 x)e^{mx}$$
>
> 3. *complex roots* $(m_1 = a + bj, m_2 = a - bj)$ of the auxiliary equation, the solution of the differential equation is
>
> $$y = e^{ax}(c_1 \cos bx + c_2 \sin bx)$$

EXAMPLE 1   ■ **Problem**

Solve $\dfrac{d^2 s}{dt^2} - 3\dfrac{ds}{dt} = 0$.

■ **Solution**

The auxiliary equation is $m^2 - 3m = 0$, or $m(m - 3) = 0$. The roots of this equation are $m = 0$ and $m = 3$. The general solution is then

$$s = c_1 e^0 + c_2 e^{3t} = c_1 + c_2 e^{3t}$$ ■

EXAMPLE 2   ▪ **Problem**

Find the particular solution for $2(d^2y/dx^2) + 5(dy/dx) - 3y = 0$ if $y = 1$ and $dy/dx = -2$ when $x = 0$.

▪ **Solution**

The auxiliary equation is

$$2m^2 + 5m - 3 = 0$$

or

$$(2m - 1)(m + 3) = 0$$

from which

$$m_1 = \tfrac{1}{2} \quad \text{and} \quad m_2 = -3$$

The general solution is then

$$y = c_1 e^{(1/2)x} + c_2 e^{-3x}$$

Since there are two arbitrary constants to be evaluated, we establish a second equation containing c_1 and c_2 by differentiating the general solution. Thus,

$$\frac{dy}{dx} = \frac{1}{2}c_1 e^{(1/2)x} - 3c_2 e^{-3x}$$

Applying the given conditions to these last two equations, we get

$$1 = c_1 + c_2 \quad \text{and} \quad -2 = \tfrac{1}{2}c_1 - 3c_2$$

Solving these two equations simultaneously yields

$$c_1 = \tfrac{2}{7} \quad \text{and} \quad c_2 = \tfrac{5}{7}$$

Therefore, the particular solution is

$$y = \tfrac{2}{7}e^{(1/2)x} + \tfrac{5}{7}e^{-3x}$$ ▪

EXAMPLE 3   ▪ **Problem**

Find the general solution for $y'' - 6y' + 9y = 0$.

▪ **Solution**

Here, the auxiliary equation is

$$m^2 - 6m + 9 = 0$$

or

$$(m - 3)^2 = 0$$

from which

$$m = 3, 3$$

Therefore, the general solution is

$$y = c_1 e^{3x} + c_2 x e^{3x} = (c_1 + c_2 x)e^{3x} \qquad \blacksquare$$

EXAMPLE 4 ■ **Problem**

Find the general solution for $\dfrac{d^2 y}{dx^2} - 2\dfrac{dy}{dx} + 5y = 0$.

■ **Solution**

The auxiliary equation is $m^2 - 2m + 5 = 0$. The roots of this equation are found by the quadratic formula to be

$$m = \frac{2 \pm \sqrt{-16}}{2} = 1 \pm 2j$$

Hence, the general solution of the given differential equation is

$$y = e^x(c_1 \cos 2x + c_2 \sin 2x)$$

where c_1 and c_2 are the arbitrary constants. $\qquad \blacksquare$

EXAMPLE 5 ■ **Problem**

Find the particular solution for $(d^2\theta/dt^2) + 4\theta = 0$ if $\theta = 3$ when $t = 0$ and $d\theta/dt = -4$ when $t = \pi/2$.

■ **Solution**

In this case, the auxiliary equation is $m^2 + 4 = 0$. The roots of this equation are

$$m = \pm 2j$$

So, the general solution is

$$\theta = c_1 \cos 2t + c_2 \sin 2t$$

Substituting $t = 0$ and $\theta = 3$ into this equation, we get

$$3 = c_1 \cos 0 + c_2 \sin 0$$

Since $\cos 0 = 1$ and $\sin 0 = 0$, this equation yields $c_1 = 3$. To use the second condition, we note that the derivative of the general solution is

$$\frac{d\theta}{dt} = -2c_1 \sin 2t + 2c_2 \cos 2t$$

Substituting $t = \pi/2$ and $d\theta/dt = -4$, we get

$$-4 = -2c_1 \sin \pi + 2c_2 \cos \pi$$

Recalling that $\cos \pi = -1$ and $\sin \pi = 0$, we get $c_2 = 2$. Therefore, the desired solution is

$$\theta = 3 \cos 2t + 2 \sin 2t \qquad \blacksquare$$

EXERCISES FOR SECTION 13.5

Find the general solution for each of Exercises 1–24.

1. $\dfrac{d^2y}{dx^2} + \dfrac{dy}{dx} = 0$

2. $\dfrac{d^2s}{dt^2} - 4s = 0$

3. $3y'' + 6y' = 0$

4. $\dfrac{d^2i}{dt^2} + \dfrac{di}{dt} - 6i = 0$

5. $\dfrac{d^2y}{dx^2} + 5\dfrac{dy}{dx} + 4y = 0$

6. $\dfrac{d^2y}{dx^2} + 4y = 0$

7. $\dfrac{d^2s}{dt^2} + 9s = 0$

8. $2y'' - 7y' + 3y = 0$

9. $\dfrac{d^2v}{dt^2} - 6\dfrac{dv}{dt} + 9v = 0$

10. $y'' + 4y' + 5y = 0$

11. $\dfrac{d^2x}{dt^2} - 2\dfrac{dx}{dt} + 2x = 0$

12. $\dfrac{d^2v}{dt^2} + 3v = 0$

13. $4y'' + 4y' + y = 0$

14. $\dfrac{d^2r}{d\theta^2} = 3\dfrac{dr}{d\theta} - 2r$

15. $\dfrac{d^2s}{dt^2} + \dfrac{2}{3}\dfrac{ds}{dt} + \dfrac{1}{9}s = 0$

16. $4y'' - y = 0$

17. $\dfrac{d^2q}{dt^2} - 3q = 0$

18. $\dfrac{d^2r}{d\theta^2} + r = 0$

19. $\dfrac{d^2y}{dx^2} + y = 0$

20. $\dfrac{d^2y}{dx^2} + 2\dfrac{dy}{dx} - y = 0$

21. $\dfrac{d^2y}{dx^2} + \dfrac{dy}{dx} - y = 0$

22. $2\dfrac{d^2y}{dx^2} - 3\dfrac{dy}{dx} - y = 0$

23. $\dfrac{d^2v}{ds^2} - 4\dfrac{dv}{ds} + 7v = 0$

24. $y'' + 8y' + 25y = 0$

In Exercises 25–31, find the particular solution of each of the differential equations that satisfies the given conditions.

25. $\dfrac{d^2y}{dx^2} - 4y = 0$; when $x = 0$, $y = 0$, and $\dfrac{dy}{dx} = 2$

26. $\dfrac{d^2s}{dt^2} - 2\dfrac{ds}{dt} - 3s = 0$; when $t = 0$, $s = 0$, and $\dfrac{ds}{dt} = -4$

27. $y'' - 8y' + 16y = 0$; when $x = 0$, $y = 0$, and $y' = 1$

28. $y'' + 16y = 0$; when $x = 0$, $y = 2$, and $y' = 0$

29. $\dfrac{d^2s}{dt^2} + 9s = 0$; when $t = \dfrac{\pi}{6}$, $s = 2$, and $\dfrac{ds}{dt} = 0$

30. $\dfrac{d^2y}{dx^2} + 3\dfrac{dy}{dx} = 0$; when $x = 0$, $y = 2$, and $\dfrac{dy}{dx} = 6$

31. $\dfrac{d^2y}{dx^2} - 6\dfrac{dy}{dx} + 10y = 0$; when $x = 0$, $y = 2$, and $\dfrac{dy}{dx} = 1$

13.6 UNDAMPED VIBRATIONS

Vibratory motion exists to some degree in all mechanical systems because of things such as noise, wind, and moving parts acting on the system. For the most part, mechanical vibration is considered undesirable. The job of the designer of a system is to understand vibratory motion and to design products that will not be affected adversely by it. For instance, the suspension system of an automobile is designed to absorb vibrations due to irregularities in roadways; booster rockets, used in launching space probes, are designed to withstand the tremendous amount of vibrational energy produced by the noise of their motors.

Perhaps the simplest system available for studying vibratory motion is an ordinary spring supporting a weight W, as shown in Figure 13.15. For our purposes, we assume the weight of the spring to be negligible. This assumption is not unrealistic since the weight of the supported object is usually much greater than that of the spring. When the weight is hanging at rest, we say it is in the **equilibrium position.** Experience tells us that a weight that is pulled down a certain distance and released will oscillate about its equilibrium position. Our objective is to describe this oscillatory motion.

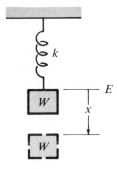

■ **Figure 13.15**

As in most mathematical discussions, a convenient thing to do is to establish a coordinate system for reference. Since the weight in Figure 13.15 is restricted to vertical movement, we take the origin as the equilibrium position and measure the vertical displacement x from this point, with the convention that x is positive when measured below the point of equilibrium and negative when above. Force is measured in pounds, mass in slugs, displacement in feet, and time in seconds.

Under the assumption that there is no force due to air resistance, the differential equation governing the motion of the mass in Figure 13.15 is given by the following formula.

FORMULA

> **Differential Equation for Undamped Vibration**
>
> $$m\frac{d^2x}{dt^2} + kx = 0 \qquad\qquad (13.10)$$
>
> where m is the mass of the weight and k is the spring constant.

EXAMPLE 1

■ **Problem**

A 2-lb weight stretches a certain spring 6 in. in coming to rest. The weight is then pulled down 3 in. and released. What is the amplitude and the period of the resulting motion?

■ **Solution**

To find the value of k, we use Hooke's law with $x = 6$ in. $= \frac{1}{2}$ ft and $F = 2$ lb. That is,

$$k = \frac{2}{1/2} = 4 \text{ lb/ft}$$

The differential equation of this system is then

$$\frac{2}{32}\frac{d^2x}{dt^2} + 4x = 0 \qquad \text{or} \qquad \frac{d^2x}{dt^2} + 64x = 0$$

The auxiliary equation is $m^2 + 64 = 0$, which has roots of $m = \pm 8j$. Therefore, the general solution is

$$x = c_1 \cos 8t + c_2 \sin 8t$$

To find the required particular solution, we note that the weight is initially pulled down 3 in. below the equilibrium position, so $x(0) = 3$ in. $= \frac{1}{4}$ ft. Furthermore, when the weight is released, its initial velocity is zero, which means that $x'(0) = 0$. Using $t = 0$ and $x = \frac{1}{4}$ in the general solution, we get $c_1 = \frac{1}{4}$. Differentiation of the general solution yields

$$x' = -8c_1 \sin 8t + 8c_2 \cos 8t$$

Substituting $t = 0$ and $x' = 0$ into this equation, we have $c_2 = 0$. The required solution is then

$$x = \tfrac{1}{4} \cos 8t$$

This equation shows that the amplitude of the oscillation is $\tfrac{1}{4}$ ft and the period is $T = 2\pi/8 = \pi/4$ sec. The graph of the solution is shown in Figure 13.16.

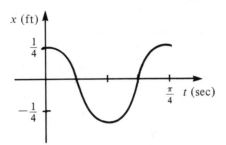

■ **Figure 13.16**   ■

EXAMPLE 2   ■ **Problem**

Find the position, the velocity, and the acceleration of the weight in Example 1 when $t = \tfrac{1}{4}$ sec.

■ **Solution**

From Example 1, we have

$$x = \tfrac{1}{4} \cos 8t$$

Differentiation of this equation with respect to t yields

$$v = \frac{dx}{dt} = -2 \sin 8t$$

and

$$a = \frac{dv}{dt} = -16 \cos 8t$$

Letting $t = \tfrac{1}{4}$ sec, we get

$$x = \tfrac{1}{4} \cos 2 \approx \tfrac{1}{4}(-0.416) \approx -0.104 \text{ ft}$$
$$v = -2 \sin 2 \approx -2(0.909) \approx -1.82 \text{ ft/sec}$$
$$a = -16 \cos 2 \approx -16(-0.416) \approx 6.66 \text{ ft/sec}^2$$

These results mean that the weight is 0.104 ft above the equilibrium position and moving upward with a velocity of 1.82 ft/sec. Furthermore, it has a downward acceleration of 6.65 ft/sec^2. Since the acceleration is opposite to the velocity, we know the weight is slowing down at this time.   ■

EXAMPLE 3   ■ **Problem**

Assume that the weight in Example 1 is given an upward velocity of 2 ft/sec. instead of simply being released. Determine the amplitude and the period of the resulting oscillation.

■ **Solution**

From Example 1, we have the differential equation

$$\frac{d^2x}{dt^2} + 64x = 0$$

from which

$$x(t) = c_1 \cos 8t + c_2 \sin 8t \tag{13.11}$$

and

$$x'(t) = -8c_1 \sin 8t + 8c_2 \cos 8t \tag{13.12}$$

In this case, the initial conditions are $x(0) = \frac{1}{4}$ ft and $x'(0) = -2$ ft/sec. Substituting the first condition into Equation (13.11) yields $c_1 = \frac{1}{4}$. Substituting the second condition into Equation (13.12) yields $c_2 = -\frac{1}{4}$. Therefore, the desired solution is

$$x(t) = \tfrac{1}{4} \cos 8t - \tfrac{1}{4} \sin 8t$$

To find the amplitude and the period of this expression, we use an alternative form for $x(t)$. From trigonometry, the sum $c_1 \cos \omega t + c_2 \sin \omega t$ can be expressed as a single cosine function by the identity

$$c_1 \cos \omega t + c_2 \sin \omega t = C \cos(\omega t - \delta)$$

where

$$C = \sqrt{c_1^2 + c_2^2} \quad \text{and} \quad \tan \delta = \frac{c_2}{c_1}$$

Thus, the solution function $x(t)$ can be expressed as

$$x(t) = C \cos(8t - \delta)$$

where

$$C = \sqrt{\left(\frac{1}{4}\right)^2 + \left(\frac{1}{4}\right)^2} \approx 0.354$$

and

$$\tan \delta = \frac{c_2}{c_1} = \frac{-1/4}{1/4} = -1 \quad \text{or} \quad \delta = -\frac{\pi}{4}$$

The particular solution to the equation can then be written as

$$x(t) = 0.354 \cos 8 \left(t + \tfrac{1}{32}\pi\right)$$

Thus, the amplitude is $A = 0.354$, and the period is $T = \frac{1}{4}\pi$. Furthermore, there is a phase shift of $-\frac{1}{32}\pi$. The graph of the function is shown in Figure 13.17.

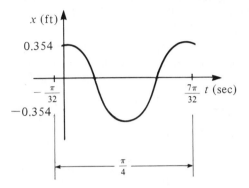

■ **Figure 13.17**

EXERCISES FOR SECTION 13.6

Exercises 1–7

1. An object weighing 32 lb is suspended from a spring with a spring constant of 4 lb/ft. The object is started from the equilibrium position with a downward velocity of 1 ft/sec.

   (a) Determine the distance-time function of the object.
   (b) Determine the amplitude and the period of the oscillation.

2. A spring with $k = 1$ lb/ft is attached to a 32-lb weight. Find the amplitude and the period of the motion of the weight if it is pulled down 3 ft below the equilibrium point and released.

3. An object weighing 2 lb is attached to a spring suspended from a fixed point. The spring was stretched 1.5 in. at the time that the weight was attached. The object is set in motion by lifting it 0.25 ft above equilibrium and releasing it.

   (a) Determine the displacement, the velocity, and the acceleration of the object as a function of time.
   (b) Find the position, the velocity, and the acceleration of the object when $t = \frac{1}{8}$ sec.

4. A spring is stretched $\frac{1}{2}$ ft by a 6-lb weight in coming to rest in its equilibrium position. The weight is pulled down $\frac{1}{3}$ ft below the equilibrium position and then given a downward velocity of 2 ft/sec.

   (a) Determine the displacement of the weight as a function of time.
   (b) Determine the amplitude, the period, and the phase shift of the motion.
   (c) Draw the graph of the displacement of the object as a function of time.

5. A spring is stretched 24 in. by a 4-lb weight in coming to rest in its equilibrium position. Determine the displacement of the weight as a function of time if it is initially lifted 0.5 ft above the equilibrium position and then given a downward velocity of 2 ft/sec.

6. A 32-lb weight hangs from a spring having a spring constant of 9 lb/ft. It is known that $x(\pi/6) = 0$ and $x'(\pi/6) = 2$ ft/sec.

   (a) Determine the displacement of the weight as a function of time.

   (b) Determine $x(0)$ and $x'(0)$.

   (c) Determine the period of the motion.

7. A steel ball suspended from a spring is set into motion and oscillates with a period of 2.5 sec. The motion is stopped, and an additional weight of 4 lb is added to the ball. The system is again set into motion, and now, it oscillates with a period of 3.5 sec. Determine the weight of the steel ball.

▰▰▰ 13.7 DAMPED VIBRATIONS

The oscillatory motion described in the preceding section would continue undiminished without some retarding force. In practice, several forces tend to dissipate the vibration. We assume the existence of a force proportional to the velocity of the object. Such a force is called a **damping force,** or more specifically, **viscous damping.** (There is empirical evidence to support this assumption provided that the velocity is not too great.)

    Mechanical systems that involve damping are usually diagrammed by connecting the mass to a *dashpot*, as shown schematically in Figure 13.18. The mechanical system is called a **spring-mass-damper system.**

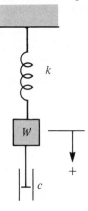

■ **Figure 13.18**

    Since the damping force opposes the movement of the mass through the medium, we denote it by

$$-c\,\frac{dx}{dt}$$

where c is called the **damping constant.** Both the spring force and the force due to the viscous friction oppose the movement. From Newton's second law,

$$m\,\frac{d^2x}{dt^2} = -c\,\frac{dx}{dt} - kx$$

or

$$m \frac{d^2x}{dt^2} + c \frac{dx}{dt} + kx = 0 \qquad \text{(13.13)}$$

The auxiliary equation written in terms of the variable r is

$$mr^2 + cr + k = 0$$

whose roots are

$$-\frac{c}{2m} \pm \frac{1}{2m} \sqrt{c^2 - 4mk}$$

Letting $\alpha = c/2m$ and $\beta = 1/2m \sqrt{c^2 - 4mk}$, we can write the roots as $-\alpha \pm \beta$.

The form of the solution depends on the damping. We distinguish three cases that come from the value of $\sqrt{c^2 - 4mk}$.

Case 1, where $c^2 < 4mk$:   In this case, β is imaginary; that is, $\beta = j\omega^*$, where

$$\omega^* = \frac{1}{2m}\sqrt{4mk - c^2}$$

Using the results of Section 13.5, we can write the general solution of (13.13) as

$$x = e^{-\alpha t}(c_1 \cos \omega^* t + c_2 \sin \omega^* t)$$

or

$$x = Ce^{-\alpha t}\cos(\omega^* t - \delta)$$

Thus, there will be oscillatory motion when the weight is released. The factor $e^{-\alpha t}$ will damp out the oscillation; that is, the amplitude of the motion approaches zero as $t \to \infty$. We say that the spring-mass-damper system is **underdamped** for this case.

The graph of the solution lies between $y = Ce^{-\alpha t}$ and $x = -Ce^{-\alpha t}$, as shown in Figure 13.19.

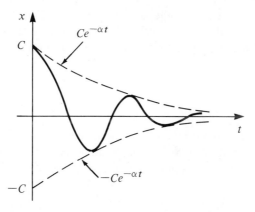

■ **Figure 13.19**

The frequency of the oscillation is $\omega^*/2\pi$ cycles per second. Note that as $c \to 0$, ω^* approaches $\sqrt{k/m}$, which is the natural frequency of the system and the frequency of harmonic oscillation. Thus, underdamped motion approaches harmonic motion as the viscous damping decreases.

Case 2, where $c^2 = 4mk$: If $c^2 = 4mk$, then $\beta = 0$, and therefore, the auxiliary equation has the double root $m = -\alpha$. Hence, the solution is

$$x = e^{-\alpha t}(c_1 + c_2 t)$$

The solution, which does not contain $\sin x$ or $\cos x$, indicates that the weight will return to equilibrium without oscillating. Since any decrease in c will cause underdamping, we say the spring-mass system is **critically damped.** Under these conditions, the weight will return to its equilibrium position in the shortest possible time without oscillation. Typical graphs of a critically damped system are shown in Figure 13.20.

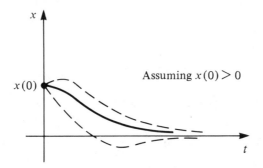

■ **Figure 13.20**

Case 3, where $c^2 > 4mk$: If $c^2 > 4\ mk$, then $c^2 - 4mk > 0$. Therefore, the auxiliary equation has two distinct real roots, $m = -\alpha \pm \beta$. The solution is

$$x = c_1 e^{-(\alpha + \beta)t} + c_2 e^{-(\alpha - \beta)t}$$

This solution is the same basic nonoscillatory form as for the critically damped case. The time required for the weight to return to its equilibrium position for this case is always greater than it is for the critically damped case. For this reason, we say the spring-mass system is **overdamped.** Figure 13.21 shows some typical overdamped motions.

EXAMPLE 1   ■ **Problem**

A spring is stretched 0.4 ft by an 8-lb weight. The 8-lb weight is pushed up 0.1 ft and released. The motion takes place in a medium that furnishes a damping force equal to twice the instantaneous velocity. Find the displacement-time equation, and draw its graph.

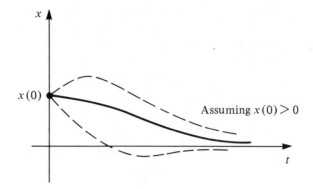

■ **Solution**

The damping force in this system is $-2\,dx/dt$, the spring constant of the spring is $8/0.4$, and the mass is $m = \frac{8}{32}$. Therefore, the differential equation is

$$\frac{8}{32}\frac{d^2x}{dt^2} = -2\frac{dx}{dt} - \frac{8}{0.4}x$$

or

$$\frac{d^2x}{dt^2} + 8\frac{dx}{dt} + 80x = 0$$

The auxiliary equation $r^2 + 8r + 80 = 0$ has roots

$$r = \frac{-8 \pm \sqrt{8^2 - 4(80)}}{2} = -4 \pm 8j$$

(The system is underdamped.) Therefore, the general solution is

$$x = e^{-4t}(c_1 \cos 8t + c_2 \sin 8t)$$

To find c_1 and c_2 use the initial conditions $x(0) = -0.1$ and $x'(0) = 0$. The derivative of the general solution is

$$x' = e^{-4t}[(8c_2 - 4c_1)(\cos 8t) - (8c_1 + 4c_2)(\sin 8t)]$$

Thus, $c_1 = -0.1$ and $c_2 = -0.05$. Using these values, we get

$$x = -e^{-4t}(0.1 \cos 8t + 0.05 \sin 8t)$$

or

$$x = -0.112e^{-4t} \cos(8t - 0.46)$$

The graph of this function is bounded above by the graph of $x = 0.112e^{-4t}$ and below by $x = -0.112e^{-4t}$, as shown in Figure 13.22. The cosine function has a period of $\pi/4$ and a phase shift of 0.058.

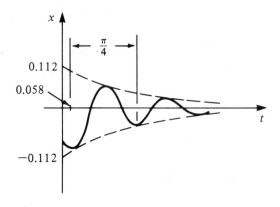

■ **Figure 13.22**   ■

EXAMPLE 2   ■ **Problem**

A spring has a spring constant of 8 lb/ft. A 4-lb weight is pulled down 0.5 ft below equilibrium and then given an initial upward velocity of 5 ft/sec. The damping force is equal to twice the instantaneous velocity. Find the equation of motion.

■ **Solution**

Here, we have

$$\frac{4}{32}\frac{d^2x}{dt^2} + 2\frac{dx}{dt} + 8x = 0$$

or

$$\frac{d^2x}{dt^2} + 16\frac{dx}{dt} + 64x = 0$$

with the initial conditions $x(0) = \frac{1}{2}$ and $x'(0) = -5$.

The auxiliary equation is $r^2 + 16r + 64 = 0$, which has the double root $r = -8$. So, the general solution for the critically damped system is

$$x = e^{-8t}(c_1 + c_2t)$$

The derivative is

$$x' = (c_2 - 8c_1 - 8c_2t)e^{-8t}$$

Therefore, we have $c_1 = \frac{1}{2}$ and $c_2 = -1$. Thus,

$$x = (\tfrac{1}{2} - t)e^{-8t}$$

The graph of this function is shown in Figure 13.23.

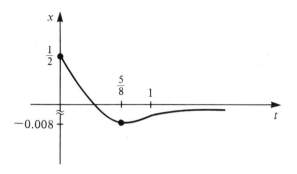

■ **Figure 13.23**   ■

EXERCISES FOR SECTION 13.7

Exercises 1–9

1. A 2-lb weight attached to a spring with a spring constant of 4 lb/ft experiences a damping force equal to its velocity. The weight is set into motion by pulling it 3 in. below the equilibrium position and releasing it.

   (a) Determine the displacement of the weight as a function of time.

   (b) Draw the graph of the displacement function.

2. Indicate the type of damping associated with each the following equations:

   (a) $\dfrac{d^2x}{dt^2} + 4\dfrac{dx}{dt} + 4x = 0$ (d) $\dfrac{d^2x}{dt^2} + 2\dfrac{dx}{dt} + 3x = 0$

   (b) $\dfrac{d^2x}{dt^2} + 10\dfrac{dx}{dt} + 9x = 0$ (e) $\dfrac{d^2x}{dt^2} + \dfrac{dx}{dt} + x = 0$

   (c) $3\dfrac{d^2x}{dt^2} + 6\dfrac{dx}{dt} + 3x = 0$

3. A spring is stretched 0.4 ft by a 4-lb weight in coming to its equilibrium position. A damping force equal to the velocity of the object acts on the system. The weight is set into motion from the equilibrium position by an initial upward velocity of 2 ft/sec.

   (a) Determine the displacement of the weight as a function of time.

   (b) Determine the displacement of the weight at the first three stops in its motion.

   (c) Draw the graph of the displacement as a function of time.

4. An object having a mass of 1 slug is suspended from a spring having a spring constant of 5 lb/ft. Assume the damping force to be equal to twice the velocity of the object. Determine the displacement of the object as a function of time if it is pulled down 24 in. below the equilibrium position and released.

5. A spring is such that a 32-lb weight stretches it 2 ft in coming to the equilibrium position. When in motion, the weight experiences a damping force equal to eight times its velocity.

The system is set into motion from the equilibrium position by a downward velocity of 1 ft/sec.

(a) Determine the displacement of the weight as a function of time.

(b) Find the maximum displacement of the weight from the equilibrium position.

(c) Draw the graph of the displacement function.

6. How does the frequency of a damped oscillation depend on the damping force?

7. Determine where the maxima and minima of an underdamped oscillation of the form $x = Ce^{-at} \sin \omega t$ occur, and compare them with the points where the graph is tangent to Ce^{-at} and $-Ce^{-at}$.

8. In the case of underdamped motion, show that the time between two successive positive maxima is $2\pi/\sqrt{\omega^2 - \omega_0^2}$.

9. Prove that in the overdamped case, the mass cannot pass through equilibrium more than once.

13.8 LAPLACE TRANSFORM

Another method of solving differential equation is based on the use of a mathematical operator called the **Laplace transform**. This approach is used in many areas of engineering and is particularly important in the study of pulse circuits in electronics.

The Laplace transform of a function $f(t)$ is defined as follows:

DEFINITION

> **Laplace Transform**   The *Laplace transform* of $f(t)$, if it exists, is denoted by $\mathcal{L}\{f(t)\}$ and is defined by
>
> $$\mathcal{L}\{f(t)\} = \int_0^\infty e^{-st} f(t)\, dt$$
>
> where the parameter s is a real variable.

Thus, the Laplace transform takes a function of t, which we denote by f, and transforms it into a function of s, which we denote by F. More generally, we will represent functions of t by lowercase letters such as f, g, or h, and we will represent their respective Laplace transforms by the corresponding capital letter F, G, or H. Therefore, we also write $\mathcal{L}\{f(t)\} = F(s)$, or

$$F(s) = \int_0^\infty e^{-st} f(t)\, dt$$

By using this notation, we make the transformation of functions of t into functions of s more explicit.

The defining equation for the Laplace transform is an improper integral, the "improperness" arising because the upper limit is unbounded. Improper integrals of this kind are defined by

$$\int_0^\infty e^{-st} f(t)\, dt = \lim_{T \to \infty} \int_0^T e^{-st} f(t)\, dt$$

Therefore, the Laplace transform of f depends on the existence of this limit.

Like the derivative and the indefinite integral, the Laplace transform is a *linear operator.* That is, if $f_1(t)$ and $f_2(t)$ have Laplace transforms, and if c_1 and c_2 are constants, then we have the following property.

PROPERTY

> ### Linearity Property of the Laplace Transform
>
> $$\mathcal{L}\{c_1 f_1(t) + c_2 f_2(t)\} = c_1\, \mathcal{L}\{f_1(t)\} + c_2\, \mathcal{L}\{f_2(t)\}$$

This property is used frequently, especially in the applications of the Laplace transform to the solution of differential equations.

We can find the Laplace transform of several elementary functions directly from the definition. For example, consider the function $f(t) = 1$ for $t \geq 0$. Then,

$$\mathcal{L}\{1\} = \int_0^\infty e^{-st}(1)\, dt = \lim_{T \to \infty} \int_0^T e^{-st}\, dt$$

To determine whether this limit exists, we must consider three separate cases involving the parameter s.

1. When $s < 0$, the exponent of e is positive for $t > 0$. Therefore,

$$\lim_{T \to \infty} \int_0^T e^{-st}\, dt = \lim_{T \to \infty} \left[\frac{e^{-st}}{-s}\right]_0^T = \lim_{T \to \infty} \left(\frac{-1}{s} e^{-sT} + \frac{1}{s}\right) = \infty$$

which means the integral diverges.

2. When $s = 0$, the integral becomes

$$\lim_{T \to \infty} \int_0^T dt = \lim_{T \to \infty} \left[t\right]_0^T = \lim_{T \to \infty} T = \infty$$

3. When $s > 0$, the exponent is negative for $t > 0$. Therefore,

$$\mathcal{L}(1) = \lim_{T \to \infty} \left(-\frac{1}{s} e^{-sT} + \frac{1}{s}\right) = \frac{1}{s}$$

The case considered here is typical. The function of s is usually a valid transform of the function of t only over some interval of s. However, this limitation (on the validity of the function of s) in no way reduces the effectiveness of Laplace transforms in applications. In fact, in most cases, the limitation on the domain of F will go almost unnoticed.

The Laplace transform of e^{kt} is derived as follows:

$$\mathcal{L}\{e^{kt}\} = \int_0^\infty e^{-st} e^{kt} \, dt = \int_0^\infty e^{-(s-k)t} \, dt = \lim_{T\to\infty} \left[\frac{-e^{-(s-k)t}}{s - k} \right]_0^T$$

This limit converges for $s > k$, so

$$\mathcal{L}\{e^{kt}\} = \lim_{T\to\infty} \left(-\frac{e^{-(s-k)T}}{s - k} + \frac{1}{s - k} \right) = \frac{1}{s - k} \qquad s > k$$

The Laplace transforms of the other elementary functions can be derived in a similar fashion. For convenience we have listed the more important ones in the accompanying table. This table, along with the linearity property, enables us to find the Laplace transform of a large number of additional functions.

Original function, $f(t) = \mathcal{L}^{-1}\{F(s)\}$	Laplace transform, $\mathcal{L}\{f(t)\} = F(s)$
1. 1	$\dfrac{1}{s}$
2. t^n, n a positive integer	$\dfrac{n!}{s^{n+1}}$
3. e^{kt}	$\dfrac{1}{s - k}$
4. $t^n e^{kt}$	$\dfrac{n!}{(s - k)^{n+1}}$
5. $\sin kt$	$\dfrac{k}{s^2 + k^2}$
6. $\cos kt$	$\dfrac{s}{s^2 + k^2}$
7. $t \sin kt$	$\dfrac{2ks}{(s^2 + k^2)^2}$
8. $t \cos kt$	$\dfrac{s^2 - k^2}{(s^2 + k^2)^2}$
9. $1 - e^{kt}$	$\dfrac{-k}{s(s - k)}$
10. $1 - \cos kt$	$\dfrac{k^2}{s(s^2 + k^2)}$

Comment: The notation $n!$ is used to denote the product of all positive integers from 1 to n. We refer to the product as *n factorial*. Thus, the value of 4 factorial is

$$4! = 4 \cdot 3 \cdot 2 \cdot 1 = 24$$

EXAMPLE 1   ■ **Problem**

Use the table to find the Laplace transform of $f(t) = 5 + \cos 2t$.

■ **Solution**

Since the Laplace transform has the linearity properties, we have

$$\begin{aligned}
\mathcal{L}\{f(t)\} &= \mathcal{L}\{5 + \cos 2t\} \\
&= 5\mathcal{L}\{1\} + \mathcal{L}\{\cos 2t\} \qquad \text{linearity properties} \\
&= 5\left(\frac{1}{s}\right) + \frac{s}{s^2 + 2^2} \qquad \text{transforms (1) and (6)} \\
&= \frac{5}{s} + \frac{s}{s^2 + 4} \qquad \text{simplifying}
\end{aligned}$$

■

EXAMPLE 2   ■ **Problem**

Use the table to find the Laplace transform of $f(t) = 5t^3$.

■ **Solution**

We have

$$\begin{aligned}
\mathcal{L}\{5t^3\} &= 5\mathcal{L}\{t^3\} \qquad \text{linearity property} \\
&= 5 \cdot \frac{3!}{s^{3+1}} \qquad \text{transform (2)} \\
&= \frac{30}{s^4} \qquad 3! = 3 \cdot 2 \cdot 1
\end{aligned}$$

■

EXAMPLE 3   ■ **Problem**

Evaluate $\mathcal{L}\{3t \sin 5t\}$.

■ **Solution**

We obtain

$$\begin{aligned}
\mathcal{L}\{3t \sin 5t\} &= 3\mathcal{L}\{t \sin 5t\} \qquad \text{linearity property} \\
&= 3 \cdot \frac{2(5)s}{(s^2 + 5^2)^2} \qquad \text{transform (7)} \\
&= \frac{30s}{(s^2 + 25)^2} \qquad \text{simplifying}
\end{aligned}$$

■

In addition to knowing how to find the Laplace transform of a function $f(t)$, you should also be able to reverse this process and reconstruct a function $f(t)$ whose Laplace transform is known. The operation is called the **inverse Laplace transform.**

DEFINITION

Inverse Laplace Transform   If there exists a function $f(t)$ such that $\mathcal{L}\{f(t)\} = F(s)$, then $f(t)$ is called the *inverse Laplace transform* of $F(s)$, and we write

$$f(t) = \mathcal{L}^{-1}\{F(s)\}$$

To find $\mathcal{L}^{-1}\{F(s)\}$, we use the table of Laplace transforms in reverse. In most cases, the given transform will not be in a form that allows the direct use of the table. Rather, the given $F(s)$ will have to be expanded into several parts, and each inverse may be found from the table.

Of fundamental importance is the underlying fact that the inverse Laplace transform has the following linearity property.

PROPERTY

Linearity Property of the Inverse Laplace Transform

$$\mathcal{L}^{-1}\{c_1 F_1(s) + c_2 F_2(s)\} = c_1 \mathcal{L}^{-1}\{F_1(s)\} + c_2 \mathcal{L}^{-1}\{F_2(s)\}$$

where c_1 and c_2 represent any constants.

EXAMPLE 4   ■ **Problem**

Evaluate $\mathcal{L}^{-1}\left\{\dfrac{2}{(s+1)^4}\right\}$ by using the table of Laplace transforms.

■ **Solution**

We have

$$\mathcal{L}^{-1}\left\{\frac{2}{(s+1)^4}\right\} = \mathcal{L}^{-1}\left\{\frac{2}{[s-(-1)]^4}\right\} \qquad \text{transform (4) requires } s - k$$

$$= \frac{1}{3}\mathcal{L}^{-1}\left\{\frac{3!}{[s-(-1)]^4}\right\} \qquad \text{transform (4) requires } n!$$

$$= \tfrac{1}{3}t^3 e^{-t} \qquad\qquad\qquad \text{inverse transform (4)} \qquad ■$$

EXAMPLE 5   ■ **Problem**

Find $\mathcal{L}^{-1}\left\{\dfrac{2s + 1}{s^2 + 4}\right\}$.

■ **Solution**

The function of which this expression is the transform is not obtainable by using the table directly. Therefore, we proceed as follows:

$$\mathcal{L}^{-1}\left\{\frac{2s + 1}{s^2 + 4}\right\} = \mathcal{L}^{-1}\left\{\frac{2s}{s^2 + 4} + \frac{1}{s^2 + 4}\right\} \qquad \frac{a + b}{c} = \frac{a}{c} + \frac{b}{c}$$

$$= 2\mathcal{L}^{-1}\left\{\frac{s}{s^2 + 4}\right\} + \mathcal{L}^{-1}\left\{\frac{1}{s^2 + 4}\right\} \qquad \text{linearity properties}$$

$$= 2 \cos 2t + \tfrac{1}{2} \sin 2t \qquad\qquad \text{inverse transforms} \\ \text{(6) and (5)} \qquad ■$$

In many cases, the given transform is a proper rational function of s. So, the table can be used only after the given transform is decomposed into its partial fractions.

EXAMPLE 6   ■ **Problem**

Evaluate $\mathcal{L}^{-1}\left\{\dfrac{s + 5}{s^2 - 2s - 3}\right\}$.

■ **Solution**

The denominator factors into $(s - 3)(s + 1)$. Hence, the partial fractions are

$$\frac{s + 5}{(s - 3)(s + 1)} = \frac{A}{s - 3} + \frac{B}{s + 1}$$

where A and B are constants to be determined. Clearing the fractions, we get

$$s + 5 = A(s + 1) + B(s - 3)$$

If $s = 3$, we have

$$8 = 4A \qquad \text{or} \qquad A = 2$$

Similarly, if $s = -1$, we have

$$4 = -4B \qquad \text{or} \qquad B = -1$$

Finally,

$$\mathcal{L}^{-1}\left\{\frac{s + 5}{s^2 - 2s - 3}\right\} = \mathcal{L}^{-1}\left\{\frac{2}{s - 3}\right\} + \mathcal{L}^{-1}\left\{\frac{-1}{s + 1}\right\} = 2e^{3t} - e^{-t} \qquad ■$$

EXAMPLE 7   ▪ **Problem**

Evaluate $\mathcal{L}^{-1}\left\{\dfrac{s^2}{(s+1)^3}\right\}$.

▪ **Solution**

The partial fractions are

$$\frac{s^2}{(s+1)^3} = \frac{A}{s+1} + \frac{B}{(s+1)^2} + \frac{C}{(s+1)^3}$$

Clearing the fractions yields

$$s^2 = A(s+1)^2 + B(s+1) + C$$

Expanding gives $s^2 = As^2 + 2As + A + Bs + B + C$. Collecting like terms, we have

$$s^2 = As^2 + (2A+B)s + (A+B+C)$$

Equating the like powers of s yields the following system of equations

$$A = 1$$
$$2A + B = 0$$
$$A + B + C = 0$$

The solution of this system is $A = 1$, $B = -2$, and $C = 1$. Making these substitutions, we have

$$\mathcal{L}^{-1}\left\{\frac{s^2}{(s+1)^3}\right\} = \mathcal{L}^{-1}\left\{\frac{1}{s+1}\right\} + \mathcal{L}^{-1}\left\{\frac{-2}{(s+1)^2}\right\} + \mathcal{L}^{-1}\left\{\frac{1}{(s+1)^3}\right\}$$

$$= e^{-t} - 2te^{-t} + \tfrac{1}{2}t^2 e^{-t} = (1 - 2t + \tfrac{1}{2}t^2)e^{-t} \qquad ■$$

EXAMPLE 8   ▪ **Problem**

Evaluate $\mathcal{L}^{-1}\left\{\dfrac{9s+14}{(s-2)(s^2+4)}\right\}$.

▪ **Solution**

Since the denominator is composed of a linear factor and an irreducible quadratic factor, the required partial fractions are

$$\frac{9s+14}{(s-2)(s^2+4)} = \frac{A}{s-2} + \frac{Bs+C}{s^2+4}$$

Clearing the fractions gives

$$9s + 14 = A(s^2 + 4) + (Bs + C)(s - 2)$$

Expanding on the right and collecting like terms yields

$$9s + 14 = (A + B)s^2 + (-2B + C)s + (4A - 2C)$$

Equating the coefficients of corresponding powers of s, we get

$$A + B = 0$$
$$-2B + C = 9$$
$$4A - 2C = 14$$

Solving this system yields $A = 4$, $B = -4$, and $C = 1$. Therefore,

$$\mathcal{L}^{-1}\left\{\frac{9s + 14}{(s - 2)(s^2 + 4)}\right\} = \mathcal{L}^{-1}\left\{\frac{4}{s - 2}\right\} + \mathcal{L}^{-1}\left\{\frac{-4s + 1}{s^2 + 4}\right\}$$

$$= 4\mathcal{L}^{-1}\left\{\frac{1}{s - 2}\right\} - 4\mathcal{L}^{-1}\left\{\frac{s}{s^2 + 4}\right\} + \mathcal{L}^{-1}\left\{\frac{1}{s^2 + 4}\right\}$$

$$= 4e^{2t} - 4\cos 2t + \tfrac{1}{2}\sin 2t \qquad \blacksquare$$

EXERCISES FOR SECTION 13.8

Evaluate the Laplace transform of the functions in Exercises 1–16, using the linearity property and the table of transforms.

1. $\mathcal{L}\{t^3\}$

2. $\mathcal{L}\{e^{-2t}\}$

3. $\mathcal{L}\{2t^4 + t^2 + 6\}$

4. $\mathcal{L}\{e^{7t} + e^{-2t}\}$

5. $\mathcal{L}\{e^{-2t} + 4e^t\}$

6. $\mathcal{L}\{3t^2 + \cos 2t\}$

7. $\mathcal{L}\{t^5 e^{-3t}\}$

8. $\mathcal{L}\{4t^3 e^{2t}\}$

9. $\mathcal{L}\{e^{3t} + \sin 3t\}$

10. $\mathcal{L}\{3\sin 4t - 2\cos 4t\}$

11. $\mathcal{L}\{2te^{-t} + e^{-t}\}$

12. $\mathcal{L}\{5 + te^{-2t} - e^{-2t}\}$

13. $\mathcal{L}\{t(t - 2)e^{3t}\}$

14. $\mathcal{L}\{(t - 2)^2 e^{4t}\}$

15. $\mathcal{L}\{t\cos 4t\}$

16. $\mathcal{L}\{5t\cos 2t\}$

Find the indicated inverse Laplace transforms in Exercises 17–38 by using the table of transforms.

17. $\mathcal{L}^{-1}\left\{\dfrac{1}{s - 3}\right\}$

18. $\mathcal{L}^{-1}\left\{\dfrac{s}{s^2 + 9}\right\}$

19. $\mathcal{L}^{-1}\left\{\dfrac{2}{s^2 + 9}\right\}$

20. $\mathcal{L}^{-1}\left\{\dfrac{s + 2}{s^2 + 1}\right\}$

21. $\mathcal{L}^{-1}\left\{\dfrac{1}{s^3}\right\}$

22. $\mathcal{L}^{-1}\left\{\dfrac{3}{s^5}\right\}$

23. $\mathcal{L}^{-1}\left\{\dfrac{2}{(s-3)^5}\right\}$

24. $\mathcal{L}^{-1}\left\{\dfrac{1}{(s+1)^3}\right\}$

25. $\mathcal{L}^{-1}\left\{\dfrac{s-3}{s^2-16}\right\}$

26. $\mathcal{L}^{-1}\left\{\dfrac{1}{s}+\dfrac{s}{s^2-9}\right\}$

27. $\mathcal{L}^{-1}\left\{\dfrac{s+1}{s^2+4s-5}\right\}$

28. $\mathcal{L}^{-1}\left\{\dfrac{s+2}{s^2-s-6}\right\}$

29. $\mathcal{L}^{-1}\left\{\dfrac{s+12}{s^2+s-6}\right\}$

30. $\mathcal{L}^{-1}\left\{\dfrac{1}{(s-2)^3}\right\}$

31. $\mathcal{L}^{-1}\left\{\dfrac{s}{(s+4)^2}\right\}$

32. $\mathcal{L}^{-1}\left\{\dfrac{3}{s(s-1)}\right\}$

33. $\mathcal{L}^{-1}\left\{\dfrac{13}{s(s-2)}\right\}$

34. $\mathcal{L}^{-1}\left\{\dfrac{3s}{(s-5)^2}\right\}$

35. $\mathcal{L}^{-1}\left\{\dfrac{3}{s(s^2+16)}\right\}$

36. $\mathcal{L}^{-1}\left\{\dfrac{s^2-9}{(s^2+9)^2}\right\}$

37. $\mathcal{L}^{-1}\left\{\dfrac{2s}{(s^2+9)^2}\right\}$

38. $\mathcal{L}^{-1}\left\{\dfrac{s}{(s^2+25)^2}\right\}$

■■■■ 13.9 SOLVING DIFFERENTIAL EQUATIONS

Laplace Transform of f' and f''

To apply the method of Laplace transforms in solving linear differential equations, we need to know the general formula for the Laplace transform of the derivative function. We derive this formula by appealing directly to the definition of the Laplace transform.

By definition of $\mathcal{L}\{f'(t)\}$, we have

$$\mathcal{L}\{f'(t)\} = \int_0^\infty e^{-st}f'(t)\,dt$$

Using integration by parts with

$u = e^{-st}$	$dv = f'(t)\,dt$
$du = -se^{-st}\,dt$	$v = f(t)$

we get

$$\mathcal{L}\{f'(t)\} = \left[f(t)e^{-st}\right]_0^\infty + s\int_0^\infty e^{-st}f(t)\,dt$$

Assuming $f(t)e^{-st}$ approaches zero as $t \to \infty$ we have the following formula.

FORMULA

> **Laplace Transform of the Derivative Function**
>
> $\mathcal{L}\{f'(t)\} = s\mathcal{L}\{f(t)\} - f(0)$

To find $\mathcal{L}\{f''(t)\}$, we let $g(t) = f'(t)$. Then,

$$\mathcal{L}\{f''(t)\} = \mathcal{L}\{g'(t)\} = s\mathcal{L}\{g(t)\} - g(0) = s\mathcal{L}\{f'(t)\} - f'(0)$$
$$= s[s\mathcal{L}\{f(t)\} - f(0)] - f'(0)$$

Expanding this result, we get the next formula.

FORMULA

> **Laplace Transform of the Second Derivative**
>
> $\mathcal{L}\{f''(t)\} = s^2\mathcal{L}\{f(t)\} - sf(0) - f'(0)$

To find $\mathcal{L}\{f'(t)\}$, we must know the value of $f(0)$. To find $\mathcal{L}\{f''(t)\}$, we must know the values of both $f(0)$ and $f'(0)$. As noted earlier, when $f(0)$ and $f'(0)$ are conditions associated with a differential equation, they are called initial conditions. A differential equation with its initial conditions is called an initial-value problem. We will use the Laplace transform to solve initial-value problems of the form

$$b_2 y'' + b_1 y' + b_0 y = f(t) \qquad y(0) = a \qquad y'(0) = d$$

where b_0, b_1, and b_2 are constant coefficients of the second-order linear differential equation, and $y(0) = a$ and $y'(0) = d$ are the initial conditions.

Some Initial-Value Problems

EXAMPLE 1   ▪ **Problem**

Use the Laplace transform to solve $y' + 2y = 0$ and $y(0) = 1$.

▪ **Solution**

We take the Laplace transform of both sides of this equation.

$$\mathcal{L}\{y' + 2y\} = \mathcal{L}\{0\}$$

$\mathcal{L}\{y'\} + 2\mathcal{L}\{y\} = \mathcal{L}\{0\}$ linearity properties

$s\mathcal{L}\{y\} - 1 + 2\mathcal{L}\{y\} = 0$ $\mathcal{L}\{y'\} = s\mathcal{L}\{y\} - y(0)$; $\mathcal{L}\{0\} = 0$

$(s + 2)\mathcal{L}\{y\} = 1$ adding 1 to both sides and factoring on the left

$$\mathcal{L}\{y\} = \frac{1}{s + 2} \qquad\qquad \text{dividing by } s + 2$$

$$y = e^{-2t} \qquad\qquad \text{inverse transform (3)} \qquad \blacksquare$$

The technique of Laplace transforms described in Example 1 is summarized in the following general outline.

PROCEDURE

Solving Initial-Value Problems by Using Laplace Transforms

1. Take the Laplace transform of both sides of the given differential equation.

2. Solve the transformed equation for the Laplace transform of the solution function. This step is purely algebraic.

3. Find the inverse transform of the expression found in step 2. This step causes the most difficulty, but experience will help you to improve your skill at finding inverse transforms.

EXAMPLE 2   ■ **Problem**

Solve the intial-value problem $y' + 3y = 3$ and $y(0) = 0$.

■ **Solution**

Taking the Laplace transform of both sides, we get

$$s\mathcal{L}\{y\} - 0 + 3\mathcal{L}\{y\} = \frac{3}{s}$$

$$(s + 3)\mathcal{L}\{y\} = \frac{3}{s} \qquad\qquad \text{factoring on the left}$$

$$\mathcal{L}\{y\} = \frac{3}{s(s + 3)} \qquad\qquad \text{dividing by } s + 3$$

$$y = 1 - e^{-3t} \qquad\qquad \text{inverse transform (9)} \qquad \blacksquare$$

EXAMPLE 3   ■ **Problem**

Solve $\dfrac{d^2x}{dt^2} + 4x = 10$, $x(0) = 1$, and $x'(0) = 2$.

■ **Solution**

Taking the Laplace transform of each side, we get

$$\mathcal{L}\left\{\frac{d^2x}{dt^2}\right\} + 4\mathcal{L}\{x\} = \frac{10}{s}$$ linearity properties: transform (1)

$$s^2\mathcal{L}\{x\} - s - 2 + 4\mathcal{L}\{x\} = \frac{10}{s}$$ $\mathcal{L}\{f''(t)\} = s^2\mathcal{L}\{f(t)\}$
$$- sf(0) - f'(0)$$

$$(s^2 + 4)\mathcal{L}\{x\} = \frac{10}{s} + s + 2$$ isolating $\mathcal{L}\{x\}$ terms

$$\mathcal{L}\{x\} = \frac{10}{s(s^2 + 4)} + \frac{s}{s^2 + 4} + \frac{2}{s^2 + 4}$$ dividing by $s^2 + 4$

$$x = \frac{5}{2}(1 - \cos 2t) + \cos 2t + \sin 2t$$ inverse transforms (10), (6), and (5)

$$= \frac{5}{2} - \frac{3}{2}\cos 2t + \sin 2t$$ simplifying

Notice that

$$\mathcal{L}^{-1}\left\{\frac{10}{s(s^2 + 4)}\right\} = \frac{5}{2}\mathcal{L}^{-1}\left\{\frac{4}{s(s^2 + 4)}\right\}$$

in order to fit inverse Laplace transform 10. ∎

EXERCISES FOR SECTION 13.9

Solve the initial-value problems in Exercises 1–14 by using the Laplace transform method.

1. $\dfrac{dy}{dt} + 3y = 0;\ y(0) = 2$

2. $y'(t) - 4y(t) = 0;\ y(0) = -1$

3. $x''(t) + 9x(t) = 0;\ x(0) = 3,\ x'(0) = 0$

4. $\dfrac{d^2y}{dt^2} + 4y = 0;\ y(0) = 1,\ y'(0) = 1$

5. $y'' + y' = 0;\ y(0) = 0,\ y'(0) = 1$

6. $\ddot{x} - 2\dot{x} = 0;\ x(0) = 0,\ \dot{x}(0) = -2\left(\textit{Note: } \dot{x} = \dfrac{dx}{dt} \text{ and } \ddot{x} = \dfrac{d^2x}{dt^2}.\right)$

7. $y''(t) + 3y'(t) + 2y(t) = 0;\ y(0) = 0,\ y'(0) = -2$

8. $\dfrac{d^2x}{dt^2} - \dfrac{dx}{dt} - 6x = 0;\ x(0) = 0,\ x'(0) = 1$

9. $\dot{x} + 2x = 4;\ x(0) = 0$

10. $y'(t) - y(t) = e^{2t};\ y(0) = 2$

11. $\dfrac{d^2y}{dt^2} + 4y = 1; \ y(0) = 0, \ y'(0) = 0$

12. $y''(t) + y(t) = 10; \ y(0) = 0, \ y'(0) = -2$

13. $y''(t) + 4y'(t) + 4y(t) = e^{-2t}; \ y(0) = 0, \ y'(0) = 0$

14. $y''(t) + 2y'(t) + y(t) = e^{-t}; \ y(0) = 0, \ y'(0) = 0$

15. A rocket weighing 96 lb is propelled by an engine that produces a thrust of 1000 lb. What is the displacement of the rocket when $t = 5$ sec if it starts from rest and $k = 1$? The equation of motion is

$$m\frac{dv}{dt} = F - kv$$

16. A spring is stretched $\frac{1}{2}$ ft by an 8-lb weight. Assume the weight is pushed up $\frac{1}{4}$ ft above equilibrium and released. Assuming there is no damping force, find the equation for the displacement of the weight.

17. Solve the spring-mass problem in Exercise 16 when the weight is given a downward velocity of 4 ft/sec.

18. What are the amplitude and the period of the oscillation described in Exercise 17?

19. Find the current in the circuit shown in the accompanying figure given that the impressed voltage is 12 V and the initial current is zero.

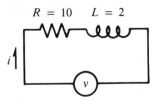

20. What is the current in the circuit if the impressed voltage varies with time according to $v = \sin 2t$? Assume $i(0) = 0$.

REVIEW EXERCISES FOR CHAPTER 13

Solve the differential equations in Exercises 1–34.

1. $y \, dx + x^{-1} \, dy = 0$

2. $(1 + x)y \, dy + y^2 \, dx = 0$

3. $e^{2x} \, dy - \sec y \, dx = 0$

4. $\dfrac{ds}{dt} = s\sqrt{2t + 3}$

5. $\dfrac{dy}{dx} = x^2 \cos^2 y$

6. $x(y + 1) \, dy + (x^2 + 3) \, dx = 0$

7. $2\,dy + \dfrac{y\,dx}{x+2} = 0$

8. $(x+3)y\,dx - (x^2 + x - 6)\,dy = 0$

9. $\dfrac{dy}{dx} = 3 + y$

10. $\dfrac{dy}{dx} + 2y = 5$

11. $y' - \dfrac{y}{x} = x^2$

12. $2y' + 6y = 5e^{-2x}$

13. $\frac{1}{2}\,ds - s\,dt = e^{2t}\,dt$

14. $\dfrac{dy}{x} + y\,dx = dx$

15. $x^2\,dy - 3xy\,dx = x\,dx$

16. $y' = \dfrac{xy}{x^2 + 2} + 3x$

17. $dy - y\tan x\,dx = \csc x\,dx$

18. $y'\tan x + y = 5\sec x$

19. $y'' - 3y' + 2y = 0$

20. $2y'' - 3y' = 0$

21. $\dfrac{d^2s}{dt^2} - 16\dfrac{ds}{dt} = 0$

22. $y'' + 5y' + 6y = 0$

23. $y'' - y' - y = 0$

24. $\dfrac{d^2v}{dt^2} + 2\dfrac{dv}{dt} - 3v = 0$

25. $y'' + 8y' + 16y = 0$

26. $y'' + 10y' + 25y = 0$

27. $\ddot{x} + 2\dot{x} + x = 0$

28. $9y'' + 6y' + y = 0$

29. $\dfrac{d^2s}{dt^2} + 9s = 0$

30. $\ddot{x} + 0.01x = 0$

31. $y'' + 4y' + 13y = 0$

32. $y'' - 6y' + 25y = 0$

33. $y'' + 2y' + 5y = 0$

34. $y'' + 2y' + 4y = 0$

35. Find the amplitude, the period, and the phase shift for $x = 3\cos 2t + \sin 2t$.

36. Find the amplitude, the period, and the phase shift for $x = \cos 3t + \sqrt{8}\sin 3t$.

 37. If a population grows at a rate proportional to the size of the population, estimate the population of a colony in the year 2000 given that the population was 500 in 1950 and 550 in 1960.

 38. The voltage applied to a series RL circuit varies with time according to $v = 0.4t$ from $t = 0$ to $t = 2$ milliseconds (msec). Find the current in the circuit at $t = 2$ if $R = 100\ \Omega$ and $L = 0.1$ H. Assume $i = 0$ when $t = 0$.

 39. Find the distance-time equation for the spring-mass system shown in the accompanying figure given that the weight is initially pulled down 4 in. and then released.

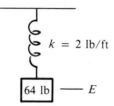

$k = 2$ lb/ft

64 lb —— E

 40. Find the distance-time equation for the spring-mass system in Exercise 39 given that it is pulled down 4 in. and then given an initial velocity of -2 ft/sec.

In Exercises 41–46, find the indicated Laplace transform.

41. $\mathcal{L}\{2 \cos \sqrt{7}t\}$

42. $\mathcal{L}\{\frac{1}{2} \sin t\}$

43. $\mathcal{L}\{3e^{5t} - 8t^2 e^{-t}\}$

44. $\mathcal{L}\{3t^4 - 9\}$

45. $\mathcal{L}\{t^5 - 3 \sin 2t\}$

46. $\mathcal{L}\{t^2 e^{-3t} + 4 \cos 5t\}$

In Exercises 47–52, find the indicated inverse Laplace transform.

47. $\mathcal{L}^{-1}\left\{\dfrac{1 - 2s}{s^2 + 9}\right\}$

48. $\mathcal{L}^{-1}\left\{\dfrac{7}{(s - 2)^4}\right\}$

49. $\mathcal{L}^{-1}\left\{\dfrac{2}{s^4} - \dfrac{1}{s^2}\right\}$

50. $\mathcal{L}^{-1}\left\{\dfrac{5}{s} + \dfrac{1}{s^3} - \dfrac{3}{s^2 + 4}\right\}$

51. $\mathcal{L}^{-1}\left\{\dfrac{2s - 7}{s^2 - 7s + 12}\right\}$

52. $\mathcal{L}^{-1}\left\{\dfrac{5}{s(s - 1)}\right\}$

Use the Laplace transform to solve the initial-value problems in Exercises 53–58.

53. $y'' + 5y' = 0; y(0) = 0, y'(0) = 1$

54. $y'' - 4y = 0; y(0) = 2, y'(0) = 0$

55. $\dot{x} - 3x = 5e^{-t}; x(0) = 0$

56. $\dfrac{dx}{dt} + 2x = 3e^{2t}; x(0) = 1$

57. $y'' + 5y = 2; y(0) = 0, y'(0) = 0$

58. $y'' + 2y = 8; y(0) = 0, y'(0) = -2$

CHAPTER 14

Vectors, Parametric Equations, and Polar Coordinates

In this chapter, we consider some applications of differential and integral calculus to functions stated in forms other than $y = f(x)$. Included in our discussion are parametric equations and polar equations, both of which are useful in describing physical problems. We begin with the concept of a vector.

14.1 VECTORS IN THE PLANE

Some quantities such as temperature, length, or volume can be described by one number. These quantities are examples of **scalar quantities.** Other quantities such as velocity, acceleration, or force must be described by giving both a magnitude and a direction. For example, a velocity of 70 mi/hr to the north is quite different from 70 mi/hr to the south.

To describe quantities that have a magnitude and an associated direction, we use (geometric) **vectors.** Usually, an arrow represents a vector quantity. Its length corresponds to the magnitude of the quantity, and the arrowhead gives its direction. Thus, in Figure 14.1, one arrow represents a velocity of 100 mi/hr north and the other a velocity of 200 mi/hr northeast.

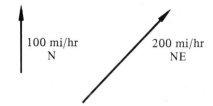

100 mi/hr
N

200 mi/hr
NE

■ **Figure 14.1**

Vectors are given in boldface to distinguish them from scalars. The length of a vector **V** is denoted by $|\,\mathbf{V}\,|$ and is always a positive quantity. The direction of a vector is given in various ways, depending on the application. Sometimes, we need to consider a vector with zero length, called the *zero vector*. It is denoted by **0** and has an undefined direction.

For mathematical purposes, two vectors are considered the same if they have the same direction and length, regardless of the location of the initial point of the vector. Thus, in Figure 14.2, all the vectors are mathematically equivalent.

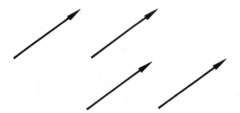

■ **Figure 14.2**

Note that this type of vector equality may not always be the kind you need. For example, in Figure 14.3, if **F** is a 10-lb force pointing down, which of the three places it is applied to does make a difference. Vectors for which you may ignore the actual point of application are said to be *free*. The physical situation shown in Figure 14.3 cannot be described by using free vectors.

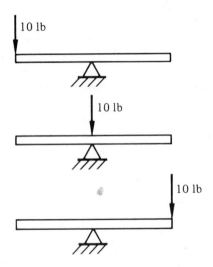

■ **Figure 14.3**

By first placing a coordinate system on the plane, we may move all the vectors so that their initial points are at the origin. See Figure 14.4. Such vectors are said to be in *standard position*. A vector at the origin represents all other vectors equivalent to it.

By referencing any vector to the origin, the terminal point of a vector will be enough to describe it. Thus, the coordinates of the terminal point completely describe the vector. Conversely, an ordered pair of numbers determines a vector whose initial point is at the origin and whose terminal point is at the point whose coordinates are given by the ordered pair of numbers.

If a vector is placed in standard position, you can find its length from the Pythagorean theorem. The direction of the vector is given as the angle it makes with the positive x-axis.

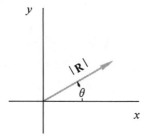

■ **Figure 14.4**

EXAMPLE 1   ■ **Problem**

Find the magnitude and direction of the vector **V** in standard position whose terminal point is at (2, 1).

■ **Solution**

Figure 14.5 shows the length of the vector to be

$$| \mathbf{V} | = \sqrt{2^2 + 1^2} = \sqrt{5}$$

The angle θ that **V** makes with the positive x-axis is

$$\theta = \text{Arctan}\,(0.5) = 26.6°$$

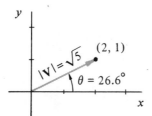

■ **Figure 14.5**   ■

EXAMPLE 2   ■ **Problem**

Figure 14.6 shows a vector **V** with $| \mathbf{V} | = 12$ and $\theta = 60°$. Find the ordered pair of numbers that determines **V**.

■ **Solution**

The ordered pair of real numbers that essentially determines the vector **V** is given by the coordinates of the point that is at the tip of **V** when **V** is in standard position. Thus, in Figure 14.6, the x-coordinate of the tip of **V** is

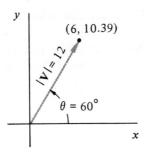

■ Figure 14.6

$$x = |\mathbf{V}| \cos \theta = 12 \cos 60° = 12\left(\frac{1}{2}\right) = 6$$

The y-coordinate is

$$y = |\mathbf{V}| \sin \theta = 12 \sin 60° = 12\left(\frac{\sqrt{3}}{2}\right) = 10.39$$

Thus, the ordered pair (6, 10.39) completely determines $\mathbf{V}$.   ■

Two basic vectors in the plane have a special notation. Figure 14.7 shows these two vectors in standard position, one ending at (1, 0), called the $\mathbf{i}$ vector, and the other ending at (0, 1), called the $\mathbf{j}$ vector. As we will see, every other vector in the plane can be expressed in terms of these two.

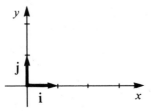

■ Figure 14.7

Scalar Multiplication

Given any vector $\mathbf{A}$, we may obtain other vectors in the same (or opposite) direction of $\mathbf{A}$ by multiplying $\mathbf{A}$ by a real number c. The resulting vector, denoted by $c\mathbf{A}$, is a vector that points in the same direction as $\mathbf{A}$ if $c > 0$ and opposite to that of $\mathbf{A}$ if $c < 0$. The magnitude of $c\mathbf{A}$ is $|c||\mathbf{A}|$. That is, it is larger than $|\mathbf{A}|$ if $|c| > 1$, and it is smaller than $|\mathbf{A}|$ if $|c| < 1$. The vector $c\mathbf{A}$ is called a **scalar multiple** of the vector $\mathbf{A}$. Figure 14.8 shows some scalar multiples of a given vector $\mathbf{A}$.

The scalar multiple of a vector obtained by multiplying a vector $\mathbf{A}$ by the reciprocal of its magnitude leads to a vector in the direction of $\mathbf{A}$ whose length is 1. This vector is the **unit vector** in the direction of $\mathbf{A}$, denoted by $\mathbf{u}_A$.

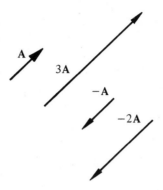

■ **Figure 14.8**

DEFINITION	**Unit Vector**

$$\mathbf{u}_A = \frac{\mathbf{A}}{|\mathbf{A}|} \qquad\qquad (14.1)$$

Vector Addition

Two vectors are added by a rule called the *parallelogram rule.*

RULE	**Addition of Vectors** To add two vectors **A** and **B**, place both with their initial points together. Then, form a parallelogram with these vectors as sides. The vector from the initial point, which is the diagonal of the parallelogram, is called the *sum* (or resultant) of **A** and **B**. Figure 14.9 depicts the resultant of two vectors.

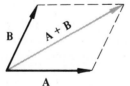

■ **Figure 14.9**

　　　　Addition of vectors is most frequently used in expressing a vector **A** as the sum of its component vectors, one in the direction of the **i** vector and one in the direction of the **j** vector. Figure 14.10 shows a vector **A** as the sum of $A_x\mathbf{i}$ and $A_y\mathbf{j}$. The quantity A_x is called the **horizontal component,** and A_y is called the **vertical component** of the vector **A**.

　　　　The numbers A_x and A_y are equal to, respectively, the *x*-axis and *y*-coordi-

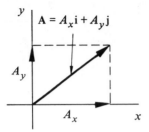

■ **Figure 14.10**

nates of the point at the terminal end of the vector **A.** From Figure 14.10, we have
the following formula.

FORMULA

Magnitude and Direction of Vector A

$$|\mathbf{A}| = \sqrt{A_x^2 + A_y^2} \qquad\qquad (14.2)$$

$$\tan \theta = \frac{A_y}{A_x} \qquad\qquad (14.3)$$

Thus, a vector **A** may ordinarily be expressed in one of the following equivalent ways:

- As the sum $A_x\mathbf{i} + A_y\mathbf{j}$;
- As the set of ordered pairs (A_x, A_y).

Addition of vectors may be removed from a geometric setting by using the
component method of representation. Note that if

$$\mathbf{A} = A_x\mathbf{i} + A_y\mathbf{j} \qquad \text{and} \qquad \mathbf{B} = B_x\mathbf{i} + B_y\mathbf{j}$$

then,

$$\mathbf{A} + \mathbf{B} = (A_x + B_x)\mathbf{i} + (A_y + B_y)\mathbf{j}$$

In words, the horizontal component of the sum of two vectors is simply the sum
of the individual horizontal components. The vertical component of the
sum is the sum of the individual vertical components.

EXAMPLE 3   ■ **Problem**

Add the vectors $\mathbf{A} = \mathbf{i} - 3\mathbf{j}$ and $\mathbf{B} = 5\mathbf{i} + \mathbf{j}$. See Figure 14.11.

■ **Solution**

We have

$$\mathbf{A} + \mathbf{B} = (\mathbf{i} - 3\mathbf{j}) + (5\mathbf{i} + \mathbf{j}) = 6\mathbf{i} - 2\mathbf{j}$$

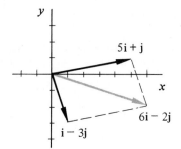

■ **Figure 14.11**   ■

There is an alternative definition for the addition of two vectors. Place the initial point of the second arrow at the terminal point of the first. Then, the resultant is the arrow whose initial point is at the initial point of the first arrow and whose terminal point is at the terminal point of the second arrow. See Figure 14.12(a). Note that this definition of a sum yields the same vector as the first definition.

■ **Figure 14.12**   (a)   (b)

For vector subtraction, reverse the procedure by placing the terminal point of **B** at the terminal point of **A**. Then **A** − **B** is drawn from the initial point of **A** to the initial point of **B**. See Figure 14.12,(b).

This alternative definition of addition can be used to add several vectors sequentially to give a general idea of the final result. Place the initial point of the second vector at the terminal point of the first, the initial point of the third at the terminal point of the second, and so on. See Figure 14.13. Since addition is a commutative operation for vectors, the order in which you perform this operation is unimportant.

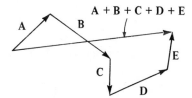

■ **Figure 14.13**

EXERCISES FOR SECTION 14.1

In Exercises 1–6, draw the vectors whose initial point is at the origin and whose terminal point is at the indicated point. Calculate the magnitude and the direction of each vector.

1. $(1, 2)$ **2.** $(-3, 2)$

3. $(-4, -4)$ **4.** $(5, 8)$

5. $(4, -3)$ **6.** $(-3, -7)$

Find the x- and y-coordinates of the terminal point of the vectors in Exercises 7–10 in standard position.

7. $|\mathbf{V}| = 10, \theta = 50°$ **8.** $|\mathbf{V}| = 0.751, \theta = 56°30'$

9. $|\mathbf{V}| = 158, \theta = 125°$ **10.** $|\mathbf{V}| = 875, \theta = 145°$

The vectors in Exercises 11–19 are given in the form $a\mathbf{i} + b\mathbf{j}$. Find the sum of **A** and **B** by adding the respective vertical and horizontal components. Draw each vector, and show the sum graphically.

11. $\mathbf{A} = 2\mathbf{i} + 2\mathbf{j}$ **12.** $\mathbf{A} = \mathbf{i} - \mathbf{j}$
     $\mathbf{B} = 3\mathbf{i} - \mathbf{j}$ $\mathbf{B} = 5\mathbf{i} + 3\mathbf{j}$

13. $\mathbf{A} = -2\mathbf{i} + 3\mathbf{j}$ **14.** $\mathbf{A} = 7\mathbf{i} + 8\mathbf{j}$
     $\mathbf{B} = -4\mathbf{i} - 5\mathbf{j}$ $\mathbf{B} = 6\mathbf{i} - 2\mathbf{j}$

15. $\mathbf{A} = -\mathbf{i} - \mathbf{j}$ **16.** $\mathbf{A} = 6\mathbf{i} + 6\mathbf{j}$
     $\mathbf{B} = 2\mathbf{i} - 2\mathbf{j}$ $\mathbf{B} = 3\mathbf{i} - 6\mathbf{j}$

17. $\mathbf{A} = -4\mathbf{i} - 6\mathbf{j}$ **18.** $\mathbf{A} = 9\mathbf{i} + 5\mathbf{j}$
     $\mathbf{B} = 4\mathbf{i} + 2\mathbf{j}$ $\mathbf{B} = -7\mathbf{i} + \mathbf{j}$

19. $\mathbf{A} = 5\mathbf{i} - 5\mathbf{j}$
     $\mathbf{B} = \mathbf{i} - 2\mathbf{j}$

The vectors in Exercises 20–27 are defined in terms of a magnitude and a direction. Find the sum of the given vectors.

20. $|\mathbf{A}| = 20, \theta_A = 15°$ **21.** $|\mathbf{A}| = 16, \theta_A = 25°$
     $|\mathbf{B}| = 25, \theta_B = 50°$ $|\mathbf{B}| = 22, \theta_B = 70°$

22. $|\mathbf{A}| = 15, \theta_A = 0°$ **23.** $|\mathbf{A}| = 9.5, \theta_A = 90°$
     $|\mathbf{B}| = 26, \theta_B = 60°$ $|\mathbf{B}| = 5.1, \theta_B = 40°$

24. $|\mathbf{A}| = 2.5, \theta_A = 35°$ **25.** $|\mathbf{A}| = 29.2, \theta_A = 15.6°$
     $|\mathbf{B}| = 3.0, \theta_B = 120°$ $|\mathbf{B}| = 82.6, \theta_B = 150°$

26. $|\mathbf{A}| = 125, \theta_A = 145°$ **27.** $|\mathbf{A}| = 550, \theta_A = 140°$
     $|\mathbf{B}| = 92, \theta_B = 215°$ $|\mathbf{B}| = 925, \theta_B = 310°$

In Exercises 28–35, use the vectors in the accompanying figure and perform the indicated additions graphically.

28. $\mathbf{A} + \mathbf{B}$ **29.** $3\mathbf{B} - \mathbf{A}$

30. $4\mathbf{A} + \mathbf{B} + 5\mathbf{C}$ **31.** $\mathbf{A} - 2\mathbf{B}$

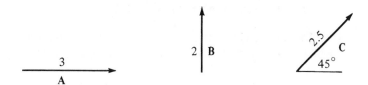

32. **C − B + 2A**   33. **2C − B − A**

34. **A + 3C − B**   35. **B − 2A + C**

 36. What are the horizontal and vertical components of the velocity of a ball that is thrown 100 ft/sec at an angle of 40° with respect to the horizontal?

 37. A plane is headed due north at 300 mi/hr. The wind is from the east at 50 mi/hr. What is the velocity of the plane?

 38. Both a vertical force of 50 lb and a horizontal force of 75 lb act through the center of gravity of an object. What single force could replace the two given forces?

 39. An object is acted upon by a force of 200 lb at an angle of 35° to the horizontal. What is the horizontal component of the force?

 40. An object is thrown vertically downward with a speed of 50 ft/sec from a plane moving horizontally with a speed of 250 ft/sec. What is the velocity of the object as it leaves the plane?

 41. A force of 60 lb acts horizontally on an object. Another force of 75 lb acts on the object at an angle of 55° with the horizontal. What is the resultant of these forces?

 42. A plane flies due west with an air speed of 150 mi/hr. The wind is from the southwest at 35 mi/hr. What is the resultant speed of the plane? In what direction is it traveling?

 43. A bullet is fired from a plane at an angle of 20° below the horizontal and in the direction the plane is moving. The bullet leaves the muzzle of the gun with a speed of 1200 ft/sec, and the plane is flying at a speed of 500 ft/sec. What is the resultant speed of the bullet?

 44. An object weighing 120 lb hangs at the end of a rope. The object is pulled sideways by a horizontal force of 30 lb. What angle does the rope make with the vertical? (*Hint:* Weight is a vector that is always considered to be acting vertically downward.)

■■■ 14.2 PARAMETRIC EQUATIONS

A convenient way to describe the position of a particle or a point in the plane is through the use of a *position vector* **R**(*t*). This vector's initial point is at the origin, and its tip is permitted to take on different coordinates depending on the value of *t*. See Figure 14.14. The tip of the vector traces out a curve, which we represent by

$$\mathbf{R}(t) = x(t)\mathbf{i} + y(t)\mathbf{j}$$

The functions $x(t)$ and $y(t)$ are called **parametric equations** of the curve; the variable t (which may represent time) is called the **parameter.**

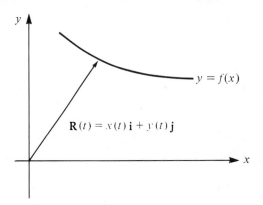

■ **Figure 14.14**

EXAMPLE 1   ■ **Problem**

Describe and sketch the curve defined by $\mathbf{R}(t) = x(t)\mathbf{i} + y(t)\mathbf{j}$, where $x = t + 3$ and $y = 2t - 5$.

■ **Solution**

The following table of values shows the ordered pairs (x, y) obtained by letting t be a range of real numbers.

t	-3	-2	-1	0	1	2	3
x	0	1	2	3	4	5	6
y	-11	-9	-7	-5	-3	-1	1

These ordered pairs may be plotted in the xy-plane and then connected by a smooth curve, as shown in Figure 14.15.

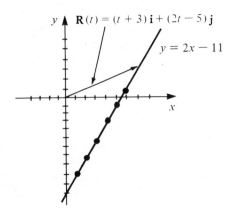

■ **Figure 14.15**

An alternative approach to sketching a curve defined by parametric equations is to **eliminate the parameter.** In this way, we obtain a representation of the curve in the more familiar form of a relation between x and y. For the parametric equations in Example 1, we solve $x = t + 3$ for t to obtain $t = x - 3$. Substituting this value for t into the equation for y, we obtain

$$y = 2(x - 3) - 5 = 2x - 11$$

In this form, we recognize the graph as a straight line, which agrees with the graph shown in Figure 14.15.

EXAMPLE 2   ■ **Problem**

Sketch the curve in the xy-plane described parametrically by $x = 2 \sin t$ and $y = 3 \cos t$.

■ **Solution**

Since $\frac{1}{2}x = \sin t$ and $\frac{1}{3}y = \cos t$, we may square both of these equations and add the corresponding sides to obtain

$$(\tfrac{1}{2}x)^2 + (\tfrac{1}{3}y)^2 = \sin^2 t + \cos^2 t = 1$$

This equation is an ellipse and is sketched in Figure 14.16.

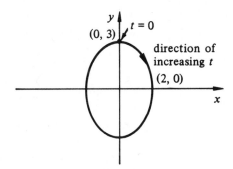

■ **Figure 14.16**

Note that the parametric representation gives a sense of direction to the curve since you can think of the point beginning at the point $(0, 3)$ when $t = 0$ and going around the curve in a clockwise sense as t increases. The point completes one revolution of the ellipse every 2π units. A curve with "direction," such as the one in this example, is said to be **oriented.**   ■

EXAMPLE 3   ■ **Problem**

Sketch the graph of the curve described by $x = \sin t$ and $y = \sin t$.

■ **Solution**

The elimination of the parameter is a simple matter of noticing that $y = x$. The graph of $y = x$ is shown in Figure 14.17 as the line that splits the first and third

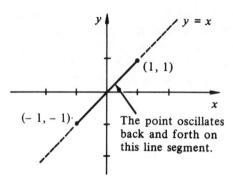

■ Figure 14.17

quadrants. However, by the nature of the parametric equations, the values of x and y are limited between 1 and -1. Thus, the point moves on the line $y = x$, but it oscillates between the points $(1, 1)$ and $(-1, -1)$.   ■

Warning: The previous example points out a general caution you should take. When eliminating the parameter, be careful that the process of elimination does not tend to include more points (or less) than the parametric equations themselves give.

EXAMPLE 4   **■ Problem**

Consider the two curves $C_1 : x = \cos^2 t, \ y = \sin^2 t$ and $C_2 : x = s, \ y = 1 - s$ (s is used as the parameter in C_2 to avoid confusion). Discuss the similarities and differences of these two curves.

■ Solution

Eliminating the parameter t in the set of equations that define C_1, we obtain $x = 1 - \sin^2 t = 1 - y$, or $x + y = 1$. Similarly, eliminating s from the equations defining C_2, we get $x + y = 1$. This result seems to imply that C_1 and C_2 are the same line. Notice, however, that the equations that define C_1 have a built-in restriction: x and y are both limited to $[0, 1]$. There is no such restriction on x and y for C_2. Thus, the curve C_1 is that part of C_2 that lies in the first quadrant. The point (x, y) oscillates back and forth on this segment when describing C_1. It traverses the entire line for curve C_2. See Figure 14.18.   ■

To obtain the points of intersection of two curves defined parametrically, we equate the x values of the two curves and then equate the y values to obtain two equations in the two unknown parametric values. These values, when substituted into the original equations, will give the points of intersection of the two curves.

EXAMPLE 5   **■ Problem**

Determine the points of intersection of $C_1 : x = 2t + 1, \ y = 4t^2 + 4t - 3$ and $C_2 : x = 3s, \ y = 3s + 2$.

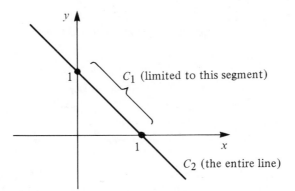

■ **Figure 14.18**

■ **Solution**

Equating the x values, we obtain

$$2t + 1 = 3s$$

Equating the y values gives

$$4t^2 + 4t - 3 = 3s + 2$$

Substituting $2t + 1$ for $3s$ in the second of these equations gives

$$4t^2 + 4t - 3 = (2t + 1) + 2$$

Solving this equation for t, we obtain $t = 1$ and $t = -\frac{3}{2}$. The corresponding values of s are $s = 1$ and $s = -\frac{2}{3}$. The points of intersection are as follows:

1. The value $s = 1$ (or $t = 1$) gives the point $(3, 5)$.
2. The value $s = -\frac{2}{3}$ (or $t = -\frac{3}{2}$) gives the point $(-2, 0)$.

These two points of intersection are shown in Figure 14.19 along with a sketch of C_1 and C_2.

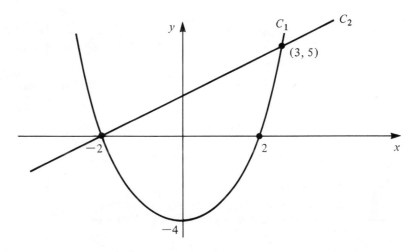

■ **Figure 14.19**
■

The parameter used to define a curve may be interpreted as time. In this case, a point of intersection may also have an additional physical meaning.

DEFINITION

> **Collision Point**   A point of intersection (x_0, y_0) of two curves C_t (parametrized by t) and C_s (parametrized by s) is called a *collision point* if $t = s$ when C_t and C_s pass through (x_0, y_0).

Thus, in Example 5, the point $(3, 5)$ is a collision point because $t = s = 1$ at that point. On the other hand, $(-2, 0)$ is not a collision point because at that point $t = -\frac{3}{2}$ and $s = -\frac{2}{3}$.

Comment: If C_t and C_s represent the paths of two objects in motion, a point of collision indicates that the two objects arrive at the same point in the plane at the same time. This collision may or may not be catastrophic, but the distinction between the paths' merely crossing and their crossing simultaneously should be clearly understood.

EXERCISES FOR SECTION 14.2

Sketch the curves described by the parametric equations in Exercises 1–10. In each case, indicate the orientation of the curve.

1. $x = \sin t, y = \cos t$

2. $x = 2 \cos t, y = \sin t$

3. $x = \cos^2 t, y = 1$

4. $x = 3t, y = 2$

5. $x = t^2, y = t^3$

6. $x = \cos t, y = \sin 2t$

7. $x = \sin t, y = \cos 2t$

8. $x = \sin t + \cos t, y = \sin t - \cos t$

9. $x = 2 \sin t + \cos t, y = 2 \cos t - \sin t$

10. $x = \sin t, y = \sin 2t$

Eliminate t from the parametric equations in Exercises 11–20 in order to obtain one equation in x and y. Sketch the curve, being careful to observe any limitations that are imposed by the parametrization.

11. $x = \sin t, y = \cos t$

12. $x = 2 \cos t, y = \sin t$

13. $x = t, y = 1 + t$

14. $x = -\cos t, y = \sin t$

15. $x = \cos t, y = \cos t$

16. $x = 1 - t^2, y = t$

17. $x = \sin^2 t, y = t$

18. $x = \cos^2 t, y = \cos^2 t$

19. $x = t - 3, y = t^2 + 1$

20. $x = 2 \sin t + \cos t, y = 2 \cos t - \sin t$

 21. The approximate path of the earth about the sun is given by $x = aR \cos 2\pi t$ and $y = bR \sin 2\pi t$, where t is the time (in years), R is the radius of the earth, and b and a are constants that are very nearly equal. Sketch this path for $a = 1$ and $b = 1.1$.

22. A piston is connected to the rim of a wheel, as shown in the accompanying figure. The point S, at a time t seconds after it has coordinates $(2, 0)$, has coordinates given by

$$x = 2 \cos 2\pi t \qquad y = 2 \sin 2\pi t$$

Find the position of the point S when $t = \frac{1}{2}, \frac{3}{4}$, and 2. Eliminate the parameter in order to find an equation relating x and y.

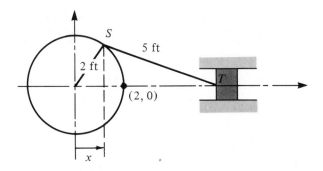

In Exercises 23–26, determine intersection/collision points. Sketch each curve.

23. $C_1 : x = 3 - 2t, y = 1 + t; C_2 : x = 2 - s, y = 2s$

24. $C_1 : x = t - 2, y = 2t; C_2 : x = s + 2, y = 1 - s$

25. $C_1 : x = t^2, y = t; C_2 : x = s + 1, y = s - 1$

26. $C_1 : x = t - 2, y = 2(3t - 5); C_2 : x = 2(s - 2), y = s$

27. The paths of two aircraft at 10,000 ft may be described approximately by the two curves $C_1 : x = \frac{16}{3} - \frac{8}{3}t, y = 4t - 5, t \geq 0$, and $C_2 : x = 2 \sin \frac{1}{2}\pi s, y = -3 \cos \frac{1}{2}\pi s, s \geq 0$. Sketch the paths of the two aircraft, and determine where the paths cross. Do the two aircraft collide?

14.3 TANGENTS TO CURVES

In this section, we examine the problem of finding a tangent to a curve defined parametrically. Then we show how the tangent vector to a curve can be interpreted as a velocity vector.

Tangent Lines

To find the slope of a tangent line, we suppose the curve C is defined by the parametric equations $x = x(t)$ and $y = y(t)$. The slope of a curve in the xy-plane is given by the value of dy/dx. To find the slope of a curve defined by a parameter t without solving for y in terms of x, we use the following formula.

FORMULA

Slope of a Curve

$$y' = \frac{dy}{dx} = \frac{dy}{dx} \cdot \frac{dt}{dt} = \frac{dy}{dt} \cdot \frac{dt}{dx} = \frac{dy/dt}{dx/dt} \qquad \textbf{(14.4)}$$

This formula gives the slope of the curve provided $dx/dt \neq 0$.

EXAMPLE 1   ■ **Problem**

Find the equation of the tangent to the ellipse

$$x = 2 \cos t \quad \text{and} \quad y = 4 \sin t$$

Determine the slope at $t = \pi/4$.

■ **Solution**

In this case, $dy/dt = 4 \cos t$, and $dx/dt = -2 \sin t$. Substituting these values into (14.4), we get

$$y' = \frac{4 \cos t}{-2 \sin t} = -2 \cot t$$

The slope of the ellipse at $t = \pi/4$ is $y' = -2 \cot \pi/4 = -2(1) = -2$. See Figure 14.20.

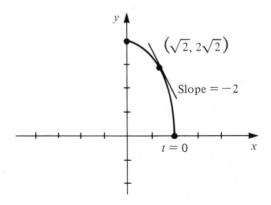

■ **Figure 14.20**

An analysis of Formula (14.4) reveals that $y' = 0$ if $dy/dt = 0$ and $dx/dt \neq 0$. Since $y' = 0$ indicates a horizontal tangent line, we may locate such points by solving $dy/dt = 0$ for t. Similarly, y' is undefined if $dx/dt = 0$, indicating a vertical tangent line. Vertical tangents are then found by solving $dx/dt = 0$ for t.

PROCEDURE

> ### Finding Horizontal and Vertical Tangents
>
> 1. To find the horizontal tangents, solve $dy/dt = 0$ for t (check dx/dt to be sure it is not equal to zero).
> 2. To find the vertical tangents, solve $dx/dt = 0$ for t.

EXAMPLE 2   ▪ **Problem**

Find the horizontal and vertical tangents to the graph of the cardioid given by the formula $C : x = 2 \cos t - \cos 2t - 1, \ y = 2 \sin t - \sin 2t$.

▪ **Solution**

To find horizontal tangents, we proceed as follows:

$$\frac{dy}{dt} = 2 \cos t - 2 \cos 2t$$

Setting $dy/dt = 0$, we get

$$2 \cos t - 2 \cos 2t = 0$$
$$\cos t - 2 \cos^2 t + 1 = 0 \qquad \cos 2t = 2 \cos^2 t - 1$$
$$(2 \cos t + 1)(\cos t - 1) = 0 \qquad \text{factoring}$$

The factor $2 \cos t + 1$ is zero for $t = 2\pi/3, \ 4\pi/3, \ \ldots$ The factor $\cos t - 1$ is zero for $t = 0, \ 2\pi, \ 4\pi, \ \ldots$ Thus, the cardioid has horizontal tangents at $t = 0, \ 2\pi/3$, and $4\pi/3$.

To find vertical tangents, we proceed as follows:

$$\frac{dx}{dt} = -2 \sin t + 2 \sin 2t$$

Setting $dx/dt = 0$, we get

$$-2 \sin t + 2 \sin 2t = 0$$
$$-\sin t + 2 \sin t \cos t = 0 \qquad \sin 2t = 2 \sin t \cos t$$
$$-\sin t(1 - 2 \cos t) = 0 \qquad \text{factoring}$$

The factor $\sin t$ is zero for $t = 0, \ \pi, \ 2\pi, \ \ldots$ The factor $1 - 2 \cos t$ is zero for $t = \pi/3, \ 5\pi/3, \ 7\pi/3. \ \ldots$ Thus, there are vertical tangents at $t = 0, \ \pi/3, \ \pi$, and $5\pi/3$. The graph in Figure 14.21 shows the horizontal and vertical tangent lines.

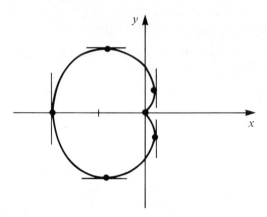

■ **Figure 14.21**   ■

EXAMPLE 3   ■ **Problem**

Find y' for the curve $C : x = 5t + 1$, $y = t^2 - 1$. At which point is the tangent parallel to the x-axis?

■ **Solution**

The slope of the curve C at any point is found by Formula (14.4) to be

$$y' = \frac{dy/dt}{dx/dt} = \frac{2t}{5}$$

Since the slope of a line parallel to the x-axis is zero, we set $y' = 0$ and solve for t. This procedure yields $y' = 0$ at $t = 0$. Thus, the point $(1, -1)$ is the point on C at which the tangent is parallel to the x-axis. See Figure 14.22.

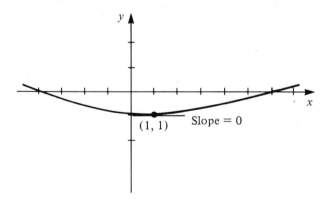

■ **Figure 14.22**   ■

The second derivative gives the **concavity of a curve** at any point on the curve. With y' as the symbol for the first derivative of y with respect to x, the second derivative y'' is given by the following formula.

FORMULA

> **The Second Derivative**
>
> $$y'' = \frac{d}{dx}(y') = \frac{dy'/dt}{dx/dt} \qquad (14.5)$$

EXAMPLE 4

■ **Problem**

Determine the concavity of the curve $C : x = t^3, y = t^2 + 1$.

■ **Solution**

We compute y' by Formula (14.4):

$$y' = \frac{(d/dt)(t^2 + 1)}{(d/dt)(t^3)} = \frac{2t}{3t^2} = \frac{2}{3t}$$

To find y'', we note that

$$\frac{dy'}{dt} = \frac{d}{dt}\left(\frac{2}{3}t^{-1}\right) = -\frac{2}{3}t^{-2}$$

and

$$\frac{dx}{dt} = 3t^2$$

Hence, by Formula (14.5), the second derivative is

$$y'' = \frac{(-2/3)t^{-2}}{3t^2} = \frac{-2}{9t^4}$$

Since t^4 is positive for all t, y'' is negative for all t. The curve is therefore concave downward. ■

Motion of a Particle

If a curve is defined parametrically and the parameter is time, then $\mathbf{R}(t)$ represents the position vector of a particle traveling on the curve. In this case, the first and second derivatives of $\mathbf{R}(t)$, given by $\mathbf{R}'(t)$ and $\mathbf{R}''(t)$, respectively, give the velocity and acceleration vectors of the particle.

FORMULA

> **Velocity and Acceleration Vectors of a Particle**
>
> $$\mathbf{R}'(t) = \mathbf{v}(t) = v_x(t)\mathbf{i} + v_y(t)\mathbf{j} = \frac{dx}{dt}\mathbf{i} + \frac{dy}{dt}\mathbf{j} \qquad (14.6)$$
>
> $$\mathbf{R}''(t) = \mathbf{a}(t) = a_x(t)\mathbf{i} + a_y(t)\mathbf{j} = \frac{d^2x}{dt^2}\mathbf{i} + \frac{d^2y}{dt^2}\mathbf{j} \qquad (14.7)$$

The quantities v_x and v_y are called the *horizontal* and *vertical components* of the velocity vector, with similar definitions for the acceleration vector.

The magnitude of the velocity is called the **speed:**

$$\left|\frac{d\mathbf{R}}{dt}\right| = \text{speed} = \sqrt{\left(\frac{dx}{dt}\right)^2 + \left(\frac{dy}{dt}\right)^2} \tag{14.8}$$

Figure 14.23 shows the path of a particle. The value of the velocity is the limit of the following quotient as Δt goes to zero:

$$\mathbf{v} = \frac{\mathbf{R}(t + \Delta t) - \mathbf{R}(t)}{\Delta t} \qquad \Delta t > 0$$

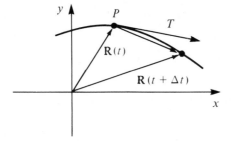

■ **Figure 14.23**

Since division by a positive real number does not affect the direction of the vector $\mathbf{R}(t + \Delta t) - \mathbf{R}(t)$, this quotient approaches a vector whose direction is tangent to the curve at P. We thus have the following result.

RULE

> **Velocity Vector** The velocity vector $\mathbf{v} = d\mathbf{R}/dt$ is tangential at every point of the curve defined by $\mathbf{R}(t)$.

EXAMPLE 5 ■ **Problem**

A particle moves according to the law $\mathbf{R}(t) = t^2\mathbf{i} + t^3\mathbf{j}$.

a. Find the velocity and acceleration vectors at $t = 2$.

b. Find the speed of the particle at $t = 2$.

c. Find a unit vector tangent to the curve at $t = 2$.

■ **Solution**

a. We have

$$v(t) = \frac{d\mathbf{R}}{dt} = 2t\mathbf{i} + 3t^2\mathbf{j} \qquad v(2) = 4\mathbf{i} + 12\mathbf{j}$$

$$a(t) = \frac{d^2\mathbf{R}}{dt^2} = 2\mathbf{i} + 6t\mathbf{j} \qquad a(2) = 2\mathbf{i} + 12\mathbf{j}$$

b. The speed is

$$\text{speed} = |\, \mathbf{v}(2)\,| = \sqrt{4^2 + 12^2} = \sqrt{160}$$

c. Since the velocity vector is tangent to the curve at any point, a unit tangent is given by

$$\frac{\mathbf{v}(2)}{|\,\mathbf{v}(2)\,|} = \frac{4\mathbf{i} + 12\mathbf{j}}{\sqrt{160}} = \frac{\mathbf{i} + 3\mathbf{j}}{\sqrt{10}}$$ ■

EXERCISES FOR SECTION 14.3

In Exercises 1–10, find the slope of the tangent lines at the point corresponding to $t = 1$.

1. $x = t + 3, y = 5t - 1$
2. $x = t^2, y = 3t + 2$
3. $x = t^2 + 2, y = t^2 - 1$
4. $x = t^3, y = 2t^2$
5. $x = e^t, y = e^{-t}$
6. $x = 2t \cos \pi t, y = \sin \pi t + t^2$
7. $x = \sin(\pi t/4), y = 2 \cos(\pi t/4)$
8. $x = e^{2t}, y = e^t$
9. $x = t, y = \ln t$
10. $x = \sqrt{t}, y = 3t + 2$

11. Find the equation of the line tangent to the curve defined by $C : x = 3 \sin \pi t + t, y = 2 \cos \pi t + t^2$ at the point corresponding to $t = 1$.

12. Find the equation of the line tangent to $C : x = e^{-t}, y = e^{-t}$ at $(1, 1)$.

13–22. Determine the concavity of each of the curves in Exercises 1–10 at the point corresponding to $t = 1$.

23. Find the points on the curve $C : x = t^2, y = t^3 - 8t$ at which the tangent line is either horizontal or vertical.

24. Find the points on the curve $C : x = t^2 - 2t, y = t^3 + 1$ at which the tangent to the curve is either vertical or horizontal.

In Exercises 25–28, find the velocity and the acceleration at the given time of a particle that moves according to the law R(t). Calculate the speed of the particle at the indicated time. Determine a unit vector that is tangent to the curve at the given time.

25. $\mathbf{R}(t) = 3t\mathbf{i} + t^2\mathbf{j}; t = 2$
26. $\mathbf{R}(t) = (t^2 + 3)\mathbf{i} + t(2t - 3)\mathbf{j}; t = \frac{1}{2}$
27. $\mathbf{R}(t) = (2 \sin \pi t)\mathbf{i} - (\cos \pi t)\mathbf{j}; t = \frac{1}{4}$
28. $\mathbf{R}(t) = (\sin^2 \pi t)\mathbf{i} + (\cos \pi t)\mathbf{j}; t = 2$

14.4 LENGTH OF ARC

Consider the problem of measuring the length of the curve shown in Figure 14.24(a). Since we cannot make a straight ruler conform to an arc, we cannot measure lengths of curves in the same way that we measure straight lines. However, as Figure 14.24(b) shows, we may use straight-line segments to approximate the length of a curve. To make such an estimate, we proceed to divide the curve from A to B into segments and connect the endpoints with straight lines to form chords. The length of a typical chord is given by the distance formula as

$$d(P_i, P_{i+1}) = \sqrt{(\Delta x_i)^2 + (\Delta y_i)^2}$$

where Δx_i is the change in the x-coordinates of P_i and P_{i+1} and Δy_i is the change in the y-coordinates of the two points.

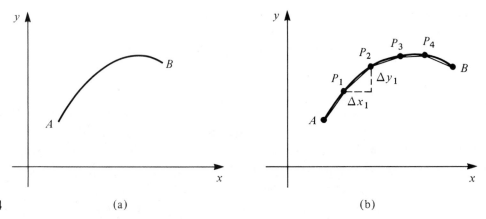

■ Figure 14.24   (a)   (b)

Using the sum of the n chord lengths to approximate the length of the curve L, we write

$$L \approx \sum_{i=1}^{n} \sqrt{(\Delta x_i)^2 + (\Delta y_i)^2}$$

Multiplying numerator and denominator by Δt_i, we write the approximating sum as

$$L \approx \sum_{i=1}^{n} \sqrt{\left(\frac{\Delta x_i}{\Delta t_i}\right)^2 + \left(\frac{\Delta y_i}{\Delta t_i}\right)^2} \cdot \Delta t_i$$

We expect this sum to get closer to the length of the curve as the number of partitions increases. Hence, we define the length L of the curve as the limit of this sum as n increases without bound, and we have the following formula for L.

FORMULA

Length of a Curve Defined Parametrically

$$L = \int_{t_A}^{t_B} \sqrt{\left(\frac{dx}{dt}\right)^2 + \left(\frac{dy}{dt}\right)^2}\, dt \qquad\qquad (14.9)$$

where t_A and t_B are values of the parameter corresponding to the points A and B on the curve.

EXAMPLE 1 ■ **Problem**

The curve traced by a point on the circumference of a circle as the circle rolls along a straight line is called a *cycloid*. See Figure 14.25. The parametric equations of a cycloid generated by a circle of radius a are

$$x = a(t - \sin t) \qquad y = a(1 - \cos t)$$

Determine the length of one arch of a cycloid with $a = 1$.

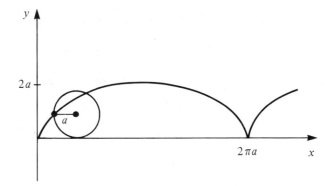

■ **Figure 14.25**

■ **Solution**

One arch of the cycloid is generated for each revolution of the generating circle, that is, as t varies from 0 to 2π. The length of the arch is then given by Formula (14.9) as

$$L = \int_0^{2\pi} \sqrt{\left(\frac{dx}{dt}\right)^2 + \left(\frac{dy}{dt}\right)^2}\, dt$$

In this case, $dx/dt = 1 - \cos t$, and $dy/dt = \sin t$. So,

$$L = \int_0^{2\pi} \sqrt{(1 - \cos t)^2 + \sin^2 t}\, dt$$

$$= \int_0^{2\pi} \sqrt{1 - 2\cos t + \cos^2 t + \sin^2 t}\, dt \qquad \text{expanding}$$

$$= \int_0^{2\pi} \sqrt{2(1 - \cos t)} \, dt \qquad\qquad \cos^2 t + \sin^2 t = 1$$

$$= \int_0^{2\pi} \sqrt{4 \sin^2 \tfrac{1}{2} t} \, dt \qquad\qquad 1 - \cos t = 2 \sin^2 \tfrac{1}{2}t$$

$$= \int_0^{2\pi} 2 \sin \tfrac{1}{2}t \, dt \qquad\qquad \text{square root}$$

$$= \left(-4 \cos \tfrac{1}{2}t\right)_0^{2\pi} = -4(-1) + 4(1) = 8 \qquad\blacksquare$$

If a curve is given in the form $y = f(x)$, then x may be treated as the parameter. If the parameter in (14.9) is x, then $dx/dx = 1$ and $dy/dx = y'$. Therefore, the formula for the length L of the curve $y = f(x)$ from $x = a$ to $x = b$ is as follows:

FORMULA

> **Length of a Curve: $y = f(x)$**
>
> $$L = \int_a^b \sqrt{1 + (y')^2} \, dx \qquad\qquad\qquad\qquad (14.10)$$

EXAMPLE 2   ■ **Problem**

Determine the length of the curve $y = x^{3/2}$ from $x = 0$ to $x = 4$. See Figure 14.26.

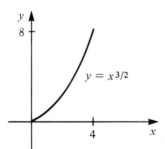

■ **Figure 14.26**

■ **Solution**

Since $y = x^{3/2}$, we obtain $y' = \tfrac{3}{2} x^{1/2}$. Thus, $1 + (y')^2 = 1 + \tfrac{9}{4} x$, and Formula (14.10) yields

$$L = \int_0^4 \left(1 + \frac{9}{4}x\right)^{1/2} dx = \frac{4}{9} \left[\frac{(1 + 9/4x)^{3/2}}{3/2}\right]_0^4$$

$$= \frac{8}{27} [(10)^{3/2} - 1] \approx 9.1 \qquad\blacksquare$$

EXERCISES FOR SECTION 14.4

In Exercises 1–10, determine the length of each curve defined by the parameter t.

1. $x = t, y = t + 3; 0 \leq t \leq 2$

2. $x = t, y = 2t^{3/2}; 0 \leq t \leq 9$

3. $x = \cos 2t, y = \sin 2t; 0 \leq t \leq \pi$

4. $x = t, y = \ln(\cos t); 0 \leq t \leq \frac{1}{6}\pi$

5. $x = \cos^3 t; y = \sin^3 t; 0 \leq t \leq \frac{1}{3}\pi$

6. $x = 3t - 2, y = t + 5; 0 \leq t \leq 4$

7. $x = t^2, y = 2t^3; 0 \leq t \leq 1$

8. $x = \cos 2t, y = \sin^2 t; 0 \leq t \leq \pi$

9. $x = 4t, y = \ln(\cos^4 t); 0 \leq t \leq \frac{1}{3}\pi$

10. $x = e^t, y = 2e^t; 0 \leq t \leq 1$

Find the length of arc of the following curves defined in xy-coordinates.

11. $y = x^{3/2}, 0 \leq x \leq 1$

12. $y = 3x + 2, 1 \leq x \leq 5$

13. $y = \sqrt{9 - x^2}, 0 \leq x \leq 3$

14. $y = \dfrac{x^4}{16} + \dfrac{1}{2x^2}, -2 \leq x \leq -1$

15. $y = \ln(\cos x), 0 \leq x \leq \frac{1}{4}\pi$

14.5 POLAR COORDINATES

The rectangular coordinate system was used exclusively in the first thirteen chapters of this book. Another coordinate system widely used in science and mathematics is the **polar coordinate system.** In this system, the position of a point is determined by specifying a distance from a given point and the direction from a given line. Actually, this concept is not new; we frequently use this system to describe the relative location of geographic points. Thus, when we say that Cincinnati is about 300 miles southeast of Chicago, we are, in fact, using polar coordinates.

To establish a frame of reference for the polar coordinate system, we begin by choosing a point O and extending a line from this point. The point O is called the **pole,** and the extended line is called the **polar axis.** The position of any point P in the plane is then determined if we know the distance OP and the angle AOP, as indicated in Figure 14.27. The directed distance OP is called the **radius vector** of P

and is denoted by *r*. The angle *AOP* is called the **vectorial angle** and is denoted by θ. The coordinates of a point *P* are then written as the ordered pair (r, θ). Notice that the radius vector is the first element and the vectorial angle is the second.

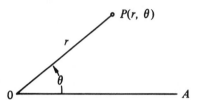

■ **Figure 14.27**

Polar coordinates, like rectangular coordinates, are regarded as signed quantities. When stating the polar coordinates of a point, we customarily use the following sign conventions.

RULE

> **Sign Conventions for Polar Coordinates**
>
> 1. The radius vector is positive when measured on the terminal side of the vectorial angle and is negative when measured in the opposite direction.
> 2. The vectorial angle is positive when generated by a counterclockwise rotation from the polar axis and negative when generated by a clockwise rotation.

The polar coordinates of a point determine the location of the point uniquely. However, the converse is not true, as we can see from Figure 14.28. Ignoring vectorial angles that are numerically greater than 360°, we have four pairs of coordinates that yield the same point. Accordingly, the pairs $(5, 60°)$, $(5, -300°)$, $(-5, 240°)$, and $(-5, -120°)$ represent the same point.

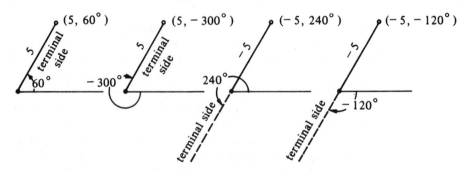

■ **Figure 14.28**

Polar coordinate paper is commercially available. As shown in Figure 14.29, this paper consists of equally spaced concentric circles with radial lines extending at equal angles through the pole. While the use of polar coordinate paper is not mandatory, you will find that it is very helpful in plotting polar curves. Several points are plotted in Figure 14.29 for illustrative purposes.

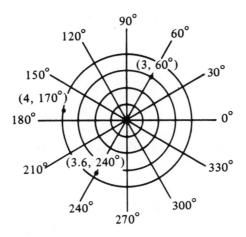

■ **Figure 14.29**

The relationship between the polar coordinates of a point and the rectangular coordinates of the same point can be found by superimposing the rectangular coordinate system on the polar coordinate system so that the origin corresponds to the pole and the positive x-axis to the polar axis. Under these circumstances, the point P shown in Figure 14.30 has both (x, y) and (r, θ) as coordinates. The desired relationship is then an immediate consequence of triangle OMP. Hence, the following equations can be used to transform a rectangular equation into a polar equation.

FORMULA

Transforming a Rectangular Equation into a Polar Equation	
$x = r \cos \theta$	(14.11)
$y = r \sin \theta$	(14.12)

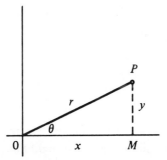

■ **Figure 14.30**

EXAMPLE 1   ■ **Problem**

Find the equation of the circle whose rectangular equation is $x^2 + y^2 = a^2$.

■ **Solution**

Substituting Equation (14.11) and (14.12) into the given equation, we have

$$r^2 \cos^2 \theta + r^2 \sin^2 \theta = a^2$$
$$r^2 (\cos^2 \theta + \sin^2 \theta) = a^2$$
$$r^2 = a^2$$
$$r = a$$ ■

To make the transformation from polar coordinates into rectangular coordinates, we use the following equations.

FORMULA

> **Transforming a Polar Equation into a Rectangular Equation**
>
> $$r^2 = x^2 + y^2 \qquad\qquad\qquad \textbf{(14.13)}$$
>
> $$\sin \theta = \frac{y}{\sqrt{x^2 + y^2}} \qquad\qquad \textbf{(14.14)}$$
>
> $$\cos \theta = \frac{x}{\sqrt{x^2 + y^2}} \qquad\qquad \textbf{(14.15)}$$

These equations are also derived from Figure 14.30.

EXAMPLE 2   ■ **Problem**

Transform the following polar equation into a rectangular equation:

$$4r \cos \theta = 1$$

■ **Solution**

Substituting Equations (14.13) and (14.15) into the given equation, we have

$$4\left(\sqrt{x^2 + y^2}\right)\left(\frac{x}{\sqrt{x^2 + y^2}}\right) = 1$$
$$4x = 1$$
$$x = \tfrac{1}{4}$$

We recognize this equation as the equation of a straight line parallel to the y-axis.   ■

EXAMPLE 3 ■ **Problem**

Show that $r = \dfrac{1}{1 - \cos \theta}$ is the polar form of a parabola.

■ **Solution**

Here, our work is simplified if we multiply both sides of the given equation by $1 - \cos \theta$ before making the substitution. Thus,

$$r - r \cos \theta = 1$$

Substituting Equations (14.13) and (14.15), we get

$$\sqrt{x^2 + y^2} - x = 1$$

Transposing x to the right and squaring both sides of the resulting equation, we get

$$x^2 + y^2 = x^2 + 2x + 1$$
$$y^2 = 2x + 1$$
$$y^2 = 2(x + \tfrac{1}{2})$$

We recognize this equation as the standard form of a parabola having its vertex at $(-\tfrac{1}{2}, 0)$ and a horizontal axis. ■

Curves in Polar Coordinates

A polar equation has a graph in the polar coordinate plane, just as a rectangular equation has a graph in the rectangular coordinate plane. To draw the graph of a polar equation, we start by assigning values to θ and finding the corresponding values of r. The desired graph is then generated by plotting the ordered pairs (r, θ) and connecting them with a smooth curve.

EXAMPLE 4 ■ **Problem**

Sketch the graph of the equation $r = 1 + \cos \theta$.

■ **Solution**

Using increments of $45°$ for θ, we obtain the following table.

θ	$0°$	$45°$	$90°$	$135°$	$180°$	$225°$	$270°$	$315°$	$360°$
r	2.00	1.71	1.00	0.29	0.00	0.29	1.00	1.71	2.00

The curve obtained by connecting these points with a smooth curve is called a *cardioid*. See Figure 14.31.

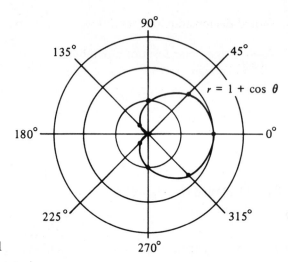

■ **Figure 14.31**

EXAMPLE 5   ■ **Problem**

Sketch the graph of the equation $r = 4 \sin \theta$.

■ **Solution**

We find the accompanying table for values of r corresponding to the indicated values of θ. Drawing a smooth curve through the plotted points, we have the *circle* shown in Figure 14.32.

θ	0	$\pi/6$	$\pi/4$	$\pi/3$	$\pi/2$	$2\pi/3$	$3\pi/4$	$5\pi/6$	π
r	0	2	$2\sqrt{2}$	$2\sqrt{3}$	4	$2\sqrt{3}$	$2\sqrt{2}$	2	0

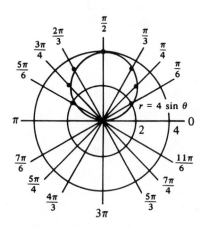

■ **Figure 14.32**

Notice that θ varies only from 0 to π radians. If we allow θ to vary from 0 to 2π radians, the graph will be traced out twice; once for $0 \le \theta \le \pi$ and again for

$\pi < \theta \leq 2\pi$. You should demonstrate this result by plotting points in the interval $\pi < \theta \leq 2\pi$. ∎

Comment: The situation in Example 5 for $\pi < \theta \leq 2\pi$ is typical. At first, you may think that the defining equation gives no ordered pairs for that interval. Actually, there is a curve. The values of r are negative, and the points are therefore reflected in the origin.

EXAMPLE 6   ▪ **Problem**

Sketch the graph of the equation $r = 1 - 2 \cos \theta$.

▪ **Solution**

Here, we will assign values to θ in increments of 30°. The ordered pairs (r, θ) are then given in the accompanying table for the interval $0 \leq \theta \leq 180°$. This curve is called a *limacon*. See Figure 14.33.

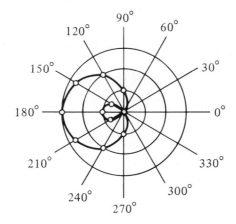

▪ **Figure 14.33**

θ	0°	30°	60°	90°	120°	150°	180°	210°
r	−1.00	−0.73	0.00	1.00	2.00	2.73	3.00	2.73

θ	240°	270°	300°	330°	360°
r	2.00	1.00	0.00	−0.73	−1.00

∎

EXAMPLE 7   ▪ **Problem**

Sketch the graph of the polar equation $r = 4 \cos 3\theta$.

▪ **Solution**

Substituting values for θ, we obtain the accompanying table. This curve is called a *three-leaved rose*. See Figure 14.34.

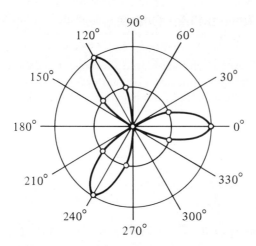

■ **Figure 14.34**

Notice that although 3θ is the argument, the curve is formed by plotting θ versus r. Also, we note that the curve retraces itself when θ assumes values greater than $180°$.

θ	$0°$	$20°$	$30°$	$40°$	$60°$	$80°$	$90°$	$100°$
3θ	$0°$	$60°$	$90°$	$120°$	$180°$	$240°$	$270°$	$300°$
r	4.00	2.00	0.00	-2.00	-4.00	-2.00	0.00	2.00

θ	$120°$	$140°$	$150°$	$160°$	$180°$
3θ	$360°$	$420°$	$450°$	$480°$	$540°$
r	4.00	2.00	0.00	-2.00	-4.00

■

Comment: We note that it is difficult to generalize the graphical forms of polar equations because slight changes in the function often produce drastic changes in the graph. For instance, the graph of $r = 2 \cos \theta$ is a circle, that of $r = 2 - 2 \cos \theta$ is a cardioid, and that of $r = \cos 2\theta$ is a four-leaved rose. See Figure 14.35.

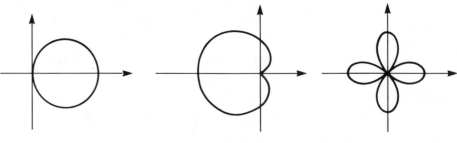

■ **Figure 14.35** $r = 2 \cos \theta$ $R = 2 - 2 \cos \theta$ $r = \cos 2\theta$

Slope of a Curve in Polar Coordinates

Suppose $r = r(\theta)$ defines a curve in polar coordinates. Then, we may consider the variable θ to be a parameter for the curve, and we may write the parametric equations as

$$x = r(\theta) \cos \theta$$
$$y = r(\theta) \sin \theta$$

These equations define the curve in terms of the parameter θ. The expression for y' may then be found by applying Formula (14.4) with $x = r(\theta) \cos \theta$ and $y = r(\theta) \sin \theta$.

FORMULA

> **Slope of a Curve in Polar Coordinates**
>
> $$y' = \frac{dy/d\theta}{dx/d\theta} = \frac{r' \sin \theta + r \cos \theta}{r' \cos \theta - r \sin \theta}$$
> (14.16)

EXAMPLE 8   ▪ **Problem**

Find y' for the curve $r = e^{2\theta}$.

▪ **Solution**

The derivative is given by Formula (14.16). Thus,

$$y' = \frac{2e^{2\theta} \sin \theta + e^{2\theta} \cos \theta}{2e^{2\theta} \cos \theta - e^{2\theta} \sin \theta} = \frac{2 \sin \theta + \cos \theta}{2 \cos \theta - \sin \theta} = \frac{2 \tan \theta + 1}{2 - \tan \theta}$$ ■

EXAMPLE 9   ▪ **Problem**

Determine the slope of the curve $r = \cos 2\theta$ at the origin.

▪ **Solution**

The curve passes through the origin when $r = 0$. Solving $\cos 2\theta = 0$, we have $2\theta = \pi/2, 3\pi/2, 5\pi/2$, and $7\pi/2$. Or in terms of θ, we get $\theta = \pi/4, 3\pi/4, 5\pi/4$, and $7\pi/4$.

The slope of the curve is given by Formula (14.16) as

$$y' = \frac{-2 \sin 2\theta \sin \theta + \cos 2\theta \cos \theta}{-2 \sin 2\theta \cos \theta - \cos 2\theta \sin \theta}$$

Since $\cos 2\theta = 0$ when the curve passes through the origin, the expression for y' can be reduced to

$$y' = \frac{-2 \sin 2\theta \sin \theta}{-2 \sin 2\theta \cos \theta} = \frac{\sin \theta}{\cos \theta} = \tan \theta$$

which is the slope at the origin. Note that $\tan \theta$ is either -1 or 1 for $\theta = \pi/4, 3\pi/4, 5\pi/4$, and $7\pi/4$. The slope of $r = \cos 2\theta$ as it passes through the origin is then either -1 or 1. See Figure 14.36.

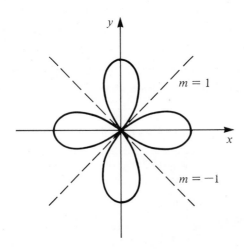

■ Figure 14.36

EXERCISES FOR SECTION 14.5

Plot the points in Exercises 1–10 on polar coordinate paper.

1. $(5, 30°)$ **2.** $(3.6, -45°)$

3. $(12, 2\pi/3)$ **4.** $(0.5, 220°)$

5. $(-7.1, 14°)$ **6.** $(-2, 7\pi/3)$

7. $(1.75, -200°)$ **8.** $(\sqrt{2}, -311°)$

9. $(-5, -30°)$ **10.** $(5, 150°)$

Convert the rectangular equations in Exercises 11–18 into equations in polar coordinates.

11. $2x + 3y = 6$ **12.** $y = x$

13. $x^2 + y^2 - 4x = 0$ **14.** $x^2 - y^2 = 4$

15. $x^2 + 4y^2 = 4$ **16.** $xy = 1$

17. $x^2 = 4y$ **18.** $y^2 = 16x$

Convert the polar equations in Exercises 19–26 into equations in rectangular coordinates.

19. $r = 5$
20. $r = \cos\theta$

21. $r = 10\sin\theta$
22. $r = 2(\sin\theta - \cos\theta)$

23. $r = 1 + 2\sin\theta$
24. $r\sin\theta = 10$

25. $r = \dfrac{5}{1 + \cos\theta}$
26. $r(1 - 2\cos\theta) = 1$

Sketch the graph of the equations in Exercises 27–40.

27. $r = 5.6$
28. $r = \sqrt{2}$

29. $\theta = \frac{1}{3}\pi$
30. $\theta = 170°$

31. $r = 2\sin\theta$
32. $r = 0.5\cos\theta$

33. $r\sin\theta = 1$
34. $r\cos\theta = -10$

35. $r = 1 + \sin\theta$
36. $r = 1 - \cos\theta$

37. $r = \sec\theta$
38. $r = -\sin\theta$

39. $r = 4\sin 3\theta$
40. $r = \sin 2\theta$

41. The radiation pattern of a particular two-element antenna is a cardioid of the form $r = 100\,(1 + \cos\theta)$. Sketch the radiation pattern of this antenna.

42. The radiation pattern of a certain antenna is given by $r = 1/(2 - \cos\theta)$. Plot this pattern.

43. By transforming the polar equation in Exercise 42 into rectangular coordinates, show that the indicated radiation pattern is elliptical.

44. The feedback diagram of a certain electronic tachometer can be approximated by the curve $r = \frac{1}{2}\theta$. Sketch the feedback diagram of this tachometer from $\theta = 0$ to $\theta = 7\pi/6$.

In Exercises 45–52, determine the slope of the polar coordinate curves at any point (r, θ).

45. $r = \sin\theta + 2$
46. $r = -\cos^2\theta$

47. $r = \dfrac{1}{\sin\theta + 2\cos\theta}$
48. $r = \dfrac{1}{3\sin\theta - \cos\theta}$

49. $r\sin\theta = 1$
50. $r\cos\theta = 1$

51. $r = 2\cos\theta - 1$
52. $r = 3\sin 2\theta + \theta$

14.6 AREA AND ARC LENGTH IN POLAR COORDINATES

Area

The area bounded by a polar curve can be found by integration. Consider the area bounded by the polar curve $r = f(\theta)$ and the radial lines $\theta = \alpha$ and $\theta = \beta$, as shown in Figure 14.37. The area of this figure can be found by dividing the area into the circular sectors and then taking the sum of the areas of these sectors as the number of sectors increases without bound.

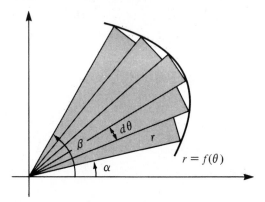

■ **Figure 14.37**

We take, as a typical element of area, a circular sector with central angle $d\theta$ and radius r. The area of any circular sector is given in elementary geometry by the formula $A = \frac{1}{2}r^2\theta$, where r is the radius of the circle and θ is the central angle of the sector. Hence, the area of our typical area element in Figure 14.37 is

$$dA = \tfrac{1}{2}r^2\,d\theta$$

Then, the total area is found from the following definite integral.

FORMULA

Area Formula for a Curve in Polar Coordinates

$$A = \frac{1}{2}\int_\alpha^\beta r^2\,d\theta \tag{14.17}$$

EXAMPLE 1   ■ **Problem**

Find the area bounded by the circle $r = 3\sin\theta$.

■ **Solution**

This area is shown in Figure 14.38. The element of area is given by

$$dA = \tfrac{1}{2}r^2\,d\theta = \tfrac{9}{2}\sin^2\theta\,d\theta$$

So, the desired area is given by

$$A = \frac{9}{2}\int_0^\pi \sin^2\theta\,d\theta = \frac{9}{2}\int_0^\pi \frac{1}{2}(1 - \cos 2\theta)\,d\theta = \frac{9}{4}\left[\theta - \frac{1}{2}\sin 2\theta\right]_0^\pi = \frac{9\pi}{4}$$

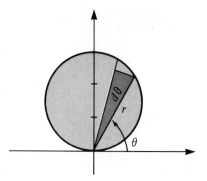

■ **Figure 14.38**   ■

EXAMPLE 2   ■ **Problem**

The radiation pattern of a two-element antenna is a cardioid of the form $r = 20(1 + \cos\theta)$, where r is measured in miles. Find the area covered by this antenna. See Figure 14.39.

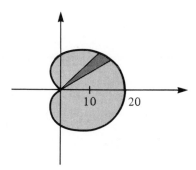

■ **Figure 14.39**

■ **Solution**

The area of this antenna pattern is given by the integral

$$A = \frac{1}{2}\int_0^{2\pi}[20(1 + \cos\theta)]^2\,d\theta = 200\int_0^{2\pi}(1 + 2\cos\theta + \cos^2\theta)\,d\theta$$

$$= 200\left[\theta + 2\sin\theta + \frac{\theta}{2} + \frac{1}{4}\sin 2\theta\right]_0^{2\pi}$$

$$= 200(2\pi + \pi) = 600\pi \approx 1884 \text{ mi}^2 \qquad\qquad ■$$

Arc Length

To conclude this section, we derive a formula for the length of arc for curves whose equations are expressed in polar coordinates. Recall from Section 14.5 that a curve in polar coordinates may be considered to be defined parametrically by the equations

$$x = r(\theta)\cos\theta \qquad y = r(\theta)\sin\theta$$

Formula (14.9) from Section 14.4 may then be used with $t = \theta$ to express the length of arc of this curve from θ_1 to θ_2. We note that

$$\frac{dx}{d\theta} = r'\cos\theta - r\sin\theta \qquad \frac{dy}{d\theta} = r'\sin\theta + r\cos\theta$$

and

$$\left(\frac{dx}{d\theta}\right)^2 = (r')^2\cos^2\theta - 2rr'\cos\theta\sin\theta + r^2\sin^2\theta$$

$$\left(\frac{dy}{d\theta}\right)^2 = (r')^2\sin^2\theta + 2rr'\cos\theta\sin\theta + r^2\cos^2\theta$$

The sum of these squares simplifies to $(r')^2 + r^2$. Using this result in Formula (14.9), we obtain the formula for length of arc for a curve in polar coordinates.

FORMULA

Length of Arc for a Curve in Polar Coordinates

$$L = \int_{\theta_1}^{\theta_2} \sqrt{(r')^2 + r^2}\, d\theta \qquad\qquad \textbf{(14.18)}$$

EXAMPLE 3   ■ **Problem**

Determine the length of the curve $r = e^\theta$ from $\theta = 0$ to $\theta = \pi/2$.

■ **Solution**

Here, $r = e^\theta$ and $r' = e^\theta$. So,

$$L = \int_0^{\pi/2} \sqrt{(r')^2 + r^2}\, d\theta = \int_0^{\pi/2} \sqrt{e^{2\theta} + e^{2\theta}}\, d\theta$$

$$= \int_0^{\pi/2} \sqrt{2e^{2\theta}}\, d\theta = \sqrt{2}\int_0^{\pi/2} e^\theta\, d\theta = \sqrt{2}(e^{\pi/2} - 1) \approx 5.4 \qquad ■$$

████████████ **EXERCISES FOR SECTION 14.6**

In Exercises 1–10, find the area enclosed by the graphs of the given equations. Sketch each graph.

1. $r = 10, \theta = 0, \theta = \pi/6$

2. $r = \sin\theta, \theta = \pi/6, \theta = \pi/4$

3. $r = \theta, \theta = 45°, \theta = 180°$

4. $r = \tan\theta, \theta = 30°, \theta = 45°$

5. $r = \cos\theta, \theta = 0, \theta = \pi/3$

6. $r = 1 + \cos\theta, \theta = 0, \theta = \pi/4$

7. $r = \cos 2\theta, \theta = 0, \theta = \pi/4$

8. $r = \theta^2, \theta = 0, \theta = \pi/2$

9. $r = \theta - \theta^2, \theta = 0, \theta = \pi/2$

10. $r = \sec\theta, \theta = 0, \theta = \pi/4$

In Exercises 11–20, find the area enclosed by the graph of the given equation. Sketch each graph.

11. $r^2 = 4 \sin 2\theta$ 12. $r^2 = \cos 2\theta$

13. $r = 7 \cos \theta$ 14. $r = 8 \sin \theta$

15. $r = 3 \cos 2\theta$ 16. $r = 2 \sin 2\theta$

17. $r = 3(1 - \sin \theta)$ 18. $r = 2 + \cos \theta$

19. $r = \sin 3\theta$ 20. $r = \cos 4\theta$

Find the length of arc of the curves defined in Exercises 21–25.

21. $r = \sin \theta, 0 \le \theta \le \pi$ 22. $r = \theta$ from $\theta = 0$ to $\theta = 2\pi$

23. $r = \sin^2 (\theta/2)$ from $\theta = 0$ to $\theta = \pi$ 24. $r = e^{-\theta}$ from $\theta = 0$ to $\theta = \pi$

25. $r = e^{2\theta}$ from $\theta = 0$ to $\theta = \pi/2$

26. The path of a certain planet as viewed from the earth is given by $r = 2 + \sin \theta$. Find the area swept out by the radius vector from the earth to the planet for $0 \le \theta \le 2\pi$.

27. The directional pattern of certain types of microphones is given by the cardioid $r = a(1 + \cos \theta)$, where a is a constant. Find the area of this pattern in terms of the constant a.

28. Find the length of the cardioid $r = 1 + \cos \theta$.

REVIEW EXERCISES FOR CHAPTER 14

1. Find the x- and y-coordinates of the terminal point of the vector $|\mathbf{V}| = 5, \theta = 130°$.

2. Find the x- and y-coordinates of the terminal point of the vector $|\mathbf{V}| = 20, \theta = 67°$.

In Exercises 3–8, find the vector sum A + B.

3. $\mathbf{A} = 3\mathbf{i} + 4\mathbf{j}$
   $\mathbf{B} = -2\mathbf{i} - 7\mathbf{j}$

4. $\mathbf{A} = \mathbf{i} + 2\mathbf{j}$
   $\mathbf{B} = -3\mathbf{i} + 4\mathbf{j}$

5. $\mathbf{A} = -\mathbf{i} - \mathbf{j}$
   $\mathbf{B} = 5\mathbf{i} + 4\mathbf{j}$

6. $\mathbf{A} = 3\mathbf{j}$
   $\mathbf{B} = -5\mathbf{i}$

7. $|\mathbf{A}| = 75, \theta_A = 23°$
   $|\mathbf{B}| = 105, \theta_B = -14°$

8. $|\mathbf{A}| = 50, \theta_A = 200°$
   $|\mathbf{B}| = 24, \theta_B = 110°$

9. Compute the vertical and horizontal components of the force shown in the accompanying figure.

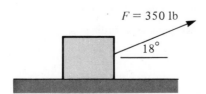

$F = 350$ lb

$18°$

10. The airplane shown in the accompanying figure is cruising at 185 knots with a heading of 220° clockwise from north. Determine the south and west components of velocity.

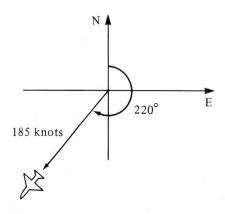

In Exercises 11–16, sketch the graph of the given parametric curve, and find a rectangular equation of the given curve. Show the orientation of the curve.

11. $x = t - 1, y = 2t + 1$

12. $x = t + 2, y = t + 3$

13. $x = t^2, y = 2 \ln t$

14. $x = e^{-t}, y = e^{2t}$

15. $x = \sin t, y = \sin 2t$

16. $x = \cos^2 t, y = \sin t$

In Exercises 17–20, find the slope of the tangent line at the point corresponding to $t = 1$.

17. $x = 2t + 1, y = t + 2$

18. $x = t - 1, y = t^2$

19. $x = t, y = t^2 + 1$

20. $x = e^{-t}, y = e^{2t}$

Find the length of each curve in Exercises 21–24 defined by the parameter t.

21. $x = 2t, y = t^{3/2}; 0 \le t \le 1$

22. $x = \sin t, y = \cos t; 0 \le t \le \pi/2$

23. $x = 2e^t, y = e^t; 0 \le t \le 1$

24. $x = \sin 2t, y = 2 \cos^2 t; 0 \le t \le \pi$

In Exercises 25–28, find the length of the arc of the curves defined in xy-coordinates.

25. $y = 3 - 2x, 1 \le x \le 4$

26. $y = 2x^{3/2}, 1 \le x \le 4$

27. $y = (4 - x)^{3/2}, 0 \le x \le 4$

28. $y = \dfrac{x^3}{6} + \dfrac{1}{2x}, 1 \le x \le 3$

29. Convert $x + 2y = 5$ into polar coordinates.

30. Convert $x^2 + 16y^2 = 4$ into polar coordinates.

31. Convert $r = 3 \cos \theta$ into rectangular coordinates.

32. Convert $r = 10/\cos \theta$ into rectangular coordinates.

In Exercises 33–38, sketch the graphs of the indicated equations.

33. $r = 4$

34. $r = \frac{1}{2} \sin \theta$

35. $r = -2 \cos \theta$

36. $r = \cos \theta - 1$

37. $r = -\sin 2\theta$

38. $r = 2 - 2 \sin \theta$

In Exercises 39–42, determine the slope of the polar coordinates curves at any point (r, θ).

39. $r = \cos \theta$

40. $r = 1 - \cos \theta$

41. $r = 2 \sin 2\theta$

42. $r = \sin^2 \theta$

In Exercises 43–46, find the length of the arc of the curves defined in polar coordinates.

43. $r = \cos \theta, 0 \le \theta \le \pi/4$

44. $r = e^{2\theta}, 0 \le \theta \le 1$

45. $r = 2\theta, 0 \le \theta \le 2\pi$

46. $r = 1 - \cos \theta, 0 \le \theta \le 2\pi$

In Exercises 47–52, find the area enclosed by the graphs of the given equations. Sketch each graph.

47. $r = 1 + \cos \theta, \theta = 0 \text{ to } \theta = \pi/3$

48. $r = 2 \sin \theta, \theta = 0 \text{ to } \theta = \pi/2$

49. $r = \sin 2\theta$

50. $r = 2(1 - \cos \theta)$

51. $r = 2 \cos 4\theta$

52. $r = 2 + \sin \theta$

CHAPTER 15

Infinite Series

The sum of the first ten integers is written as

$$1 + 2 + 3 + 4 + 5 + 6 + 7 + 8 + 9 + 10$$

This sum is equal to 55. Sums of a finite number of terms, as in this example, can always be computed, although it might take some time. Have you ever wondered what it means to sum an infinite number of terms? Is the sum of an infinite number of terms infinitely large? The answers to these questions are contained in this chapter. Although this topic may seem to be of little practical value, there are many physical concepts that require our understanding of infinite sums.

15.1 SEQUENCES

The half-life of a radioactive substance is the time it takes for half of the original atoms to undergo radioactive transformation. For radium, the half-life is 1620 years. That is, if we start with 100 g of radium, we will have 50 g at the end of 1620 years, 25 g after 3240 years, 12.5 g after 4860 years, and so on. The mass of radium remaining after successive half-lives may then be written as

$$100, 50, 25, 12.5$$

Of course, this succession of half-lives can be extended to include as many half-life values as we choose. Notice that the mass after each half-life is one half its previous value, so successive half-lives can also be written as

$$100, \frac{100}{2^1}, \frac{100}{2^2}, \frac{100}{2^3}, \ldots, \frac{100}{2^n}, \ldots$$

This example shows how a mathematical quantity called a **sequence** arises naturally in scientific work. In mathematics, a sequence of terms is a special type of function defined only for the positive integers. In the example, the general rule for the sequence is given by

$$a_n = \frac{100}{2^n}$$

where each term of the sequence can be obtained by simply substituting successive integer values for n. Normally, the general rule is either given or can be inferred from the information given.

EXAMPLE 1   The sequence defined by the rule $a_n = 2n - 1$ gives the terms $a_1 = 1$, $a_2 = 3$, $a_3 = 5$, and so on. Thus, the indicated sequence is

$$1, 3, 5, \ldots$$

Inspection of this sequence reveals that any two succeeding terms differ by a constant. Sequences that have the property $a_{n+1} - a_n = d$ are called **arithmetic sequences.** The given sequence is therefore arithmetic with $d = 2$. ■

EXAMPLE 2   A sequence is called **geometric** if the ratio of any two succeeding terms is a constant, that is, if $a_{n+1}/a_n = r$. The sequence $1, \frac{1}{2}, \frac{1}{4}, \frac{1}{8}, \ldots$ is geometric with $r = \frac{1}{2}$. Inspection of this sequence suggests that the general term is $a_n = 1/2^{n-1}$. ■

EXAMPLE 3   The sequence defined by the rule $a_n = (-1)^n$ has the terms

$$a_1 = -1 \qquad a_2 = 1 \qquad a_3 = -1$$

and so on. ■

EXAMPLE 4   The sequence defined by the rule $S_n = \sum_{k=1}^{n} (\frac{1}{2})^{k-1}$ has the terms

$$S_1 = 1 \qquad S_2 = 1 + \frac{1}{2} \qquad S_3 = 1 + \frac{1}{2} + \frac{1}{4}$$

and so on. ■

EXAMPLE 5   ■ **Problem**

Find the general term of the sequence $-\frac{1}{2}, 0, \frac{1}{4}, \frac{2}{3}, \ldots$.

■ **Solution**

We observe that the numerator of the first term is -1, of the second term is 0, of the third term is 1, of the fourth term is 2. This observation suggests that the numerator of the general term is specified by $n - 2$. The denominators appear to be one more than the number of the term, so $n + 1$ represents the denominator of the general term. The expression

$$a_n = \frac{n - 2}{n + 1}$$

is therefore the general term of the sequence.   ■

When one is working with sequences, products of successive integers beginning with 1 occur quite often. To handle this feature, we introduce the notation $n!$ to denote the product

$$n! = 1 \times 2 \times 3 \times \cdots \times (n - 1) \times n$$

We refer to this product as n **factorial.** Thus, the value of 4 factorial is

$$4! = 1 \times 2 \times 3 \times 4 = 24$$

Because of circumstances that arise in formulating the general term of a sequence, it is convenient to define factorial zero to be equal to one; that is,

$$0! = 1$$

EXAMPLE 6   ■ **Problem**

Find the general term of the sequence $\dfrac{x}{1}, \dfrac{x^2}{1}, \dfrac{x^3}{1 \cdot 2}, \dfrac{x^4}{1 \cdot 2 \cdot 3}, \ldots$

■ **Solution**

The numerator of the general term is obviously x^n. The denominator can be described by $(n - 1)!$, since the denominator of the first term is $0!$, of the second term is $1!$, of the third term is $2!$, and so forth. Hence, the general term is given by

$$a_n = \frac{x^n}{(n - 1)!}$$ ■

EXAMPLE 7   ■ **Problem**

Find the general term of the sequence $\dfrac{1}{\sqrt{3}}, \dfrac{-1}{\sqrt{5}}, \dfrac{1}{\sqrt{7}}, \dfrac{-1}{\sqrt{9}}, \ldots$

■ **Solution**

To account for the alternate positive and negative terms of the sequence, we use $(-1)^{n+1}$. This term is positive when n is odd and negative when n is even. We also observe that the denominator can be expressed by $\sqrt{2n + 1}$. The general term is therefore given by

$$a_n = \frac{(-1)^{n+1}}{\sqrt{2n + 1}}$$ ■

■■■■■   **EXERCISES FOR SECTION 15.1**

In Exercises 1–8, write the first four terms of each sequence (n a positive integer).

1. $a_n = \dfrac{1}{2n - 1}$

2. $a_n = \dfrac{n}{2^{n-1}}$

3. $a_n = \dfrac{2^n}{(n - 1)!}$

4. $a_n = \dfrac{x^{n-1}}{\sqrt{2n + 1}}$

5. $a_n = \dfrac{(-1)^n}{n!}$

6. $a_n = (-1)^{n+1} n^2$

7. $S_n = \displaystyle\sum_{k=1}^{n} \dfrac{1}{k}$

8. $S_n = \displaystyle\sum_{k=1}^{n} \dfrac{1}{k^2}$

Discover, by inspection, the general term of the indicated sequence.

9. $2, 4, 6, 8, \ldots$

10. $1, 8, 27, 64, \ldots$

11. $1, 2, \dfrac{3}{2!}, \dfrac{4}{3!}, \ldots$

12. $1, \dfrac{1}{2}, \dfrac{1}{4}, \dfrac{1}{8}, \ldots$

13. $1, \dfrac{1}{3!}, \dfrac{1}{5!}, \dfrac{1}{7!}, \ldots$

14. $1, \dfrac{1}{4}, \dfrac{1}{9}, \dfrac{1}{16}, \ldots$

15. $\dfrac{1}{3}, \dfrac{2}{5}, \dfrac{3}{7}, \dfrac{4}{9}, \ldots$

16. $\dfrac{1}{\sqrt{2}}, \dfrac{1}{\sqrt{3}}, \dfrac{1}{\sqrt{4}}, \dfrac{1}{\sqrt{5}}, \ldots$

17. $\dfrac{x^2}{2}, \dfrac{x^4}{4}, \dfrac{x^6}{8}, \dfrac{x^8}{16}, \ldots$

18. $x, \dfrac{x^3}{3!}, \dfrac{x^5}{5!}, \dfrac{x^7}{7!}, \ldots$

19. $1, x, \dfrac{x^2}{2!}, \dfrac{x^3}{3!}, \ldots$

20. $\dfrac{e}{\sqrt{2}}, \dfrac{e^2}{\sqrt{4}}, \dfrac{e^3}{\sqrt{6}}, \dfrac{e^4}{\sqrt{8}}, \ldots$

21. $1, -\frac{1}{2}, \frac{1}{3}, -\frac{1}{4}, \ldots$

22. $\frac{1}{3}, -\frac{1}{5}, \frac{1}{7}, -\frac{1}{9}, \ldots$

23. $1, 1 + \frac{1}{3}, 1 + \frac{1}{3} + \frac{1}{5}, 1 + \frac{1}{3} + \frac{1}{5} + \frac{1}{7}, \ldots$

24. $\frac{1}{2}, \frac{1}{2} + \frac{1}{4}, \frac{1}{2} + \frac{1}{4} + \frac{1}{6}, \frac{1}{2} + \frac{1}{4} + \frac{1}{6} + \frac{1}{8}, \ldots$

■■■■■   **15.2 CONVERGENCE AND DIVERGENCE**

When we worked with general functions of the form $f(x)$, we were concerned with limiting values. With sequences, our main concern is with asymptotic behavior for large values of n.

The sequence $1, \frac{1}{2}, \frac{1}{3}, \ldots$ is specified by the general term $a_n = 1/n$. Since the general term is close to zero for large n, we say that the limit of the sequence is zero. Similarly, since $a_n = 3 - 1/\sqrt{n}$ is close to 3 when n is large, we say that the limit of the sequence defined by this general term is 3. Accordingly, we make the following definition of the limit of a sequence.

DEFINITION

Limit of a Sequence   Let a_n be the nth term of a sequence. If a_n approaches some specific number L as n increases without bound, then L is said to be the *limit of the sequence.* We write

$$\lim_{n \to \infty} a_n = L$$

In this case, the sequence is said to be *convergent*, and it *converges to L*; otherwise, the sequence is said to be *divergent*.

EXAMPLE 1   The sequence $a_n = 1/(n + 1)$ converges to zero, since

$$\lim_{n \to \infty} a_n = \lim_{n \to \infty} \frac{1}{n + 1} = 0$$

■

EXAMPLE 2   The sequence $a_n = n/(n + 1)$ converges to 1, since

$$\lim_{n \to \infty} a_n = \lim_{n \to \infty} \frac{n}{n + 1} = \lim_{n \to \infty} \frac{1}{1 + (1/n)} = 1$$

■

EXAMPLE 3   The sequence $a_n = n^2$ diverges, since

$$\lim_{n \to \infty} a_n = \lim_{n \to \infty} n^2 = \infty$$

■

EXAMPLE 4   The sequence $a_n = (-1)^n$ diverges, since $\lim_{n \to \infty} (-1)^n$ does not exist. The terms of the sequence alternate between 1 and -1.   ■

Examples 3 and 4 are important because they show that a sequence can diverge for two basic reasons:

1. The values of the terms may increase or decrease without bound as n increases without bound.
2. The general term may fail to approach a definite limit as n increases without bound.

Sometimes, even though the general rule is given, to determine the convergence or divergence of a sequence is very difficult. This difficulty arises especially when the general rule is given in terms of a sum.

EXAMPLE 5   ■ **Problem**

Show that the sequence $1, 1 + \frac{1}{2}, 1 + \frac{1}{2} + \frac{1}{4}, \ldots$ converges to 2.

■ **Solution**

This sequence was described in Example 4 of Section 15.1, so the general term is

$$S_n = \sum_{k=1}^{n} \left(\frac{1}{2}\right)^{k-1}$$

However, S_n is not in a form where we can easily take the limit as n increases without bound. The key to this problem is to recognize S_n as the sum of n terms of a geometric progression with $a_1 = 1$ and $r = \frac{1}{2}$.

To find the sum of n terms of a geometric sequence, we let

$$S_n = a_1 + a_1 r + \cdots + a_1 r^{n-1}$$

Multiplying both sides of this equation by the common ratio r yields

$$rS_n = a_1 r + a_1 r^2 + \cdots + a_1 r^n$$

Subtracting the left side of the second equation from the first equation and equating it to the difference of the right sides of the two equations, we get

$$S_n(1 - r) = a_1(1 - r^n)$$

from which, if $r \neq 1$,

$$S_n = \frac{1 - r^n}{1 - r} a_1$$

Using this formula with $a_1 = 1$ and $r = \frac{1}{2}$, we can write S_n as

$$S_n = \sum_{k=1}^{n} \left(\frac{1}{2}\right)^{k-1} = \frac{1 - (1/2)^n}{1 - (1/2)} = 2\left[1 - \left(\frac{1}{2}\right)^n\right]$$

Since

$$\lim_{n \to \infty} S_n = \lim_{n \to \infty} 2\left[1 - \left(\frac{1}{2}\right)^n\right] = 2$$

the sequence converges to 2.   ■

EXERCISES FOR SECTION 15.2

In Exercises 1–18, the general term of the sequence is given. Write the first four terms of the sequence, and then investigate the convergence or divergence of each sequence.

1. $a_n = \dfrac{2}{n}$

2. $a_n = 3 - \dfrac{2}{n}$

3. $a_n = 1 + \dfrac{5}{n}$

4. $a_n = \dfrac{2}{n + 2}$

5. $a_n = \dfrac{n}{n + 3}$

6. $a_n = \dfrac{n + 1}{2n - 1}$

7. $a_n = n + \dfrac{1}{n}$

8. $a_n = \sqrt{n}$

9. $a_n = \ln(n + 1)$

10. $a_n = \dfrac{1}{\ln n}$

11. $a_n = 5 + (-1)^n$

12. $a_n = \dfrac{n^2 - 2}{n^2 + 2}$

13. $a_n = \dfrac{(-1)^n}{n}$

14. $a_n = \dfrac{3n}{n^2 + 3}$

15. $a_n = \dfrac{n^3}{n^2 + 5}$

16. $a_n = (-1)^n + \dfrac{1}{n}$

17. $S_n = \displaystyle\sum_{k=1}^{n} \left(\dfrac{1}{3}\right)^k$

18. $S_n = \displaystyle\sum_{k=1}^{n} \left(\dfrac{1}{4}\right)^{k-1}$

19. The current in an electric circuit is measured after each minute and found to be approximated by

$$i_n = (1 + e^{-n})(10)$$

The limit of this value is the *steady-state current*. What is the steady-state current?

20. The height of an electronic "bouncing ball" is described by the sequence

$$h_n = \dfrac{n + 2}{3n + 5}$$

What is the limiting value of the height?

15.3 INFINITE SERIES AND THE INTEGRAL TEST

We are frequently interested in the sum of the terms of a sequence. When the sequence is finite, the problem of finding the sum consists simply of finding the algebraic sum of all of the terms in the sequence. To find such a sum is always possible, although sometimes impractical. However, the problem is not so simple when we are dealing with an infinite sequence, since no matter how many terms we sum, there are always more terms remaining to be summed. It is soon apparent that the "sum" of an infinite number of terms of a sequence does not exist in the sense that we ordinarily use for sum. We therefore define what we mean by an *infinite sum*.

DEFINITION

Infinite Series   An expression of the form

$$a_1 + a_2 + a_3 + \cdots + a_n + \cdots = \sum_{n=1}^{\infty} a_n \qquad (15.1)$$

is called an *infinite series*.

This definition only gives a name to an endless summation; it does not give it a meaning.

We can arrive at a reasonable meaning for an infinite series by associating with the series in Equation (15.1) a sequence $S_1, S_2, \ldots . S_n$ that is defined by

$$S_1 = a_1$$
$$S_2 = a_1 + a_2$$
$$S_3 = a_1 + a_2 + a_3$$
$$\begin{array}{c} . \\ . \\ . \end{array}$$
$$S_n = a_1 + a_2 + a_3 + \cdots + a_n$$

The sequence $S_1, S_2, S_3, \ldots$ is called the **sequence of partial sums** for the series $a_1 + a_2 + a_3 + \cdots$. Each term of the sequence of partial sums is said to be a finite series of terms of the original series.

EXAMPLE 1   ■ **Problem**

Find the sequence of partial sums for the series

$$\sum_{k=1}^{\infty} k = 1 + 2 + 3 + \cdots$$

■ **Solution**

The sequence of partial sums for this series is

$$S_1 = 1$$
$$S_2 = 1 + 2$$
$$S_3 = 1 + 2 + 3$$
$$\begin{array}{c} . \\ . \\ . \end{array}$$
$$S_n = 1 + 2 + 3 + \cdots + n$$ ■

As with any sequence, we are interested in whether a sequence of partial sums converges or diverges. For a sequence of partial sums, we must determine the existence or nonexistence of

$$\lim_{n \to \infty} S_n$$

or, another way to write the same thing, of

$$\lim_{n \to \infty} \sum_{k=1}^{n} a_k \tag{15.2}$$

The limit in (15.2) has come to be written $\sum_{k=1}^{\infty} a_k$, and it is called the sum of the infinite series. You must realize that $\sum_{k=1}^{\infty} a_k$ is not the sum of infinitely many terms but that it is merely a notation for the limit of the partial sums of the series $a_1 + a_2 + a_3 + \cdots$. If the limit in (15.2) exists, the infinite series is said to converge; otherwise, it is said to diverge.

EXAMPLE 2   ■ **Problem**

Find the sum of the infinite series $\displaystyle\sum_{k=1}^{\infty} \left(\frac{1}{2}\right)^{k-1}$ if it exists.

■ **Solution**

We resist the temptation to think of adding infinitely many terms together by first writing the partial sum

$$S_n = \sum_{k=1}^{n} \left(\frac{1}{2}\right)^{k-1}$$

Then, as in Example 5 of the previous section, we can write

$$S_n = \sum_{k=1}^{n} \left(\frac{1}{2}\right)^{k-1} = \frac{1 - (1/2)^n}{1 - (1/2)} = 2\left[1 - \left(\frac{1}{2}\right)^n\right]$$

Therefore,

$$\lim_{n\to\infty} S_n = \lim_{n\to\infty} 2[1 - (\tfrac{1}{2})^n] = 2$$

■

EXAMPLE 3   ■ **Problem**

Show that the sum of the infinite series $\sum_{k=1}^{\infty} (-1)^{k+1}$ does not exist.

■ **Solution**

Here,

$$S_n = 1 - 1 + 1 - 1 + \cdots + (-1)^{n+1}$$

The value of S_n depends on whether n is odd or even. When n is odd, $S_n = 1$; and when n is even, $S_n = 0$. Consequently, $\lim_{n\to\infty} S_n$ does not exist, and the series diverges.■

If a series converges, its nth term must become arbitrarily small as n becomes large. Conversely, if the nth term does not become small as n becomes large, then the series must diverge. We state this conclusion in the form of a test for divergence.

TEST

> **Divergence Test** If a_n denotes the nth term of an infinite series, and if
>
> $$\lim_{n \to \infty} a_n \neq 0$$
>
> then the infinite series diverges.

EXAMPLE 4 ■ **Problem**

Use the divergence test to show that the series

$$2 + \frac{3}{2} + \frac{4}{3} + \cdots + \frac{n + 1}{n} + \cdots$$

diverges.

■ **Solution**

We must show that the nth term does not approach zero as n becomes large. Thus,

$$\lim_{n \to \infty} \frac{n + 1}{n} = \lim_{n \to \infty} 1 + \frac{1}{n} = 1 \neq 0$$

which means the series diverges. ■

A common mistake that beginners make is to misinterpret the test for divergence and assume they can use it as a test for convergence. But as the following example shows, the fact that the nth term goes to zero does not necessarily mean that the series converges.

EXAMPLE 5 ■ **Problem**

By examining the sequence of partial sums, show that the series $\sum_{k=1}^{\infty} \frac{1}{\sqrt{k}}$ diverges.

■ **Solution**

The general term of this series, which is $a_n = 1/\sqrt{n}$, is easily seen to go to zero as n increases without bound. However, the partial sum

$$S_n = \frac{1}{\sqrt{1}} + \frac{1}{\sqrt{2}} + \frac{1}{\sqrt{3}} + \cdots + \frac{1}{\sqrt{n}}$$

increases without bound as n increases. To see why, note that the smallest term in the sum is the last. Therefore,

$$S_n > \frac{1}{\sqrt{n}} + \frac{1}{\sqrt{n}} + \frac{1}{\sqrt{n}} + \cdots + \frac{1}{\sqrt{n}} = n\left(\frac{1}{\sqrt{n}}\right) = \sqrt{n}$$

Since $\sqrt{n}$ clearly increases as n increases, so must S_n. Thus, the series diverges.   ■

A test for convergence, called the **integral test,** uses the convergence or divergence of the improper integral of a function to tell precisely what the infinite series does. Consider a series $\sum_{n=1}^{\infty} a_n$ of positive terms decreasing in size so that $\lim_{n\to\infty} a_n = 0$. Assume the terms of the series are represented graphically as in Figure 15.1. Let $f(x)$ be a continuous function that agrees with the terms of the series on the positive integers. Then, the graph of $f(x)$ would look something like the smooth curve in Figure 15.1.

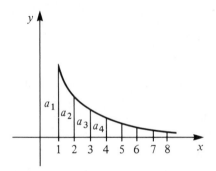

■ **Figure 15.1**

The basis of the integral test is suggested by the diagram. By comparing the area under the curve $y = f(x)$ from 1 to n with the sum $a_1 + a_2 + \cdots + a_n$, we can show that the series can converge only if the area under the curve approaches a limit as $n \to \infty$; otherwise, the series will diverge.

TEST

Integral Test   Let $f(n)$ denote the general term of the series of positive terms

$$a_1 + a_2 + a_3 + \cdots + a_n + \cdots$$

If the function $f(x)$ is always positive and decreasing for $x \geq 1$, then the series converges if the integral

$$\int_1^{\infty} f(x)\, dx$$

exists, and it diverges if it does not exist.

The function $f(x)$ referred to in the integral test is formed by simply replacing n with x in the expression for the general term of the series.

Note that the integral test allows us to discover whether or not a series is convergent, but it does not yield the numerical value to which the series converges. This limitation, however, is not a severe one since, in practice, a finite number of terms of a convergent series is used to approximate the sum.

Comment: The sum of the first n terms of an infinite series is called a **truncation** of the series.

EXAMPLE 6   ■ **Problem**

Consider the series

$$1 + \frac{1}{2^2} + \frac{1}{3^2} + \cdots + \frac{1}{n^2} + \cdots$$

a.   Use the integral test to show that the series converges.

b.   Approximate the value of the sum by using the first 50 terms of the series.

■ **Solution**

a.   The general term of this series is $f(n) = 1/n^2$, so $f(x) = 1/x^2$. Since $1/x^2$ is positive and decreasing for $x \geq 1$, the integral test applies, and we have

$$\int_1^\infty \frac{dx}{x^2} = \lim_{b \to \infty} \int_1^b x^{-2}\, dx = \lim_{b \to \infty} \left[-\frac{1}{x} \right]_1^b = \lim_{b \to \infty} \left(-\frac{1}{b} + 1 \right) = 1$$

Since the integral exists, the given series converges.

b.   The sum of the first 50 terms of this series is found, using the BASIC program in the exercise set, to be

$$1 + \frac{1}{2^2} + \frac{1}{3^2} + \cdots + \frac{1}{50^2} \approx 1.62513$$

Methods beyond the scope of this book show that the given series converges to $\pi^2/6 \approx 1.644934$.   ■

EXAMPLE 7   ■ **Problem**

Prove that the following series converges:

$$1 + \frac{1}{3^2} + \frac{1}{5^2} + \frac{1}{7^2} + \cdots$$

■ **Solution**

The general term for this series is $f(n) = 1/(2n - 1)^2$. So, $f(x) = 1/(2x - 1)^2$. Throughout the interval $x \geq 1$, the function $f(x) = 1/(2x - 1)^2$ is positive and decreases as x increases. The required conditions are satisfied, so the integral test may be used. This test yields the integral

$$\int_1^\infty \frac{dx}{(2x - 1)^2} = \lim_{b \to \infty} \int_1^b \frac{dx}{(2x - 1)^2}$$

$$= \lim_{b \to \infty} \left[-\frac{1}{2(2x - 1)} \right]_1^b$$

$$= \lim_{b \to \infty} \left[-\frac{1}{2(2b - 1)} + \frac{1}{2} \right] = \frac{1}{2}$$

Since this integral exists, the given series converges. Keep in mind that the integral test only establishes convergence; it does not yield the value to which the series converges. Thus, the numerical value of the integral plays no essential role in the test. ■

EXAMPLE 8   ■ **Problem**

Using the integral test, show that the series

$$\sum_{n=1}^\infty \frac{1}{n} = 1 + \frac{1}{2} + \frac{1}{3} + \cdots$$

diverges. This series is known as the **harmonic series.**

■ **Solution**

Here, the general term is given by $f(n) = 1/n$, so $f(x) = 1/x$. The function $f(x) = 1/x$ is positive and decreasing as x increases. Therefore, the integral test is applicable to this series, Thus,

$$\int_1^\infty \frac{dx}{x} = \lim_{p \to \infty} \int_1^p \frac{dx}{x} = \lim_{p \to \infty} \left[\ln x \right]_1^p$$

$$= \lim_{p \to \infty} (\ln p) = \infty$$

Since this integral does not exist, the harmonic series diverges.   ■

EXERCISES FOR SECTION 15.3

In Exercises 1–10, use the divergence test to determine whether the given infinite series diverges. For those for which the divergence test fails, use the integral test.

1. $\displaystyle\sum_{n=1}^{\infty} \frac{n}{n+1}$

2. $\displaystyle\sum_{n=1}^{\infty} \frac{n^2+2}{n^2}$

3. $\displaystyle\sum_{n=1}^{\infty} \frac{2n+5}{n}$

4. $\displaystyle\sum_{n=1}^{\infty} \frac{e^{10}}{e^n}$

5. $\displaystyle\sum_{n=1}^{\infty} \frac{2}{e^n}$

6. $\displaystyle\sum_{n=1}^{\infty} \frac{1}{1+n^2}$

7. $\displaystyle\sum_{n=1}^{\infty} \ln n$

8. $\displaystyle\sum_{n=1}^{\infty} \frac{2}{n+1}$

9. $\displaystyle\sum_{n=1}^{\infty} \frac{2+3n}{3+n}$

10. $\displaystyle\sum_{n=1}^{\infty} \frac{1+e^{-n}}{2+e^{-n}}$

Each of the series in Exercises 11–14 is a geometric series. Use (15.2) and the definition of convergence to tell whether the series is convergent or divergent. If it is convergent, find its sum.

11. $\displaystyle\sum_{n=1}^{\infty} \left(\frac{1}{3}\right)^{n-1}$

12. $\displaystyle\sum_{n=1}^{\infty} \left(\frac{3}{2}\right)^{n-1}$

13. $\displaystyle\sum_{n=1}^{\infty} \frac{3}{2^{n-1}}$

14. $\displaystyle\sum_{n=1}^{\infty} \left(\frac{4}{5}\right)^{n-1}$

Using the integral test, investigate the convergence of the series in Exercises 15–24.

15. $1 + \dfrac{1}{2^3} + \dfrac{1}{3^3} + \dfrac{1}{4^3} + \cdots + \dfrac{1}{n^3} + \cdots$

16. $\dfrac{1}{2} + \dfrac{1}{4} + \dfrac{1}{6} + \dfrac{1}{8} + \cdots + \dfrac{1}{2n} + \cdots$

17. $1 + \dfrac{1}{\sqrt{2}} + \dfrac{1}{\sqrt{3}} + \dfrac{1}{\sqrt{4}} + \cdots + \dfrac{1}{\sqrt{n}} + \cdots$

18. $1 + \dfrac{1}{2^{3/2}} + \dfrac{1}{3^{3/2}} + \dfrac{1}{4^{3/2}} + \cdots + \dfrac{1}{n^{3/2}} + \cdots$

19. $\dfrac{1}{e} + \dfrac{1}{e^2} + \dfrac{1}{e^3} + \dfrac{1}{e^4} + \cdots + \dfrac{1}{e^n} + \cdots$

20. $\dfrac{1}{3^2} + \dfrac{1}{5^2} + \dfrac{1}{7^2} + \dfrac{1}{9^2} + \cdots + \dfrac{1}{(2n+1)^2} + \cdots$

21. $\dfrac{1}{2} + \dfrac{2}{5} + \dfrac{3}{10} + \dfrac{4}{17} + \cdots + \dfrac{n}{n^2+1} + \cdots$

22. $\dfrac{1}{2} + \dfrac{1}{5} + \dfrac{1}{10} + \dfrac{1}{17} + \cdots + \dfrac{1}{n^2+1} + \cdots$

23. $\dfrac{e}{1+e} + \dfrac{e^2}{1+e^2} + \dfrac{e^3}{1+e^3} + \dfrac{e^4}{1+e^4} + \cdots + \dfrac{e^n}{1+e^n} + \cdots$

24. $\dfrac{1}{4} + \dfrac{2}{25} + \dfrac{3}{100} + \cdots + \dfrac{n}{(1+n^2)^2} + \cdots$

25. The series

$$\sum_{n=1}^{\infty} \frac{1}{n^p} = 1 + \frac{1}{2^p} + \frac{1}{3^p} + \cdots + \frac{1}{n^p} + \cdots$$

is called the p series. Use the integral test to prove that (a) the series converges for $p > 1$, and (b) the series diverges for $p = 1$ and $p < 1$.

Use the following BASIC program to calculate the nth partial sum of each infinite series in this exercise set.

```
10   REM  SUM OF INFINITE SERIES
20   HOME
30   PRINT "THIS PROGRAM WILL FIND THE N TH"
40   PRINT "PARTIAL SUM OF AN INFINITE SERIES"
50   PRINT "WHERE N IS SUPPLIED BY THE USER."
60   PRINT
70   PRINT "THE N TH TERM IN THIS PROGRAM IS 1/N^2"
80   PRINT
90   PRINT "DO YOU WISH TO CHANGE THE N TH TERM? Y/N"
100  INPUT A$
110  IF A$ = "Y" THEN 310
120  PRINT "HOW MANY TERMS DO YOU WISH TO SUM?"
130  INPUT K
140  DEF  FN F(N) = 1 / N ^ 2
150  PRINT
160  HTAB 12: FLASH : PRINT "PATIENCE PLEASE"
170  PRINT
180  HTAB 12: PRINT "CALCULATING!!!": NORMAL
190  PRINT
200 S = 0
210  FOR I = 1 TO K
220 S = S +  FN F(I)
230  NEXT I
240  PRINT "FOR N = ";K;" THE SUM = ";S
250  PRINT
260  PRINT "DO YOU WISH TO RUN THIS PROGRAM AGAIN? Y/N"
270  INPUT B$
280  HOME
290  IF B$ = "Y" THEN 90
300  GOTO 340
310  PRINT "TYPE"
320  INVERSE : PRINT "130 DEF FN F(N) = (YOUR FUNCTION) <RET>"
330  PRINT "RUN 110 <RET>": NORMAL
340  END
```

15.4 OTHER TESTS FOR CONVERGENCE

The Ratio Test

Another test for convergence is the so-called **ratio test**. The ratio test is based on a comparison of the given series with some geometric series. In the ratio test, we determine the ratio of two succeeding terms *in the limit*, which, for a geometric series, is always the same. The following example shows how to determine the convergence or divergence of a general geometric series.

EXAMPLE 1   ■ **Problem**

Determine the values of R for which the geometric series $\sum_{k=1}^{\infty} R^k$ is convergent.

■ **Solution**

We write the sequence of partial sums

$$S_n = \sum_{k-1}^{n} R^k$$

By the formula for the sum of a geometric series,

$$S_n = \frac{1 - R^{n+1}}{1 - R} = \frac{1}{1 - R} - \frac{R^{n+1}}{1 - R} \qquad R \neq 1$$

Since R is a fixed number, we have

$$\lim_{n \to \infty} S_n = \frac{1}{1 - R} \qquad \text{if} \qquad |R| < 1$$

$$\lim_{n \to \infty} S_n = \infty \qquad \text{if} \qquad |R| > 1$$

If $R = 1$, then the original series becomes $\sum_{k=1}^{\infty} (1)^k$, which obviously diverges.   ■

The ratio test tells us that the same techniques we used to determine the convergence or divergence of a geometric series hold for nongeometric series, with one exception: when the limiting value of the ratio is 1.

TEST

> **The Ratio Test**   Consider the infinite series
>
> $$a_1 + a_2 + a_3 + \cdots + a_n + a_{n+1} + \cdots$$
>
> of constant terms, with either mixed or like signs. Let a_n and a_{n+1} represent any two consecutive terms of the series, and take the limit of the absolute value of the ratio of a_{n+1} to a_n as n increases. Denoting this limit by

$$R = \lim_{n \to \infty} \left| \frac{a_{n+1}}{a_n} \right|$$

we have the following results:

1. When $R < 1$, the series is convergent.
2. When $R > 1$, the series is divergent.
3. When $R = 1$, the test fails.

If the ratio test yields $R = 1$, we cannot make a decision about whether the series is convergent or divergent. In this case, the series must be investigated by another test, such as the integral test.

EXAMPLE 2   ▪ **Problem**

Use the ratio test to investigate the convergence or divergence of the series

$$-\frac{1}{2} + \frac{2}{2^2} - \frac{3}{2^3} + \cdots + \frac{(-1)^n n}{2^n} + \cdots$$

▪ **Solution**

Here, $a_n = [(-1)^n n]/(2^n)$, and $a_{n+1} = [(-1)^{n+1}(n+1)]/2^{n+1}$. The test ratio is then given by

$$\left| \frac{a_{n+1}}{a_n} \right| = \left| \frac{[(-1)^{n+1}(n+1)/2^{n+1}]}{(-1)^n n/2^n} \right| = \frac{n+1}{2^{n+1}} \cdot \frac{2^n}{n} = \frac{n+1}{2n} = \frac{1}{2} + \frac{1}{2n}$$

Taking the limit as $n \to \infty$, we have

$$R = \lim_{n \to \infty}\left(\frac{1}{2} + \frac{1}{2n}\right) = \frac{1}{2}$$

Since $R < 1$, we conclude that the given series converges.   ▪

EXAMPLE 3   ▪ **Problem**

Use the ratio test to investigate the convergence or divergence of the series

$$1 + \frac{1}{2^2} + \frac{1}{3^2} + \cdots + \frac{1}{n^2} + \cdots$$

▪ **Solution**

Here, $a_n = 1/n^2$ and $a_{n+1} = 1/(n+1)^2$. So,

$$R = \lim_{n \to \infty} \left[\frac{1/(n + 1)^2}{1/n^2} \right] = \lim_{n \to \infty} \left(\frac{n^2}{n^2 + 2n + 1} \right)$$

$$= \lim_{n \to \infty} \left(\frac{1}{1 + 2/n + 1/n^2} \right) = 1$$

Since $R = 1$, the test fails to determine whether the series is convergent or divergent. However, the series can be shown to be convergent by applying the integral test with $f(x) = 1/x^2$. Thus,

$$\int_1^\infty \frac{dx}{x^2} = \lim_{b \to \infty} \int_1^b x^{-2} \, dx = \lim_{b \to \infty} \left[-\frac{1}{x} \right]_1^b = \lim_{b \to \infty} \left(-\frac{1}{b} + 1 \right) = 1$$

Therefore, since the integral exists, the given series converges.   ∎

EXAMPLE 4   ■ **Problem**

Use the ratio test to show that the following series is divergent:

$$\frac{1!}{\sqrt{2}} + \frac{2!}{\sqrt{3}} + \frac{3!}{\sqrt{4}} + \cdots + \frac{n!}{\sqrt{n + 1}} + \cdots$$

■ **Solution**

The nth term of this series is $a_n = n!/\sqrt{n + 1}$, and the $(n + 1)$st term is $a_{n+1} = (n + 1)!/\sqrt{n + 2}$. So,

$$\frac{a_{n+1}}{a_n} = \frac{(n + 1)!}{\sqrt{n + 2}} \cdot \frac{\sqrt{n + 1}}{n!} = \frac{(n + 1)!}{n!} \cdot \sqrt{\frac{n + 1}{n + 2}}$$

The test ratio can be simplified by noting that

$$\frac{(n + 1)!}{n!} = \frac{(n + 1) \times n \times (n - 1) \times \cdots \times 2 \times 1}{n \times (n - 1) \times \cdots \times 2 \times 1} = n + 1$$

Making this simplification and at the same time dividing the numerator and denominator of the radicand by n, we have

$$R = \lim_{n \to \infty} \left[(n + 1) \sqrt{\frac{1 + (1/n)}{1 + (2/n)}} \right] = \infty$$

Since $R > 1$, the series is divergent.   ∎

EXAMPLE 5   ■ **Problem**

Use the ratio test to show that the following alternating series is divergent:

$$\frac{2}{3^2} - \frac{2^2}{5^2} + \frac{2^3}{7^2} - \frac{2^4}{9^2} + \cdots + \frac{(-1)^{n+1}2^n}{(2n + 1)^2} + \cdots$$

■ **Solution**

Here,

$$|a_n| = \frac{2^n}{(2n + 1)^2} \quad \text{and} \quad |a_{n+1}| = \frac{2^{n+1}}{[2(n + 1) + 1]^2}$$

$$= \frac{2^{n+1}}{(2n + 3)^2}$$

Therefore,

$$R = \lim_{n \to \infty} \left| \frac{a_{n+1}}{a_n} \right| = \lim_{n \to \infty} \frac{2^{n+1}}{(2n + 3)^2} \cdot \frac{(2n + 1)^2}{2^n} = \lim_{n \to \infty} 2 \cdot \left(\frac{2n + 1}{2n + 3} \right)^2$$

$$= \lim_{n \to \infty} 2 \cdot \left(\frac{2 + 1/n}{2 + 3/n} \right)^2 = 2$$

Since $R > 1$, we conclude that the series is divergent.   ■

The Alternating Series Test

A series whose terms are alternately positive and negative is called an **alternating series**. There is a rather easy test to determine the convergence of this kind of series.

TEST

> **The Alternating Series Test**   Let
>
> $$a_1 - a_2 + a_3 - a_4 + \cdots + a_n - a_{n+1} + \cdots$$
>
> represent an alternating series in which the absolute value of each term is eventually less than the preceding term; that is, eventually,
>
> $$|a_{n+1}| < |a_n|$$
>
> Then, the series will be convergent if
>
> $$\lim_{n \to \infty} a_n = 0$$

EXAMPLE 6   ■ **Problem**

Show that the alternating series $1 - 1/2^2 + 1/3^2 - 1/4^2 + \cdots$ is convergent.

■ **Solution**

Here, $|a_n| = 1/n^2$ and $|a_{n+1}| = 1/(n + 1)^2$. Therefore,

$$|a_{n+1}| < |a_n|$$

for all values of n. Furthermore,

$$\lim_{n \to \infty} \frac{1}{n^2} = 0$$

and the series is convergent. ∎

EXAMPLE 7 ■ **Problem**

Show that the series $\displaystyle\sum_{n=1}^{\infty} \frac{(-1)^n}{n}$ converges.

■ **Solution**

The absolute value of each term of this series is less than the previous one, and $\lim\limits_{n \to \infty} 1/n = 0$. Hence, the series converges. ∎

EXERCISES FOR SECTION 15.4

Use the ratio test to investigate the convergence or divergence of the indicated series. If the ratio test fails, apply either the integral test or the alternating series test.

1. $1 + \dfrac{1}{2!} + \dfrac{1}{3!} + \dfrac{1}{4!} + \cdots$

2. $\dfrac{1}{2} - \dfrac{2}{2^2} + \dfrac{3}{2^3} - \dfrac{4}{2^4} + \cdots$

3. $\dfrac{3}{2} + \dfrac{4}{2^2} + \dfrac{5}{2^3} + \dfrac{6}{2^4} + \cdots$

4. $\displaystyle\sum_{n=1}^{\infty} \dfrac{(-1)^{n+1}\, 3n}{(n+1)^2}$

5. $1 - \dfrac{1}{2} + \dfrac{1}{3} - \dfrac{1}{4} + \cdots$

6. $1 - \dfrac{1}{3!} + \dfrac{1}{5!} - \dfrac{1}{7!} + \cdots$

7. $\displaystyle\sum_{n=1}^{\infty} \dfrac{n}{(n-1)!}$

8. $\displaystyle\sum_{n=1}^{\infty} \dfrac{(-1)^n n!}{e^n}$

9. $\dfrac{1}{e^{10}} + \dfrac{2}{e^{10}} + \dfrac{3}{e^{10}} + \dfrac{4}{e^{10}} + \cdots$

10. $\dfrac{1}{\pi} - \dfrac{2}{\pi^2} + \dfrac{3}{\pi^3} - \dfrac{4}{\pi^4} + \cdots$

11. $\displaystyle\sum_{n=1}^{\infty} \dfrac{(-1)^{n+1}}{\sqrt{n}}$

12. $\displaystyle\sum_{n=1}^{\infty} \dfrac{\sqrt{n-1}}{(n-1)!}$

13. $\displaystyle\sum_{n=1}^{\infty} \dfrac{\sqrt{2n}}{n!}$

14. $\displaystyle\sum_{n=1}^{\infty} \dfrac{1}{\sqrt{n}}$

15. $\displaystyle\sum_{n=1}^{\infty} \dfrac{5}{n^{2/3}}$

16. $\displaystyle\sum_{n=1}^{\infty} \dfrac{n-1}{n+1}$

15.5 POWER SERIES

So far, we have considered only series of constant terms. Since some of the more important types of infinite series involve terms that are functions of a variable, we wish to extend our discussion to include series of this type. The most important series involving variable terms is known as a **power series**.

DEFINITION

> **Power Series**   With x denoting the variable quantity, a *power series* is given by
>
> $$c_0 + c_1x + c_2x^2 + c_3x^3 + \cdots + c_nx^n + \cdots$$
>
> where $c_0, c_1, c_2, \ldots c_n, \ldots$ are constant coefficients and are independent of the value of x.

In dealing with power series, we must again consider the problem of convergence and divergence. However, the problem is now one of finding the values of the variable x for which the series will converge. We find that a power series may converge for all values of x, or it may converge for some values of x but not for others. The set of values of x for which a power series converges is usually referred to as the **interval of convergence** of the series. In this section, we will determine the interval of convergence of a power series by applying the ratio test to the given power series.

EXAMPLE 1   ■ **Problem**

Find the interval of convergence of the series

$$x + 2x^2 + 3x^3 + \cdots + nx^n + \cdots$$

■ **Solution**

Here, $a_n = nx^n$, and $a_{n+1} = (n + 1)x^{n+1}$. So,

$$R = \lim_{n \to \infty} \left| \frac{(n + 1)x^{n+1}}{nx^n} \right| = \lim_{n \to \infty} \frac{n + 1}{n} \cdot |x| = \lim_{n \to \infty} \left(1 + \frac{1}{n}\right) \cdot |x| = |x|$$

The ratio test reveals that the given series is convergent for all values of x for which $|x| < 1$, that is, for the interval

$$-1 < x < 1$$

We also know from the ratio test that the given series is divergent for $|x| > 1$, but we do not know what happens when $|x| = 1$. The endpoints of the interval of convergence—in this case, $x = \pm 1$—must be examined separately.

If $x = 1$, the given series becomes

$$1 + 2 + 3 + \cdots + n + \cdots$$

which is clearly a divergent series.

If $x = -1$, the given series becomes the alternating series

$$-1 + 2 - 3 + 4 + \cdots + (-1)^n n + \cdots$$

The terms of this series increase without bound as n increases without bound, and therefore, the series is divergent.

Since the given series diverges at both $x = 1$ and $x = -1$, we conclude that the interval of convergence for this series is

$$-1 < x < 1 \qquad\qquad\qquad\qquad\qquad\qquad\qquad\qquad \blacksquare$$

EXAMPLE 2 ■ **Problem**

Find the interval of convergence of the series

$$x + \frac{x^2}{\sqrt{2}} + \frac{x^3}{\sqrt{3}} + \cdots + \frac{x^n}{\sqrt{n}} + \cdots$$

■ **Solution**

Here, $a_n = x^n/\sqrt{n}$, and $a_{n+1} = x^{n+1}/\sqrt{n+1}$. So,

$$R = \lim_{n \to \infty} \left| \frac{x^{n+1}}{\sqrt{n+1}} \cdot \frac{\sqrt{n}}{x^n} \right| = \lim_{n \to \infty} \left| \left(\frac{n}{n+1} \right)^{1/2} \right| \cdot |x| = |x|$$

Therefore, the series converges for $|x| < 1$, that is, $-1 < x < 1$. The ratio test fails at $x = 1$ and $x = -1$, so we examine these points separately.

If $x = 1$, the given series becomes

$$1 + \frac{1}{\sqrt{2}} + \frac{1}{\sqrt{3}} + \cdots + \frac{1}{\sqrt{n}} + \cdots$$

which can be tested for convergence by using the integral test with $f(x) = 1/\sqrt{x}$. Thus,

$$\int_1^\infty \frac{dx}{\sqrt{x}} = \lim_{t \to \infty} \int_1^t x^{-1/2}\, dx = \lim_{t \to \infty} \left[2x^{1/2} \right]_1^t = \lim_{t \to \infty} (2t^{1/2} - 2) = \infty$$

Since the limit does not exist, the given series diverges for $x = 1$.

If $x = -1$, the given series becomes

$$-1 + \frac{1}{\sqrt{2}} - \frac{1}{\sqrt{3}} + \cdots + \frac{(-1)^n}{\sqrt{n}} + \cdots$$

which is an alternating series with $|a_n| = 1/\sqrt{n}$ and $|a_{n+1}| = 1/\sqrt{n+1}$. Since $|1/\sqrt{n+1}| < |1/\sqrt{n}|$ for all positive n, and

$$\lim_{n\to\infty} \frac{1}{\sqrt{n}} = 0$$

the given series will be convergent for $x = -1$.

Summarizing the result, we see that the interval of convergence for the given series is

$$-1 \le x < 1$$

■

EXAMPLE 3   ■ **Problem**

Find the interval of convergence for the series

$$1 + x + \frac{x^2}{2!} + \frac{x^3}{3!} + \cdots + \frac{x^n}{n!} + \cdots$$

■ **Solution**

Here, $|a_n| = |x^n/n!|$, and $|a_{n+1}| = |x^{n+1}/(n+1)!|$. So,

$$R = \lim_{n\to\infty} \left| \frac{x^{n+1}}{(n+1)!} \cdot \frac{n!}{x^n} \right| = \lim_{n\to\infty} \left(\frac{1}{n+1} \right) \cdot |x| = 0$$

Since $R < 1$ for all finite values of x, we conclude that the interval of convergence is

$$-\infty < x < \infty$$

■

EXERCISES FOR SECTION 15.5

Find the interval of convergence of each of the given power series.

1. $x - x^2 + x^3 - x^4 + \cdots$

2. $2x + 3x^2 + 4x^3 + 5x^4 + \cdots$

3. $\sum_{n=1}^{\infty} \frac{x^{n-1}}{n!}$

4. $\sum_{n=1}^{\infty} \frac{x^n}{n}$

5. $\frac{1}{2} + \frac{x}{3} + \frac{x^2}{4} + \frac{x^3}{5} + \cdots$

6. $\sum_{n=1}^{\infty} \frac{(-1)^{n+1}x^n}{\sqrt{n}}$

7. $\sum_{n=1}^{\infty} \frac{(-1)^{n+1}x^{2n-1}}{(2n-1)!}$

8. $\sum_{n=1}^{\infty} \frac{nx^{2n+3}}{(n+1)^2}$

9. $1 - \frac{x^2}{2!} + \frac{x^4}{4!} - \frac{x^6}{6!} + \cdots$

10. $\frac{x}{2} + \frac{x^2}{2^2} + \frac{x^3}{2^3} + \frac{x^4}{2^4} + \cdots$

11. $1 - x + \frac{x^2}{2^2} - \frac{x^3}{3^2} + \cdots$

12. $\sum_{n=1}^{\infty} 2^n x^{n-1}$

13. $\displaystyle\sum_{n=1}^{\infty} 3^{n-1}x^n$

14. $\dfrac{x^2}{2} - \dfrac{x^4}{4} + \dfrac{x^6}{6} - \dfrac{x^8}{8} + \cdots$

15. $\displaystyle\sum_{n=1}^{\infty} \dfrac{2^n x^n}{n!}$

16. $\displaystyle\sum_{n=1}^{\infty} (-1)^{n+1} \dfrac{x^{2n-1}}{2n-1}$

17. $1 + \dfrac{x}{e} + \dfrac{x^2}{e^2} + \dfrac{x^3}{e^3} + \cdots$

18. $\displaystyle\sum_{n=1}^{\infty} \dfrac{x^n}{n2^n}$

■ REVIEW EXERCISES FOR CHAPTER 15

Write the general term of each sequence in Exercises 1–4.

1. $3, 5, 7, 9, \ldots$

2. $1, 4, 9, 16, \ldots$

3. $\dfrac{1}{2}, -\dfrac{1}{4}, \dfrac{1}{6}, -\dfrac{1}{8}, \ldots$

4. $-1, \dfrac{1}{3}, -\dfrac{1}{5}, \dfrac{1}{7}, \ldots$

In Exercises 5–10, write the first four terms of the sequence, and determine whether or not it is convergent.

5. $a_n = \dfrac{3}{n+1}$

6. $a_n = \dfrac{2n}{n+1}$

7. $a_n = \dfrac{(-1)^n}{\sqrt{n}}$

8. $a_n = \dfrac{n^2+1}{n}$

9. $a_n = \dfrac{n^2}{2+n}$

10. $a_n = (\tfrac{1}{2})^n$

In Exercises 11–14, use the integral test or the divergence test to determine the convergence or divergence of the given series.

11. $1 + \dfrac{1}{3} + \dfrac{1}{5} + \dfrac{1}{7} + \cdots$

12. $1 + \dfrac{1}{\sqrt[3]{2}} + \dfrac{1}{\sqrt[3]{3}} + \dfrac{1}{\sqrt[3]{4}} + \cdots$

13. $1 + \dfrac{1}{8} + \dfrac{1}{27} + \dfrac{1}{64} + \cdots$

14. $1 + e^{-1} + e^{-2} + e^{-3} + \cdots$

Use the ratio test or the divergence test to determine the convergence or divergence of the series in Exercises 15–20.

15. $\dfrac{1}{2!} + \dfrac{1}{4!} + \dfrac{1}{6!} + \dfrac{1}{8!} + \cdots$

16. $\dfrac{1}{2!} + \dfrac{2}{3!} + \dfrac{3}{4!} + \dfrac{4}{5!} + \cdots$

17. $1 + \dfrac{2}{e} + \dfrac{3}{e^2} + \dfrac{4}{e^3} + \cdots$

18. $1 + \dfrac{\sqrt{2}}{2!} + \dfrac{\sqrt{3}}{3!} + \dfrac{\sqrt{4}}{4!} + \cdots$

19. $\displaystyle\sum_{n=1}^{\infty} \dfrac{(2n-1)^2}{2^n}$

20. $\displaystyle\sum_{n=1}^{\infty} \dfrac{n+1}{(n+2)!}$

Test the series in Exercises 21–24 for convergence by using the alternating series test, if possible.

21. $1 - \dfrac{3}{2} + \dfrac{4}{3} - \dfrac{5}{4} + \cdots$

22. $1 - \dfrac{1}{\sqrt{3}} + \dfrac{1}{\sqrt{5}} - \dfrac{1}{\sqrt{7}} + \cdots$

23. $\dfrac{2}{3} - \dfrac{3}{3^2} + \dfrac{4}{3^3} - \dfrac{5}{3^4} + \cdots$

24. $1 - \dfrac{2}{e} + \dfrac{3}{e^2} - \dfrac{4}{e^3} + \cdots$

Find the interval of convergence for each of the given power series.

25. $1 + x + x^2 + x^3 + \cdots$

26. $x - 2x^2 + 3x^3 - 4x^4 + \cdots$

27. $\dfrac{x}{2} - \dfrac{x^2}{5} + \dfrac{x^3}{10} - \dfrac{x^4}{17} + \cdots$

28. $1 + \dfrac{x}{\pi} + \dfrac{x^2}{\pi^2} + \dfrac{x^3}{\pi^3} + \cdots$

CHAPTER 16

Expansion of Functions

In trigonometry, we learn how to find the values of the trigonometric functions for some special angles such as 0°, 30°, 45°, 60°, and 90°. The text then points out that the values of the trigonometric functions for other angles are available in table form. Have you ever wondered how the values in the trigonometric tables were computed? For instance, how would you find sin 39° without a table? Surprising as it may seem, tables of trigonometric functions of angles are constructed by using power series of the form $a_0 + a_1x + a_2x^2 + a_3x^3 + \cdots$.

16.1 MACLAURIN SERIES

An infinite sum of terms of the form

$$a_0 + a_1x + a_2x^2 + a_3x^3 + \cdots$$

where $a_0, a_1, a_2, \ldots$ are constant coefficients of the variable x, is called a **power series** as we learned in Chapter 15. In this section, we explore the connection between functions and power series. The relationship between a function f and its power series is illustrated by the following example.

Consider the function

$$f(x) = \frac{1}{1 - x}$$

If we divide $1 - x$ into 1, we get

$$\frac{1}{1 - x} = 1 + x + x^2 + x^3 + \cdots$$

where the terms on the right continue indefinitely. Thus, the fifth term is x^4, the next x^5, and so on. We call $1 + x + x^2 + x^3 + \cdots$ the **power series expansion** of the function $1/(1 - x)$. We can show, by the methods of Chapter 15, that this series converges for those values of x in the interval $-1 < x < 1$, called the **interval of convergence** of the power series. The power series is said to be **convergent** on this interval. Moreover, the power series converges to $1/(1 - x)$ on this interval. We say that the power series **represents the function**.

The fact that $f(x) = 1/(1 - x)$ can be represented by a power series leads us to ask whether or not other functions can be represented in this way. We restrict our discussion to functions that can be represented by a power series over some interval of convergence.

To find a power series expansion for a function $f(x)$, we assume that it has a power series expansion of the form

$$f(x) = a_0 + a_1 x + a_2 x^2 + a_3 x^3 + a_4 x^4 + \cdots + a_n x^n + \cdots \qquad \textbf{(16.1)}$$

where $a_0, a_1, a_2, a_3, \ldots$ are constants to be determined. If the power series is to represent $f(x)$, we must be able to evaluate these coefficients. By setting $x = 0$ in (16.1), we have

$$a_0 = f(0)$$

And therefore, a_0 is the constant obtained by evaluating the function at $x = 0$.

To evaluate the remaining coefficients, we assume that $f(x)$ can be differentiated as many times as we wish. The first derivative gives

$$f'(x) = a_1 + 2a_2 x + 3a_3 x^2 + 4a_4 x^3 + \cdots + na_n x^{n-1} + \cdots$$

Letting $x = 0$, we have

$$a_1 = f'(0)$$

Similarly,

$$f''(x) = 2 \cdot 1a_2 + 3 \cdot 2a_3 x + 4 \cdot 3a_4 x^2 + \cdots + n(n - 1)a_n x^{n-2} + \cdots$$

When $x = 0$, we get

$$2 \cdot 1a_2 = f''(0) \qquad \text{or} \qquad a_2 = \frac{f''(0)}{2 \cdot 1} = \frac{f''(0)}{2!}$$

Differentiating again, we obtain

$$f'''(x) = 3 \cdot 2 \cdot 1a_3 + 4 \cdot 3 \cdot 2a_4 x + 5 \cdot 4 \cdot 3a_5 x^2 + \cdots$$

Letting $x = 0$, we have

$$3 \cdot 2 \cdot 1a_3 = f'''(0) \qquad \text{or} \qquad a_3 = \frac{f'''(0)}{3 \cdot 2 \cdot 1} = \frac{f'''(0)}{3!}$$

By continuing this process, we see that each of the coefficients $a_0, a_1, a_2, \ldots$ can be evaluated, and that the general term a_n is given by

$$a_n = \frac{f^{(n)}(0)}{n!} \qquad (16.2)$$

where

$$f^{(0)}(x) = f(x)$$

Substituting these values into Equation (16.1), we conclude that if a function $f(x)$ has a power series expansion, it will be of the form

$$f(x) = f(0) + f'(0)x + \frac{f''(0)x^2}{2!} + \frac{f'''(0)x^3}{3!} + \cdots + \frac{f^{(n)}(0)x^n}{n!} + \cdots$$

which can be written more compactly as follows:

FORMULA

> **Maclaurin Series Expansion of $f(x)$**
>
> $$f(x) = \sum_{n=0}^{\infty} \frac{f^{(n)}(0)x^n}{n!} \qquad (16.3)$$

Observe that Equation (16.3) assumes that the function and all of its derivatives exist at $x = 0$.

EXAMPLE 1   ■ **Problem**

Using Equation (16.3), verify the Maclaurin series of $f(x) = \dfrac{1}{1 - x}$.

■ **Solution**

Evaluating the function and its successive derivatives at $x = 0$, we have

$$
\begin{array}{ll}
f(x) = (1 - x)^{-1} & f(0) = 1 \\
f'(x) = 1(1 - x)^{-2} & f'(0) = 1 \\
f''(x) = 2(1 - x)^{-3} & f''(0) = 2 \\
f'''(x) = 6(1 - x)^{-4} & f'''(0) = 6
\end{array}
$$

Substituting these values into Equation (16.3), we obtain the Maclaurin series expansion:

$$\frac{1}{1 - x} = \frac{1 \cdot x^0}{0!} + \frac{1 \cdot x^1}{1!} + \frac{2 \cdot x^2}{2!} + \frac{6 \cdot x^3}{3!} + \cdots$$

$$= 1 + x + x^2 + x^3 \cdots$$

This result agrees with the series expansion obtained earlier. The interval of convergence is $(-1, 1)$.   ■

EXAMPLE 2   ■ **Problem**

Find the Maclaurin series expansion of the function $f(x) = \sin x$.

■ **Solution**

Here,

$$
\begin{aligned}
f(x) &= \sin x & f(0) &= 0 \\
f'(x) &= \cos x & f'(0) &= 1 \\
f''(x) &= -\sin x & f''(0) &= 0 \\
f'''(x) &= -\cos x & f'''(0) &= -1 \\
f^{(4)}(x) &= \sin x & f^{(4)}(0) &= 0 \\
f^{(5)}(x) &= \cos x & f^{(5)}(0) &= 1
\end{aligned}
$$

Substituting these values into Equation (16.3), we have

$$f(x) = \sin x = x - \frac{x^3}{3!} + \frac{x^5}{5!} - \cdots + \frac{(-1)^n x^{2n+1}}{(2n+1)!} + \cdots$$

This series converges for $-\infty < x < \infty$.   ■

EXAMPLE 3   ■ **Problem**

Find the first four nonzero terms of the Maclaurin series expansion of $f(x) = \sqrt{x+1}$.

■ **Solution**

Writing $f(x)$ in exponential form, we have

$$
\begin{aligned}
f(x) &= (x+1)^{1/2} & f(0) &= 1 \\
f'(x) &= \tfrac{1}{2}(x+1)^{-1/2} & f'(0) &= \tfrac{1}{2} \\
f''(x) &= -\tfrac{1}{4}(x+1)^{-3/2} & f''(0) &= -\tfrac{1}{4} \\
f'''(x) &= \tfrac{3}{8}(x+1)^{-5/2} & f'''(0) &= \tfrac{3}{8}
\end{aligned}
$$

Thus, by Equation (16.3),

$$\sqrt{x+1} = 1 + \frac{x}{2} - \frac{x^2}{8} + \frac{3x^3}{48} - \cdots$$

■

There are four Maclaurin series that are very common. They are listed here for reference.

FORMULA

Geometric Series

$$\frac{1}{1 - x} = 1 + x + x^2 + x^3 + \cdots + x^n + \cdots$$

See Example 1. The interval of convergence is $-1 < x < 1$.

Sine Series

$$\sin x = x - \frac{x^3}{3!} + \frac{x^5}{5!} - \cdots + \frac{x^{2n+1} (-1)^n}{(2n + 1)!} + \cdots$$

See Example 2. The interval of convergence is $-\infty < x < \infty$.

Cosine Series

$$\cos x = 1 - \frac{x^2}{2!} + \frac{x^4}{4!} - \cdots + \frac{(-1)^n x^{2n}}{(2n)!} + \cdots$$

See Exercise 1. The interval of convergence is $-\infty < x < \infty$.

Exponential Series

$$e^x = 1 + x + \frac{x^2}{2!} + \frac{x^3}{3!} + \cdots + \frac{x^n}{n!} + \cdots$$

See Exercise 2. The interval of convergence is $-\infty < x < \infty$.

EXERCISES FOR SECTION 16.1

In Exercises 1–4, find the first three nonzero terms of the Maclaurin series expansion of the given functions, and write the general term.

1. $f(x) = \cos x$

2. $f(x) = e^x$

3. $f(x) = \ln (1 + x)$

4. $f(x) = \ln(1 - x)$

In the remaining exercises, find the first three nonzero terms of the Maclaurin series expansion of the given function.

5. $f(x) = \sin 3x$

6. $f(x) = \sqrt{1 + 4x}$

7. $f(x) = \dfrac{1}{(1 + x)^2}$

8. $f(x) = \dfrac{1}{\sqrt{1 + x}}$

9. $f(x) = \text{Arcsin } x$

10. $f(x) = \text{Arctan } x$

11. $f(x) = e^{-x}$

12. $f(x) = \tan x$

13. $f(x) = \sec x$

14. $f(x) = e^{\sin x}$

15. $f(x) = \ln \cos x$

16. $f(x) = e^{x^2}$

17. $f(x) = x^2$

18. $f(x) = x^3 + 3x + 2$

19. $f(x) = (x - 1)^2$

20. $f(x) = \dfrac{x - 1}{x + 1}$

■■■■ 16.2 OPERATIONS WITH POWER SERIES

Substitution

Once a Maclaurin series expansion of a function $f(x)$ is known, the series for $f(ax)$ may be obtained by substitution.

EXAMPLE 1　**a.**　The Maclaurin series for $1/(1 + 2x)$ may be obtained from the series for $1/(1 - x)$ by substituting $-2x$ for x. Thus,

$$\frac{1}{1 + 2x} = \frac{1}{1 - (-2x)} = 1 + (-2x) + (-2x)^2 + (-2x)^3 + \cdots$$
$$= 1 - 2x + 4x^2 - 8x^3 + \cdots$$

b.　The Maclaurin series for $\sin 3x$ may be obtained by substituting $3x$ for x in the sine series. Thus,

$$\sin 3x = 3x - \frac{(3x)^3}{3!} + \frac{(3x)^5}{5!} - \cdots = 3x - \frac{9}{2}x^3 + \frac{81}{40}x^5 - \cdots \qquad ■$$

EXAMPLE 2　■ **Problem**

Find the Maclaurin series expansion for $\dfrac{1}{3 + 2x}$.

■ **Solution**

First, we write the given function as

$$\frac{1}{3 + 2x} = \frac{1}{3[1 + (2/3)x]}$$

Now, we use the geometric series with $\left(-\frac{2}{3}x\right)$ substituted for x:

$$\frac{1}{3 + 2x} = \frac{1}{3}\left(1 - \frac{2}{3}x + \frac{4}{9}x^2 - \frac{8}{27}x^3 + \cdots\right) \qquad ■$$

As we pointed out earlier, we may wish to know where a given power series is convergent. One reason we need this information is that many of the operations that we know to be valid for finite summations can be extended immediately to infinite series of terms provided that the series is convergent. The operations of differentiation and integration are such operations.

Differentiation of Power Series

A convergent power series may be differentiated term by term, and the resulting series will be convergent. Not only will this series be convergent, but it will have the same interval of convergence as the series from which it was obtained, except perhaps for the endpoints.

EXAMPLE 3   ■ **Problem**

Use the expansion for $\sin x$ to derive the expansion for $\cos x$.

■ **Solution**

Beginning with

$$\sin x = x - \frac{x^3}{3!} + \frac{x^5}{5!} - \cdots$$

we differentiate both sides to obtain

$$\cos x = 1 - \frac{3x^2}{3!} + \frac{5x^4}{5!} - \cdots = 1 - \frac{x^2}{2!} + \frac{x^4}{4!} - \cdots$$ ■

Integration of Power Series

A convergent power series may be integrated term by term, and the resulting series will have the same interval of convergence, except perhaps for the endpoints.

EXAMPLE 4   ■ **Problem**

Use the expansion for $\dfrac{1}{1 + x^2}$ to derive the expansion for Arctan x.

■ **Solution**

The expansion for $1/(1 + x^2)$ is given by

$$\frac{1}{1 + x^2} = 1 - x^2 + x^4 - x^6 + \cdots$$

The integral of the left side yields

$$\int_0^x \frac{dv}{1 + v^2} = \left[\text{Arctan } v \right]_0^x = \text{Arctan } x$$

Where the indefinite integral is evaluated from 0 to x to can avoid the necessity of adding a constant of integration to each term. Therefore, if we integrate both sides, we have

$$\text{Arctan } x = x - \frac{x^3}{3} + \frac{x^5}{5} - \cdots$$

■

EXERCISES FOR SECTION 16.2

In Exercises 1–15, write the infinite series expansion of the given function by using one of the Maclaurin series obtained in Section 16.1 or 16.2.

1. $\sin 2x$

2. $e^{\pi x}$

3. $\cos \frac{1}{2}x$

4. $\sin x^2$

5. $\text{Arcsin } 3x$

6. $\sin(x - a)$

7. $\cos\left(x - \frac{\pi}{6}\right)$

8. $\tan(x + \theta)$

9. $\dfrac{1}{1 - x^2}$

10. $\dfrac{1}{2 + x}$

11. $\dfrac{1}{3x + 1}$

12. $\dfrac{1}{2 + x^3}$

13. $\dfrac{2}{2 - 5x}$

(*Hint*: For Exercises 14 and 15, first use a partial-fractions decomposition.)

14. $\dfrac{1}{(x - 4)(x - 1)}$

15. $\dfrac{1}{(x + 1)(x + 2)}$

In Exercises 16–20, use the given Maclaurin series to derive the Maclaurin series of the expression in brackets.

16. $\tan x = x + \dfrac{x^3}{3} + \dfrac{2x^5}{15} + \cdots$, $[\sec^2 x]$

17. $\ln(1 + x) = x - \dfrac{x^2}{2} + \dfrac{x^3}{3} - \dfrac{x^4}{4} + \cdots$, $\left[\dfrac{1}{1 + x}\right]$

18. $\tan x = x + \dfrac{x^3}{3} + \dfrac{2x^5}{15} + \cdots$, $[\ln \cos x]$

19. $\cos x = 1 - \dfrac{x^2}{2!} + \dfrac{x^4}{4!} - \dfrac{x^6}{6!} + \cdots$, $[\sin x]$

20. $\dfrac{1}{1 + x} = 1 - x + x^2 - x^3 + x^4 - \cdots$, $\left[\left(\dfrac{1}{1 + x}\right)^2\right]$

■ 16.3 APPROXIMATION BY TRUNCATED SERIES

If a series expansion represents a function $f(x)$, then we can expect a good approximation to the value of $f(x)$ by using the first few terms of the series.

EXAMPLE 1   ■ **Problem**

Evaluate $\sin x$ when $x = 0.5$.

■ **Solution**

Using the first three terms of the series expansion of $\sin x$, we have

$$\sin 0.5 = 0.5 - \frac{(0.5)^3}{3!} + \frac{(0.5)^5}{5!} = 0.5 - 0.020833 + 0.000260$$
$$= 0.47943$$

This approximation can be shown to be accurate to five decimal places.   ■

EXAMPLE 2   ■ **Problem**

Use four terms of the Maclaurin series expansion for e^x to approximate $e^{0.1}$.

■ **Solution**

We have

$$e^{0.1} = 1 + 0.1 + \frac{(0.1)^2}{2!} + \frac{(0.1)^3}{3!}$$
$$= 1.00000 + 0.10000 + 0.00500 + 0.00017 = 1.10517$$

This answer is accurate to the indicated number of decimal places.   ■

EXAMPLE 3   ■ **Problem**

Find the value of $\int_0^1 \cos x^2 \, dx$.

■ **Solution**

This integral cannot be evaluated by the elementary methods of Chapter 10. However, the value of the definite integral can be approximated by employing power series. Using the power series expansion for $\cos x^2$, we have

$$\int_0^1 \cos x^2 \, dx = \int_0^1 \left[1 - \frac{x^4}{2!} + \frac{x^8}{4!} - \cdots \right] dx = \left[x - \frac{x^5}{10} + \frac{x^9}{216} - \cdots \right]_0^1$$

Considering the first three terms of the series, we get

$$\int_0^1 \cos x^2 \, dx \approx \left[x - \frac{x^5}{10} + \frac{x^9}{216} \right]_0^1 = 1.000 - 0.100 + 0.005 = 0.905$$ ■

EXERCISES FOR SECTION 16.3

In Exercises 1–12, calculate the value of the given expression by using the first three terms of its Maclaurin series expansion.

1. $\cos 0.1$
2. $\tan 0.5$
3. $\sin 5°$
4. $\cos 10°$
5. $e^{0.2}$
6. $\ln 1.2$
7. $\ln 0.9$
8. $e^{1.1}$
9. $\text{Arctan } 0.1$
10. $\sin 0.3$
11. $\sec^2 0.1$
12. $\sin(0.1)^2$

In Exercises 13–18, evaluate the indicated definite integrals by using three terms of the Maclaurin series expansion of the integrand.

13. $\displaystyle\int_0^1 \sin x^2 \, dx$
14. $\displaystyle\int_0^1 \ln \cos x \, dx$
15. $\displaystyle\int_0^1 e^{x^2} \, dx$
16. $\displaystyle\int_0^1 \text{Arctan } x \, dx$
17. $\displaystyle\int_0^1 \tan x^2 \, dx$
18. $\displaystyle\int_0^{0.3} e^{4x^2} \, dx$

19. Maclaurin series are particularly useful in working with digital computers. By using Maclaurin series, a programmer can express a trigonometric function as a series of algebraic terms that can be evaluated by the machine. Write the first few terms of the Maclaurin series for $f(x) = \sin(\sin x)$.

20. Evaluate the function in Exercise 19 at $x = \frac{1}{2}$. Use three terms of the series.

21. The displacement of a simple harmonic oscillator varies according to $S = \sin 2\pi t$. Write the first three terms of a series that could be used to evaluate this function on a digital computer.

22. What is the displacement of the oscillator in Exercise 21 when $t = 0.05$? Use the first two terms of the series expansion for this purpose.

16.4 TAYLOR SERIES

When we use a truncated Maclaurin series to estimate values of $f(x)$, the accuracy is dependent on the value of x. Generally, the larger the value of x, the more terms will be required for the same accuracy. For instance, when $x = 0.5$, three terms of the Maclaurin sine series are needed to obtain an answer accurate to five decimal places; five terms of the expansion are needed for the same accuracy when we compute $\sin 1.0$.

For large values of x, the Maclaurin series may be useless for computational purposes since too many terms might be required to attain reasonable accuracy. To circumvent this problem, we use an expansion of a function about the point $x = a$. This type expansion, which is called a **Taylor series**, has the form

$$f(x) = b_0 + b_1(x - a) + b_2(x - a)^2 + b_3(x - a)^3 + \cdots \qquad (16.4)$$

where a is some constant and $b_0, b_1, b_2, \ldots$ are the Taylor coefficients to be determined.

Assuming that (16.4) may be differentiated term by term any number of times, we get the following equations:

$$f'(x) = b_1 + 2b_2(x - a) + 3b_3(x - a)^2 + 4b_4(x - a)^3 + \cdots$$
$$f''(x) = 2b_2 + 3 \cdot 2b_3(x - a) + 4 \cdot 3b_4(x - a)^2 + \cdots$$
$$f'''(x) = 3 \cdot 2b_3 + 4 \cdot 3 \cdot 2b_4(x - a) + \cdots$$

and so forth. The coefficients $b_0, b_1, b_2, \cdots$ can now be evaluated by letting $x = a$ in these equations. When $x = a$, any term containing $x - a$ as a factor will be zero. The Taylor coefficients are then

$$b_0 = f(a) \qquad b_1 = f'(a)$$
$$b_2 = \frac{f''(a)}{2!} \qquad b_3 = \frac{f'''(a)}{3!}$$

and so on. Therefore, if $f(x)$ can be represented by a power series such as (16.4), that series will have the following form.

FORMULA

Taylor Series Expansion of $f(x)$

$$f(x) = \sum_{n=0}^{\infty} \frac{f^{(n)}(a)}{n!} (x - a)^n \qquad (16.5)$$

Note that the Taylor series simplifies to the Maclaurin series when $a = 0$.

EXAMPLE 1   ■ **Problem**

Write the Taylor series expansion of $\sin x$ about the point $a = \dfrac{\pi}{3}$.

■ **Solution**

In this problem, we have

$$f(x) = \sin x \qquad f\left(\frac{\pi}{3}\right) = \frac{\sqrt{3}}{2}$$

$$f'(x) = \cos x \qquad f'\left(\frac{\pi}{3}\right) = \frac{1}{2}$$

$$f''(x) = -\sin x \qquad f''\left(\frac{\pi}{3}\right) = -\frac{\sqrt{3}}{2}$$

$$f'''(x) = -\cos x \qquad f'''\left(\frac{\pi}{3}\right) = -\frac{1}{2}$$

Substituting the values into (16.5), we get

$$\sin x = \frac{\sqrt{3}}{2} + \frac{1}{2}\left(x - \frac{\pi}{3}\right) + \frac{(-\sqrt{3}/2)[x - (\pi/3)]^2}{2!}$$

$$+ \frac{(-1/2)[x - (\pi/3)]^3}{3!} + \cdots$$

$$= \frac{\sqrt{3}}{2} + \frac{1}{2}\left(x - \frac{\pi}{3}\right) - \frac{\sqrt{3}}{4}\left(x - \frac{\pi}{3}\right)^2 - \frac{1}{12}\left(x - \frac{\pi}{3}\right)^3 + \cdots \qquad \blacksquare$$

EXAMPLE 2   ■ **Problem**

Evaluate sin 33° by use of the Taylor series expansion of sin x about the point $a = \frac{\pi}{6}$.

■ **Solution**

We choose to expand the Taylor series abut $a = \pi/6$ (that is, 30°), because 30° is the closest angle to 33° for which we know the values of sine and cosine without using tables. Thus,

$$f(x) = \sin x \qquad f\left(\frac{\pi}{6}\right) = \frac{1}{2}$$

$$f'(x) = \cos x \qquad f'\left(\frac{\pi}{6}\right) = \frac{\sqrt{3}}{2}$$

$$f''(x) = -\sin x \qquad f''\left(\frac{\pi}{6}\right) = -\frac{1}{2}$$

and

$$\sin x = \frac{1}{2} + \frac{\sqrt{3}}{2}\left(x - \frac{\pi}{6}\right) - \frac{1}{4}\left(x - \frac{\pi}{6}\right)^2 - \cdots$$

To find the value of sin 33°, we convert 33° into radians and substitute this value into the Taylor series. Thus,

$$x = 33° = 33 \cdot \frac{\pi}{180} = \frac{11\pi}{60} \text{ rad}$$

Using the first three terms of the Taylor series, we obtain

$$\sin 33° = \frac{1}{2} + \frac{\sqrt{3}}{2}\left(\frac{11\pi}{60} - \frac{\pi}{6}\right) - \frac{1}{4}\left(\frac{11\pi}{60} - \frac{\pi}{6}\right)^2 = \frac{1}{2} + \frac{\sqrt{3}}{2}\left(\frac{\pi}{60}\right) - \frac{1}{4}\left(\frac{\pi}{60}\right)^2$$

$$\approx 0.50000 + (0.86603)(0.05236) - (0.25000)(0.05236)^2 = 0.5446$$

This answer is accurate to the indicated number of decimal places.   ■

EXAMPLE 3   ■ **Problem**

Use the Taylor series derived in Example 2 to evaluate sin 27°.

■ **Solution**

In this problem,

$$x = 27° = 27 \cdot \frac{\pi}{180} = \frac{3\pi}{20} \text{ rad}$$

so

$$\left(x - \frac{\pi}{6}\right) = \left(\frac{3\pi}{20} - \frac{\pi}{6}\right) = -\frac{\pi}{60} \approx -0.05236$$

Therefore, the value of sin 27° to four decimal places is given by

$$\sin 27° = 0.50000 - (0.86603)(0.05236) - (0.25000)(0.05236)^2 = 0.4540$$

   ■

EXERCISES FOR SECTION 16.4

In Exercises 1–16, find the Taylor series expansion of the given function about the indicated point.

1. $\cos x, a = \pi/3$

2. $\cos x, a = 30°$

3. $\sin x, a = 45°$

4. $\tan x, a = \pi/4$

5. $e^x, a = 1$

6. $\ln x, a = 1$

7. $\dfrac{1}{1 - x}, a = 2$

8. $\dfrac{1}{x^2}, a = 1$

9. $x^2, a = 1$

10. $2x + 3, a = -2$

11. $x^2 - 8x + 2, a = 1$

12. $\dfrac{1}{3x + 1}, a = 2$

13. $\dfrac{1}{2x + 5}, a = 1$

14. $x^2 + 2x + 1, a = -1$

15. $\dfrac{1}{x(x - 1)}, a = -1$

16. $\dfrac{1}{(x + 1)(x + 2)}, a = 1$

In the remaining exercises, evaluate the given expressions by using Taylor series expansions found in Exercises 1–5. Use three terms of the series.

17. $\cos 55°$ **18.** $\sin 48°$

19. $\tan 42°$ **20.** $\cos 62°$

21. $e^{1.01}$ **22.** $e^{0.98}$

■■■■■ 16.5 FOURIER SERIES

Another method of representing functions is by a trigonometric series of the form

$$f(x) = a_0 + \sum_{k=1}^{\infty} a_k \cos kx + \sum_{k=1}^{\infty} b_k \sin kx \tag{16.6}$$

If the coefficients are properly chosen, this series is called the **Fourier series** of $f(x)$ valid on the interval $(-\pi, \pi)$. More generally, the Fourier representation of $f(x)$ on $(-L, L)$ is written as follows:

FORMULA

> **Fourier Series Expansion of $f(x)$**
>
> $$f(x) = a_0 + \sum_{k=1}^{\infty} a_k \cos \frac{k\pi}{L} x + \sum_{k=1}^{\infty} b_k \sin \frac{k\pi}{L} x \tag{16.7}$$

The class of functions that can be represented by a Fourier series includes almost any that you are likely to encounter in applications. In particular, Fourier series can be used to represent functions even when their graphs have sharp corners or even finite discontinuities. This feature is *not* true for Taylor or Maclaurin series.

The Fourier series of a function converges to the function at every point on the interval $(-L, L)$ where the function is continuous. At finite discontinuities, the series converges to a value equal to the *average* of the two extremes of the discontinuity. Outside $(-L, L)$, the Fourier series converges to the periodic extension. Therefore, the Fourier series is especially useful as a representation of periodic functions.

The analysis of many mechanical and electrical systems requires the use of periodic waves that are not simple sinusoids. Figure 16.1 depicts a few of the commonly used waves. Each wave can be conveniently represented by the infinite trigonometric Fourier series. In this case, each Fourier coefficient is called a **harmonic** and thus gives a measure of how much each sine and cosine wave contributes to the total function. We next give the formulas for the Fourier coefficients on the general interval $(-L, L)$. In this way, we can expand any function with period $2L$.

FORMULA

Fourier Coefficients   If $f(x)$ is to be represented on $(-L, L)$ by its Fourier series, then we choose the Fourier coefficients as follows:

$$a_0 = \frac{1}{2L} \int_{-L}^{L} f(x)\, dx$$

$$a_k = \frac{1}{L} \int_{-L}^{L} f(x) \cos \frac{k\pi}{L} x\, dx$$

$$b_k = \frac{1}{L} \int_{-L}^{L} f(x) \sin \frac{k\pi}{L} x\, dx$$

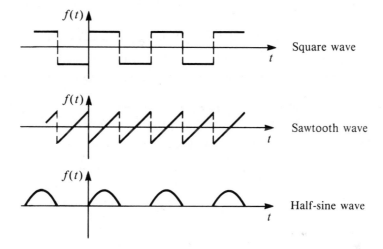

Square wave

Sawtooth wave

Half-sine wave

■ **Figure 16.1**

EXAMPLE 1   ■ **Problem**

Find the Fourier series expression of $f(x) = x$, $-\pi < x < \pi$, and periodic with period 2π. See Figure 16.2.

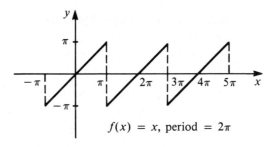

$f(x) = x$, period $= 2\pi$

■ **Figure 16.2**

■ **Solution**

Using the formulas for the Fourier coefficients with $L = \pi$, we have

$$a_0 = \frac{1}{2\pi} \int_{-\pi}^{\pi} x \, dx = \frac{1}{2\pi} \left[\frac{x^2}{2} \right]_{-\pi}^{\pi} = 0$$

Also,

$$a_k = \frac{1}{\pi} \int_{-\pi}^{\pi} x \cos kx \, dx$$

Using integration by parts, we obtain

$$a_k = \frac{1}{\pi} \left[\frac{x \sin kx}{k} + \frac{1}{k^2} \cos kx \right]_{-\pi}^{\pi}$$

$$= \frac{1}{\pi} \left[\frac{\pi \sin k\pi}{k} + \frac{1}{k^2} \cos k\pi + \frac{\pi \sin(-k\pi)}{k} - \frac{1}{k^2} \cos(-k\pi) \right] = 0$$

The quantity in the brackets is zero because $\cos(-k\pi) = \cos k\pi$ and $\sin(-k\pi) = -\sin k\pi$.

Finally, the value of b_k is given by

$$b_k = \frac{1}{\pi} \int_{-\pi}^{\pi} x \sin kx \, dx$$

Using integration by parts, we get

$$b_k = \frac{1}{\pi} \left[-\frac{x \cos kx}{k} + \frac{1}{k^2} \sin kx \right]_{-\pi}^{\pi}$$

$$= \frac{1}{\pi} \left[-\frac{\pi \cos k\pi}{k} + \frac{1}{k^2} \sin k\pi + \frac{(-\pi) \cos(-k\pi)}{k} - \frac{1}{k^2} \sin(-k\pi) \right]$$

$$= -\frac{2 \cos k\pi}{k} = (-1)^{k+1} \left[\frac{2}{k} \right]$$

The sign of b_k is alternately positive and negative because $\cos k\pi = 1$ when k is even, and $\cos k\pi = -1$ when k is odd.

Therefore, the Fourier series for the given function is

$$f(x) = 2 \sum_{k=1}^{\infty} \frac{(-1)^{k+1}}{k} \sin kx$$

$$= 2 \sin x - \sin 2x + \frac{2}{3} \sin 3x - \frac{1}{2} \sin 4x + \cdots \qquad ■$$

Using this series, other Fourier series may be obtained without doing all the computations.

EXAMPLE 2   ▪ **Problem**

Find the Fourier series of the function $g(x)$ whose graph is shown in Figure 16.3.

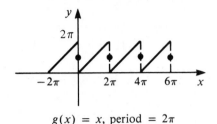

▪ **Figure 16.3** $g(x) = x$, period $= 2\pi$

▪ **Solution**

Note that $g(x)$ and $f(x)$ of Example 1 are related by

$$g(x) = \pi + f(x - \pi)$$

Hence, using the series for $f(x)$ with x replaced by $x - \pi$, the Fourier series of $g(x)$ is

$$g(x) = \pi + 2 \sum_{k=1}^{\infty} \frac{(-1)^{k+1}}{k} \sin k(x - \pi) = \pi + 2 \sum_{k=1}^{\infty} \frac{(-1)^{k+1}(-1)^k \sin kx}{k}$$

$$= \pi - 2 \sum_{k=1}^{\infty} \frac{\sin kx}{k} = \pi - 2 \sin x - \sin 2x - \frac{2}{3} \sin 3x - \cdots$$

Figure 16.3 shows the function to which the Fourier series converges. At the endpoint of each interval, the series converges to the average of the two extremes, that is, to $\frac{1}{2}[2\pi + 0] = \pi$.   ▪

EXAMPLE 3   ▪ **Problem**

Find the Fourier series for the square wave

$$f(x) = \begin{cases} -1, & \text{if } -1 \le x < 0 \\ 1, & \text{if } 0 < x < 1 \end{cases}$$

and periodic with period 2. See Figure 16.4.

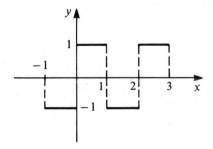

▪ **Figure 16.4**

■ **Solution**

Figure 16.4 shows the function to which the series converges. At the points of discontinuity, the average of the extremes is zero. Using the formulas for a_k and b_k with $L = 1$, we have

$$a_0 = \frac{1}{2} \int_{-1}^{0} (-1) \, dx + \frac{1}{2} \int_{0}^{1} (1) \, dx = 0$$

$$a_k = \int_{-1}^{0} (-1) \cos k\pi x \, dx + \int_{0}^{1} (1) \cos k\pi x \, dx$$

$$= \left[- \frac{\sin k\pi x}{k\pi} \right]_{-1}^{0} + \left[\frac{\sin k\pi x}{k\pi} \right]_{0}^{1} = 0$$

$$b_k = \int_{-1}^{0} (-1) \sin k\pi x \, dx + \int_{0}^{1} (1) \sin k\pi x \, dx$$

$$= \frac{1}{k\pi} \left(\left[\cos k\pi x \right]_{-1}^{0} - \left[\cos k\pi x \right]_{0}^{1} \right)$$

$$= \frac{1}{k\pi}(1 - \cos k\pi - \cos k\pi + 1) = \frac{2}{k\pi} (1 - \cos k\pi)$$

Therefore,

$$b_k = \frac{4}{k\pi}, \qquad \text{if } k \text{ is odd}$$

$$b_k = 0, \qquad \text{if } k \text{ is even}$$

We see that the Fourier series will contain only odd harmonics, as follows:

$$f(x) = \frac{4}{\pi} \sum_{k=0}^{\infty} \frac{\sin(2k + 1)\pi x}{2k + 1}$$

$$= \frac{4}{\pi} \left(\sin \pi x + \frac{1}{3} \sin 3\pi x + \frac{1}{5} \sin 5\pi x + \cdots \right)$$

■

Comment: Figure 16.5 shows how the first three partial sums of the Fourier series of the square wave in Example 3 approximates the square wave. Notice that around the points of discontinuity, the approximating curve tends to *overshoot*, a tendency known as *Gibbs' phenomenon.*

In passing, we note that the function in Example 3 might be more easily defined on some other interval, say $[0, 2L]$, rather than $[-L, L]$. In such a case, merely determine (from the periodicity conditon) what the definition of the function is on $[-L, 0]$ and proceed as before.

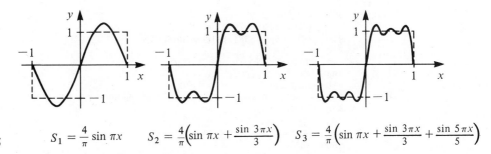

■ **Figure 16.5** $S_1 = \dfrac{4}{\pi}\sin \pi x$ $S_2 = \dfrac{4}{\pi}\left(\sin \pi x + \dfrac{\sin 3\pi x}{3}\right)$ $S_3 = \dfrac{4}{\pi}\left(\sin \pi x + \dfrac{\sin 3\pi x}{3} + \dfrac{\sin 5\pi x}{5}\right)$

Use the following BASIC program to draw the Fourier approximation of the square wave in Example 3 for $n = 1$ to $n = 10$.

```
10   HOME
20   REM  FOURIER APPROXIMATION TO SQUARE WAVE
30   REM  THIS PROGRAM WORKS BEST WITH A COLOR MONOTOR
40   PRINT "THIS PROGRAM GRAPHS THE SQUARE WAVE"
50   PRINT "AND THE FOURIER APPROXIMATION, NAMELY"
60   PRINT "THE SUM OF 4*SIN((2K+1)X)/(2K+1)"
70   PRINT "FOR K = 0 TO N, FOR A VALUE OF N"
80   PRINT "SUPPLIED BY THE USER."
90   PRINT
100  PRINT "ENTER A VALUE FOR N"
110  PRINT "VALUES BETWEEN 1 & 10 ARE BEST"
120  INPUT N
130  HGR
140  HCOLOR= 2
150  REM  AXES
160  HPLOT 1,80 TO 239,80
170  HPLOT 120,0 TO 120,160
180  HCOLOR= 5
190  REM  SQUARE WAVE
200  HPLOT 0,120 TO 120,120
210  HPLOT 120,40 TO 240,40
220  REM  PLOT THE APPROXIMATION
230  HCOLOR= 1
240  PI = 3.14159
250  FOR X1 = 1 TO 239
260  X = 2 * PI * (X1 - 120) / 239.5
270  Y = 0
280  FOR K = 0 TO N
290  Y = Y +  SIN ((2 * K + 1) * X) / (2 * K + 1)
300  NEXT K
310  Y = 4 * Y / PI
320  Y1 =  INT (80 - 40 * Y)
330  IF X1 <  > 1 THEN 360
340  HPLOT X1,Y1
350  GOTO 370
360  HPLOT  TO X1,Y1
370  NEXT X1
380  REM  RE-INSERT Y-AXIS
```

```
390   HCOLOR= 2
400   HPLOT 120,0 TO 120,160
410   REM  RE-INSERT SQUARE WAVE
420   HCOLOR= 5
430   HPLOT 0,120 TO 120,120
440   HPLOT 120,40 TO 239,40
450   VTAB 21
460   PRINT "DO YOU WISH TO RUN THE PROGRAM AGAIN? Y/N"
470   INPUT A$
480   TEXT : HOME
490   IF A$ = "Y" THEN 90
500   END
```

EXAMPLE 4   ■ **Problem**

In electronics, a half-wave rectifier is a device that allows electric current to flow in only one direction. Consequently, if a simple alternating current is applied to the rectifier, current will flow during only half of the cycle. Find the Fourier series of the half-wave rectification of $i = \sin t$, which is depicted in Figure 16.6. This current may be expressed mathematically as

$$i = \begin{cases} \sin t, & \text{if } 0 \le t < \pi \\ 0, & \text{if } \pi \le t \le 2\pi \end{cases}$$

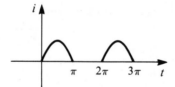

■ **Figure 16.6**

■ **Solution**

We have

$$a_0 = \frac{1}{2\pi} \int_0^\pi \sin t \, dt = \frac{1}{2\pi} \left[-\cos t \right]_0^\pi$$

$$= \frac{1}{2\pi}(1 + 1) = \frac{1}{\pi}$$

$$a_n = \frac{1}{\pi} \int_0^\pi \sin t \cos nt \, dt$$

$$= -\frac{1}{2\pi} \left[\frac{\cos(1 - n)t}{1 - n} + \frac{\cos(1 + n)t}{1 + n} \right]_0^\pi \qquad \text{(Table of Common Integrals;}$$
$$\text{Formula 45)}$$

$$= -\frac{1}{2\pi} \left[\frac{\cos(1 - n)\pi}{1 - n} + \frac{\cos(1 + n)\pi}{1 + n} - \frac{1}{1 - n} - \frac{1}{1 + n} \right]$$

This formula is valid for all n except $n = 1$. For $n > 1$, the value of a_n depends on whether n is odd or even. When n is *odd*,

$$a_n = -\frac{1}{2\pi}\left(\frac{1}{1-n} + \frac{1}{1+n} - \frac{1}{1-n} - \frac{1}{1+n}\right) = 0$$

When n is *even*,

$$a_n = -\frac{1}{2\pi}\left(-\frac{1}{1-n} - \frac{1}{1+n} - \frac{1}{1-n} - \frac{1}{1+n}\right)$$

$$= \frac{1}{\pi}\left(\frac{1}{1-n} + \frac{1}{1+n}\right)$$

$$= \frac{2}{\pi(1-n^2)}$$

When $n = 1$, the integral becomes

$$a_1 = \frac{1}{\pi}\int_0^\pi \sin t \cos t \, dt = \frac{1}{\pi}\left[\frac{1}{2}\sin^2 t\right]_0^\pi = 0$$

Finally, the value of b_n is found by evaluating

$$b_n = \frac{1}{\pi}\int_0^\pi \sin t \sin nt \, dt$$

For $n > 1$,

$$b_n = \frac{1}{2\pi}\left[\frac{\sin(1-n)t}{1-n} - \frac{\sin(1+n)t}{1+n}\right]_0^\pi \qquad \text{(Table of Common Integrals;}\\ \text{Formula 44)}$$

$$= \frac{1}{2\pi}\left[\frac{\sin(1-n)\pi}{1-n} - \frac{\sin(1+n)\pi}{1+n}\right] = 0$$

For the separate case $n = 1$, the integral becomes

$$b_1 = \frac{1}{\pi}\int_0^\pi \sin t \sin t \, dt = \frac{1}{\pi}\int_0^\pi \sin^2 t \, dt$$

$$= \frac{1}{2\pi}\left[t - \frac{\cos 2t}{2}\right]_0^\pi = \frac{1}{2}$$

Using these values of a_0, a_n, and b_n, we get

$$i = \frac{1}{\pi} + \frac{1}{2}\sin t - \frac{2}{\pi}\left(\frac{1}{3}\cos 2t + \frac{1}{15}\cos 4t + \cdots\right)$$

as the Fourier series of the half-wave rectification of $\sin t$.   ■

EXERCISES FOR SECTION 16.5

In Exercises 1–16, sketch several periods of the given function and find its Fourier series expansion.

1. $f(x) = 2x, \ -\pi \le x < \pi$

2. $f(x) = \frac{1}{2}x^2, \ -\pi \le x \le \pi$

3. $f(x) = \frac{1}{4}x, \ -1 \le x \le 1$

4. $f(x) = \sin\frac{1}{2}x, \ -\pi \le x \le \pi$

5. $f(x) = x^2, -1 \le x \le 1$

6. $f(x) = x + \pi, -4 \le x \le 4$

7. $f(x) = \begin{cases} 0, & \text{if } -2 \le x < 0 \\ 1, & \text{if } 0 \le x < 2 \end{cases}$

8. $F(s) = \begin{cases} 0, & \text{if } -\pi \le s < 0 \\ 1, & \text{if } 0 \le s < \dfrac{\pi}{2} \\ 2, & \text{if } \pi/2 \le s < \pi \end{cases}$

9. $g(x) = \begin{cases} -1, & \text{if } -\pi \le x < 0 \\ 3, & \text{if } 0 \le x < \pi \end{cases}$

10. $v = \begin{cases} 0, & \text{if } -\pi \le t < 0 \\ t, & \text{if } 0 \le t < \pi \end{cases}$

11. $f(x) = \sin x, -\dfrac{\pi}{2} \le x \le \dfrac{\pi}{2}$

12. $i = \begin{cases} 1, & \text{if } -\pi \le t < 0 \\ 2t, & \text{if } 0 \le t < \pi \end{cases}$

13. $s(x) = \begin{cases} 0, & \text{if } -\pi \le x < -\frac{1}{2}\pi \\ -x, & \text{if } -\frac{1}{2}\pi \le x < 0 \\ x, & \text{if } 0 \le x < \frac{1}{2}\pi \\ 0, & \text{if } \frac{1}{2}\pi \le x \le \pi \end{cases}$

14. $p(x) = \begin{cases} x, & \text{if } -\pi \le x < 0 \\ \pi, & \text{if } 0 \le x < \pi \end{cases}$

15. $h(x) = \begin{cases} 0, & \text{if } -\pi \le x < -\frac{1}{2}\pi \\ x, & \text{if } -\frac{1}{2}\pi \le x < \frac{1}{2}\pi \\ 0, & \text{if } \frac{1}{2}\pi \le x \le \pi \end{cases}$

16. $f(x) = \cos^2 x, -\pi \le x \le \pi$

17. An engineer wishes to use the exponential wave $v = e^t$, $0 \le t \le 2\pi$, in the frequency control circuit of a radar. Find the Fourier series that represents this wave.

18. The forcing function being applied to a particular spring-mass system is given by $F = t^2 - t$ in the interval $0 \le t \le 2\pi$. Derive the Fourier series expansion for this function.

19. In fatigue-testing a specimen, the deforming force was applied alternately at two different levels. The applied force is given by

$$f(x) = \begin{cases} 1, & \text{if } 0 \le t \le \pi \\ 2, & \text{if } \pi \le t < 2\pi \end{cases}$$

What is the Fourier series that can be used to synthesize the necessary force?

20. The output of a full-wave rectifier is shown in the accompanying figure when the input is a simple sinusoidal current. Find the Fourier series of the output if one period is described by

$$i = \begin{cases} \sin t, & \text{if } 0 \le t < \pi \\ -\sin t, & \text{if } \pi \le t \le 2\pi \end{cases}$$

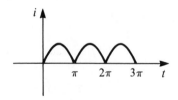

REVIEW EXERCISES FOR CHAPTER 16

In Exercises 1–10, find the first three nonzero terms of the Maclaurin series *directly*, that is, by using the formulas for the Maclaurin coefficients.

1. $(x + 1)^{1/3}$

2. $(x + 4)^{1/2}$

3. $\ln \sin(x + \frac{1}{6}\pi)$

4. $\tan 2x$

5. $\ln(x + 2)$

6. $x^2 + 5x + 1$

7. $\text{Arcsin } 3x$

8. $x^3 - 2$

9. $e^{\cos x}$

10. $\dfrac{x + 3}{x - 1}$

In Exercises 11–15, find the first three nonzero terms of the Taylor series expansion about the indicated point.

11. $x^2 + 5x + 1, a = 2$

12. $\cos x, a = \pi/6$

13. $\tan 2x, a = \pi/8$

14. $x^2 + 3x + 5, a = 1$

15. $\dfrac{1}{3x - 5}, a = 1$

Use known Maclaurin series to obtain series for Exercises 16–25.

16. $\sin \frac{1}{2}x$

17. $\cos 3x$

18. $\sin x^2$

19. $\dfrac{1}{1 + 3x^2}$

20. $\dfrac{2 + 3x}{x + 5}$

21. $\dfrac{x - 2}{(x + 1)(x - 3)}$

22. $\dfrac{1}{x^2 + 4x + 3}$

23. $\ln(3 + x)$

24. e^{2x}

25. e^{x^2}

In Exercises 26–30, approximate the given quantity by using the first three terms of an appropriate expansion.

26. $\sin 0.2$

27. $\ln 0.98$

28. $\cos 91°$

29. $\text{Arctan } 1.2$

30. $e^{1.1}$

In Exercises 31–35, find the Fourier series of the given function.

31. $f(x) = \sin^2 x$ on $(-\pi, \pi)$

32. $f(x) = |x|$ on $(-\pi, \pi)$

33. $f(x) = 1 + \sin x$ on $(-\pi, \pi)$

34. $f(x) = 3 + 2\cos 3x - \sin 20x$ on $(-\pi, \pi)$

35. $f(x) = 1 - |x|$ on $(-1, 1)$

36. Approximate the value of

$$\int_0^1 \frac{e^x - 1}{x} \, dx$$

by using a suitable Maclaurin expansion truncated at three nonzero terms.

37. Find the Fourier series expansion of

$$F(x) = \begin{cases} -1, & \text{if } -\pi \le x \le 0 \\ 1, & \text{if } 0 \le x \le \pi \end{cases}$$

Evaluate the sum of the first five terms at $x = \pi$.

APPENDIX I:
Approximate Numbers

SIGNIFICANT DIGITS

If you count each card in a deck of 52 playing cards, you can say there are exactly 52 cards in the deck. However, if, when using a meterstick, you measure the distance between two points to be 3.5 cm, you realize that 3.5 is only an approximation to the precise answer and that the accuracy of the measurement depends on the accuracy of the measuring device being used. Most calculations in the physical sciences involve measurements that are accurate to some specified number of digits. The digits of a number that are known to be accurate are called **significant digits**. Zeros that are required to locate the decimal point are not considered to be significant digits.

EXAMPLE 1   The number 5.793 has four significant digits. The number 20.781 has five significant digits. The number 0.000059 has two significant digits. The number 0.08300 has four significant digits.   ■

EXAMPLE 2   ■ **Problem**

How many significant digits does 9480 have?

■ **Solution**

A number like 9480 is difficult to categorize unless we know something about the number. Thus, it could represent the exact number of cards in a computer program. Or it could represent a measurement accurate to either three or four significant digits. Sometimes, we use scientific notation for numbers like this one to avoid confusion. Thus,

$$9.48 \times 10^3$$

is used to indicate three significant digits, while

$$9.480 \times 10^3$$

indicates four significant digits.   ■

Note that the last significant digit of an approximate number is not completely accurate. For instance, if you measure a length of wire to be 2.56 cm, you realize that the length could be anywhere between 2.555 and 2.565 cm since it has been obtained by estimation.

The two numbers 28,500 and 0.285 are both **accurate** to three significant digits. However, 0.285 is a more precise measurement than 28,500. The decimal position of the last significant digit of a number determines its **precision**. Both accuracy and precision are important concepts when making computations with approximate numbers.

EXAMPLE 3

The measurements 395 cm and 0.0712 cm are both accurate to three significant digits. But 395 cm is precise to the nearest 1 cm, and 0.0712 cm is precise to the nearest 0.0001 cm.

The number 497.3 is more accurate than 0.025 but less precise. ∎

The process of writing a given number to a specified number of significant digits is called **rounding off**. There are several popular schemes for rounding off numbers. The following method is used in many calculators and computers.

PROCEDURE

Rounding Off

1. If the last digit is less than 5, drop the digit and use the remaining digits. For example, 8.134 rounds off as 8.13 to three significant digits.

2. If the last digit is 5 or greater, drop the digit and increase the last remaining digit by 1. For example, 0.0225 rounds off as 0.023 to two significant digits.

Another method used to reduce the digits in an approximate number is **truncation**. In this scheme, the unwanted digits are simply dropped or truncated. Thus, the numbers 23.157 and 23.154 both truncate to the four-digit number 23.15. Likewise, under the truncation scheme, $\frac{2}{3}$ to seven decimal places is carried as 0.6666666, not as 0.6666667. Check your calculator to see if it rounds off or truncates.

EXAMPLE 4

∎ **Problem**

Round off each of the following numbers to three significant digits:

a. 18.89

b. 0.0003725

c. 99430

d. 4.996

∎ **Solution**

a. 18.89 is rounded off to 18.9.

b. 0.0003725 is rounded off to 0.000373.

c. 99430 is rounded off to 99400.

d. 4.996 is rounded off to 5.00.   ∎

OPERATIONS WITH APPROXIMATE NUMBERS

There is a great temptation, especially when you are using a calculator, to write the answer to arithmetic calculations to as many digits as the calculator will display. In doing so, you make the answer seem more accurate or precise than it really is. For instance, writing

$$8.4 \times 12.137 = 101.9508$$

implies that the product is accurate to seven significant digits, when the numbers being multiplied are only accurate to two and five significant digits, respectively. To avoid this problem, we adopt the following conventions when performing arithmetic operations on approximate numbers.

PROCEDURE

> **Precision and Accuracy**
>
> **1.** When adding or subtracting approximate numbers, express the result with the precision of the least precise number. For example,
>
> $$0.74 + 0.0515 - 0.3329 = 0.4586$$
>
> is rounded off to 0.46.
>
> **2.** When multiplying or dividing approximate numbers, express the result with the accuracy of the least accurate number. For example,
>
> $$(1.93)(13.77) = 26.5761$$
>
> is rounded off to 26.6.
>
> **3.** When finding the root of an approximate number, express the root with the same accuracy as the number. For example,
>
> $$\sqrt{29.14} = 5.398$$

EXAMPLE 5 **a.** The answer to $R = 1.9(63.21) + 4.9072$ should have two significant digits.

b. The answer to $3.005 \sqrt{2}$ should have four significant digits.

c. The answer to $y = 3 - \sqrt{29}$, where 3 and $\sqrt{29}$ are exact numbers can be written to as many significant digits as desired since there are no approximate numbers being used.   ∎

EXAMPLE 6   ■ **Problem**

The length of a rectangle is measured with a meterstick to be 95.7 cm, and the width is measured with a micrometer to be 8.433 cm. What is the area of the rectangle?

■ **Solution**

The area is given by

$$A = 8.433 \times 95.7 = 807.0381 \text{ cm}^2$$

Since the length has only three significant digits, the area must be rounded off to three significant digits. Therefore, $A = 807 \text{ cm}^2$.   ■

EXERCISES

In Exercises 1–10, indicate the number of significant digits in the given numbers.

1. 3.37

2. 2.002

3. 812.0

4. 6161

5. 0.03

6. 0.000215

7. 0.40

8. 57.001

9. 500.0

10. 0.06180

In Exercises 11–20, round off the given number to three significant digits.

11. 9818

12. 72267

13. 54.745

14. 1.002

15. 0.06583

16. 2435

17. 39.75

18. 0.4896

19. 0.9997

20. 900,498

In Exercises 21–32, perform the indicated operations, and round off the answer to the appropriate number of significant digits.

21. 23.45(0.91669)

22. 4.7(54.75)

23. 0.5782 + 1.34 + 0.0057

24. 50.68 + 9.666 − 24.059

25. 2.9(3.57 + 10.28) + 25.0

26. 0.20 + 3.86(0.127 − 0.097)

27. $\sqrt{2.4^2 + 0.9^2}$

28. $2.176 \sqrt{3}$

29. $\dfrac{25(0.9297)}{0.0102}$

30. $\dfrac{5.0887(2.20)}{8813}$

31. $\dfrac{0.9917(771.33)}{\sqrt{30}}$

32. $\sqrt{2.14^2 + 3.9^2}$

APPENDIX II:
Table of Common Integrals

TABLE OF COMMON INTEGRALS

(1) $\int \dfrac{x\,dx}{ax + b} = \dfrac{x}{a} - \dfrac{b}{a^2}\ln(ax + b) + C$

(2) $\int \dfrac{x\,dx}{(ax + b)^2} = \dfrac{b}{a^2(ax + b)} + \dfrac{1}{a^2}\ln(ax + b) + C$

(3) $\int x(ax + b)^n\,dx = \dfrac{x(ax + b)^{n+1}}{a(n + 1)} - \dfrac{(ax + b)^{n+2}}{a^2(n + 1)(n + 2)} + C$

(4) $\int \dfrac{dx}{x(ax + b)} = \dfrac{1}{b}\ln\dfrac{x}{ax + b} + C$

(5) $\int \dfrac{dx}{x(ax + b)^2} = \dfrac{1}{b(ax + b)} + \dfrac{1}{b^2}\ln\dfrac{x}{ax + b} + C$

(6) $\int \dfrac{dx}{a^2 - x^2} = \dfrac{1}{2a}\ln\dfrac{a + x}{a - x} + C$

(7) $\int \dfrac{dx}{(ax^2 + b)^2} = \dfrac{x}{2b(ax^2 + b)} + \dfrac{1}{2b}\int\dfrac{dx}{ax^2 + b}$

(8) $\int \dfrac{dx}{x(ax^2 + b)} = \dfrac{1}{2b}\ln\dfrac{x^2}{ax^2 + b} + C$

(9) $\int x\sqrt{ax + b}\,dx = \dfrac{2x}{3a}(ax + b)^{3/2} - \dfrac{4}{15a^2}(ax + b)^{5/2} + C$

(10) $\int \dfrac{x\,dx}{\sqrt{ax + b}} = \dfrac{2x}{a}(ax + b)^{1/2} - \dfrac{4}{3a^2}(ax + b)^{3/2} + C$

(11) $\int \sqrt{a^2 - x^2}\,dx = \tfrac{1}{2}x\sqrt{a^2 - x^2} + \tfrac{1}{2}a^2\,\text{Arcsin}\,\dfrac{x}{a} + C$

(12) $\int \sqrt{x^2 \pm a^2}\,dx = \tfrac{1}{2}x\sqrt{x^2 \pm a^2} \pm \tfrac{1}{2}a^2\ln(x + \sqrt{x^2 \pm a^2}) + C$

(13) $\int \dfrac{dx}{\sqrt{x^2 \pm a^2}} = \ln(x + \sqrt{x^2 \pm a^2}) + C$

(14) $\int \dfrac{dx}{x\sqrt{a^2 \pm x^2}} = \dfrac{1}{a}\ln\dfrac{x}{a + \sqrt{a^2 \pm x^2}} + C$

TABLE (*Continued*)

---

(15) $\displaystyle\int \frac{dx}{x\sqrt{x^2 - a^2}} = -\frac{1}{a} \operatorname{Arcsin} \frac{a}{x} + C$

(16) $\displaystyle\int \frac{\sqrt{x^2 - a^2}}{x} \, dx = \sqrt{x^2 - a^2} + a \operatorname{Arcsin} \frac{a}{x} + C$

(17) $\displaystyle\int x^2\sqrt{a^2 - x^2} \, dx = -\tfrac{1}{4}x(a^2 - x^2)^{3/2} + \tfrac{1}{8}a^2 x\sqrt{a^2 - x^2} + \tfrac{1}{8}a^4 \operatorname{Arcsin} \frac{x}{a} + C$

(18) $\displaystyle\int x^3\sqrt{a^2 - x^2} \, dx = \tfrac{1}{5}(a^2 - x^2)^{5/2} - \tfrac{1}{3}a^2(a^2 - x^2)^{3/2} + C$

(19) $\displaystyle\int x^2\sqrt{x^2 \pm a^2} \, dx = \tfrac{1}{4}x(x^2 \pm a^2)^{3/2} \pm \tfrac{1}{8}a^2 x\sqrt{x^2 \pm a^2}$

$$- \tfrac{1}{8}a^4 \ln(x + \sqrt{x^2 \pm a^2}) + C$$

(20) $\displaystyle\int \frac{x^2 \, dx}{\sqrt{a^2 - x^2}} = -\tfrac{1}{2}x\sqrt{a^2 - x^2} + \tfrac{1}{2}a^2 \operatorname{Arcsin} \frac{x}{a} + C$

(21) $\displaystyle\int \frac{x^3 \, dx}{\sqrt{a^2 - x^2}} = -x^2\sqrt{a^2 - x^2} - \tfrac{2}{3}(a^2 - x^2)^{3/2} + C$

(22) $\displaystyle\int \frac{x^2 \, dx}{\sqrt{x^2 \pm a^2}} = \tfrac{1}{2}x\sqrt{x^2 \pm a^2} \pm \tfrac{1}{2}a^2 \ln(x + \sqrt{x^2 \pm a^2}) + C$

(23) $\displaystyle\int \frac{dx}{\sqrt{2ax - x^2}} = 2 \operatorname{Arcsin} \sqrt{\frac{x}{2a}} + C$

(24) $\displaystyle\int \sin^2 x \, dx = \tfrac{1}{2}x - \tfrac{1}{4}\sin 2x + C$

(25) $\displaystyle\int \cos^2 x \, dx = \tfrac{1}{2}x + \tfrac{1}{4}\sin 2x + C$

(26) $\displaystyle\int \sin^n x \, dx = -\frac{\sin^{n-1} x \cos x}{n} + \frac{n-1}{n} \int \sin^{n-2} x \, dx$

(27) $\displaystyle\int \cos^n x \, dx = \frac{1}{n} \cos^{n-1} x \sin x + \frac{n-1}{n} \int \cos^{n-2} x \, dx$

(28) $\displaystyle\int \cos^m x \sin^n x \, dx = \frac{\cos^{m-1} x \sin^{n+1} x}{m+n} + \frac{m-1}{m+n} \int \cos^{m-2} x \sin^n x \, dx$

(29) $\displaystyle\int \cos^m x \sin^n x \, dx = -\frac{\sin^{n-1} x \cos^{m+1} x}{m+n} + \frac{n-1}{m+n} \int \cos^m x \sin^{n-2} x \, dx$

(30) $\displaystyle\int \tan^2 x \, dx = \tan x - x + C$

TABLE (*Continued*)

(31) $\int \cot^2 x \, dx = -\cot x - x + C$

(32) $\int \tan^n x \, dx = \dfrac{\tan^{n-1} x}{n-1} - \int \tan^{n-2} x \, dx$

(33) $\int \cot^n x \, dx = -\dfrac{\cot^{n-1} x}{n-1} - \int \cot^{n-2} x \, dx$

(34) $\int \sec^3 x \, dx = \frac{1}{2} \sec x \tan x + \frac{1}{2} \ln(\sec x + \tan x) + C$

(35) $\int \csc^3 x \, dx = -\frac{1}{2} \csc x \cot x + \frac{1}{2} \ln(\csc x - \cot x) + C$

(36) $\int \sec^n x \, dx = \dfrac{\tan x \sec^{n-2} x}{n-1} + \dfrac{n-2}{n-1} \int \sec^{n-2} x \, dx$

(37) $\int \csc^n x \, dx = -\dfrac{\cot x \csc^{n-2} x}{n-1} + \dfrac{n-2}{n-1} \int \csc^{n-2} x \, dx$

(38) $\int x \sin x \, dx = \sin x - x \cos x + C$

(39) $\int x \cos x \, dx = \cos x + x \sin x + C$

(40) $\int x^n \sin x \, dx = -x^n \cos x + n \int x^{n-1} \cos x \, dx$

(41) $\int x^n \cos x \, dx = x^n \sin x - n \int x^{n-1} \sin x \, dx$

(42) $\int x \sin^n x \, dx = \dfrac{(\sin^{n-1} x)(\sin x - nx \cos x)}{n^2} + \dfrac{n-1}{n} \int x \sin^{n-2} x \, dx$

(43) $\int x \cos^n x \, dx = \dfrac{(\cos^{n-1} x)(\cos x + nx \sin x)}{n^2} + \dfrac{n-1}{n} \int x \cos^{n-2} x \, dx$

(44) $\int \sin mx \sin nx \, dx = \dfrac{\sin(m-n)x}{2(m-n)} - \dfrac{\sin(m+n)x}{2(m+n)} + C$

(45) $\int \sin mx \cos nx \, dx = -\dfrac{\cos(m-n)x}{2(m-n)} - \dfrac{\cos(m+n)x}{2(m+n)} + C$

(46) $\int \cos mx \cos nx \, dx = \dfrac{\sin(m-n)x}{2(m-n)} + \dfrac{\sin(m+n)x}{2(m+n)} + C$

(47) $\int x e^{ax} \, dx = \dfrac{e^{ax}}{a^2} (ax - 1) + C$

(48) $\int x^2 e^{ax} \, dx = \dfrac{e^{ax}}{a^3} (a^2 x^2 - 2ax + 2) + C$

(49) $\int x^n e^{ax} \, dx = \dfrac{x^n e^{ax}}{a} - \dfrac{n}{a} \int x^{n-1} e^{ax} \, dx$

TABLE (*Concluded*)

---

(50) $\displaystyle\int e^{ax} \sin mx \, dx = \frac{e^{ax}(a \sin mx - m \cos mx)}{m^2 + a^2} + C$

(51) $\displaystyle\int e^{ax} \cos mx \, dx = \frac{e^{ax}(m \sin mx + a \cos mx)}{m^2 + a^2} + C$

(52) $\displaystyle\int \ln x \, dx = x \ln x - x + C$

(53) $\displaystyle\int x^n \ln x \, dx = x^{n+1}\left[\frac{\ln x}{n + 1} - \frac{1}{(n + 1)^2}\right] + C$

GLOSSARY

ABSCISSA The x-coordinate of a point in the rectangular plane.

ABSOLUTE MAXIMUM The maximum value of a function on a given interval.

ABSOLUTE VALUE The magnitude of a real number, denoted by $|a|$. Algebraically, $|a|$ is given by a if a is positive and by $-a$ if a is negative.

ACCELERATION The rate of change of velocity with respect to time ($a = dv/dt$).

AMPERE A unit of electrical current (abbreviated A).

AMPLITUDE The maximum deviation of sine and cosine waves from their mean values.

ANGLE OF ELEVATION The smallest acute angle made by a line of sight to an object above the horizontal.

ANGLE OF INCLINATION The smallest positive angle between a straight line and the positive x-axis.

ANTIDERIVATIVE A function $f(x)$ whose derivative is $f'(x)$. Also called the indefinite integral or integral.

ANTIDIFFERENTIATION The process of finding the function having a given derivative. Also called integration.

AREA UNDER A CURVE The area bounded by a curve and the x-axis over an interval $[a, b]$.

ARITHMETIC SEQUENCE A sequence of numbers that are equally spaced.

ASYMPTOTES Lines that are approached arbitrarily closely by the graph of a function for large values of the variables. *Vertical asymptotes* occur when a function becomes unbounded.

BENDING MOMENT The sum of moments of the external forces on either side of a particular cross section of a beam. Used in analyzing deflections in beams.

BINOMIAL COEFFICIENT The coefficient in the expansion of the binomial $(x + y)^n$, given by

$$C_{n,k} = \frac{n!}{(n - k)!k!}$$

BINOMIAL THEOREM The rule for expanding expressions of the form $(1 + x)^n$.

BOUNDED FUNCTION A function whose range values lie between two finite values.

CALCULUS A word meaning a method of calculation.

CANTILEVER BEAM A beam supported at only one end; for example, a diving board.

CAPACITOR An electrical device for storing electric charge.

CARTESIAN COORDINATE SYSTEM   A coordinate system in which the axes are mutually perpendicular lines

CENTROID OF AN AREA   The point at which all of the area of a plane figure is considered to be concentrated.

CHAIN RULE   A rule stating that the derivative of a function formed by the composition of two elementary functions is equal to the product of the derivatives of the composing functions.

COEFFICIENT   Usually refers to the numerical factor of a variable. For example, in the term $5x^2$, 5 is the coefficient of x^2.

COFUNCTION   Trigonometric functions that have equal values for complementary angles, such as sine and cosine.

COMMON LOGARITHMS   Logarithms to the base 10.

COMPLEMENTARY ANGLES   Two angles whose sum is $90°$ (or $\frac{1}{2}\pi$ radian)

COMPLETING THE SQUARE   The process of adding a constant to an expression of the form $x^2 + ax$ to form a perfect square. The constant to be added is $(\frac{1}{2}a)^2$ to make the perfect square $(x + \frac{1}{2}a)^2$.

COMPLEX NUMBER   A number of the form $a + bj$, where $j = \sqrt{-1}$.

COMPONENTS OF A VECTOR   Projections of a vector onto specified lines such as the x- and y-axes.

CONCAVE DOWN   A portion of a graph that opens downward.

CONCAVE UP   A portion of a graph that opens upward.

CONIC SECTION   A curve formed by cutting a right circular cone with a plane; namely, a circle, an ellipse, a hyperbola, or a parabola.

CONVERGENCE   The existence of a limit of a sequence, series , or improper integral.

COORDINATE   An association of a number with a point. A point in the plane has two coordinates, and a point in three dimensions has three coordinates.

COSINE OF AN ANGLE   The ratio of the x-coordinate of a point in the plane divided by the distance from the origin to the point. In terms of the sides of a right triangle, $\cos\theta = $ adjacent/hypotenuse.

CRITICALLY DAMPED MOTION   Motion characterized by the equation

$$y(t) = e^{-at}(c_1 + c_2 t)$$

CRITICAL VALUES   Values of the variable for which the derivative of the function is zero.

CYCLE   The shortest segment of a graph that includes one period of a periodic function.

DECIBEL   A unit for measuring sound intensity. The measure is $10 \log_{10}(I/I_0)$, where I_0 is the minimum intensity detectable by the human ear.

DEFINITE INTEGRAL   A number that is denoted and defined by

$$\int_a^b f(x)\, dx = \lim_{n \to \infty} \sum_{i=1}^{n} f(x_i)\, \Delta x$$

DEGREE   A unit of measure of an angle equal to $\frac{1}{360}$ of a revolution.

DEPENDENT SYSTEM   A system of equations that has more than one simultaneous solution.

DERIVATIVE   The limit of the difference quotient of $f(x)$ as Δx approaches zero, or

$$f'(x) = \lim_{\Delta x \to 0} \frac{f(x + \Delta x) - f(x)}{\Delta x}$$

DIFFERENCE QUOTIENT   The expression $[f(a + h) - f(a)]/h$. The difference quotient leads to the notion of the derivative in calculus.

DIFFERENTIAL   A small change in a variable, denoted by dx.

DIFFERENTIAL CALCULUS   The study and calculation of derivatives.

DIFFERENTIAL EQUATION   An equation containing a derivative or differential term.

DIRECT PROPORTIONALITY   A situation for two quantities in which one is a constant multiple of the other. Also called direct variation.

DISTANCE   The straight-line measure between two points in the plane. The formula for distance is $d = \sqrt{(x_2 - x_1)^2 + (y_2 - y_1)^2}$.

DIVERGENCE   The failure of a limit of a sequence, series, or improper integral to exist.

DOMAIN   The set of numbers that a variable can represent. When the word *domain* is used with the concept of a function, the implied limitation is to the largest permissible set of real numbers.

DOUBLE INTEGRAL   An integral of the form $\int_a^b \int_c^d f(x, y)\, dx\, dy$.

DYNE   A unit of force (abbreviated dyn).

ELECTRIC CURRENT   The time rate of change of electric charge past a point $(i = dq/dt)$.

ELLIPSE   The set of all points in a plane the sum of whose distance from two fixed points in the plane is constant. A conic section obtained by cutting the cone with a plane not parallel to the base, the axis, or the sides of the cone.

EQUATION   A mathematical statement that two expressions are equal. Equations are usually considered "open" in that they are true for some numbers and false for others.

EVEN FUNCTION   A function for which $f(x) = f(-x)$. The graph of an even function is symmetric with respect to the y-axis.

EXPANDING   A mathematical term referring to performing an indicated product.

EXPONENT   A special mathematical notation to indicate repeated multiplication of the same number. Also called the power of a number.

EXPONENTIAL EQUATION   An equation in which the variable occurs in the exponent.

EXPONENTIAL FUNCTION   A function of the form $y(x) = a^x$. Exponential functions are used to describe electric current, decomposition of uranium, and a wide variety of other physical phenomena.

FACTORIAL   A mathematical abbreviation for the product of integers 1 through n. The symbol $n!$ is read "n factorial."

FARAD   A unit of electrical capacitance (abbreviated F). Usually given in microfarads (10^{-6}).

FOOT-POUND   The unit of work in the British engineering system of units (abbreviated ft-lb). Usually associated with mechanical systems.

FOURIER SERIES   An infinite series representation of $f(x)$ of the form

$$f(x) = a_0 + \sum_{k=1}^{\infty} a_k \cos \frac{k\pi}{L} x + \sum_{k=1}^{\infty} b_k \sin \frac{k\pi}{L} x$$

FREQUENCY   A term used to describe the number of times a periodic function repeats itself in each unit interval of time.

FUNCTION   A rule of correspondence between x and y in which there is exactly one value of y for each value of x.

GEOMETRIC SEQUENCE   A sequence of numbers in which succeeding numbers are always in the same ratio.

GRAPH   The set of points in the plane that correspond to the ordered pairs of a function or a relation.

HALF-LIFE   The time required for a radioactive substance to decay to one-half its original mass.

HARMONIC MOTION   Any motion that can be described by $c_1 \cos bt + c_2 \sin bt$.

HENRY   A unit of electric inductance (abbreviated H).

HORSEPOWER   The unit of power in the British engineering system of units. Usually associated with mechanical systems.

HYPERBOLA   The set of all points in a plane the difference of whose distances from two fixed points in the plane is constant. A conic section obtained by cutting the cone by a plane inclined so that it cuts both nappes.

HYPOTENUSE   The side opposite the right angle in a right triangle.

IDENTITY   An equation that is true for all permissible values of the variable.

IMPLICIT DIFFERENTIATION   The process of finding the derivative of an implicit function.

IMPLICIT FUNCTION   A function stated in the form $g(x, y) = 0$.

IMPROPER INTEGRAL   An integral having either infinite limits or a discontinuity within the interval of integration

INDEFINITE INTEGRAL   A function denoted and defined by

$$F(x) = \int_a^x f(t)\, dt$$

INDUCTANCE COIL   An electrical device that impedes the growth or decay of current.

INTEGRAL CALCULUS   The study and calculation of definite integrals

INTEGRAND   The function being integrated.

INTEGRATION BY PARTS   The use of the formula $\int u\,dv = uv - \int v\,du$.

INTERPOLATION   A method of estimating a value between two given values.

INTERVAL   The set of numbers between two given real numbers a and b. If a and b are not included, the interval is *open*; if a and b are both included, the interval is *closed*.

INVERSE LAPLACE TRANSFORM   The process of finding the function whose Laplace transform is given.

INVERSE PROPORTIONALITY   A situation for two quantities in which one is a constant times the reciprocal of the other. Also called inverse variation.

INVERSE TRIGONOMETRIC FUNCTIONS   The set of functions in which the value of the trigonometric function is specified and the corresponding angle or number must be determined.

ITERATED INTEGRAL   An integral of the form

$$\int_a^b \int_{f(x)}^{g(x)} F(x, y)\,dy\,dx$$

JOULE   The unit of work in the international system of units (abbreviated J).

KILOGRAM   The unit of mass in the international system of units (abbreviated kg).

KINEMATICS   The study of motion without regard to the forces causing the motion.

KINETIC ENERGY   The capacity of a moving object to do work.

LAPLACE TRANSFORM   Denoted and defined by

$$\mathcal{L}\{f(t)\} = \int_0^\infty e^{-st} f(t)\,dt$$

LIMIT   The number approached by $f(x)$ as x approaches some number a or increases without bound.

LIMITS OF INTEGRATION   The extreme points of the interval over which the integration takes place.

LINEAR FUNCTION   A function of the form $y = ax + b$. Also called a linear equation.

LOGARITHMIC FUNCTION   A function described by

$$y = \log_a x$$

This function is the inverse of the exponential function; that is, $y = \log_a x$ means $x = a^y$. The two most important kinds of logarithms are common logarithms (base 10) and natural logarithms (base e).

LOGARITHMIC EQUATIONS   Equations that involve the use of logarithms.

MACH NUMBER   The ratio of the velocity of an object to that of sound.

MACLAURIN SERIES   A power series representation of a function of the form

$$f(x) = \sum_{n=0}^{\infty} \frac{f^{(n)}(0)x^n}{n!}$$

MEAN VALUE OF A FUNCTION   A value given by

$$\bar{f}(x) = \frac{1}{b - a} \int_a^b f(x)\, dx$$

MEAN VALUE THEOREM   A theorem stating that there is a rectangle of width $b - a$ and height $f(x)$ that has the same area as the area under the curve from $x = a$ to $x = b$.

METER   The unit of distance in the international system of units (abbreviated m).

MOMENT OF A FORCE   Another expression for torque.

MOMENT OF AN AREA   The product of the area times the perpendicular distance from the centroid of the area to an axis of rotation.

MOMENT OF INERTIA   The product of area times the square of the distance from its centroid to an axis of rotation.

NATURAL LOGARITHM   Logarithms to the base e. Written $\ln x$.

NORMAL   A line that is perpendicular to another line.

NUMBER LINE   The geometric association of every real number with a point on a line.

NUMBER SYSTEM   Some particular set of numbers, such as the integers or the rational, real, or complex numbers.

ODD FUNCTION   A function for which $f(x) = -f(-x)$. The graph of an odd function is symmetric with respect to the origin.

OHM   A unit of electric resistance (abbreviated Ω).

ORDERED PAIR   A pair of elements in which a distinction is made between a *first* and a *second* element.

ORDINATE   The y-coordinate of a point in the plane.

PARABOLA   A U-shaped curve that is the graph of a quadratic function. The set of all points in a plane equidistant from a fixed point and a fixed line. A conic section obtained from the cone by cutting it with a plane parallel to one element in the side of the cone.

PARALLELOGRAM RULE   The rule for adding vectors.

PARAMETRIC EQUATIONS   Two equations that describe the behavior of x and y in terms of some third parameter t.

PARTIAL DERIVATIVE   The process of finding the derivative of a function of two or more variables.

PARTIAL-FRACTION DECOMPOSITION   A method of decomposing a rational function into a sum of functions each of which has but one factor in the denominator.

PERIOD   The phenomenon of repeating values of a function on a cyclic basis. For example, the sine and cosine functions repeat their values every 2π units.

PHASE SHIFT   A quantity that describes the horizontal shift of a function from its normal state. If a function is shifted to the right, the phase shift is positive; a shift to the left is considered negative.

POINT OF INFLECTION   A point on a graph at which the graph changes its sense of concavity.

POLAR COORDINATES   A system of coordinates utilizing the distance from the origin and the angle of the line segment to locate points in the plane.

POLYNOMIAL   Any expression involving only constant terms or variable terms with positive exponents.

POUND   The unit of force in the British engineering system of units (abbreviated lb).

POWER   The time rate of doing work ($p = dw/dt$).

POWER SERIES   An infinite sum of the form $a_0 + a_1x + a_2x^2 + a_3x^3 + \cdots + a_nx^n + \cdots$.

PRESSURE   The applied force divided by the area of application.

PRINCIPAL VALUES   The values to which the trigonometric functions are limited to make a one-to-one matching of the domain and range elements. For example, the principal values of $\sin x$ are $-\frac{1}{2}\pi \le x \le \frac{1}{2}\pi$.

PROJECTILE   An object moving through the atmosphere without a propelling force, for example, a ball or a bullet.

PYTHAGOREAN RELATION OF TRIGONOMETRY   The relation $\sin^2\theta + \cos^2\theta = 1$.

PYTHAGOREAN THEOREM   For a right triangle, $a^2 + b^2 = c^2$, where c is the hypotenuse.

QUADRATIC EQUATION   An equation of the form $ax^2 + bx + c = 0$.

QUADRATIC FORMULA   A formula for the roots of the quadratic equation expressed as

$$x = \frac{-b \pm \sqrt{b^2 - 4ac}}{2a}$$

RADIAN   The measure of an angle obtained by taking the ratio of arc length of a circle to the corresponding radius.

RADICAL   The symbol $\sqrt{}$, which refers to taking the root of a number. For example, $\sqrt{a}$ is a square root; $\sqrt[3]{a}$ is a cube root.

RANGE   The set of values of a function corresponding to the elements in its domain.

RATIONAL FUNCTION   The ratio of two polynomial functions.

RATIONALIZING   An algebra process whereby the denominator of a fraction is expressed free of radicals.

RC **CIRCUIT**   An electric circuit containing a resistor and a capacitor.

RECIPROCAL   If a is a number or expression, then $1/a$ is its reciprocal.

RECTIFIED SINE WAVE   A function such as $f(x) = |\sin x|$, which is called the full-wave, rectified sine wave.

RELATION   A rule of correspondence between x and y in which there may be more than one value of y for a given x.

RELATIVE MAXIMUM   A peak in the graph of $f(x)$ that is characterized by $f'(x) = 0$ and $f''(x) < 0$.

RELATIVE MINIMUM   A valley in the graph of $f(x)$ that is characterized by $f'(x) = 0$ and $f''(x) > 0$.

RESISTOR   An electrical device that impedes the flow of current.

RIGHT TRIANGLE   A triangle that contains a right angle.

RL CIRCUIT   An electric circuit containing a resistor and an inductance coil.

SCALAR QUANTITIES   Physical quantities such as temperature, length, and volume that can be described by *one* number.

SCIENTIFIC NOTATION   A special notation in which real numbers are expressed as the product of a number between 1 and 10 and a power of 10. The scientific form of 5280 is 5.280×10^3.

SECANT LINE   A straight line that passes through two distinct points of a curve.

SEPARABLE DIFFERENTIAL EQUATION   A first-order differential equation of the form $f(x)\,dx + g(y)\,dy = 0$.

SEQUENCE   A set of numbers in a fixed order.

SERIES   The sum of the terms of a sequence.

SHEAR   The rate of change of bending moment ($s = dm/dx$).

SINE OF AN ANGLE   The ratio of the y-coordinate of a point in the plane to the distance from the origin to the point. In terms of the sides of a triangle, $\sin \theta =$ opposite/hypotenuse.

SINUSOID   The graph of the sine function. Also called a sine wave.

SLANT RANGE   The line of sight distance between an observer and an elevated object.

SLOPE   A measure of the inclination of a line, defined as the ratio of the difference in the ordinates divided by the difference of the corresponding abscisses of any two points.

SLUG   The unit of mass in the British engineering system of units (weight/g).

SOLUTION   Those numbers that satisfy mathematical open sentences such as equations or inequalities.

STANDARD POSITION   A location of an angle in the Cartesian plane by placing the vertex at the origin and the initial side along the positive x-axis.

STRESS   The internal force per unit area of an object.

SUMMATION NOTATION   The use of the Greek letter Σ to represent a sum of terms.

TANGENT LINE   A straight line drawn so that it touches a curve at only one point without necessarily cutting through it.

TANGENT OF AN ANGLE   The ratio of the y-coordinate of a point in the plane to the x-coordinate of the point. In terms of the sides of a right triangle, $\tan \theta =$ opposite/adjacent.

TAYLOR SERIES   A series representation of $f(x)$ given by

$$f(x) = \sum_{n=0}^{\infty} \frac{f^{(n)}(a)}{n!}(x - a)^n$$

TENSILE STRENGTH   A measure of the greatest load in the direction of length that a given substance can bear without tearing apart.

TERMINAL VELOCITY   The maximum velocity reached by an object moving through a resisting medium such as air or water.

TORQUE   A measure of the tendency of a force to cause rotation. Given by the product of the force and the distance to a point of rotation. Also called *moment*.

TRANSLATION   A process of shifting the location of a graph in the plane without changing its shape.

TRANSVERSE WAVE   A wave in which the particles of the wave move perpendicular to the direction of motion of the wave.

TRIGONOMETRIC IDENTITY   An expression involving trigonometric functions that is true for all permissible values of the variable.

TRIGONOMETRIC RATIOS   The six ratios of sides of a right triangle.

TRUNCATED SERIES   If $\sum_{n=1}^{\infty} a_n$ is a series, then $\sum_{n=1}^{N} a_n$ is a truncated series.

UNBOUNDED FUNCTIONS   Functions whose range values are *not* contained between two finite numbers.

UNIT VECTOR   A vector of unit length.

VECTOR   A quantity requiring both a magnitude and a direction for its description. Velocity and force are examples of vectors.

VELOCITY   The time rate of change of displacement ($v = ds/dt$).

VOLT   A unit of electric voltage (abbreviated V).

VOLUME OF REVOLUTION   A solid figure obtained by revolving an area under a segment of a curve about a fixed line.

WATT   A unit of power in the international system of units (abbreviated W). Usually associated with electric power.

WORK   The product of applied force and distance moved.

ZERO OF A FUNCTION   Any number in the domain of the function for which the corresponding range value is zero.

ANSWERS FOR ODD-NUMBERED EXERCISES

SECTIONS 1.1–1.2

1. 4   **3.** 4   **5.** 7.1   **7.** 6

9.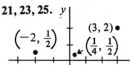

11.   **13.**

15.   **17.**   **19.**

21, 23, 25. 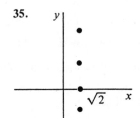   **27.** I, IV   **29.** III   **31.** 0

33. 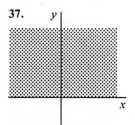   **35.** 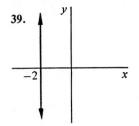   **37.**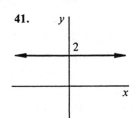

39.   **41.**

43. $\sqrt{2}$; $(-\frac{1}{2}, \frac{7}{2})$ **45.** $\frac{5}{4}$; $(\frac{1}{2}, -\frac{1}{8})$ **47.** 8.2; (3.35, 4.55)

49. $2\sqrt{4 + \sqrt{3}} \approx 4.8$; $(\frac{1}{2}(2 - \sqrt{3}), -\frac{9}{2})$ **51.** $2\sqrt{x + y}$ **53.** $2|y|$ **55.** $(0, -2)$

SECTION 1.3

1. function   **3.** function   **5.** function   **7.** function   **9.** not a function

11. function   **13.** not a function   **15.** function   **17.** not a function

19. D: all reals; R: all reals   **21.** D: all reals except -3; R: all reals except 0

23. D: all nonpositive real numbers; R: all nonnegative reals

25. (a) 10   (b) $3\pi + 1$ (c) $3z + 1$ (d) $3(x - h) + 1$ (e) 3   (f) all reals   (g) 31

27.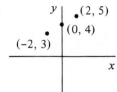

29.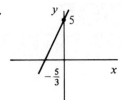

31.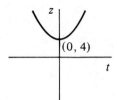

33.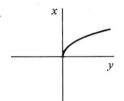

35.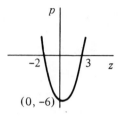

37.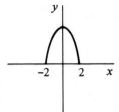

39.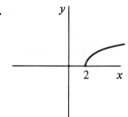

41.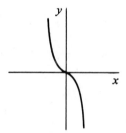

43.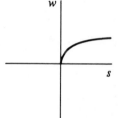

45.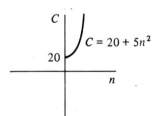

47. function   **49.** not a function   **51.** not a function

SECTION 1.4

1. $m = \frac{1}{2}$ **3.** $m = -\frac{9}{8}$ **5.** $m = \sqrt{481}/12 \approx 1.83$ **7.** 5 A   **9.** 29.67 ft/sec^2
11. $x - 2y = -8$ **13.** $7x + y = 33$ **15.** $x + 5y = 16$ **17.** $2y - 3x = 1$
19. $5y - 2x = 10$ **21.** $m = \frac{2}{3}, b = -\frac{5}{3}$ **23.** $m = -1, b = 2$ **25.** $m = \frac{1}{2}, b = -2$
27. $m = \frac{1}{2}, b = \frac{5}{2}$ **29.** $m = 0, b = 5$ **31.** $m = \frac{1}{5}, b = \frac{7}{5}$ **33.** $2x - 3y = -4$
35. $x + y = 8$ **37.** $2x + 3y = 8$ **39.** $7x + 5y = 0$ **41.** $3x + y = -5$
45. $I = V/12$ **47.** \$225   **49.** \$151.11 per month

SECTION 1.5

1. D: all x
R: $f(x) \geq -4$
zeros: $x = -2, 2$
intercept: $(0, -4)$
vertex: $(0, -4)$

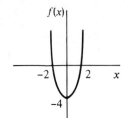

3. D: all x
R: $y \geq 1$
no zeros
intercept: $(0, 1)$
vertex: $(0, 1)$

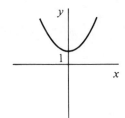

5. D: all x
R: $f(x) \leq 9$
zeros: $x = -3, 3$
intercept: $(0, 9)$
vertex: $(0, 9)$

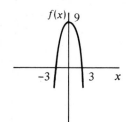

7. D: all t
R: $x(t) \geq -\frac{9}{4}$
zeros: $t = 0, 3$
intercept: $(0, 0)$
vertex: $(\frac{3}{2}, -\frac{9}{4})$

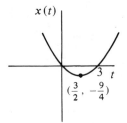

9. D: all x
R: $y \geq -\frac{1}{4}$
zeros: $x = 2, 3$
intercept: $(0, 6)$
vertex: $(\frac{5}{2}, -\frac{1}{4})$

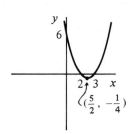

11. D: all x
R: $y \geq \frac{2}{3}$
no zeros
intercept: $(0, 1)$
vertex: $(-\frac{1}{3}, \frac{2}{3})$

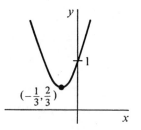

13. D: all x
R: $y \leq \frac{9}{4}$
zeros: $x = -1, 2$
intercept: $(0, 2)$
vertex: $(\frac{1}{2}, \frac{9}{4})$

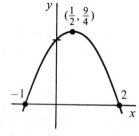

15. D: all x
R: $f(x) \geq -\frac{49}{8}$
zeros: $x = -1, \frac{5}{2}$
intercept: $(0, -5)$
vertex: $(\frac{3}{4}, -\frac{49}{8})$

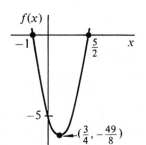

17. D: all x
R: $y \leq \frac{13}{4}$
zeros: $x = (1 \pm \sqrt{13})/2$
intercept: $(0, 3)$
vertex: $(\frac{1}{2}, \frac{13}{4})$

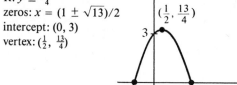

19. D: all x
R: $G(x) \geq \frac{3}{4}$
no zeros
intercept: $(0, 1)$
vertex: $(-\frac{1}{2}, \frac{3}{4})$

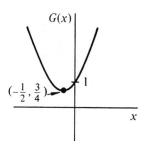

21. $y = x^2 - 2x$ **23.** $y = x^2 - 2x + 1$ **25.** $y = -x^2 - 4x - 3$ **27.** $y = \frac{2}{3}x^2 - \frac{4}{3}x - 2$

29. $y = -\frac{1}{10}x^2 - \frac{3}{10}x + 1$ **31.** symmetric with respect to the x-axis

33.

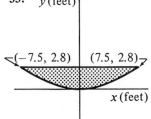

35.

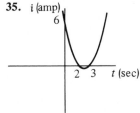

37. $v_{max} = 500$ ft/sec at $t = 4$ sec
$v = 0$ at $t = 4 + 2\sqrt{5} \approx 8.47$ sec

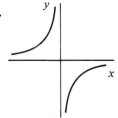

SECTION 1.6

1.

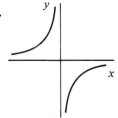

3.

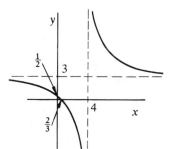

5.

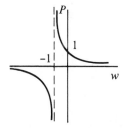

7.

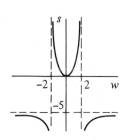

9.

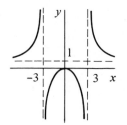

11.

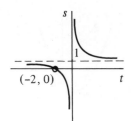

13.

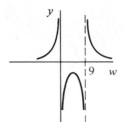

15.

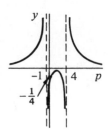

17.

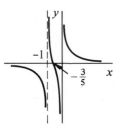

19.

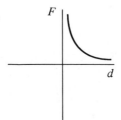

21.

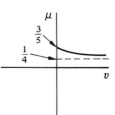

SECTION 1.7

1.

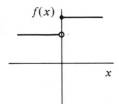

3.

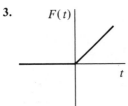

5.

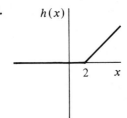

7.

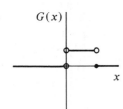

9.

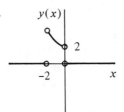

11.

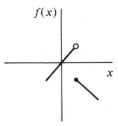

13.

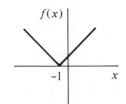

15.

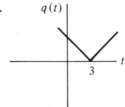

17.

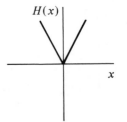

19.

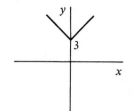

21.

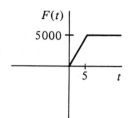

23.

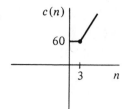

25. let $x = 2, y = -3$

27.

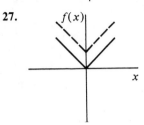

REVIEW EXERCISES FOR CHAPTER 1

1. 9   **3.** 2   **5.** 3   **7.**

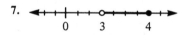

9.

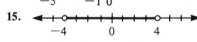

11.

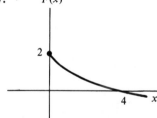

13.

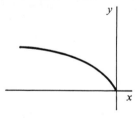

15.

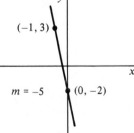

17.

19.

21. $\sqrt{89}$; $(\frac{1}{2}, 3)$ **23.** function   **25.** not a function   **27.** function   **29.** not a function

31. function   **33.** D: all reals; R: all nonnegative reals   **35.** D: all reals ≥ 3; R: all nonnegative reals

37.

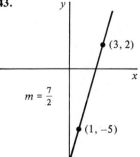

39.

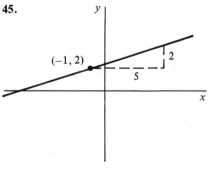

41.

43.

45.

47. $-\frac{6}{5}$ **49.** $y + 2x - 5 = 0$ **51.** $4y + x - 15 = 0$ **53.** $4x + 2y = 5$ **55.** $3x + y = 1$

57. D: all real x
R: $y \geq -\frac{81}{8}$

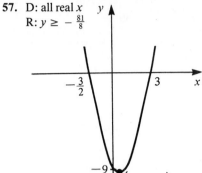

59.

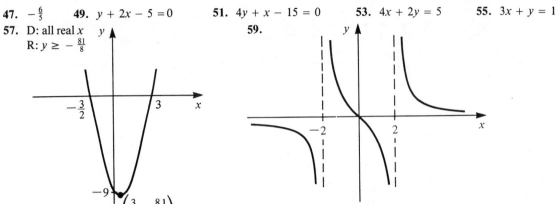

61.

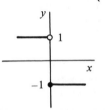

63.

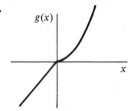

65.

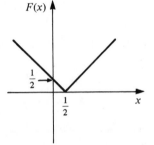

67.

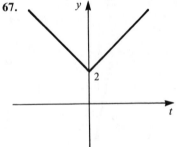

69.

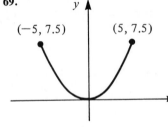

71.

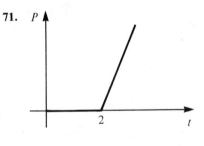

SECTION 2.1

1. $\frac{9}{7}$ **3.** 2   **5.** 1   **7.** -3 **9.** $\frac{4}{3}$ **11.** 4   **13.** 16   **15.** $\frac{1}{2}$
17. -1 **19.** -9 **23.** 7   **25.** 2   **27.** 1   **29.** 2.718   **31.** 12

SECTION 2.2

1. 0   **3.** $\frac{3}{2}$ **5.** $\frac{1}{3}$ **7.** $-\infty$ **9.** 1   **11.** ∞ **13.** 0   **15.** 0
17. ∞ **19.** $\frac{5}{4}$ **21.** ∞ **23.** ∞ **25.** V/R

SECTION 2.3

1. -6 **3.** 2 **5.** -2 **7.** 0 **9.** 12 **11.** $-\frac{1}{4}$ **13.** $\frac{1}{2}$
15. $2x - y = 1$ **17.** 9 ft/sec **19.** 6 ft/sec **21.** $-\frac{1}{4}$ ft/sec **23.** $11A$ **25.** $0.2A$

27. $-20A$ **29.** $a(t_0) = \lim\limits_{t \to t_0} \dfrac{v(t) - v(t_0)}{t - t_0}$ **31.** -1.4 cm/sec

SECTION 2.4

1. $\frac{2}{3}x$ **3.** $2x + 3$ **5.** $x + \frac{1}{3}$ **7.** $6x + 5$ **9.** $-5/x^2$ **11.** $-4/x^3$
13. $3x^2 - 2$ **15.** $-1/(x + 5)^2$ **17.** $-\frac{2}{3}; 0; \frac{2}{3}$ **19.** $3.2; 4$ **21.** $-2/(t - 2)^2$ ft/sec

REVIEW EXERCISES FOR CHAPTER 2

1. 27 **3.** 6 **5.** 0 **7.** 0 **9.** 1 **11.** 7 **13.** $\frac{1}{6}$ **15.** ∞
17. $\frac{1}{3}$ **19.** does not exist **21.** ∞ **23.** 5 **25.** 31 ft/sec **27.** x **29.** $2x + 5$
31. $3t^2 + 1$ **33.** $-1/(x + 1)^2$

SECTION 3.1

1. $9x^8$ **3.** $36x^{11}$ **5.** $-2t^{-3}$ **7.** $2x^{-1/2}$ **9.** 0 **11.** 3 **13.** $3 + 8\phi$
15. $-3/x^4$ **17.** $\frac{3}{2}t^{1/2}$ **19.** $(\sqrt{3}/2)t^{-1/2}$ **21.** $(5at^{3/2}/2) - (3bt^{-2/5}/5)$ **23.** $-\frac{4}{3}s^{-4/3} - \frac{5}{2}s^{-3/2}$
25. $4x^3 - 6x^2 + 10x - 7$ **27.** $-3/(2\sqrt{2t^5})$ **29.** 1 **31.** $2x - 4$ **33.** $\frac{3}{2}t^{-1/2} - \frac{4}{3}t^{-2/3}$
35. $-\frac{3}{2}t^{-3/2} + \frac{3}{2}t^{1/2}$ **37.** $2.34t^{1.6}$ **39.** $-2.4w^{-1.2} + 0.8w^{-0.2}$ **41.** $60x + 4y = 47$
43. $2x + y + 6 = 0$ **45.** $x = 0$ and $x = -6$ **47.** no real values **49.** $18 + 2\pi(3)^{\pi - 1}$

SECTION 3.2

1. $-15(4 - 3x)^4$ **3.** $8t(t^2 + 4)^3$ **5.** $1/(5x + 1)^{4/5}$ **7.** $\dfrac{8t - 3}{2(4t^2 - 3t)^{1/2}}$

9. $-6(3t + 1)^{-3}$ **11.** $\dfrac{-30}{(2 + 5x)^4}$ **13.** $\dfrac{-3x}{(x^2 + 1)^{3/2}}$ **15.** $\left(\dfrac{3}{x^2}\right)\left(2 - \dfrac{1}{x}\right)^2$

17. $\dfrac{\pi}{4\sqrt{y}(1 + \pi\sqrt{y})}$ **19.** $\dfrac{4x - 3}{3(4 + 3x - 2x^2)^{4/3}}$ **21.** $(2 - 3x)^{-5/3}(3x + 8)$ **23.** $\frac{9}{2}(3t + 4)^{1/2}$

25. $1/(4L\sqrt{Tu})$ **27.** $-(3t + 2)^{-4/3}$ **29.** $dG/du = -E/[2(1 + \mu)^2]$ **31.** $s = 6(3x - 2)$
33. $s = -6x(10 - x^2)^2$

SECTION 3.3

1. $12x + 7$ **3.** $\dfrac{12}{(3 - 2x)^2}$ **5.** $t^2(t + 1)(5t + 3)$ **7.** $\dfrac{-24s}{(s^2 - 4)^2}$

9. $(3x + 2)^2(5x + 1)^3(105x + 49)$ **11.** $\dfrac{12z(4z + 2)^2}{(3z + 1)^3}$ **13.** $(-2)\dfrac{(1 - x)(4 - x)}{(4 - x^2)^2}$

15. $\dfrac{-1}{5m^{4/5}(m - 1)^{6/5}}$ **17.** $\dfrac{-4t}{3(t^2 - 1)^{4/3}(t^2 + 1)^{2/3}}$ **19.** $\dfrac{-6(5x + 4)}{(2x - 1)^4(3x + 5)^3}$

21. $\frac{3}{2}, -\frac{13}{6}$ **23.** $0, 2$ **25.** $29x - 12y = 25$

27. $v(t) = \dfrac{5t^2 + 12}{(t^2 + 4)^{2/3}}$ cm/sec; $s(2) = 12$ cm, $v(2) = 8$ cm/sec **29.** $\dfrac{3T^2(2T - 3)}{2(T^2 - T)^{3/2}}$

SECTION 3.4

1. $30x - 126x^5$ **3.** $-\frac{1}{4}t^{-3/2}$ **5.** $13.77x^{0.7}$ **7.** $(\frac{8}{3})w^{-7/3} + 108w^{-5}$

9. $(\frac{9}{16})t^{-5/4}$ **11.** $-(2x - 5)^{-3/2}$ **13.** $90(3x + 7)^{-3}$ **15.** $30x(x^3 - 8)^3(7x^3 - 8)$

17. $\dfrac{2(2t^2 - 1)}{(t^2 + 1)^{5/2}}$ **19.** $60(5x + 4)(5x + 2)$ **21.** 2 **23.** 0

25.

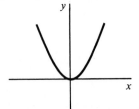

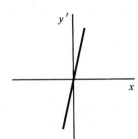

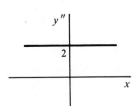

27.

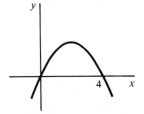

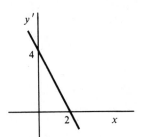

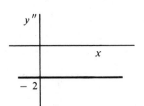

SECTION 3.5

1. $-x/y$ **3.** $-4x^3/3y^2$ **5.** $-(1 + 2xy^2)/(1 + 2x^2y)$ **7.** $-y^{1/3}/x^{1/3}$

9. $-(3x^2 + y)/(x + 3y^2)$ **11.** $-y^2/x^2$ **13.** $1/[2(1 + y)]$ **15.** $-16/y^3$

17. $\dfrac{(y - 1)(2x + 6y + 6)}{(6y + x)^3}$ **19.** $\dfrac{8y(6x + y)}{(4x + y)^3}$ **21.** $7x + 8y = 38$ **23.** $x - 2y = -28$

25. $-\frac{1}{12}$

REVIEW EXERCISES FOR CHAPTER 3

1. $10x^4 - 9x^2 + 1$ **3.** $\frac{1}{2}x^{-1/2} + (\frac{4}{3})x^{-2/3}$ **5.** $-45t^{-4} + (\frac{8}{5})t^{-6/5}$

7. $(\frac{7}{3})w^{4/3} - 21w^6 + 14w$ **9.** $(6 - x)/x^3$ **11.** $-21(2 - 3x)^6$ **13.** $-6x(x^2 + 2)^{-4}$

15. $(\frac{2}{3})(1 - 2t)(t - t^2)^{-2/3}$ **17.** $-12(3x + 7)^{-3/2}$ **19.** $(2t + 5)^2(24t + 15)$ **21.** $8x^3(x^4 - 1)$

23. $3(x^2 + 7x)^2(3x^3 + 1)^3(18x^4 + 105x^3 + 2x + 7)$ **25.** $\dfrac{s^4(2s + 25)}{(s + 5)^4}$ **27.** $\dfrac{4 - 2t}{(5t - 3)^{4/5}(3t + 1)^{4/3}}$

29. $(2t - 1)(5t + 4)^2(240t^2 - 34t - 17)$ **31.** $40x^3 - 18x; 120x^2 - 18$

33. $-\frac{1}{4}x^{-3/2} - (\frac{8}{9})x^{-5/3}; (\frac{3}{8})x^{-5/2} + (\frac{40}{27})x^{-8/3}$ **35.** $180t^{-5} - (\frac{48}{25})t^{-11/5}; -900t^{-6} + (\frac{528}{125})t^{-16/5}$

37. $(\frac{28}{9})w^{1/3} - 126w^5 + 14; (\frac{28}{27})w^{-2/3} - 630w^4$ **39.** $\dfrac{2x - 18}{x^4}; \dfrac{72 - 6x}{x^5}$ **41.** $\dfrac{(3x^2 - y^2)}{(2xy + 3)}$

43. $y' = 1 - x; 2, 0$ **45.** $8x + 5y = 13$ **47.** 42 cm/sec   **49.** $\dfrac{3t + 2}{2\sqrt{t + 1}}$

SECTION 4.1

1. (a) $v = 100 - 32t$ (b) 68 ft/sec; 36 ft/sec; 4 ft/sec   **3.** $\frac{23}{3}$ ft/sec; $\frac{1}{9}$ ft/sec^2 **5.** 15 cm/sec^2

7. $-12(2t + 1)^{-3}$ ft-lb/sec   **9.** 6500 ft   **11.** $a(t) = -0.4t(50 - 0.4t^2)^{-1/2}$

13. $v = 3.568$ cm/sec; $a = 0.115$ cm/sec^2 **15.** $P = 1.5(2t + 1)^{-1/2}$

17. (a) $v = \frac{3}{2}t^2(t^3 + 8)^{-1/2}$ cm/sec; $a = \frac{3}{4}t(t^3 + 32)(t^3 + 8)^{-3/2}$ cm/sec^2 **19.** 0.35 cm/sec   **21.** 6.4 sec

(b) $s(2) = 4$ cm; $v(2) = \frac{3}{2}$ cm/sec; $a(2) = \frac{15}{16}$ cm/sec^2

SECTION 4.2

1. $17.1t^2$ **3.** $0.96t^{-0.7} - 8.58t^{0.3}$ **5.** $-48/(2t - 1)^{2.6}$ **7.** $-(0.1t^{-0.9} + 0.5t^{-1.1})$

9. 3.12 A   **11.** $-360t^2(4t^3 + 5)$ **13.** 2.22 V   **15.** 12,800 W   **17.** 0.4 A   **19.** $21t$ V

21. 159.7 V   **23.** $5/6\sqrt[3]{2}$ W ≈ 0.66 W

SECTION 4.3

1. 100 mi/hr   **3.** 1.0 cm^2/min   **5.** 0.75 dyn/sec   **7.** 600 psi/sec   **9.** 1.1 Ω/min

11. 11.8 ft/sec   **13.** 26.3   **15.** 0.0318 ft^3/sec   **17.** 0.3 psi/sec   **19.** $-\frac{2}{3}$ **21.** $\frac{2}{3}$ cm^2/sec

SECTION 4.4

1. at $x = 3$, minimum of $y = -4$ **3.** at $t = 0$, maximum of $i = 2$; at $t = 2$, minimum of $i = -2$

5. no maximum or minimum   **7.** at $x = 1$, minimum of $p = -2$; at $x = -1$, maximum of $p = 2$

9. at $x = -4$, minimum of $y = -78$ **11.** at $r = 1$, minimum of $m = 0$

13. no maximum or minimum

15. at $x = \sqrt{10}$, minimum of $y = 2\sqrt{10}$; at $x = -\sqrt{10}$, maximum of $y = -2\sqrt{10}$

17. maximum at $x = 4$ ($y = 16$); minimum at $x = 10$ ($y = -20$)

19. maximum at $x = -2$ ($y = 4$); minimum at $x = -4$ ($y = -16$)

21. maximum at $x = 3$ ($y = \frac{5}{2}$); minimum at $x = 0$ ($y = 1$); relative maximum at $x = 1$ ($y = \frac{11}{6}$);

relative minimum at $x = 2$ ($y = \frac{5}{3}$)

23. maximum at $x = 8$ ($y = 2$); minimum at $x = 0$ ($y = 0$)   **25.** 7812.5 ft   **27.** $\frac{13}{4}$ W

SECTION 4.5

1. no relative extremes   **3.** relative minimum at $x = -\frac{3}{2}$ **5.** no relative extremes

7. relative maximum at $x = 2$; relative minimum at $x = 4$

9. relative minimum at $x = 2$; relative maximum at $x = -2$ **11.** relative minimum at $x = \sqrt[3]{2}$

13. relative minimum at $x = -\frac{1}{2}$

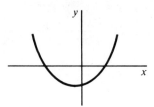

15. relative maximum at $x = -2$;
relative minimum at $x = 0$;
point of inflection at $x = -1$

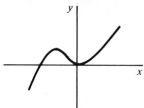

17. relative maximum at $x = 1$;
relative minimum at $x = 2$;
point of inflection at $x = \frac{3}{2}$

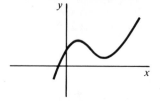

19. point of inflection at $x = 0$

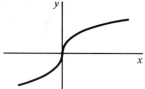

21. relative minimum at $x = 0$

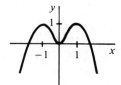

23. $t = 3\sqrt{10}/10$

SECTION 4.6

1.

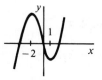

3.

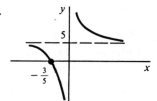

5.

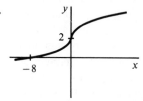

7.

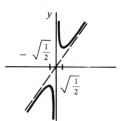

9.

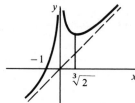

11.

13.

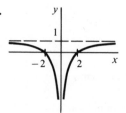

15.

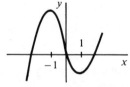

17.

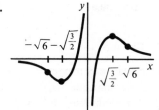

19.

21.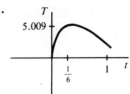

SECTION 4.7

1. at $t = 2$, $v = 4$ ft/sec   **3.** at $t = 3$, $i = 13$ A   **5.** at $t = 0.2$, $i = 0.12$ A
7. at $t = \frac{1}{4}$, $v = \frac{7}{4}$ in./sec is a maximum   **9.** 15, 15   **11.** 4, 2   **13.** a 2×2 square
15. 62,500 ft^2 **17.** $b = 5.76$ in.; $h = 8.12$ in.   **19.** $r = 6.8$ cm and $h = 13.6$ cm
21. $A = 16$ square units   **23.** $d = 3$ units

SECTION 4.8

1. $8x\,dx$ **3.** $-\frac{1}{2}x^{-3/2}\,dx$ **5.** $(\sqrt{3}/2)t^{-1/2}\,dt$ **7.** $\frac{8}{3}b^{-1/3}\,db$
9. $(-3p^{-4} + 2p^{-3} - 4)\,dp$ **11.** 30   **13.** 0.0083   **15.** -6.8 **17.** 0.512 A
19. 1.14×10^{-3} C   **21.** 312   **23.** 187.5 cm^3 **25.** 7860 ft^3 **27.** 0.0318 in. increase
29. -0.0318 in. decrease in both radii

REVIEW EXERCISES FOR CHAPTER 4

1. $a = 4 - 10t$ cm/sec^2 **3.** $s = 0.031$; $v = 0.166$; $a = 0.565$ **5.** 0.0000833 A
7. $6(2t + 1)(t^2 + t)^2$ **9.** 2.56 A   **11.** 90 cm^2/sec   **13.** 3.02 ft/sec
15. relative minimum at $x = 16$;   **17.** relative maximum at $x = -2/\sqrt{3}$;   **19.** at $t = 1$ sec, $a = 3$ ft/sec^2
relative maximum at $x = 0$;   relative minimum at $x = 2/\sqrt{3}$;
point of inflection $x = 8$ point of inflection $x = 0$

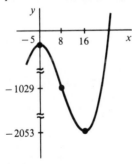

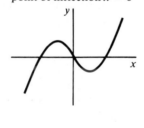

21. construct circle with entire L **23.** $\left(\dfrac{1}{2\sqrt{x}} + \dfrac{3}{4\sqrt[4]{x^3}}\right) dx$ **25.** $(-\frac{5}{3})(1 - 5x)^{-2/3}\,dx$
27. 6.28×10^{-5} in.^{3}

SECTION 5.1

1. $3x + C$ **3.** $3\left(\dfrac{z^2}{2}\right) + C$ **5.** $\dfrac{x^4}{4} + \dfrac{x^3}{3} + C$ **7.** $\dfrac{3x^8}{8} + \dfrac{3x^{10}}{5} + C$

9. $\dfrac{3p^{4/3}}{4} + C$ **11.** $\dfrac{2\sqrt{3}}{3}\phi^{3/2} + C$ **13.** $-\dfrac{x^{-4}}{4} + C$ **15.** $\dfrac{x^3}{3} - \dfrac{x^5}{5} + C$

17. $\dfrac{-\theta^{-2}}{2} + 2\theta^{-1} + C$ **19.** $\dfrac{x^3}{3} - 3x^2 + 9x + C$ **21.** $\dfrac{x^3}{3} + 2x - x^{-1} + C$

23. $\frac{6}{11}x^{11/3} - \frac{9}{8}x^{8/3} + \frac{3}{5}x^{5/3} + C$ **25.** $\dfrac{x^4}{4} - \dfrac{x^{-2}}{2} + C$ **27.** $\dfrac{t^{13}}{13} + C$

29. $\frac{3}{5}\phi^{5/3} + \frac{24}{19}\phi^{19/12} + \frac{2}{3}\phi^{3/2} + C$ **31.** $\dfrac{8p^{9/8}}{9} - \dfrac{24p^{19/24}}{19} + C$ **33.** $\dfrac{x^{1.35}}{1.35} + C$

35. $17x + \frac{4}{3}x^{0.75} + C$ **37.** $\dfrac{x^{e+1}}{e+1} + C$ **39.** $y = \frac{3}{2}x^2 + \frac{1}{2}$ **41.** $s = \frac{3}{4}r^{4/3} - 14$

43. $p = \frac{2}{3}t^{3/2} + 2t^{1/2} - \frac{25}{3}$

SECTION 5.2

1. $\frac{1}{8}(2x + 1)^4 + C$ **3.** $-\frac{1}{20}(1 - 5x)^4 + C$ **5.** $\frac{1}{6}(3 + 4p)^{3/2} + C$ **7.** $\frac{1}{10}(x^2 + 1)^5 + C$

9. $\frac{3}{8}(z^2 + 2z)^{4/3} + C$ **11.** $\dfrac{1}{1 - s} + C$ **13.** $\frac{3}{8}(2x + 4)^{4/3} + C$ **15.** $-\frac{1}{12}(2 + 4x)^{-3} + C$

17. $\frac{8}{3}(r^3 + 1)^{1/2} + C$ **19.** $\frac{2}{3}(4 + \sqrt{x})^3 + C$ **21.** $\dfrac{x^5}{5} + \dfrac{8x^3}{3} + 16x + C$

23. $\frac{1}{6}(x^2 - 3)^3 + C$ **25.** $\frac{1}{7}x^7 + \frac{6}{5}x^5 + 4x^3 + 8x + C$ **27.** $y = \frac{1}{12}(3x + 2)^4 + \frac{11}{3}$

29. $y = \frac{4}{3}(\frac{1}{2}t + 1)^{3/2} - \frac{32}{3}$ **31.** If $u = 2x + 3$, then $du = 2\,dx$, not $du = x\,dx$.   **33.** $\dfrac{1}{5 - t} + C$

35. $y = (x^2 + 1)^{1/2} + C$

SECTION 5.3

1. $v = \frac{2}{3}t^{3/2}$ **3.** 11 sec   **5.** 2000 ft   **7.** $s = \frac{1}{3}(t^3 + 1)$ ft   **9.** $t = 1.62$ sec
11. 32 ft/sec   **13.** 15.6 ft-lb   **15.** 0.558   **17.** $v = 300t - t^2 + 3000$ m/sec
19. $s = 0.115(5.1t + 1)^{1.7} + 13.15$ ft   **21.** $w = \frac{200}{3}(0.01t + 1)^{3/2} - \frac{200}{3}$

SECTION 5.4

1. 0.3 C   **3.** $i = \frac{-1}{18}(3t + 1)^{3/2} + \frac{5}{9}$ **5.** $v = \frac{2}{3}t^3 + 6t^2 + 18t + 6$ **7.** 1920 J
9. $\frac{7}{20}t^4 + \frac{7}{25}t^3 + t + 2$ **11.** $\frac{1}{8}t^4 + 250t^3 - \frac{1}{5}t^2 - \frac{1997}{10}t + 1$ **13.** -0.105 A
15. $v = 20t^2 + 1000t + 33.3t^3$ **17.** 9.5 V   **19.** $40t^2 + \frac{58}{5}t + \frac{121}{5}$

REVIEW EXERCISES FOR CHAPTER 5

1. $\frac{1}{2}x^2 - x + C$ **3.** $\frac{5}{2}x^2 - \frac{1}{4}x^4 + C$ **5.** $\frac{1}{2}x^{1/2} + C$ **7.** $\frac{3}{2}r^{2/3} + 3r^{5/3} + C$
9. $\frac{2}{3}x^{3/2} - \frac{72}{5}x^{5/4} + 81x + C$ **11.** $\frac{1}{9}(x^3 + 1)^3 + C$ **13.** $\frac{1}{10}(x^2 + 1)^5 + C$
15. $\frac{5}{96}(16x + 3)^{6/5} + C$ **17.** $\frac{-1}{45}(5t^3 - 7)^{-12} + C$ **19.** $\frac{2}{5}\sqrt{5x - 7} + C$
21. $y = 3x^2 - 5x - 6$ **23.** $\frac{99}{2}$ ft, 42 ft/sec, 23 ft/sec^2 **25.** $s = \frac{4}{15}t^{5/2} + 10t$ ft
27. $v = -2t^2 + t,\ s = \frac{-2}{3}t^3 + \frac{1}{2}t^2 + 5$ **29.** $i = \frac{1}{12}t^4 + \frac{1}{5}t^3 + \frac{13}{2}t^2 + \frac{2}{3}t$
31. $v = 20t^{3/2} + \frac{3}{2}t^{1/2} + \frac{1}{5}t^{5/2} + 5$

SECTION 6.1

1. 1.093 **3.** 55 **5.** 45 **7.** 20 **9.** 10 **11.** $x_1 + x_2 + x_3 + x_4 + x_5$
13. $\sum_{k=1}^{4} k$ **15.** $\sum_{k=1}^{4} k^2$ **17.** $\sum_{k=1}^{5} 2^k$ **19.** true **21.** true **23.** true **25.** 1,625,625

SECTION 6.2

1. 5.25 **3.** 9.5 **5.** 1.93 **7.** 0.79125 **9.** 5.38 **11.** (a) 0.437 (b) 0.938
13. (a) 6.75 (b) 10.75 **15.** 210 ft **17.** (a) 140 ft (b) 300 ft **19.** (a) 175 ft (b) 255 ft
21. 18.25 cm **23.** (a) 20.125 cm (b) 15.625 cm **25.** 60.04 m

SECTION 6.3

1. $\frac{64}{3}$ **3.** $\frac{4}{3}$ **5.** 6 **7.** 200 **9.** 3π **11.** 8 **13.** $\frac{2}{3}$ **15.** $\frac{46}{3}$
17. $\frac{62}{3}$ **19.** $\frac{10}{3}$ cm **21.** $\frac{4}{3}$

SECTION 6.4

1. $4x^2$ **3.** $2\sqrt{t}$ **5.** $\sin t$ **7.** $(3 - x)^{-1}$ **9.** 1 **11.** $\frac{9}{2}$ **13.** $\frac{5}{6}$
15. $\frac{16}{3}$ **17.** 46 **19.** $\frac{9}{4}$ **21.** $-\frac{64}{7}$ **23.** 1 **25.** $\frac{8}{3}\sqrt{2}(4 - \sqrt{2})$ **27.** 0
29. $\frac{32}{3}\sqrt{2} - \frac{2}{3}$ **31.** $\frac{3}{20}$ **33.** $\frac{26}{3}$ **35.** 8 **37.** $\frac{16}{3}$ **39.** $\frac{4}{5}$ **41.** 99 ft

SECTION 6.5

9. 3 **11.** 7, $\sqrt{\frac{7}{3}}$ **13.** 4 **15.** 3 **17.** $\frac{1}{4}$ **19.** $(4\sqrt{2} - 2)/3\sqrt{2}$
21. (a) $v_{min} = 0$ (b) $v_{max} = 10$ ft/sec (c) $v_{avg} = 4.33$ ft/sec **23.** $s = 8.67$ ft **25.** $\frac{10}{3}$ A

SECTION 6.6

1. 27 **3.** 0.446 **5.** 10.15 **7.** 8.86 **9.** 11.38 **11.** 7.21 **13.** 2.83 cm
15. 0.25 ft/sec^2 **17.** 22 gal/min **19.** 75.565 ft-lb

REVIEW EXERCISES FOR CHAPTER 6

1. 10.5 **3.** 1.5 **5.** 19.125 **7.** 36 **9.** 57 **11.** 8 **13.** $\frac{17}{2}$ **15.** $\frac{8}{3}$
17. 20 **19.** -32 **21.** $\frac{9}{350}$ **23.** $\frac{13}{10}$ **25.** $\frac{14}{3}$ **27.** $\frac{28}{3}$ **29.** 9 **31.** $\frac{129}{4}$
33. $\frac{153}{2}$ cm **35.** $\frac{14}{3}, \frac{7}{3}$ **37.** 37.5

SECTION 7.1

1. 5 **3.** $\frac{17}{4}$ **5.** $\frac{75}{4}$ **7.** 12 **9.** $\frac{34}{3}$ **11.** $\frac{3}{2}$ **13.** $\frac{2}{5}$ **15.** $\frac{21}{2}$
17. $\frac{21}{4}$ **19.** $\frac{62}{5}$ **21.** 0

SECTION 7.2

1. $\frac{4}{3}$ **3.** $\frac{4}{15}$ **5.** $\frac{32}{3}$ **7.** $\frac{1}{3}$ **9.** $\frac{32}{3}$ **11.** $\frac{125}{6}$ **13.** $\frac{125}{6}$ **15.** 9

17. 2   **19.** $\frac{375}{4}$ **21.** $\frac{80}{3}$ **23.** $\frac{33}{4}$

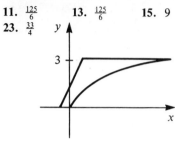

SECTION 7.3

1. $M_x = \frac{17}{2}$, $M_y = \frac{7}{2}$ **3.** $M_x = 4$, $M_y = 12$ **5.** $\left(\frac{7}{18}, \frac{17}{18}\right)$ **7.** $(2, 2)$

9. $\left(\frac{4}{9}, \frac{8}{3}\right)$ **11.** $\left(\frac{68}{30}, \frac{79}{30}\right)$ **13.** $\left(3, \frac{24}{5}\right)$ **15.** $\left(\frac{9}{8}, \frac{27}{5}\right)$ **17.** $\left(-2, \frac{12}{5}\right)$

19. $\left(\frac{3}{2}, \frac{14}{5}\right)$ **21.** $\left(\frac{1}{3}, \frac{5}{3}\right)$ **23.** $(2, 1)$

SECTION 7.4

1. 8π **3.** 72π **5.** $\frac{128}{7}\pi$ **7.** $\frac{1296}{5}\pi$ **9.** $\frac{200}{3}\pi$ **11.** $\frac{384}{7}\pi$ **13.** $\frac{3}{5}\pi$

15. $\frac{512}{15}\pi$ **17.** $\frac{576}{5}\pi$ **19.** $\frac{8}{3}\pi$ **21.** 9π **23.** 8π **25.** 8π **27.** $\frac{96}{5}\pi$

29. $\frac{81}{2}\pi$ **31.** $\frac{8}{3}\pi$ **33.** 64π **35.** 64π

SECTION 7.5

1. 32   **3.** $\frac{1215}{4}$ **5.** $\frac{2}{7}$; $\frac{2}{15}$ **7.** $\frac{63}{20}$ **9.** $\frac{32}{7}$ **11.** $\frac{3}{7}$; $\frac{1}{5}$ **13.** $\frac{7}{10}$

15. $\frac{108}{5}\pi\rho$; $108\pi\rho$; 5   **17.** $8\pi\rho$; $\frac{64}{5}\pi\rho$; $\frac{8}{5}$ **19.** $\frac{81}{2}\pi\rho$; $243\pi\rho$; 6   **21.** $\frac{8}{3}\pi\rho$; $\frac{64}{15}\pi\rho$; $\frac{8}{5}$

SECTION 7.6

1. 2268 ft-lb   **3.** 50 in.-lb   **5.** $\frac{256}{3}$ in.-lb   **7.** 8333π ft-lb   **9.** $20{,}250\pi$ ft-lb

11. 4725π ft-lb   **13.** $25{,}125\pi/8$ ft-lb   **15.** $91{,}100\pi$ ft-lb   **17.** 6000π ft-lb

19. $15{,}200\pi$ ft-lb   **21.** $546{,}000\pi$ ft-lb   **23.** 4500 lb   **25.** 375,000 lb

REVIEW EXERCISES FOR CHAPTER 7

1. 3.5   **3.** 1.5; average height $=0$ **5.** $\frac{14}{3}$ **7.** $\frac{125}{6}$ **9.** $\left(\frac{255}{84}, \frac{1023}{210}\right)$

11. $\left(\frac{12}{7}, \frac{213}{35}\right)$ **13.** $\frac{208}{3}\pi$ **15.** 40π **17.** $\frac{27}{2}\pi$ **19.** $\frac{3}{5}\pi$ **21.** $\frac{1}{5}$; $\frac{4}{21}$ **23.** $\frac{243}{5}\pi\rho$; $\frac{18}{5}$

25. 28 in.-lb   **27.** $404{,}625\pi/32$ ft-lb   **29.** $253{,}125/4$ lb   **31.** 15.4 J

SECTION 8.1

1. -8.55 **3.** $4, -1$ **5.** -0.465 **7.** 1000   **9.** 4   **11.** 2   **13.** 1.188

15.

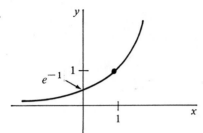

17.

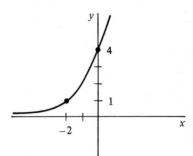

19.

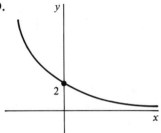

21.

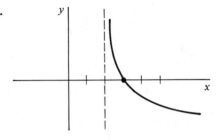

23.

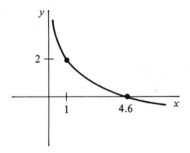

25. mirror image in the y-axis **27.** the same **29.** mirror image in the x-axis
31. translated 3 units left **33.** mirror image in the y-axis **35.** $3 \log x$ **37.** $\log_2 (2x^4)$
39. $\log_5[x^3/(2x - 3)]$ **41.** $x > -1$ **43.** $x > 0$ **45.** 2.77 years

47.

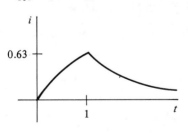

49.

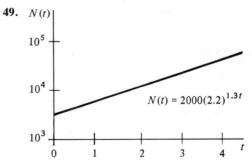

51. 1.77 mg

53.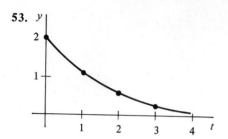

SECTION 8.2

1. $\dfrac{\log e}{x}$ **3.** $\dfrac{2}{2t-1}$ **5.** $\dfrac{2}{p}$ **7.** $\dfrac{3(1-2x)}{x-x^2}$ **9.** $\dfrac{12-5x}{x(4-x)}$ **11.** $\dfrac{3t^2}{2(t^3-1)}$

13. $\dfrac{-4r(r^2+2)}{r^4-1}$ **15.** $\dfrac{12\ln^3(3T+2)}{3T+2}$ **17.** $\dfrac{2R\ln(R^2-2)-2R^3/(R^2-2)}{\ln^2(R^2-2)}$

19. $\dfrac{3V^2}{3V+2}+2V\ln(3V+2)$ **21.** $\dfrac{-100}{r}$ **23.** $\dfrac{-42}{p}$ **25.** $\dfrac{3t(2-t^3)}{(t^3+1)^2}$ ft/sec^2

27. $\dfrac{-3}{1+3T}$ psi/degree   **29.** at $t=1$ sec

SECTION 8.3

1. $2e^{2x}$ **3.** $\dfrac{2(2x-1)}{(2x-3)^{2/3}(2x+3)^{1/3}}$ **5.** $2t(1+t)e^{2t}$ **7.** $-15e^{-3t}$

9. $\dfrac{-2x}{(x^2+1)^{1/2}(x^2-1)^{3/2}}$ **11.** $(L^2+1)^{-1/2}\left[\dfrac{8L(L^2+1)}{4L^2-5}+L\ln(4L^2-5)\right](4L^2-5)^{(L^2+1)1/2}$

13. $\dfrac{24x^2+36x+13}{(4x+3)^2}$

SECTION 8.4

1. $2e^{2t}$ **3.** $8xe^{x^2}$ **5.** $-12e^{-3x}$ **7.** $\dfrac{1}{2\sqrt{x}}e^{\sqrt{x}}$ **9.** $T^2e^T(T+3)$

11. $\dfrac{e^v(v-2)}{v^3}$ **13.** $\dfrac{-2e^r}{(e^r-1)^2}$ **15.** $\dfrac{3e^{3x}}{2\sqrt{2+e^{3x}}}$ **17.** $e^t\left(\dfrac{1}{t}+\ln t\right)$ **19.** e^{-x}

21. $2e^{t^2}(2t^2+1)$ **23.** $2te^{t^2}(2t^2+3)$ **25.** $\dfrac{V}{L}e^{-Rt/L}$ **27.** $e^x(x+1)$ **29.** $-kp_0e^{-kz}$

31. $3e^{-2}=0.405$ lb

SECTION 8.5

1. $2\ln x+C$ **3.** $\frac{1}{2}\ln 3$ **5.** $\frac{1}{2}\ln(t^2+4)+C$ **7.** $\ln(e^x+1)+C$

9. $-\dfrac{1}{2(x^2-3)}+C$ **11.** $\frac{1}{3}\ln^3 x+C$ **13.** $\ln(\ln x)+C$ **15.** $\ln(e^x+e^{-x})+C$

17. $\dfrac{\ln(2+3\ln I)}{3}+C$ **19.** $-\dfrac{1}{1+\ln z}+C$ **21.** $v=-u\ln(m_0-qt)-gt+u\ln m_0$

23. $\frac{1}{2}\ln 2=0.347$ lb-sec   **25.** $10+\ln 6=11.79$ cm^3 **27.** $\frac{5}{6}\ln 13=2.14$ cm   **29.** $\frac{1}{2}\ln x$

SECTION 8.6

1. $2e^x + C$ **3.** $-\dfrac{1}{2^x \ln 2} + C$ **5.** $\frac{1}{2}e^{2x} + \frac{1}{5}e^{5x} + C$ **7.** $\frac{1}{2}e^{s2} + C$ **9.** $\frac{1}{3}e^{x3} + \frac{1}{4}x^4 + C$

11. $2e^{(y+1)^{1/2}} + C$ **13.** $\frac{2}{3}(1 - e^{-x})^{3/2} + C$ **15.** $\frac{1}{2}e^{2x} + 2x - \frac{1}{2}e^{-2x} + C$ **17.** $3e^{s/3} + C$

19. $-e^{-x} + \frac{1}{2}e^{2x} + C$ **21.** $-Ve^{-t/RC} + K$ **23.** $2e^4 - 2 = 107.2$ mi

25. $(e^4 - 1)/2 = 26.8$ mi **27.** $e^3 + 99 = 119.1$ **29.** 20.6 min **31.** $(1 - e^{-\pi s})/s$

REVIEW EXERCISES FOR CHAPTER 8

1. $\dfrac{1}{x + 2}$ **3.** $\dfrac{8x[\ln(x^2 - 3)]^3}{x^2 - 3}$ **5.** $10te^{5t2}$ **7.** $-5^{-x} \ln 5$ **9.** $2x + 4$

11. $\dfrac{e^{3x}}{\sqrt{x^2 + 4}}(3x^2 + x + 12)$ **13.** $4 \ln(2t - 5) + C$ **15.** $\frac{1}{3}e^{3x-1} + C$ **17.** $\ln(\ln 2x) + C$

19. $\frac{1}{2}e^{x2} + C$ **21.** $10x(x^2 - 1)^4$ **23.** $y = \ln \dfrac{1}{3 - 2x}$ **25.** 44.5 g **27.** 1.35

29. $\ln 3$

SECTION 9.1

1. $5/\sqrt{29},\ -2/\sqrt{29},\ -5/2,\ -2/5,\ -\sqrt{29}/2,\ \sqrt{29}/5$ **3.** $-3/5,\ -4/5,\ 3/4,\ 4/3,\ -5/4,\ -5/3$

5. (a) $180/\pi° \approx 57.28°$ (b) $68.75°$ (c) $30°$ (d) $572.8°$ (e) $-114.6°$

7. $A = 20$, period $= 2\pi/7$

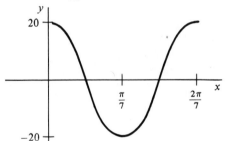

9. $A = 0.3$, period $= \pi$

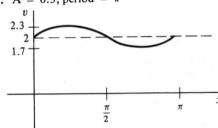

11. asymptotes: $x = -5\pi/16,\ 3\pi/16$; phase shift $= -\pi/16$; period $= \pi/2$

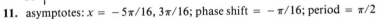

13. $A = 1$; phase shift $= \pi/3$; period $= 2\pi$

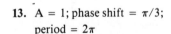

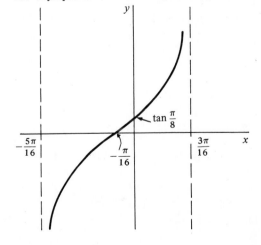

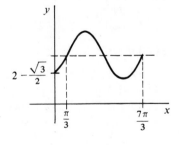

15. asymptotes: $x = -\frac{1}{4}\pi, \frac{7}{4}\pi$; phase shift $= -\frac{1}{4}\pi$; period $= 2\pi$

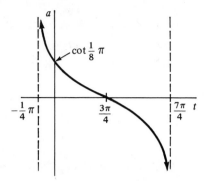

17. $\sec\theta$ **19.** $\sec x$ **21.** $\cot x$ **23.** $\cot x$ **25.** $-\tan^2 x$ **27.** $\sin x$ **29.** 1

SECTION 9.2

1. $\frac{1}{2}\pi\cos\frac{1}{2}\pi t$ **3.** $-6\sin 3t$ **5.** $-\frac{1}{2}\sin(t+1)$ **7.** $3T^2\cos T^3$ **9.** $-e^t\sin e^t$
11. $-3\sin t\cos^2 t$ **13.** $\cos 2x/\sqrt{\sin 2x}$ **15.** $e^t(\cos t + \sin t)$
17. $(-\ln x)(\sin 2x) + (1/2x)\cos 2x$ **19.** $5\cot 5s$ **21.** $-2\sin 2te^{\cos 2t}$ **23.** $-\sin x\cos(\cos x)$
25. $-4x^2\sin x^2 + 2\cos x^2$ **27.** $-W\sin\theta$ **29.** $-L\omega I_m\cos(\omega t - \phi)$
31. $-(\sigma_x - \sigma_y)\cos 2\phi - 2T\sin 2\phi$

SECTION 9.3

1. $3\sec^2 3x$ **3.** $2\pi\sec\pi t\tan\pi t$ **5.** $-\csc^2(2t+1)$ **7.** $2x\sec x^2\tan x^2$
9. $(3\csc 3\sqrt{x}\cot 3\sqrt{x})/\sqrt{x}$ **11.** $6\tan^2 2x\sec^2 2x$ **13.** $-(1+\cot 2t)^{-1/2}\csc^2 2t$
15. $e^{2t}\sec\pi t(2 + \pi\tan\pi t)$ **17.** $3x^2\tan x^3$ **19.** $r^2(3\tan 2r + 2r\sec^2 2r)$ **21.** $8\sec^2 2x\tan 2x$
25. 4   **27.** 1196 cm/sec   **29.** $-N\csc^2 E$

SECTION 9.4

1. $\pi/6$ **3.** $\pi/4$ **5.** $5\pi/6$ **7.** $2\pi/3$ **9.** 0   **11.** $-\pi/3$ **13.** $\frac{12}{13}$
15. $2/\sqrt{5}$ **17.** $\sqrt{15}/4$ **19.** $\sqrt{15}/8$ **21.** 1.211   **23.** 1.399   **25.** 1.407
27. 0.459   **29.** -1.010 **31.** 0.9711   **33.** no solution   **35.** 2.5722   **37.** -4.2556
39. 0.7800

SECTION 9.5

1. $3/\sqrt{1-9t^2}$ **3.** $2/(4+y^2)$ **5.** $1/(2\sqrt{t-t^2})$ **7.** $3/(9+x^2)$
9. $e^t/(1+e^{2t})$ **11.** $1/(2\sqrt{1-x}) + \text{Arcsin}\sqrt{x}/(2\sqrt{x})$ **13.** $e^{-t}[2/(1+4t^2) - \text{Arctan } 2t]$
15. $1/[4\sqrt{\phi-\phi^2}(\text{Arccos}\sqrt{\phi})^{3/2}]$ **17.** $\text{Arctan } x$ **19.** 1.101   **21.** $-X/(R^2+X^2)$
23. $-T/(\omega^2+T^2)$ **25.** -0.025 rad/sec

SECTION 9.6

1. $-\frac{1}{3}\cos 3x + C$ **3.** $\frac{1}{2}\ln\sec 2t + C$ **5.** $\ln\sin\omega t + C$ **7.** $3\sec x^2 + C$
9. $\cos e^{-x} + C$ **11.** $2\ln(\csc\frac{1}{2}x - \cot\frac{1}{2}x) + C$ **13.** $-(1/\omega)(\csc\omega t) + C$ **15.** $\frac{1}{6}\sin^6\theta + C$

17. $\frac{1}{3}\sec^3\phi + C$ **19.** $\frac{1}{2}e^{\sin 2\phi} + C$ **21.** $-1/(1 + \tan x) + C$
23. $\omega t + 2\ln(\sec \omega t + \tan \omega t) + \tan \omega t + C$ **25.** 8.98   **27.** 289 ft-lb   **29.** πh
31. 153.5

SECTION 9.7

1. $(\theta/2) + \frac{1}{4}\sin 2\theta + C$ **3.** $(\phi/2) - \frac{1}{12}\sin 6\phi + C$ **5.** $\frac{1}{10}\sin^5 2t - \frac{1}{14}\sin^7 2t + C$
7. $\frac{1}{4}\sec^4\theta + C$ **9.** $\sin \omega t - \frac{1}{3}\sin^3\omega t + C$ **11.** $(t/2) - \frac{1}{4}\sin 2t + \ln\sin t + C$
13. $\ln\sin x - \frac{1}{2}\sin^2 x + C$ **15.** $\theta - 3\cot\theta - \frac{1}{3}\cot^3\theta + C$ **17.** $-\frac{1}{2}\cos x^2 + \frac{1}{6}\cos^3 x^2 + C$
19. $\frac{1}{9}$ **21.** 0.071 J

SECTION 9.8

1. $\frac{1}{2}\text{Arctan } y/2 + C$ **3.** $\frac{1}{3}\text{Arcsin }\sqrt{3}v + C$ **5.** $\sqrt{6}/6\,\text{Arctan }(\sqrt{6}x/2) + C$
7. $\frac{1}{2}\text{Arctan }(x + 2)/2 + C$ **9.** $\frac{1}{6}\ln(3z^2 + 5) + C$ **11.** $\text{Arcsin }(x - 2)/2 + C$
13. $\frac{1}{6}\text{Arctan } 2(y + 1)/3 + C$ **15.** $\frac{1}{2}\ln(x^2 + 1) + 2\text{Arctan } x + C$
17. $(\sqrt{3}/15)\text{Arctan }(\sqrt{3}/3)y + C$ **19.** $\frac{1}{2}\ln(x^2 + 2x + 3) + (1/\sqrt{2})\text{Arctan }[(x + 1)/\sqrt{2}] + C$
21. $v = \frac{1}{3}\text{Arctan } 3t + 100$ **23.** 0.197 A   **25.** $x = (1/k)\text{Arcsin } ky/c + C$

REVIEW EXERCISES FOR CHAPTER 9

1. $2x\cos(x^2 + 1)$ **3.** $-6\csc^2(3x + 2)$ **5.** $\tan^3(2x - 1) + 6x\tan^2(2x - 1)\sec^2(2x - 1)$
7. $e^t\cot e^t$ **9.** $12/\sqrt{1 - 16t^2}$ **11.** $-1/[(2 - x^2)\sqrt{1 - x^2}]$ **13.** $30x\sec^5 3x^2\tan 3x^2$
15. $\frac{1}{6}\sin 6t + C$ **17.** $\frac{1}{6}\ln(\sec 3x^2) + C$ **19.** $\frac{1}{2}x - \frac{1}{20}\sin 10x + C$ **21.** $\frac{1}{2}(\ln\sin x)^2 + C$
23. $\frac{2}{3}\sin^{3/2} x + C$ **25.** $-\frac{2}{3}\cos^{3/2} x + \frac{2}{7}\cos^{7/2} x + C$ **27.** $\frac{1}{3}\text{Arctan }(3x/2) + C$
29. $\frac{1}{6}\text{Arctan }[(s + 8)/6] + C$ **31.** 6.31   **33.** $5\pi\sqrt{2}/2 = 11.1$ ft/sec   **35.** $\pi/2$ **37.** 1

SECTION 10.1

1. $(x + 2)^{3/2}[(6x - 8)/15] + C$ **3.** $\frac{2}{15}(t - 3)^{3/2}(3t + 16) + C$
5. $-\frac{2}{5}\sqrt{1 - s}(8 + 4s + 3s^2) + C$ **7.** $(1/\sqrt{17})\text{Arctan }(2\sqrt{y} - 1/\sqrt{17}) + C$
9. $\ln(\sqrt{x} + 1)^2 + C$ **11.** $\frac{3}{28}(x - 1)^{4/3}(4x + 3) + C$ **13.** $\frac{6}{5}(2t + 1)^{1/4}(t - 2) + C$
15. $\frac{3}{160}(2x + 1)^{5/3}(10x - 3) + C$ **17.** $(x/2)(4 - x^2)^{1/2} + 2\text{Arcsin }\frac{1}{2}x + C$
19. $\ln[x + (9 + x^2)^{1/2}] + C$ **21.** $\frac{1}{5}(1 - x^2)^{5/2} - \frac{1}{3}(1 - x^2)^{3/2} + C$ **23.** $z/(9\sqrt{9 + z^2}) + C$
25. $(x/\sqrt{9 - x^2}) - \text{Arcsin }(x/3) + C$ **27.** 80.2 J   **29.** $v = 2\sqrt{t} - 2\text{Arctan }\sqrt{t}$ **31.** $\pi/4$

SECTION 10.2

1. $-x\cos x + \sin x + C$ **3.** $-e^{-t}(1 + t) + C$ **5.** $(2 - \theta^2)(\cos\theta) + 2\theta\sin\theta + C$
7. $\frac{1}{4}z^4\ln z - \frac{1}{16}z^4 + C$ **9.** $y\text{Arctan } y - \frac{1}{2}\ln(y^2 + 1) + C$
11. $x\text{Arccos } 3x - \frac{1}{3}\sqrt{1 - 9x^2} + C$ **13.** $-\frac{1}{18}(3t^2 - 2)^{-3} + C$ **15.** $\frac{2}{3}x(2x - 5)^{3/4} - \frac{4}{21}(2x - 5)^{7/4} + C$
17. $\frac{1}{2}e^{\phi^2}(\phi^2 - 1) + C$ **19.** $\frac{3}{10}e^{-x}(\sin 3x - \frac{1}{3}\cos 3x) + C$ **21.** $-1 + \ln 4$ **23.** 0.657
25. $y = \frac{2}{3}x^{3/2}(\ln x - \frac{2}{3}) + \frac{4}{9}$ **27.** 0.438   **29.** 2.51 J

SECTION 10.3

1. $\ln\sqrt[3]{(x + 5)^2(x - 1)} + C$ **3.** $\ln\dfrac{x}{(x + 1)} + C$ **5.** $\frac{1}{2}\ln z + 2\ln(z + 5) - \frac{1}{2}\ln(z + 2) + C$
7. $\ln\sqrt{(x + 1)(x + 5)} + C$ **9.** $-6\ln y + 7\ln(y - 1) + \dfrac{2}{(y - 1)} + C$

11. $\ln \dfrac{s}{(s+1)} + \dfrac{1}{s+1} + C$ **13.** $\ln \dfrac{(w-1)^2}{(w+1)(3-2w)^{3/2}} + C$ **15.** $2\ln(x-1) - \dfrac{1}{x-1} + C$

17. $\ln t + \dfrac{\sqrt{2}}{2}\operatorname{Arc\,tan}\dfrac{\sqrt{2}}{2}t + C$ **19.** $\frac{1}{2}\ln(2x+3) - \dfrac{1}{\sqrt{3}}\operatorname{Arc\,tan}\dfrac{x+1}{\sqrt{3}} + C$

21. $\ln(x-1) - \dfrac{1}{x-1} + \operatorname{Arc\,tan} x + C$ **23.** $\ln 3 = 1.0986$ **25.** $1 + \ln 2$

SECTION 10.4

1. $\frac{1}{2}$ **3.** 4   **5.** diverges   **7.** diverges   **9.** diverges   **11.** diverges
13. $\pi/2$ **15.** $-\frac{3}{2} + \frac{3}{2}\sqrt[3]{4}$ **17.** 0   **19.** $8\sqrt{2}/3$ **21.** 1   **23.** 2

REVIEW EXERCISES FOR CHAPTER 10

1. $\frac{3}{32}(2x-5)^{8/3} + \frac{3}{4}(2x-5)^{5/3} + C$ **3.** $\ln[x + (4+x^2)^{1/2}] + C$
5. $\frac{1}{2}x^2\operatorname{Arctan} x^2 - \frac{1}{4}\ln(1+x^4) + C$ **7.** $\frac{2}{5}(x-3)^{5/2} + 2(x-3)^{3/2} + C$
9. $(1/\pi)\sin \pi x - x\cos \pi x + C$ **11.** $\frac{1}{2}(\ln 3x)^2 + C$ **13.** $\tan x \sin x + \cos x + C$
15. $-\frac{2}{3}\ln(x+1) + \frac{5}{3}\ln(x+4) + C$ **17.** $\frac{1}{2}\ln(x-1) + \frac{1}{4}\ln(x^2+1) + \frac{1}{2}\operatorname{Arctan} x + C$
19. $\ln(x+1) - \ln(x-3) - [4/(x-3)] + C$ **21.** diverges   **23.** $\frac{1}{9}$ **25.** $2\sqrt{3}$
27. $\frac{1}{2}(\frac{1}{2}\pi - 1)$ cm

SECTION 11.1

1. focus: $(-2, 0)$; dir.: $x = 2$; r.c.: $(-2, 4), (-2, -4)$ **3.** focus: $(0, \frac{3}{2})$; dir.: $y = -\frac{3}{2}$; r.c.: $(-3, \frac{3}{2}), (3, \frac{3}{2})$
5. focus: $(-4, 0)$; dir.: $x = 4$; r.c.: $(-4, 8), (-4, -8)$ **7.** focus: $(\frac{3}{4}, 0)$; dir.: $x = -\frac{3}{4}$; r.c.: $(\frac{3}{4}, \frac{3}{2}), (\frac{3}{4}, -\frac{3}{2})$
9. focus: $(-\frac{1}{2}, 0)$; dir.: $x = \frac{1}{2}$; r.c.: $(-\frac{1}{2}, 1), (-\frac{1}{2}, -1)$ **11.** $x^2 = 8y$ **13.** $y^2 = 6x$ **15.** $x^2 = -4y$
17. $x^2 = y$ **19.** $(0, 0), (-2, 1)$ **21.** $(0.09, -1.05)$ **23.** $x^2 = 1000y$
25.

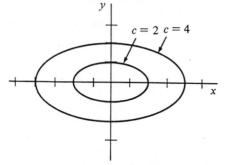

SECTION 11.2

1. ellipse; $a = 5, b = \sqrt{5}$; foci $(0, \pm 2\sqrt{5})$ **3.** ellipse; $a = 2, b = 1$; foci $(0, \pm\sqrt{3})$
5. circle; center $(0, 0)$; $r = 3$ **7.** ellipse; $a = \sqrt{12}, b = \sqrt{8}$; foci $(\pm 2, 0)$
9. ellipse; $a = 1, b = \frac{2}{3}$; foci $(0, \pm\frac{1}{3}\sqrt{5})$ **11.** $9x^2 + 16y^2 = 144$ **13.** $100x^2 + 9y^2 = 225$
15. $9x^2 + 25y^2 = 225$ **17.** $15x^2 + 16y^2 = 240$ **19.** $4x^2 + 21y^2 = 25$
21. $(1.3, -1.9), (-1.3, 1.9)$ **23.** $(0.94, 0.88), (-0.94, 0.88)$ **25.** $9x^2 + 100y^2 = 225$
27.   **29.**

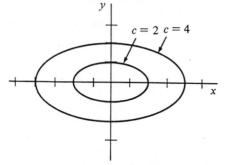
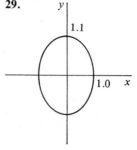

SECTION 11.3

1. $a = 4, b = 4$; foci($\pm \sqrt{32}, 0$)   **3.** $a = 3, b = 2$; foci ($\pm \sqrt{13}, 0$)

5. $a = 5, b = 2$; foci ($0, \pm \sqrt{29}$)   **7.** $a = \frac{5}{2}, b = \frac{5}{4}$; foci ($\pm \frac{5}{4}\sqrt{5}, 0$)   **9.** $a = 1, b = 1$; foci ($\pm \sqrt{2}, 0$)

11. $9x^2 - 16y^2 = 144$ **13.** $4y^2 - x^2 = 4$ **15.** $13x^2 - 36y^2 = 117$ **17.** $4y^2 - x^2 = 64$

19. $y^2 - 10x^2 = 9$ **23.**

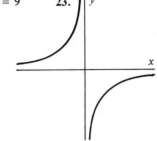

SECTION 11.4

1. $8x = y^2 - 2y + 25$ **3.** $12y = -13 + 2x - x^2$ **5.** $4x = y^2 - 4y + 8$

7. $7x^2 + 16y^2 - 28x - 32y - 68 = 0$ **9.** $16x^2 + y^2 - 16x + 8y + 4 = 0$

11. $3x^2 + 4y^2 + 24x - 24y + 72 = 0$ **13.** $64y^2 - 49x^2 - 98x - 256y = 577$

15. $7x^2 - 9y^2 - 28x + 18y = 44$ **17.** $x^2 - 4y^2 + 4x - 16y - 16 = 0$ **19.** $(x - h)^2 + y^2 = r^2$

21. $y^2 = 4a(x - h)$ **23.** $(x - r)^2 + y^2 = r^2$

27. $\dfrac{(x - 2)^2}{4} - \dfrac{(y - 3)^2}{5} = 1$

25.

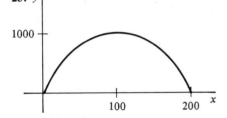

SECTION 11.5

1. $(x + 2)^2 + (y + 3)^2 = 3^2$ **3.** $\dfrac{(x + 1)^2}{4} + \dfrac{(y + 1)^2}{9} = 1$ **5.** $\dfrac{(x - 2)^2}{25} - \dfrac{y^2}{25} = 1$

7. $(x - 3)^2 = 3(y + 3)$ **9.** $(y - \frac{1}{2})^2 = -\frac{1}{2}(x - \frac{3}{2})$ **11.** $\dfrac{(x + 1)^2}{1} - \dfrac{(y - 1)^2}{4} = 1$

13. $\dfrac{(x - 1)^2}{4} + \dfrac{(y + 2)^2}{9} = 1$ **15.** $\dfrac{(y - \frac{1}{2})^2}{1/4} - \dfrac{x^2}{1/2} = 1$ **17.** $\dfrac{(x + 2)^2}{7} - \dfrac{y^2}{7} = 1$

19. $(x + \frac{5}{2})^2 = y - \frac{3}{4}$ **21.** ellipses   **23.** parabola; $y^2 = 10x + 25$

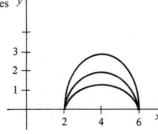

REVIEW EXERCISES FOR CHAPTER 11

1. parabola; vertex $(0, 0)$; focus $(0, -\frac{3}{4})$

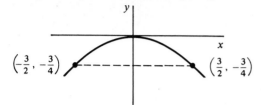

3. circle; center $(0, -2)$; radius $= 3$

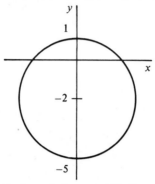

5. hyperbola; center $(-1, -1)$;
   $a = 1, b = 1, c = \sqrt{2}$

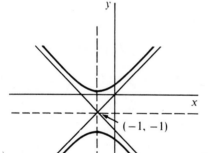

7. hyperbola; center $(0, 0)$;
   $a = 10, b = 10, c = \sqrt{2}$

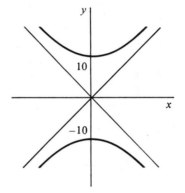

9. ellipse; center $(-\frac{3}{2}, 0)$;
   $a = \sqrt{\frac{23}{2}} \approx 3.4, b = \sqrt{\frac{23}{4}} \approx 2.4, c = \sqrt{\frac{23}{4}} \approx 2.4$

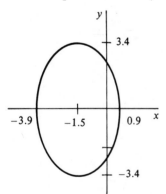

11. $x^2 = 8(y - 2)$ **13.** $(x + 1)^2 + (y - 2)^2 = 64$ **15.** $\dfrac{x^2}{16} + \dfrac{y^2}{12} = 1$

17. $\dfrac{(x-3)^2}{16} - \dfrac{(y-4)^2}{64} = 1$ **19.** 9 ft from center   **21.** $(0.67, -0.33)$ and $(0.5, 0)$

SECTION 12.1

1. 19   **3.** $4 + \pi^2$ **5.** t^2 **7.** $4t + t^2$ **9.** 4

11. 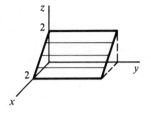   **13.** 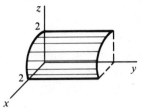   **15.**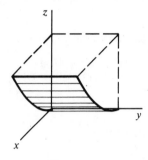

17. 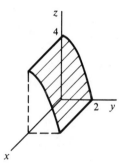   **19.** 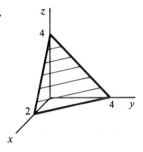   **21.**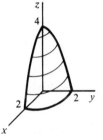

SECTION 12.2

1. 9   **3.** 20   **5.** $5 + y - y^3$ **7.** $8 + 2y - y^3$ **9.** $z_x = 3x^2y^3 + 2x; z_y = 3x^3y^2$

11. $f_x = 2x \cos t; f_t = -x^2 \sin t$ **13.** $G_x = ye^{xy}; G_y = xe^{xy}$ **15.** $F_u = 6v^2; F_v = 12uv$

17. $z_x = \cos x \cos y; z_y = -\sin x \sin y$ **19.** $f_x = \dfrac{e^x}{\sin y}; f_y = \dfrac{-e^x \cos y}{\sin^2 y}$

21. $A_r = \text{Arctan } x; A_x = \dfrac{r}{1 + x^2}$ **23.** $v_x = \dfrac{y}{2\sqrt{xy}} + 3; v_y = \dfrac{x}{2\sqrt{xy}}$

25. $H_r = 2s \sec^2 rs \tan rs; H_s = 2r \sec^2 rs \tan rs$ **27.** $z_x = \dfrac{y}{2\sqrt{1 + xy}}; z_y = \dfrac{x}{2\sqrt{1 + xy}}$

29. $z_{xx} = 20x^3y^2 - \frac{1}{4}x^{-3/2}; z_{yy} = 2x^5; z_{xy} = z_{yx} = 10yx^4$

31. $f_{xx} = 2 \tan y; f_{yy} = 2x^2 \sec^2 y \tan y; f_{xy} = f_{yx} = 2x \sec^2 y$

33. $h_{ss} = -t^2 \sin st + 2; h_{tt} = -s^2 \sin st; h_{st} = h_{ts} = -st \sin st + \cos st$

35. $F_{zz} = 2y \sec^2 zy(zy \tan zy + 1); F_{yy} = 2z^3 \sec^2 zy \tan zy; F_{zy} = F_{yz} = 2z \sec^2 zy(zy \tan zy + 1)$

37. $z_{xx} = -x^{-2}; z_{yy} = 2; z_{xy} = z_{yx} = 0$

39. $z_{xxx} = 60x^2y^3 - 24x; z_{yyy} = 6x^5; z_{xyy} = z_{yxy} = z_{yyx} = 30x^4y; z_{xxy} = z_{xyx} = z_{yxx} = 60x^3y^2$

51. $D_A = 4/P; D_P = -4A/P^2$ **53.** $W_v = wv/g$ **55.** 2.5×10^{-2} mhos

SECTION 12.3

1. $2xy^3 \, dx + 3x^2y^2 \, dy$ **3.** $(x + 1)e^x \, dx + y^{-1} \, dy$ **5.** $\sin y \, dx + (e^y + x \cos y) \, dy$

7. $ye^{xy} \sec^2 e^{xy}\, dx + (2y \sec y^2 \tan y^2 + xe^{xy} \sec^2 e^{xy})\, dy$ **9.** $2(x + y)(dx + dy)$

11. $2 \sin t \cos^4 t - 3 \sin^3 t \cos^2 t$ **13.** $2t(t^2 + 1)e^{t^2} + (1/t)$ **15.** $2t \sin(\ln t) + 1 + t \cos(\ln t)$

17. 0.25 in.^{2} **19.** $dV = \frac{2}{3}\pi rh\, dr + \frac{1}{3}\pi r^2\, dh$ **21.** $dP = \rho v\, dv + \frac{1}{2}v^2\, d\rho$

23. $dv = 32 \sin \alpha\, dt + 32t \cos \alpha\, d\alpha$ **25.** 80 ft/sec, 0.29 ft/sec   **27.** $\dfrac{dV}{dt} = 2\pi rh\dfrac{dr}{dt} + \pi r^2 \dfrac{dh}{dt}$

29. -11.600 in.^{3}/sec   **31.** (a) $22°$ (b) 750 psi/sec

SECTION 12.4

1. $\frac{1}{3}x^3 + xy + G(y)$ **3.** $xy + F(x)$ **5.** $\frac{1}{4}x^2 y^2 + F(x) + G(y)$

7. $\frac{1}{3}x^3 + (2/y)e^{xy} + G(y)$ **9.** $(-\cos xy)/x^2 + yF_1(x) + F_2(x)$ **11.** 9 **13.** 0

15. $\frac{63}{2}$ **17.** $\frac{16}{3}$ **19.** 2 **21.** $\frac{33}{10}$ **23.** $\frac{317}{60}$ **25.** $\pi/4$

SECTION 12.5

1. $\frac{9}{2}$ **3.** $(-25 + 32\sqrt{2})/24$ **5.** $\frac{9}{2}$ **7.** $\frac{16}{3}$ **9.** $\frac{1}{10}$ **11.** $\frac{9}{2}$

SECTION 12.6

1. 6 **3.** $\frac{256}{3}$ **5.** $\frac{136}{15}$ **7.** $\frac{9}{2}$ **9.** $\frac{1}{15}$ **11.** $\frac{1}{12}$ **13.** 8 lb

SECTION 12.7

1. $\frac{4}{3}$ **3.** 51 **5.** $\frac{5}{42}$ **7.** $2 \ln 10$ **9.** $\frac{8}{3}$

REVIEW EXERCISES FOR CHAPTER 12

1. -26 **3.**   **5.** $f_x = -3x^2 y^4 + 2x; f_y = -4x^3 y^3$

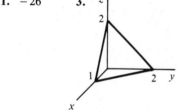

7. $z_x = 2e^{2x} \cos 3y; z_y = -3e^{2x} \sin 3y$ **9.** $f_t = \cos 2st - 2st \sin 2st; f_s = -2t^2 \sin 2st$

11. $f_{xx} = 20x^3; f_{yy} = 2x; f_{yx} = f_{xy} = 2y$

13. $z_{xx} = 2y(\sec x)(\sec^2 x + \tan^2 x); z_{yy} = 0; z_{xy} = z_{yx} = 2 \sec x \tan x$

15. $2x \sin 2y\, dx + 2x^2 \cos 2y\, dy$ **17.** $dx + 3e^{3y}\, dy$ **19.** $4e^{4t} + 6 \sin 3t \cos 3t$ **21.** $\frac{27}{2}$

23. 2 **25.** $\frac{9}{2}$ **27.** $\frac{99}{4}$ **29.** $\frac{16}{3}$ **31.** $\frac{500}{3}$ **33.** $\frac{4}{15}$

SECTION 13.1

13. $y = \ln(x + 2) + C$ **15.** $y = -\frac{1}{2}\cos 2x + C$ **17.** $y = \frac{2}{5}(x^2 + 3)^{5/2} + C$

19. $y = \frac{2}{3}(x + 3)^{3/2} + C$ **21.** $y = \frac{1}{2}\text{Arctan}\,\frac{1}{2}x + C$ **23.** $y = \frac{1}{3}x \sin 3x + \frac{1}{9}\cos 3x + C$

25. $y = x^3 + 1$ **27.** $M = -\frac{1}{2}(3 + 2x)^{-1} + \frac{61}{6}$

SECTION 13.2

1. $xy = c$ **3.** $x^2 + y^2 = c^2; y = 0$ **5.** $1 + s^2t^2 = cs^2$ **7.** $\sqrt{x^2 - 2} = c(1 - y)$
9. $2 \operatorname{Arctan} x + y^2 = c$ **11.** $\ln y = c + e^{-x}(x + 1)$ **13.** $e^y - e^{-x} = c$
15. $(1/\sqrt{5}) \operatorname{Arctan} (y/\sqrt{5}) + \frac{1}{2} \ln (x^2 - 1) = C$ **17.** $y^2 + \sec^2 x = c$ **19.** $y = \dfrac{c}{x}e^{-1/x}$
21. $(x - 4)(y + 2) = -4$ **23.** $\sqrt{9 + x^2} + \ln y = 5$ **25.** $\operatorname{Arctan} x + \operatorname{Arctan} y = \frac{7}{12}\pi$
27. 0.0023 A   **29.** 311 years   **31.** $15{,}238$ years   **33.** (a) $64.7°$ (b) 50.7 min
35. $x^2 + 2y^2 = e$, ellipse with center at origin and horizontal major axis

SECTION 13.3

1. xy **3.** $\frac{1}{2}x^2y^2$ **5.** $y \ln x$ **7.** $y \sin x$ **9.** $y(x^2 + 1)^{-1}$ **11.** $y = 2 + ce^{-2x}$
13. $y = -\frac{3}{4} + ce^{4x}$ **15.** $6xy - 2x^3 - 3x^2 = c$ **17.** $q = -\frac{1}{4}(2t + 1) + ce^{2t}$
19. $s(t^2 + 1) = 3t + c$ **21.** $y = -2 + c\sqrt{x^2 + 1}$ **23.** $v = (r + c) \cos r$
25. $y = ce^x - e^{-x}$ **27.** $s = (t/2) + (c/t)$ **29.** $y = xe^x - ex$ **31.** $y = 5 \csc x - 10 \cot x$

SECTION 13.4

1. $v = 5(t - 1 + e^{-t})$ **3.** $v = 2000 - 400\sqrt{25 - t}$ **5.** 420 ft/sec
7. (a) $v = 320(1 - e^{-t/10})$ (b) $v = [16(1 + 0.9e^{120-4t})]/(1 - 0.9e^{120-4t})$
9. (a) $v = (16 - 4\sqrt{3})t$ (b) $x = (8 - 2\sqrt{3})t^2$ (c) 45.4 ft/sec, 113.4 ft
11. (a) $Q = 30 - 25e^{-t/5}$ lb   (b) 30 lb
13. (a) $Q = 4t + 100 - 22{,}500\sqrt{2}(2t + 50)^{-3/2}$ lb   (b) 51.5 lb
15. $Q = 300(1 - e^{-0.05t}) - 15te^{-0.05t}$ **17.** 8.7 years   **19.** $i = 0.6(1 - e^{-100t})$
21. $i = \begin{cases} \frac{1}{10}(2t - 1 + e^{-2t}), & \text{if } 0 \leq t \leq 2 \\ 0.4 - 5.4e^{-2t}, & \text{if } t > 2 \end{cases}$ **23.** $i = \begin{cases} 3[1 - (1 + t)^{-2}], & \text{if } 0 \leq t \leq 1 \\ 3 - 0.75e^{-(t-1)}, & \text{if } t > 1 \end{cases}$
25. $q = VC(1 - e^{-t/RC})$

SECTION 13.5

1. $y = C_1 + C_2e^{-x}$ **3.** $y = C_1 + C_2e^{-2x}$ **5.** $y = C_1e^{-x} + C_2e^{-4x}$
7. $s = C_1 \cos 3t + C_2 \sin 3t$ **9.** $v = C_1e^{3t} + C_2te^{3t}$ **11.** $x = e^t(C_1 \cos t + C_2 \sin t)$
13. $y = C_1e^{-(1/2)x} + C_2xe^{-(1/2)x}$ **15.** $s = C_1e^{-(1/3)t} + C_2te^{-(1/3)t}$
17. $q = C_1e^{\sqrt{3}t} + C_2e^{-\sqrt{3}t}$ **19.** $y = C_1 \cos \frac{1}{3}x + C_2 \sin \frac{1}{3}x$
21. $y = C_1e^{[(-1+\sqrt{5})/2]x} + C_2e^{[(-1-\sqrt{5})/2]x}$
23. $v = e^{2s}(C_1 \cos \sqrt{3}s + C_2 \sin \sqrt{3}s)$ **25.** $y = \frac{1}{2}e^{2x} - \frac{1}{2}e^{-2x}$ **27.** $y = xe^{4x}$
29. $s = 2 \sin 3t$ **31.** $y = e^{3x}(2 \cos x - 5 \sin x)$

SECTION 13.6

1. (a) $x = \frac{1}{2} \sin 2t$ (b) A $= \frac{1}{2}$ ft, period $= \pi$ sec
3. (a) $x = -\frac{1}{4} \cos 16t, v = 4 \sin 16t, a = 64 \cos 16t$ (b) 0.104 ft, $v = 3.64$ ft/sec, $a = -26.6$ ft/sec^2
5. $x = -\frac{1}{2} \cos 4t + \frac{1}{2} \sin 4t$ **7.** 4.16 lb

SECTION 13.7

1. (a) $x = \frac{1}{4}(1 + 8t)e^{-8t}$

(b)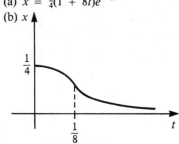

3. (a) $x = -0.25e^{-4t} \sin 8t$

(b) -0.114 ft; 0.024 ft; -0.005 ft

(c)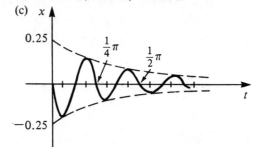

5. (a) $x = te^{-4t}$

(b) $1/(4e)$ ft

(c)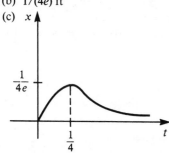

7. maxima and minima occur at $t = (1/\omega) \text{Arctan}(\omega/a) + (n\pi/\omega)$; $n = 0, 1, 2, \ldots$; tangents to Ce^{-at} and Ce^{at} occur at $t = (\pi/2\omega) + (n\pi/\omega)$; $n = 0, 1, 2, \ldots$

SECTION 13.8

1. $\dfrac{6}{s^4}$ **3.** $\dfrac{6s^4 + 2s^2 + 48}{s^5}$ **5.** $\dfrac{5s + 7}{(s + 2)(s - 1)}$ **7.** $\dfrac{120}{(s + 3)^6}$ **9.** $\dfrac{s(s + 3)}{(s - 3)(s^2 + 9)}$

11. $\dfrac{s + 3}{(s + 1)^2}$ **13.** $\dfrac{2(4 - s)}{(s - 3)^3}$ **15.** $\dfrac{s^2 - 4}{(s^2 + 4)^2}$ **17.** e^{3t} **19.** $\frac{2}{3} \sin 3t$ **21.** $\frac{1}{2}t^2$

23. $\frac{1}{12}t^4e^{3t}$ **25.** $\frac{1}{8}(e^{4t} + 7e^{-4t})$ **27.** $\frac{2}{3}e^{-5t} + \frac{1}{3}e^t$ **29.** $\frac{14}{5}e^{2t} - \frac{9}{5}e^{-3t}$ **31.** $(1 - 4t)e^{-4t}$

33. $-\frac{13}{2}(1 - e^{2t})$ **35.** $\frac{3}{16}(1 - \cos 4t)$ **37.** $\frac{1}{3}t \sin 3t$

SECTION 13.9

1. $y = 2e^{-3t}$ **3.** $x = 3 \cos 3t$ **5.** $y = 1 - e^{-x}$ **7.** $y = 2(e^{-2t} - e^{-t})$

9. $x = 2(1 - e^{-2t})$ **11.** $y = \frac{1}{4}(1 - \cos 2t)$ **13.** $y = \frac{1}{2}t^2e^{-2t}$ **15.** 2567 ft

17. $x = \frac{1}{2} \sin 8t - \frac{1}{4} \cos 8t$ **19.** $i = 1.2(1 - e^{-5t})$

REVIEW EXERCISES FOR CHAPTER 13

1. $(x^2/2) + \ln y = c$ **3.** $\sin y - \frac{1}{2}e^{2x} = c$ **5.** $\tan y - \frac{1}{3}x^3 = c$ **7.** $y^2(x + 2) = c$

9. $y = -3 + ce^x$ **11.** $y = \frac{1}{2}x^3 + cx$ **13.** $s = (t + c)e^{2t}$ **15.** $y = -\frac{1}{3} + cx^3$

17. $y \cos x = \ln \sin x + c$ **19.** $y = c_1 e^x + c_2 e^{2x}$ **21.** $s = c_1 + c_2 e^{16t}$

23. $y = c_1 e^{(1+\sqrt{5})x/2} + c_2 e^{(1-\sqrt{5})x/2}$ **25.** $y = (c_1 + c_2 x)e^{-4x}$ **27.** $x = (c_1 + c_2 t)e^{-t}$

29. $s = c_1 \cos 3t + c_2 \sin 3t$ **31.** $y = e^{-2x}(c_1 \cos 3x + c_2 \sin 3x)$

33. $y = e^{-x}(c_1 \cos 2x + c_2 \sin 2x)$ **35.** $A = \sqrt{10}$, period $= \pi$, phase shift $= \frac{1}{2} \text{Arctan} \frac{1}{3}$

37. 805 **39.** $x = \frac{1}{3} \cos t$ **41.** $\dfrac{2s}{s^2 + 7}$ **43.** $\dfrac{3}{s - 5} - \dfrac{16}{(s + 1)^3}$ **45.** $\dfrac{5!}{s^6} - \dfrac{6}{s^2 + 4}$

47. $\frac{1}{3} \sin 3t - 2 \cos 3t$ **49.** $\frac{1}{3} t^3 - t$ **51.** $e^{3t} + e^{4t}$ **53.** $\frac{1}{5}(1 - e^{-5t})$ **55.** $\frac{5}{4}(e^{3t} - e^{-t})$

57. $\frac{2}{5}(1 - \cos\sqrt{5}t)$

SECTION 14.1

1. 2.24, 63.4° **3.** 5.65, −135° **5.** 5, −36.9° **7.** (6.43, 7.66)

9. (−90.63, 129.43) **11.** $5\mathbf{i} + \mathbf{j}$ **13.** $-6\mathbf{i} - 2\mathbf{j}$ **15.** $\mathbf{i} - 3\mathbf{j}$ **17.** $-4\mathbf{j}$ **19.** $6\mathbf{i} - 7\mathbf{j}$

21. 35.19, 51.2° **23.** 13.36, 73.0° **25.** 65.58, 131.4° **27.** 395.1, −64.0°

29. **31.** 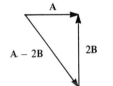 **33.**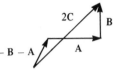

35. 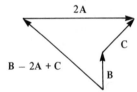 **37.** 304 mi/hr, 80.5° N of W **39.** 163.8 lb **41.** 11.99 lb, 30.8°

43. 1678.6 ft/sec, 14.1° below the horizontal

SECTION 14.2

1. 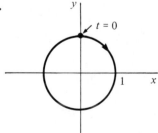 **3.** 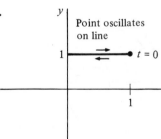 **5.**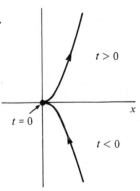

7.

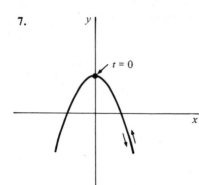

9.

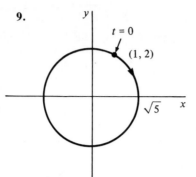

11. $x^2 + y^2 = 1$

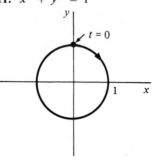

13. $y = x + 1$

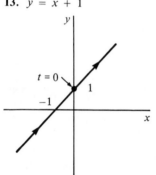

15. $y = x,\ -1 \le x \le 1$

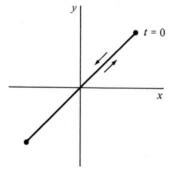

17. $x = \sin^2 y$

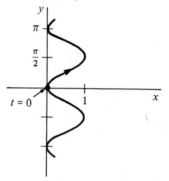

19. $(y - 1) = (x + 3)^2$

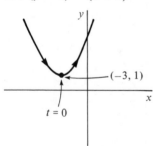

21.

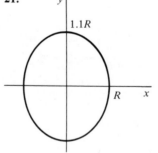

23.

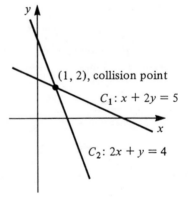

25.

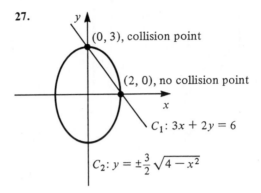

$C_2: x - y = 2$

$C_1: x = y^2$

$(4, 2)$, no collision point

$(1, 1)$, no collision point

27.

$(0, 3)$, collision point

$(2, 0)$, no collision point

$C_1: 3x + 2y = 6$

$C_2: y = \pm\dfrac{3}{2}\sqrt{4 - x^2}$

SECTION 14.3

1. 5   **3.** 1   **5.** $-e^{-2}$ **7.** -2 **9.** 1   **11.** $y + 1 = [2/(1 - 3\pi)](x - 1)$

13. not concave   **15.** not concave   **17.** concave up   **19.** concave down

21. concave down   **23.** horizontal tangent: $t = \pm\sqrt{8/3}$; vertical tangent: $t = 0$

25. $\mathbf{v} = 3\mathbf{i} + 2t\mathbf{j}$; $v(2) = 3\mathbf{i} + 4\mathbf{j}$; $\mathbf{a} = 2\mathbf{j}$; $\mathbf{a}(2) = 2\mathbf{j}$; $|\mathbf{v}(2)| = 5$; $\mathbf{v}(2)/|\mathbf{v}(2)| = \frac{3}{5}\mathbf{i} + \frac{4}{5}\mathbf{j}$

$= \mathbf{a}(\tfrac{1}{4})\sqrt{2}\,\pi^2\,\mathbf{i} + \tfrac{1}{2}\sqrt{2}\,\pi^2\mathbf{j}$; $\mathbf{v}(\tfrac{1}{4}) = \pi\sqrt{\tfrac{5}{2}}$; $\mathbf{v}(\tfrac{1}{4})\,|\mathbf{v}(\tfrac{1}{4})/| = (2/\sqrt{5})\mathbf{i} + (1/\sqrt{5})\mathbf{j}$

27. $\mathbf{v} = 2\pi\cos\pi t\mathbf{i} + \pi\sin\pi t\mathbf{j}$; $v(\tfrac{1}{4}) = \pi\sqrt{2}\,\mathbf{i} + \tfrac{1}{2}\pi\sqrt{2}\,\mathbf{j}$; $\mathbf{a} = -2\pi^2\sin\pi t\mathbf{i} + \pi^2\cos\pi t\mathbf{j}$;

SECTION 14.4

1. $2\sqrt{2}$ **3.** 2π **5.** $\frac{9}{8}$ **7.** $\frac{2}{27}(10^{3/2} - 1)$ **9.** $4\ln(\sqrt{3} + 2)$ **11.** $\frac{8}{27}[(\frac{13}{4})^{3/2} - 1]$

13. $\frac{3}{2}\pi$ **15.** $\ln(\sqrt{2} + 1)$

SECTION 14.5

1.

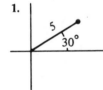

3.

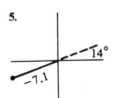

5.

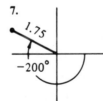

7.

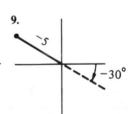

9.

11. $r(2\cos\theta + 3\sin\theta) = 6$ **13.** $r = 4\cos\theta$ **15.** $r^2(\cos^2\theta + 4\sin^2\theta) = 4$

17. $r\cos^2\theta = 4\sin\theta$ **19.** $x^2 + y^2 = 25$ **21.** $x^2 + y^2 = 10y$

23. $x^2 + y^2 = (x^2 + y^2)^{1/2} + 2y$ **25.** $y^2 = 25 - 10x$

27.

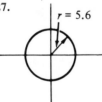

$r = 5.6$

29.

$\dfrac{\pi}{3}$

31.

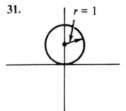

$r = 1$

33.

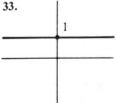

1

35.

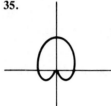

37.

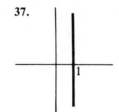

39.

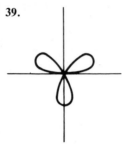

41.

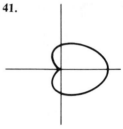

43. $3x^2 + 4y^2 - 2x - 1 = 0$ **45.** $\dfrac{2\cos\theta(1 + \sin\theta)}{\cos^2\theta - \sin^2\theta - 2\sin\theta}$ **47.** -2 **49.** 0

51. $\dfrac{-2\sin^2\theta + \cos\theta(2\cos\theta - 1)}{\sin\theta(1 - 4\cos\theta)}$

SECTION 14.6

1. $25\pi/3$ **3.** $31\pi^3/128$ **5.** $\frac{1}{4}[(\pi/3) + (\sqrt{3}/4)]$ **7.** $\pi/16$
9. $(\pi^3/960)(20 - 15\pi + 3\pi^2)$ **11.** 4 **13.** $49\pi/4$ **15.** $9\pi/2$
17. $27\pi/2$ **19.** $\frac{1}{2}\pi$ **21.** π **23.** 2 **25.** $(\sqrt{5}/2)(e^\pi - 1)$ **27.** $3a^2\pi/2$

REVIEW EXERCISES FOR CHAPTER 14

1. $x = -3.2, y = 3.8$ **3.** $\mathbf{i} - 3\mathbf{j}$ **5.** $4\mathbf{i} + 3\mathbf{j}$ **7.** $171, 1.3°$
9. $F_x = 332.9\,\text{lb}, F_y = 108.2\,\text{lb}$
11. $y = 2x + 3$ **13.** $y = \ln x$ **15.** $y = 2x\sqrt{1 - x^2}$

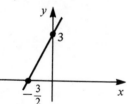
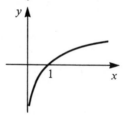
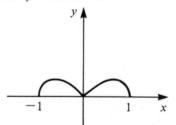

17. $y' = \frac{1}{2}$ **19.** $y' = 2$ **21.** $\frac{61}{27}$ **23.** $\sqrt{5}(e - 1)$ **25.** $3\sqrt{5}$ **27.** $\frac{1}{27}(40^{3/2} - 8)$
29. $r = 5/(\cos\theta + 2\sin\theta)$ **31.** $x^2 + y^2 = 3x$
33.   **35.**   **37.**

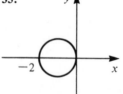
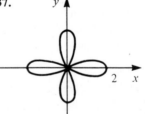

39. $y' = \dfrac{\cos^2\theta - \sin^2\theta}{-2\sin\theta\cos\theta} = -\cot 2\theta$ **41.** $y' = \dfrac{4\cos 2\theta\sin\theta + \sin 2\theta\cos\theta}{4\cos 2\theta\cos\theta - \sin 2\theta\sin\theta}$ **43.** $\frac{1}{4}\pi$

45. $2\pi\sqrt{1 + 4\pi^2} + \ln(2\pi + \sqrt{1 + 4\pi^2})$ **47.** $\frac{1}{2}[\frac{1}{2}\pi + \frac{9}{8}\sqrt{3}]$ **49.** π **51.** 2π

SECTION 15.1

1. $1, \frac{1}{3}, \frac{1}{5}, \frac{1}{7}$ **3.** $2, 4, 4, \frac{8}{3}$ **5.** $-1, 1/2!, -1/3!, 1/4!$
7. $1 + \frac{1}{2}, 1 + \frac{1}{2} + \frac{1}{3}, 1 + \frac{1}{2} + \frac{1}{3} + \frac{1}{4}$ **9.** $a_n = 2n$ **11.** $a_n = n/(n - 1)!$
13. $a_n = 1/(2n - 1)!$ **15.** $a_n = n/(2n + 1)$ **17.** $a_n = x^{2n}/2^n$
19. $a^n = x^{n-1}/(n - 1)!$ **21.** $a_n = (-1)^{n+1}/n$ **23.** $a_n = \sum_{k=1}^{n} \frac{1}{2k - 1}$

SECTION 15.2

1. converges to 0   **3.** converges to 1   **5.** converges to 1   **7.** diverges
9. diverges   **11.** diverges   **13.** converges to 0   **15.** diverges   **17.** converges to $\frac{1}{2}$
19. 10

SECTION 15.3

1. diverges   **3.** diverges   **5.** converges   **7.** diverges   **9.** diverges
11. converges to $\frac{3}{2}$ **13.** converges to 6   **15.** converges   **17.** diverges   **19.** converges
21. diverges   **23.** diverges

SECTION 15.4

1. converges   **3.** converges   **5.** converges   **7.** converges   **9.** diverges
11. converges   **13.** converges   **15.** diverges

SECTION 15.5

1. $-1 < x < 1$ **3.** all x **5.** $-1 \le x < 1$ **7.** all x **9.** all x
11. $-1 \le x \le 1$ **13.** $-\frac{1}{3} < x < \frac{1}{3}$ **15.** all x **17.** $-e < x < e$

REVIEW EXERCISES FOR CHAPTER 15

1. $a_n = 2n + 1$ **3.** $a_n = (-1)^{n+1}/2n$ **5.** $\frac{3}{2}, \frac{3}{3}, \frac{3}{4}, \frac{3}{5}$; converges to 0
7. $-1, 1/\sqrt{2}, -1/\sqrt{3}, 1/\sqrt{4}$; converges to 0   **9.** $\frac{1}{3}, \frac{4}{4}, \frac{9}{5}, \frac{16}{6}$; diverges   **11.** diverges
13. converges   **15.** converges   **17.** converges   **19.** converges   **21.** diverges
23. converges   **25.** $-1 < x < 1$ **27.** $-1 \le x \le 1$

SECTION 16.1

1. $1 - \dfrac{x^2}{2!} + \dfrac{x^4}{4!} - \cdots + \dfrac{(-1)^n x^{2n}}{(2n)!} + \cdots$ **3.** $x - \dfrac{x^2}{2} + \dfrac{x^3}{3} - \cdots + \dfrac{(-1)^n x^{n+1}}{n + 1} + \cdots$

5. $3x - \dfrac{9x^3}{2} + \dfrac{81x^5}{40} - \cdots$ **7.** $1 - 2x + 3x^2 - \cdots$ **9.** $x + \dfrac{x^3}{6} + \dfrac{3x^5}{40} + \cdots$

11. $1 - x + \dfrac{x^2}{2!} - \dfrac{x^3}{3!} + \cdots$ **13.** $1 + \dfrac{x^2}{2} + \dfrac{5x^4}{24} + \cdots$ **15.** $-\dfrac{x^2}{2} - \dfrac{x^4}{12} - \dfrac{x^6}{45} - \cdots$

17. x^2 **19.** $x^2 - 2x + 1$

SECTION 16.2

1. $2x - \dfrac{4x^3}{3} + \dfrac{4x^5}{15} - \cdots$ **3.** $1 - \dfrac{x^2}{8} + \dfrac{x^4}{384} - \cdots$ **5.** $3x + \dfrac{(3x)^3}{6} + \dfrac{3(3x)^5}{40} + \cdots$

7. $1 - \dfrac{(x - \pi/6)^2}{2!} + \dfrac{(x - \pi/6)^4}{4!} - \cdots$ **9.** $1 + x^2 + x^4 + x^6 + \cdots$

11. $1 - 3x + 9x^2 - 27x^3 + \cdots$ **13.** $1 + \frac{5}{2}x + \frac{25}{4}x^2 + \frac{125}{8}x^3 + \cdots$

15. $\frac{1}{2} - \frac{3}{4}x + \frac{7}{8}x^2 - \frac{15}{16}x^3 + \cdots$ **17.** $1 - x + x^2 - x^3 + \cdots$ **19.** $x - \dfrac{x^3}{3!} + \dfrac{x^5}{5!} - \cdots$

SECTION 16.3

1. 0.9950 **3.** 0.08729 **5.** 1.22 **7.** -0.1053 **9.** 0.099669 **11.** 1.010067

13. 0.31028 **15.** 1.43333 **17.** 0.39307

19. $\left(x - \dfrac{x^3}{3!} + \dfrac{x^5}{5!}\right) - \dfrac{1}{3!}\left(x - \dfrac{x^3}{3!} + \dfrac{x^5}{5!}\right)^3 + \dfrac{1}{5!}\left(x - \dfrac{x^3}{3!} + \dfrac{x^5}{5!}\right)^3$ **21.** $2\pi t - \dfrac{(2\pi t)^3}{3!} + \dfrac{(2\pi t)^5}{5!} - \cdots$

SECTION 16.4

1. $\dfrac{1}{2} - \dfrac{\sqrt{3}(x - \pi/3)}{2} - \dfrac{(x - \pi/3)^2}{4} + \cdots$ **3.** $\dfrac{\sqrt{2}}{2}\left[1 + (x - \pi/4) - \dfrac{(x - \pi/4)^2}{2!} - \cdots\right]$

5. $e\left[1 + \dfrac{(x - 1)}{1!} + \dfrac{(x - 1)^2}{2!} + \cdots\right]$ **7.** $-1 + (x - 2) - (x - 2)^2 + (x - 2)^3 - \cdots$

9. $(x - 1)^2 + 2(x - 1) + 1$ **11.** $-5 - 6(x - 1) + (x - 1)^2$

13. $\frac{1}{7}[1 - \frac{2}{7}(x - 1) + \frac{4}{49}(x - 1)^2 - \frac{8}{343}(x - 1)^3 + \cdots]$

15. $\frac{1}{2} + \frac{3}{4}(x + 1) + \frac{7}{8}(x + 1)^2 + \frac{15}{16}(x + 1)^3 + \cdots$ **17.** 0.5754 **19.** 0.9008 **21.** 2.749

SECTION 16.5

1. $4\sin x - 2\sin 2x + \frac{4}{3}\sin 3x - \cdots$ **3.** $\dfrac{1}{2\pi}\left(\sin \pi x - \dfrac{\sin 2\pi x}{2} + \dfrac{\sin 3\pi x}{3} - \dfrac{\sin 4\pi x}{4} + \cdots\right)$

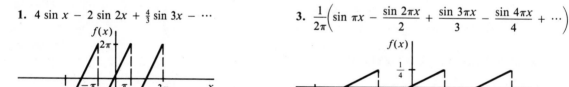

5. $\dfrac{1}{3} + \dfrac{4}{\pi^2}\left(-\cos \pi x + \dfrac{\cos 2\pi x}{4} - \dfrac{\cos 3\pi x}{9} + \dfrac{\cos 4\pi x}{16} - \cdots\right)$

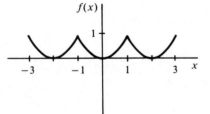

7. $\frac{1}{2} + (2/\pi)(\sin \frac{1}{2}\pi x + \frac{1}{3}\sin \frac{3}{2}\pi x + \frac{1}{5}\sin \frac{5}{2}\pi x + \cdots)$

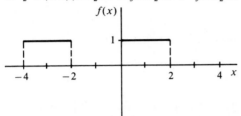

9. $1 + (8/\pi)(\sin x + \frac{1}{3}\sin 3x + \frac{1}{5}\sin 5x + \cdots)$

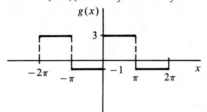

11. $(8/\pi)(\frac{1}{3}\sin 2x - \frac{2}{15}\sin 4x + \frac{3}{35}\sin 6x - \frac{4}{63}\sin 8x + \cdots)$

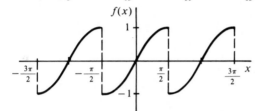

13. $\frac{1}{8}\pi + \frac{1}{\pi}\left[(\pi - 2)\cos x - \cos 2x - \left(\frac{\pi}{3} + \frac{2}{9}\right)\cos 3x + \left(\frac{\pi}{5} - \frac{2}{25}\right)\cos 5x - \frac{1}{9}\cos 6x - \cdots\right]$

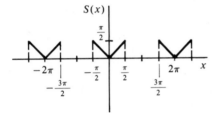

15. $(1/\pi)(2\sin x + \frac{1}{2}\pi\sin 2x - \frac{2}{9}\sin 3x - \frac{1}{4}\pi\sin 4x + \cdots)$

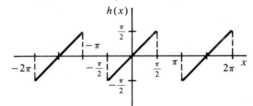

17. $\dfrac{e^{2\pi} - 1}{2\pi} + \dfrac{1}{\pi}\displaystyle\sum_{n=1}^{\infty}\left[\dfrac{e^{2\pi} - 1}{1 + n^2}\cos nt + \dfrac{n(1 - e^{2\pi})}{1 + n^2}\sin nt\right]$ **19.** $\frac{3}{2} - (2/\pi)[\sin x + \frac{1}{3}\sin 3x + \frac{1}{5}\sin 5x + \cdots]$

REVIEW EXERCISES FOR CHAPTER 16

1. $1 + \frac{1}{4}x - \frac{1}{6}x^2$ **3.** $-\ln 2 + \sqrt{3}x - 2x^2$ **5.** $\ln 2 - \frac{1}{2}x - \frac{1}{8}x^2$ **7.** $3x - \frac{9}{2}x^3 - \frac{1}{40}x^5$

9. $1 - \frac{1}{2}x^2 - \frac{1}{24}x^4$ **11.** $2 + 9(x - 2) + \frac{15}{2}(x - 2)^2$ **13.** $1 + 4(x - \frac{1}{8}\pi) + 8(x - \frac{1}{8}\pi)^2$

15. $-[1 + \frac{3}{2}(x - 1) + \frac{9}{4}(x - 1)^2]$ **17.** $1 - \frac{9}{2}x^2 + \frac{81}{24}x^4 - \frac{729}{720}x^6 + \cdots$

19. $1 - 3x^2 + 9x^4 - 27x^6 + 81x^8 - \cdots$ **21.** $\frac{2}{3}(1 - \frac{4}{3}x + \frac{8}{9}x^2 - \frac{28}{27}x^3 + \frac{80}{81}x^4 - \cdots)$

23. $\frac{1}{3}x - \frac{1}{18}x^2 + \frac{1}{81}x^3 - \frac{1}{324}x^4 - \cdots$ **25.** $1 + x^2 + \frac{1}{2}x^4 + \frac{1}{6}x^6 + \frac{1}{24}x^8 + \cdots$ **27.** -0.020203
29. 1.12166   **31.** $\frac{1}{2}(1 - \cos 2x)$ **33.** $1 + \sin x$
35. $\frac{1}{2} + (4/\pi^2)(\cos \pi x + \frac{1}{9}\cos 3\pi x + \frac{1}{25}\cos 5\pi x + \cdots)$
37. $(4/\pi)(\sin x + \frac{1}{3}\sin 3x + \frac{1}{5}\sin 5x + \frac{1}{7}\sin 7x + \cdots)$; at $x = \pi$, the sum of the first five terms is zero.

APPENDIX

1. three   **3.** four   **5.** one   **7.** two   **9.** four   **11.** 9820   **13.** 54.7
15. 0.0658   **17.** 39.8   **19.** 1.00   **21.** 21.50   **23.** 1.92   **25.** 65   **27.** 3.1
29. 2300   **31.** 139.6

INDEX

ANALYTIC GEOMETRY

Straight line: slope $= m = \dfrac{y_2 - y_1}{x_2 - x_1}$; $\quad y = mx + b$; $\quad y - y_1 = m(x - x_1)$

Circle: $\quad x^2 + y^2 = r^2$; $\quad (x - h)^2 + (y - k)^2 = r^2$

Parabola: *Vertical axis:* $\qquad x^2 = 4ay$; $\quad (x - h)^2 = 4a(y - k)$

            Horizontal axis: $\qquad y^2 = 4ax$; $\quad (y - k)^2 = 4a(x - h)$

Ellipse: *Vert. major axis:* $\qquad \dfrac{x^2}{b^2} + \dfrac{y^2}{a^2} = 1$; $\quad \dfrac{(x - h)^2}{b^2} + \dfrac{(y - k)^2}{a^2} = 1$

          Hor. major axis: $\qquad \dfrac{x^2}{a^2} + \dfrac{y^2}{b^2} = 1$; $\quad \dfrac{(x - h)^2}{a^2} + \dfrac{(y - k)^2}{b^2} = 1$

Hyperbola: *Vert. transverse axis:* $\dfrac{y^2}{a^2} - \dfrac{x^2}{b^2} = 1$; $\quad \dfrac{(y - k)^2}{a^2} - \dfrac{(x - h)^2}{b^2} = 1$

          Hor. transverse axis: $\dfrac{x^2}{a^2} - \dfrac{y^2}{b^2} = 1$; $\quad \dfrac{(x - h)^2}{a^2} - \dfrac{(y - k)^2}{b^2} = 1$

DERIVATIVE FORMULAS

1. $\dfrac{d}{dx}(c) = 0$

2. $\dfrac{d}{dx}(x) = 1$

3. $\dfrac{d}{dx}[af(x)] = a\dfrac{d}{dx}[f(x)]$

4. $\dfrac{d}{dx}[f(x) \pm g(x)] = \dfrac{d}{dx}[f(x)] \pm \dfrac{d}{dx}[g(x)]$

5. $\dfrac{d}{dx}[f(x)g(x)] = f(x)g'(x) + g(x)f'(x)$

6. $\dfrac{d}{dx}\left[\dfrac{f(x)}{g(x)}\right] = \dfrac{g(x)f'(x) - f(x)g'(x)}{[g(x)]^2}$

7. $\dfrac{d}{dx}[x^n] = nx^{n-1}$

8. $\dfrac{d}{dx}[u^n] = nu^{n-1}\dfrac{du}{dx}$

9. $\dfrac{d}{dx}[\log_a u] = \dfrac{1}{u}\dfrac{du}{dx}\log_a e$

10. $\dfrac{d}{dx}[\ln u] = \dfrac{1}{u}\dfrac{du}{dx}$

11. $\dfrac{d}{dx}[a^u] = a^u\dfrac{du}{dx}\ln a$

12. $\dfrac{d}{dx}[e^u] = e^u\dfrac{du}{dx}$

13. $\dfrac{d}{dx}[\sin u] = \cos u\dfrac{du}{dx}$

14. $\dfrac{d}{dx}[\cos u] = -\sin u\dfrac{du}{dx}$